Mathematics and Visualization

For further volumes:
http://www.springer.com/series/4562

We dedicate this book to our families.

To Doris, Sonja, Johannes, Christian,
Jonathan and Dominik,
Christa and Gerhard

To Rita and Ariel with love

To Rena, Monica and Sevasti,
Agelis and Sevasti

Preface

Shape understanding remains one of the most intriguing problems in computer vision and human perception. This book is a collection of chapters on shape analysis, by experts in the field, highlighting several viewpoints, including modeling and algorithms, in both discrete and continuous domains. It is a summary of research presentations and discussions on these topics at a Dagstuhl workshop in April 2011.

The content is grouped into three main areas:

Part I – Discrete Shape Analysis
Part II – Partial Differential Equations for Shape Analysis
Part III – Optimization Methods for Shape Analysis

The chapters contain both new results and tutorial sections that survey various areas of research.

It was a pleasure for us to have had the opportunity to collaborate and exchange scientific ideas with our colleagues who participated in the Dagstuhl Workshop on Shape Analysis and subsequently contributed to this collection. We hope that this book will promote new research and further collaborations.

Cottbus, Haifa and Athens

Michael Breuß
Alfred Bruckstein
Petros Maragos

Acknowledgments

We would like to express our thanks to the many people who supported the publication of this book.

First of all, we would like to thank the staff of Schloss Dagstuhl for their professional help in all aspects. The breathtaking Dagstuhl atmosphere was the basis that made our workshop such a unique and successful meeting.

This book would never have attained its high level of quality without a rigorous peer-review process. Each chapter has been reviewed by at least two researchers in one or more stages. We would like to thank Alexander M. Bronstein, Oliver Demetz, Jean-Denis Durou, Laurent Hoeltgen, Yong Chul Ju, Margret Keuper, Reinhard Klette, Jan Lellmann, José Alberto Iglesias Martínez, Pascal Peter, Luis Pizarro, Nilanjan Ray, Christian Rössl, Christian Schmaltz, Simon Setzer, Sibel Tari, Michael Wand, Martin Welk, Benedikt Wirth, and Laurent Younes for their dedicated and constructive help in this work.

Moreover, we would like to thank the editors of the board of the Springer series *Mathematics and Visualization* for the opportunity to publish this book at an ideal position in the scientific literature. We are also grateful to Ruth Allewelt from Springer-Verlag for her practical and very patient support.

Finally, we would like to thank Anastasia Dubrovina for producing the nice cover image for the book.

Contents

List of Contributors

Gilad Adiv Rafael, Haifa, Israel

Carlo Arcelli Institute of Cybernetics "E. Caianiello", Pozzuoli (Naples), Italy

Shai Avidan Department of Electrical Engineering, Tel Aviv University, Tel Aviv, Israel

Egil Bae Department of Mathematics, University of California, Los Angeles, USA

Michael Breuß Applied Mathematics and Computer Vision Group, BTU Cottbus, Cottbus, Germany

Faculty of Mathematics and Computer Science, Mathematical Image Analysis Group, Saarland University, Saarbrücken, Germany

Alexander M. Bronstein School of Electrical Engineering, Faculty of Engineering, Tel Aviv University, Tel Aviv, Israel

Michael M. Bronstein Faculty of Informatics, Institute of Computational Science, Universita della Svizzera Italiana, Lugano, Switzerland

Alfred Bruckstein Department of Computer Science Technion-Israel Institute of Technology Haifa, Israel

Elisabetta Carlini Dipartimento di Matematica "G. Castelnuovo", Sapienza – Università di Roma, Roma, Italy

Edward Castillo Department of Radiation Oncology, University of Texas MD Anderson Cancer Center, Houston, USA

Department of Computational and Applied Mathematics, Rice University, Houston, USA

Ritwik Chaudhuri University of North Carolina, Chapel Hill, USA

Lidija Čomić Faculty of Technical Sciences, University of Novi Sad, Novi Sad, Serbia

Daniel Cremers Department of Computer Science, Institut für Informatik, TU München, Garching bei München, Germany

James N. Damon University of North Carolina, Chapel Hill, USA

Leila De Floriani Department of Computer Science, University of Genova, Genova, Italy

Shay Dekel Department of Electrical Engineering, Tel Aviv University, Tel Aviv, Israel

Kimon Drakopoulos Massachusetts Institute of Technology, Cambridge, USA

Anastasia Dubrovina Technion, Israel Institute of Technology, Haifa, Israel

Maurizio Falcone Dipartimento di Matematica "G. Castelnuovo", Sapienza – Università di Roma, Roma, Italy

Adriano Festa Dipartimento di Matematica "G. Castelnuovo", Sapienza – Università di Roma, Roma, Italy

P. Thomas Fletcher University of Utah, Salt Lake City, USA

Dibyendusekhar Goswami University of North Carolina, Chapel Hill, USA

Arne Grumpe Image Analysis Group, Dortmund University of Technology, Dortmund, Germany

Stephan Huckemann Institute for Mathematical Stochastics, University of Göttingen, Göttingen, Germany

Federico Iuricich Department of Computer Science, University of Genova, Genova, Italy

Sungkyu Jung University of Pittsburgh, Pittsburgh, USA

Ron Kimmel Department of Computer Science, Technion, Haifa, Israel

Jian Liang Department of Mathematics, University of California, Irvine, USA

Roee Litman School of Electrical Engineering, Tel Aviv University, Tel Aviv, Israel

Petros Maragos School of Electrical and Computer Engineering, National Technical University of Athens, Athens, Greece

J.S. Marron Department of Statistics and Operations Research, University of North Carolina, Chapel Hill, USA

Roberto Mecca Dipartimento di Matematica "G. Castelnuovo", Sapienza – University of Rome, Rome, Italy

Serena Morigi Department of Mathematics-CIRAM, University of Bologna, Bologna, Italy

Claudia Nieuwenhuis Department of Computer Science, Institut für Informatik, TU München, Garching bei München, Germany, nieuwenhuis@in.tum.de

Martin R. Oswald Department of Computer Science, Institut für Informatik, TU München, Garching bei München, Germany

Pascal Peter Faculty of Mathematics and Computer Science, Mathematical Image Analysis Group, Saarland University, Saarbrücken, Germany

Stephen M. Pizer University of North Carolina, Chapel Hill, USA

Dan Raviv Technion, Israel Institute of Technology, Haifa, Israel

Paul L. Rosin School of Computer Science & Informatics, Cardiff University, Cardiff, UK

Guy Rosman Department of Computer Science, Technion, Haifa, Israel

Marco Rucci Department of Mathematics-CIRAM, University of Bologna, Bologna, Italy

Gabriella Sanniti di Baja Institute of Cybernetics "E. Caianiello", CNR, Pozzuoli (Naples), Italy

Shachar Shem-Tov Technion – Israel Institute of Technology, Haifa, Israel

Luca Serino Institute of Cybernetics "E. Caianiello", CNR, Pozzuoli (Naples), Italy

Nir Sochen Department of Applied Mathematics, Tel Aviv University, Tel Aviv, Israel

Xue-Cheng Tai Department of Mathematics, University of Bergen, Norway

Sibel Tari Middle East Technical University, Ankara, Turkey

Eno Töppe Department of Computer Science, Institut für Informatik, TU München, Garching bei München, Germany

Silvia Tozza Dipartimento di Matematica "G. Castelnuovo", Sapienza – University of Rome, Rome, Italy

Jared Vicory University of North Carolina, Chapel Hill, USA

Christian Wöhler Image Analysis Group, Dortmund University of Technology, Dortmund, Germany

Alon Wolf Department of Mechanical Engineering, Technion – Israel Institute of Technology, Haifa, Israel

Jing Yuan Computer Science Department, University of Western Ontario, Canada

Hongkai Zhao Department of Mathematics, University of California, Irvine, USA

Xiaojie Zhao University of North Carolina, Chapel Hill, USA

Steven W. Zucker Computer Science, Biomedical Engineering and Applied Mathematics, Yale University, New Haven, USA

Joviša Žunić Computer Science, University of Exeter, Exeter, UK

Mathematical Institute Serbian Academy of Sciences and Arts, Belgrade, Serbia

Part I
Discrete Shape Analysis

Chapter 1
Modeling Three-Dimensional Morse and Morse-Smale Complexes

Lidija Čomić, Leila De Floriani, and Federico Iuricich

Abstract Morse and Morse-Smale complexes have been recognized as a suitable tool for modeling the topology of a manifold M through a decomposition of M induced by a scalar field f defined over M. We consider here the problem of representing, constructing and simplifying Morse and Morse-Smale complexes in 3D. We first describe and compare two data structures for encoding 3D Morse and Morse-Smale complexes. We describe, analyze and compare algorithms for computing such complexes. Finally, we consider the simplification of Morse and Morse-Smale complexes by applying coarsening operators on them, and we discuss and compare the coarsening operators on Morse and Morse-Smale complexes described in the literature.

1.1 Introduction

Topological analysis of discrete scalar fields is an active research field in computational topology. The available data sets defining the fields are increasing in size and in complexity. Thus, the definition of compact topological representations for scalar fields is a first step in building analysis tools capable of analyzing effectively large data sets. In the continuous case, Morse and Morse-Smale complexes have been recognized as convenient and theoretically well founded representations for modeling both the topology of the manifold domain M, and the behavior of a scalar field f over M. They segment the domain M of f into regions associated with critical points of f, which encode the features of both M and f.

L. Čomić (✉)
Faculty of Technical Sciences, University of Novi Sad, Trg D. Obradovića 6, Novi Sad, Serbia
e-mail: comic@uns.ac.rs

L. De Floriani · F. Iuricich
Department of Computer Science, University of Genova, via Dodecaneso 35, Genova, Italy
e-mail: deflo@disi.unige.it; federico.iuricich@unige.it

M. Breuß et al. (eds.), *Innovations for Shape Analysis*, Mathematics and Visualization,
DOI 10.1007/978-3-642-34141-0_1, © Springer-Verlag Berlin Heidelberg 2013

Morse and Morse-Smale complexes have been introduced in computer graphics for the analysis of 2D scalar fields [5, 20], and specifically for terrain modeling and analysis, where the domain is a region in the plane, and the scalar field is the elevation function [14,39]. Recently, Morse and Morse-Smale complexes have been considered as a tool to analyze also 3D functions [21, 24]. They are used in scientific visualization, where data are obtained through measurements of scalar field values over a volumetric domain, or through simulation, such as the analysis of mixing fluids [8]. With an appropriate selection of the scalar function, Morse and Morse-Smale complexes are also used for segmenting molecular models to detect cavities and protrusions, which influence interactions between proteins [9, 35]. Morse complexes of the distance function have been used in shape matching and retrieval.

Scientific data, obtained either through measurements or simulation, is usually represented as a discrete set of vertices in a 2D or 3D domain M, together with function values given at those vertices. Algorithms for extracting an approximation of Morse and Morse-Smale complexes from a sampling of a (continuous) scalar field on the vertices of a simplicial complex Σ triangulating M have been extensively studied in 2D [1,6,9,13,20,37,39]. Recently, some algorithms have been proposed for dealing with scalar data in higher dimensions [11,19,21,26,27].

Although Morse and Morse-Smale complexes represent the topology of M and the behavior of f in a much more compact way than the initial data set at full resolution, simplification of these complexes is a necessary step for the analysis of noisy data sets. Simplification is achieved by applying the *cancellation operator* on f [33], and on the corresponding Morse and Morse-Smale complexes. In 2D [6, 20, 24, 39, 43], a cancellation eliminates critical points of f, reduces the incidence relation on the Morse complexes, and eliminates cells from the Morse-Smale complexes. In higher dimensions, surprisingly, a cancellation may introduce cells in the Morse-Smale complex, and may increase the mutual incidences among cells in the Morse complex.

Simplification operators, together with their inverse refinement ones, form a basis for the definition of a multi-resolution representation of Morse and Morse-Smale complexes, crucial for the analysis of the present-day large data sets. Several approaches for building such multi-resolution representations in 2D have been proposed [6,7,15]. In higher dimensions, such hierarchies are based on a progressive simplification of the initial full-resolution model.

Here, we briefly review the well known work on extraction, simplification, and multi-resolution representation of Morse and Morse-Smale complexes in 2D. Then, we review in greater detail and compare the extension of this work to three and higher dimensions. Specifically, we compare the data structure introduced in [25] for encoding 3D Morse-Smale complexes with a 3D instance of the dimension-independent data structure proposed in [11] for encoding Morse complexes. We review the existing algorithms for the extraction of an approximation of Morse and Morse-Smale complexes in three and higher dimensions. Finally, we review and compare the two existing approaches in the literature to the simplification of the topological representation given by Morse and Morse-Smale complexes, without changing the topology of M. The first approach [24] implements a cancellation operator defined for Morse functions [33] on the corresponding Morse-Smale complexes. The second approach [10] implements only a well-behaved subset of cancellation operators, which still forms a basis for the set of operators that modify Morse and Morse-Smale complexes on M in a topologically consistent manner. These operators also form a basis for the definition of a multi-resolution representation of Morse and Morse-Smale complexes.

1.2 Background Notions

We review background notions on Morse theory and Morse complexes for C^2 functions, and some approaches to discrete representations for Morse and Morse-Smale complexes.

Morse theory captures the relationships between the topology of a manifold M and the critical points of a scalar (real-valued) function f defined on M [33, 34]. An *n-manifold* M without boundary is a topological space in which each point p has a neighborhood homeomorphic to $\mathbb{R}^n$. In an n-manifold with boundary, each point p has a neighborhood homeomorphic to $\mathbb{R}^n$ or to a half-space $\mathbb{R}^n_+ = \{(x_1, x_2, \ldots, x_n) \in \mathbb{R}^n : x_n \geq 0\}$ [30].

Let f be a C^2 real-valued function (scalar field) defined over a manifold M. A point $p \in M$ is a *critical point* of f if and only if the gradient $\nabla f = (\frac{\partial f}{\partial x_1}, \ldots, \frac{\partial f}{\partial x_n})$ (in some local coordinate system around p) of f vanishes at p. Function f is said to be a *Morse function* if all its critical points are non-degenerate (the Hessian matrix $Hess_p f$ of the second derivatives of f at p is non-singular). For a Morse function f, there is a neighborhood of each critical point $p = (p_1, p_2, \ldots, p_n)$ of f, in which $f(x_1, x_2, \ldots, x_n) = f(p_1, p_2, \ldots, p_n) - x_1^2 - \ldots - x_i^2 + x_{i+1}^2 + \ldots + x_n^2$ [34]. The number i is equal to the number of negative eigenvalues of $Hess_p f$, and is called the *index* of critical point p. The corresponding eigenvectors point in the directions in which f is decreasing. If the index of p is i, $0 \leq i \leq n$, p is called an *i-saddle*. A 0-saddle is called a *minimum*, and an n-saddle is called a *maximum*. Figure 1.1 illustrates a neighborhood of a critical point in three dimensions.

An *integral line* of a function f is a maximal path that is everywhere tangent to the gradient ∇f of f. It follows the direction in which the function has the maximum increasing growth. Two integral lines are either disjoint, or they are the same. Each integral line starts at a critical point of f, called its *origin*, and ends at another critical point, called its *destination*. Integral lines that converge to a critical point p of index i cover an i-cell called the *stable (descending) cell* of p. Dually, integral lines that originate at p cover an $(n-i)$-cell called the *unstable (ascending) cell* of p. The descending cells (or manifolds) are pairwise disjoint, they cover M, and the boundary of every cell is a union of lower-dimensional cells. Descending cells decompose M into a cell complex Γ_d, called the *descending Morse complex* of f on M. Dually, the ascending cells form the *ascending Morse complex* Γ_a of f on M. Figures 1.2a, b and 1.3a, b show an example of a descending and dual ascending Morse complex in 2D and 3D, respectively.

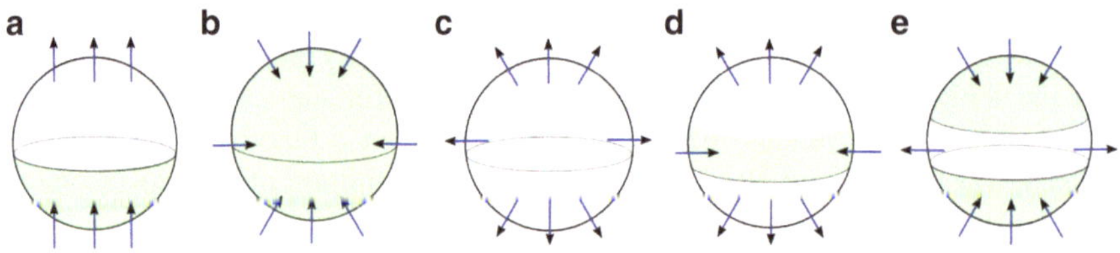

Fig. 1.1 Classification of non-degenerate critical points in the 3D case. *Arrowed lines* represent *integral lines*, *green regions* contain points with the lower function value. (**a**) A regular point, (**b**) a local maximum, (**c**) a local minimum, (**d**) a 1-saddle and (**e**) a 2-saddle

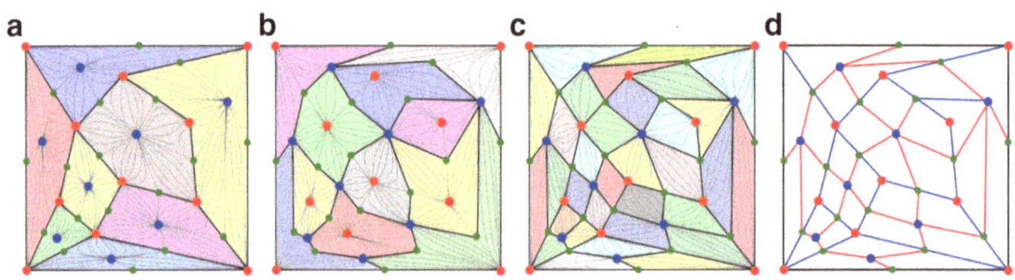

Fig. 1.2 A portion of (**a**) a descending Morse complex; (**b**) the dual ascending Morse complex; (**c**) the Morse-Smale complex; (**d**) the 1-skeleton of the Morse-Smale complex in 2D

A Morse function f is called a *Morse-Smale function* if and only if each non-empty intersection of a descending and an ascending cell is transversal. This means that each connected component of the intersection (if it exists) of the descending i-cell of a critical point p of index i, and the ascending $(n-j)$-cell of a critical point q of index j, $i \geq j$, is an $(i-j)$-cell. The connected components of the intersection of descending and ascending cells of a Morse-Smale function f decompose M into a *Morse-Smale complex*. If f is a Morse-Smale function, then there is no integral line connecting two different critical points of f of the same index. Each 1-saddle is connected to exactly two (not necessarily distinct) minima, and each $(n-1)$-saddle is connected to exactly two (not necessarily distinct) maxima. The 1-skeleton of the Morse-Smale complex is the subcomplex composed of 0-cells and 1-cells. It plays an important role in the applications, as it is often used as a graph-based representation of the Morse and Morse-Smale complexes. Figure 1.2c in 2D and Fig. 1.3c in 3D illustrate the Morse-Smale complex corresponding to the ascending and descending Morse complexes of Figs. 1.2a, b and 1.3a, b, respectively. Figure 1.2d shows the 1-skeleton of the Morse-Smale complex in Fig. 1.2c.

The first approaches to develop a discrete version of Morse theory aimed at a generalization of the notion of critical points (maxima, minima, saddles) to the case of a scalar field f defined on the vertices of a simplicial complex Σ triangulating a 2D manifold (surface) M. This generalization was first done in [2] in 2D, and has been used in many algorithms [20, 36, 39]. The classification of critical points is done based on the f value at a vertex p, and the vertices in the link $Lk(p)$ of p.

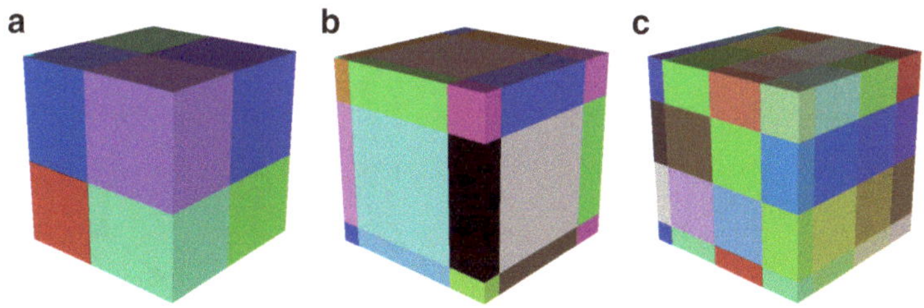

Fig. 1.3 A portion of (**a**) a descending and (**b**) ascending 3D Morse complex, and (**c**) the corresponding Morse-Smale complex, defined by a function $f(x, y, z) = \sin x + \sin y + \sin z$

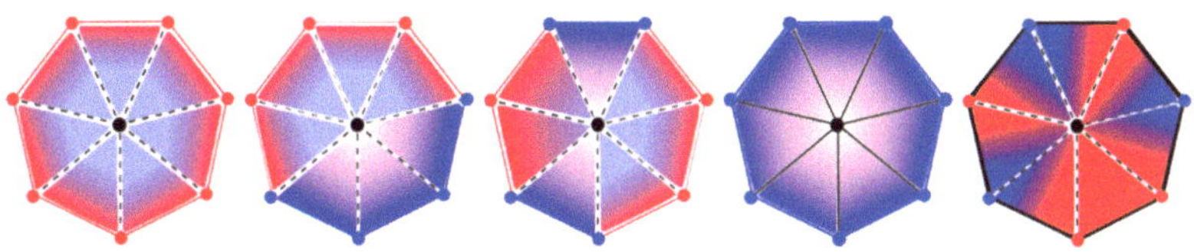

Fig. 1.4 The classification of a vertex based on the function values of the vertices in its link (minimum, regular point, simple saddle, maximum, 2-fold saddle). The lower link Lk^- is marked in *blue*, the upper link is *red*

The link $Lk(p)$ of each vertex p of Σ can be decomposed into three sets, $Lk^+(p)$, $Lk^-(p)$, and $Lk^{\pm}(p)$. The *upper link* $Lk^+(p)$ consists of the vertices $q \in Lk(p)$ with higher f value than $f(p)$, and of edges connecting such vertices. The *lower link* $Lk^-(p)$ consists of the vertices with lower f value than $f(p)$, and of edges connecting such vertices. The set $Lk^{\pm}(p)$ consists of *mixed edges* in $Lk(p)$, each connecting a vertex with higher f value than $f(p)$ to a vertex with lower f value than $f(p)$. If the lower link $Lk^-(p)$ is empty, then p is a minimum. If the upper link $Lk^+(p)$ is empty, then p is a maximum. If the cardinality of $Lk^{\pm}(p)$ is $2+2m(p)$, then p is a saddle with multiplicity $m(p) \geq 1$. Otherwise, p is a regular point. The classification of a vertex based on these rules is illustrated in Fig. 1.4.

There have been basically two approaches in the literature to extend the results of Morse theory and represent Morse and Morse-Smale complexes in the discrete case. One approach, called *Forman theory* [22], considers a discrete Morse function (*Forman* function) defined on all cells of a cell complex. The other approach, introduced in [20] in 2D, and in [21] in 3D, provides a combinatorial description, called a *quasi-Morse-Smale complex*, of the Morse-Smale complex of a scalar field f defined at the vertices of a simplicial complex.

The main purpose of Forman theory is to develop a discrete setting in which almost all the main results from smooth Morse theory are valid. This goal is achieved by considering a function F defined on all cells, and not only on the vertices, of a cell complex Γ. Function F is a Forman function if for any i-cell σ, all the $(i-1)$-cells on the boundary of σ have a lower F value than $F(\sigma)$, and all the $(i+1)$-cells in the co-boundary of σ have a higher F value than $F(\sigma)$, with at most

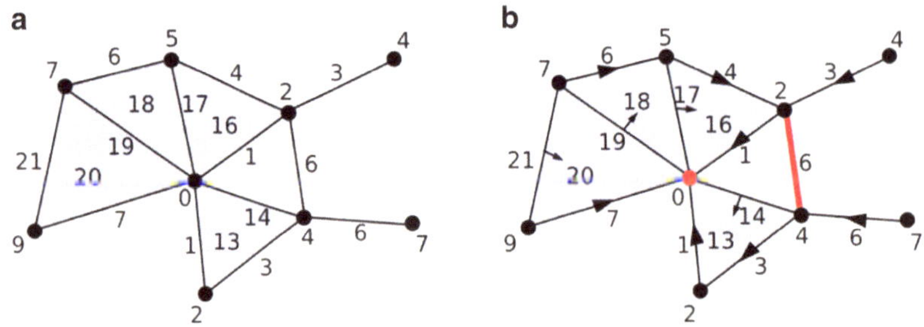

Fig. 1.5 (**a**) Forman function F defined on a 2D simplicial complex, and (**b**) the corresponding discrete gradient vector field. Each simplex is labelled by its F value

one exception. If there is such an exception, it defines a pairing of cells of Γ, called a *discrete* (or *Forman*) *gradient vector field* V. Otherwise, i-cell σ is a critical cell of index i. Similar to the smooth Morse theory, critical cells of a Forman function can be cancelled in pairs. In the example in Fig. 1.5a, a Forman function F defined on a 2D simplicial complex is illustrated. Each simplex is labelled by its function value. Figure 1.5b shows the Forman gradient vector field defined by Forman function F in Fig. 1.5a. Vertex labelled 0 and edge labelled 6 are critical simplexes of F.

Forman theory finds important applications in computational topology, computer graphics, scientific visualization, molecular shape analysis, and geometric modeling. In [32], Forman theory is used to compute the homology of a simplicial complex with manifold domain, while in [9], it is used for segmentation of molecular surfaces. Forman theory can be used to compute Morse and Morse-Smale complexes of a scalar field f defined on the vertices of a simplicial or cell complex, by extending scalar field f to a Forman function F defined on all cells of the complex [12, 27, 31, 38].

The notion of a quasi-Morse-Smale complex in 2D and 3D has been introduced in [20, 21] with the aim of capturing the combinatorial structure of a Morse-Smale complex of a Morse-Smale function f defined over a manifold M. In 2D, a quasi-Morse-Smale complex is defined as a complex whose 1-skeleton is a tripartite graph, since the set of its vertices is partitioned into subsets corresponding to critical points (minima, maxima, and saddles). A vertex corresponding to a saddle has four incident edges, two of which connect it to vertices corresponding to minima, and the other two connect it to maxima. Each region (2-cell of the complex) is a quadrangle whose vertices are a saddle, a minimum, a saddle, and a maximum. In 3D, vertices of a quasi-Morse-Smale complex are partitioned into four sets corresponding to critical points. Each vertex corresponding to a 1-saddle is the extremum vertex of two edges connecting it to two vertices corresponding to minima, and dually for 2-saddles and maxima. Each 2-cell is a quadrangle, and there are exactly four 2-cells incident in each edge connecting a vertex corresponding to a 1-saddle to a vertex corresponding to a 2-saddle.

1.3 Related Work

In this section, we review related work on topological representations of 2D scalar fields based on Morse or Morse-Smale complexes. We concentrate on three topics relevant to the work presented here, namely: computation, simplification and multi-resolution representation of Morse and Morse-Smale complexes.

Several algorithms have been proposed in the literature for decomposing the domain of a 2D scalar field f into an approximation of a Morse or a Morse-Smale complex. For a review of the work in this area see [4]. Algorithms for decomposing the domain M of field f into an approximation of a Morse, or of a Morse-Smale, complex can be classified as *boundary-based* [1, 6, 20, 37, 39], or *region-based* [9, 13]. Boundary-based algorithms trace the integral lines of f, which start at saddle points and converge to minima and maxima of f. Region-based methods grow the 2D cells corresponding to minima and maxima of f, starting from those critical points.

One of the major issues that arise when computing a representation of a scalar field as a Morse, or as a Morse-Smale, complex is the over-segmentation due to the presence of noise in the data sets. Simplification algorithms eliminate less significant features from these complexes. Simplification is achieved by applying an operator called *cancellation*, defined in Morse theory [33]. It transforms a Morse function f into Morse function g with fewer critical points. Thus, it transforms a Morse-Smale complex into another, with fewer vertices, and it transforms a Morse complex into another, with fewer cells. It enables also the creation of a hierarchical representation. A cancellation in 2D consists of collapsing a maximum-saddle pair into a maximum, or a minimum-saddle pair into a minimum. Cancellation is performed in the order usually determined by the notion of *persistence*. Intuitively, persistence measures the importance of the pair of critical points to be cancelled, and is equal to the absolute difference in function values between the paired critical points [20]. In 2D Morse-Smale complexes, the cancellation operator has been investigated in [6, 20, 39, 43]. In [15], the cancellation operator in 2D has been extended to functions that may have multiple saddles and macro-saddles (saddles that are connected to each other).

Due to the large size and complexity of available scientific data sets, a multi-resolution representation is crucial for their interactive exploration. There have been several approaches in the literature to multi-resolution representation of the topology of a scalar field in 2D [6, 7, 15]. The approach in [6] is based on a hierarchical representation of the 1-skeleton of a Morse-Smale complex, generated through the cancellation operator. It considers the 1-skeleton at full resolution and generates a sequence of simplified representations of the complex by repeatedly applying a cancellation operator. In [7], the inverse anticancellation operator to the cancellation operator in [6] has been defined. It enables a definition of a dependency relation between refinement modifications, and a creation of a multi-resolution model for 2D scalar fields. The method in [15] creates a hierarchy of graphs (generalized critical nets), obtained as a 1-skeleton of an overlay of ascending and descending Morse complexes of a function with multiple saddles and saddles that are connected to each other. Hierarchical watershed approaches have been developed to cope with the increase in size of both 2D and 3D images [3].

There have been two attempts in the literature to couple the multi-resolution topological model provided by Morse-Smale complexes with the multi-resolution model of the geometry of the underlying simplicial mesh. The approach in [6] first creates a hierarchy of Morse-Smale complexes by applying cancellation operators to the full-resolution complex, and then, by Laplacian smoothing, it constructs the smoothed function corresponding to the simplified topology. The approach in [16] creates the hierarchy by applying half-edge contraction operator, which simplifies the geometry of the mesh. When necessary, the topological representation corresponding to the simplified coarser mesh is also simplified. The data structure encoding the geometrical hierarchy of the mesh, and the data structure encoding the topological hierarchy of the critical net are interlinked. The hierarchical critical net is used as a topological index to query the hierarchical representation of the geometry of the simplicial mesh.

1.4 Representing Three-Dimensional Morse and Morse-Smale Complexes

In this section, we describe and compare two data structures for representing the topology and geometry of a scalar field f defined over the vertices of a simplicial complex Σ with manifold domain in 3D. The topology of scalar field f (and of its domain Σ) is represented in the form of Morse and Morse-Smale complexes. The two data structures encode the topology of the complexes in essentially the same way, namely in the form of a graph, usually called an *incidence graph*. The difference between the two data structures is in the way they encode the geometry: the data structure in [11] (its 3D instance) encodes the geometry of the 3-cells of the descending and ascending complexes; the data structure in [25] encodes the geometry of the ascending and descending 3-, 2- and 0-cells in the descending and ascending Morse complexes, and that of the 1-cells in the Morse-Smale complexes.

1.4.1 A Dimension-Independent Compact Representation for Morse Complexes

The incidence-based representation proposed in [11] is a dual representation for the ascending and the descending Morse complexes Γ_a and Γ_d. The topology of both complexes is represented by encoding the immediate boundary and co-boundary relations of the cells in the two complexes in the form of a *Morse Incidence Graph (MIG)*. The Morse incidence graph provides also a combinatorial representation of

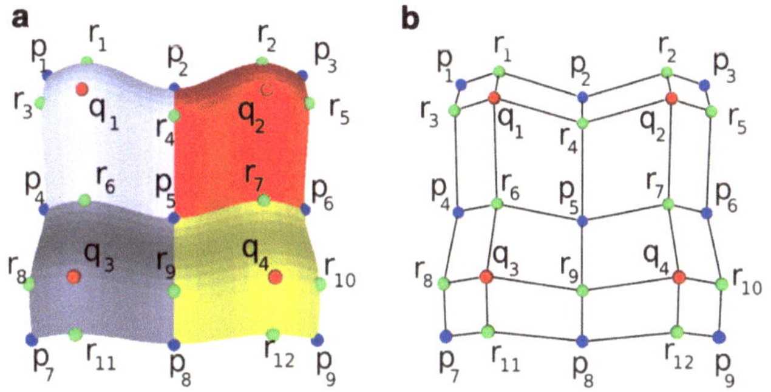

Fig. 1.6 (**a**) Ascending 2D Morse complex of function $f(x, y) = \sin x + \sin y$ and (**b**) the corresponding Morse incidence graph

the 1-skeleton of a Morse-Smale complex. In the discrete case the Morse incidence graph is coupled with a representation for the underlying simplicial mesh Σ. The two representations (of the topology and of the geometry) are combined into the *incidence-based data structure*, which is completely dimension-independent. This makes it suitable also for encoding Morse complexes in higher dimensions, e.g. 4D Morse complexes representing time-varying 3D scalar fields.

A *Morse Incidence Graph (MIG)* is a graph $\mathbb{G} = (\mathbb{N}, \mathbb{A})$ in which:

- The set of nodes $\mathbb{N}$ is partitioned into $n+1$ subsets $\mathbb{N}_0, \mathbb{N}_1, \ldots, \mathbb{N}_n$, such that there is a one-to-one correspondence between the nodes in $\mathbb{N}_i$ (*i-nodes*) and the i-cells of Γ_d, (and thus the $(n-i)$-cells of Γ_a);
- There is an arc joining an i-node p with an $(i+1)$-node q if and only if the corresponding cells p and q differ in dimension by one, and p is on the boundary of q in Γ_d, (and thus q is on the boundary of p in Γ_a);
- Each arc connecting an i-node p to an $(i+1)$-node q is labelled by the number of times i-cell p (corresponding to i-node p) in Γ_d is incident to $(i+1)$-cell q (corresponding to $(i+1)$-node q) in Γ_d.

In Fig. 1.6, we illustrate a 2D ascending complex, and the corresponding incidence graph of function $f(x, y) = \sin x + \sin y$. In the ascending complex, cells labeled p are 2-cells (corresponding to minima), cells labeled r are 1-cells (corresponding to saddles), and cells labeled q are 0-cells (corresponding to maxima).

The data structure for encoding the *MIG* $\mathbb{G} = (\mathbb{N}, \mathbb{A})$ is illustrated in Fig. 1.7. The nodes and the arcs of $\mathbb{G}$ are encoded as two lists. Recall that each node in the graph corresponds to a critical point p of f and to a vertex in the Morse-Smale complex. When p is an extremum, the corresponding element in the list of nodes contains three fields, G_0, G_n and A. The geometry of the extremum (its coordinates) is stored in field G_0, and the geometry of the associated n-cell (ascending n-cell of a minimum, or a descending n-cell of a maximum), which is the list of n-simplexes forming the corresponding n-cell in the ascending or descending complex, is stored in field G_n. The list of the pointers to the arcs incident in the extremum is stored in

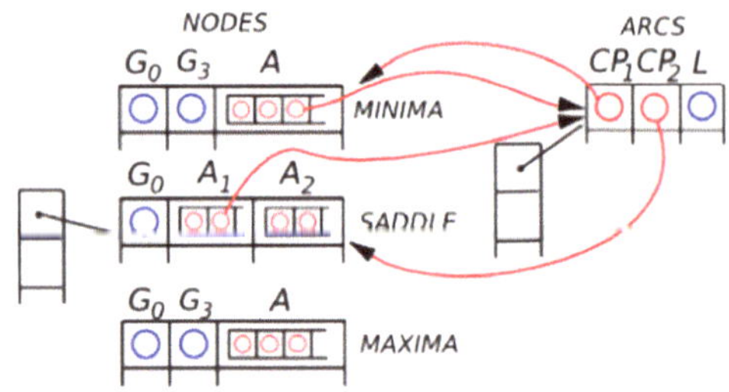

Fig. 1.7 Dimension-independent data structure for storing the incidence graph. The nodes corresponding to i-saddles are stored in lists, as are the arcs. Each element in the list of nodes stores the geometry of the corresponding critical point, and the list of pointers to arcs incident in the node. A node corresponding to an extremum stores also a list of pointers to the n-simplexes in the corresponding n-cell associated with the extremum. Each element in the list of arcs stores pointers to its endpoints, and a label indicating its multiplicity

field A. If p is a maximum (n-saddle), these arcs connect p to $(n-1)$-saddles. If p is a minimum (0-saddle), they connect p to 1-saddles. When p is not an extremum, element in the node list contains fields G_0, A_1 and A_2. The geometry of i-saddle p (its coordinates) is stored in field G_0. A list of pointers to the arcs connecting i-saddle p to $(i+1)$-saddles and to $(i-1)$-saddles is stored in fields A_1 and A_2, respectively.

Each arc in the *MIG* corresponds to integral lines connecting two critical points of f, which are the endpoints of the arc. Each element in the list of arcs has three fields, CP_1, CP_2 and L. If the arc connects an i-saddle to an $(i+1)$-saddle, then CP_1 is a pointer to the i-saddle, and CP_2 is a pointer to the $(i+1)$-saddle. The label of the arc (its multiplicity) is stored in field L.

The manifold simplicial mesh Σ discretizing the graph of the scalar field is encoded in a data structure which generalizes the indexed data structure with adjacencies, commonly used for triangular and tetrahedral meshes [17]. It stores the 0-simplexes (vertices) and n-simplexes explicitly plus some topological relations, namely: for every n-simplex σ, the $n+1$ vertices of σ; for every n-simplex σ, the $n+1$ n-simplexes which share an $(n-1)$-simplex with σ; for every 0-simplex, one n-simplex incident in it.

The vertices and n-simplexes are stored in two arrays. In the array of vertices, for each vertex its Cartesian coordinates are encoded, and the field value associated with it. In the array of n-simplexes, with each n-simplex σ of the underlying mesh Σ the indexes of the minimum and of the maximum node in $\mathbb{G}$ are associated such that σ belongs to the corresponding ascending n-cell of the minimum, and descending n-cell of the maximum.

The resulting data structure is completely dimension-independent, since both the encoding of the mesh and of the graph are independent of the dimension of the mesh and of the algorithm used for the extraction of Morse complexes. The only geometry is the one of the maximal cells in the two Morse complexes, from which the geometry of all the other cells of the Morse complexes can be extracted. The geometry of these cells can be computed iteratively, from higher to lower

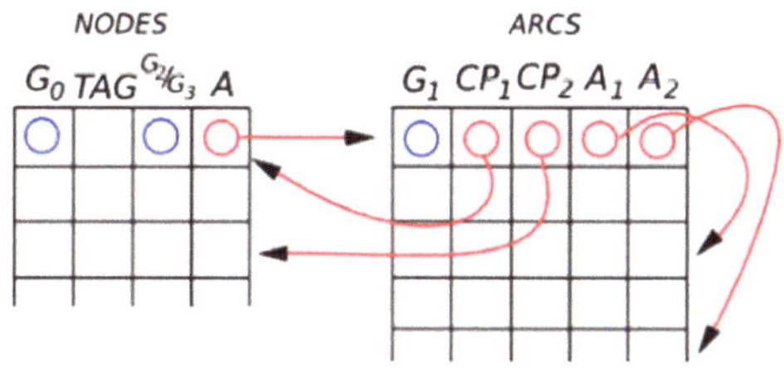

Fig. 1.8 Dimension-specific data structure for storing the incidence graph. Nodes and arcs are stored in lists. Each element in the list of nodes stores the geometry of the corresponding critical point, tag indicating the index of the critical point, geometry of the associated 2- or 3-cell in the Morse complex, and a pointer to one incident arc. Each element in the list of arcs stores the geometry of the arc, two pointers to its endpoints, and two pointers to the next arcs incident in the two endpoints

dimensions, by searching for the k-simplexes that are shared by $(k + 1)$-simplexes belonging to different $(k + 1)$-cells.

The incidence-based data structure encodes also the topology of the Morse-Smale complex. The arcs in the graph (i.e., pairs of nodes connected through the arc) correspond to 1-cells in the Morse-Smale complex. Similarly, pairs of nodes connected through a path of length k correspond to k-cells in the Morse-Smale complex. The geometry of these cells can be computed from the geometry of the cells in the Morse complex through intersection. For example, the intersection of ascending n-cells corresponding to minima and descending n-cells corresponding to maxima defines n-cells in the Morse-Smale complex.

1.4.2 A Dimension-Specific Representation for 3D Morse-Smale Complexes

In [25] a data structure for 3D Morse-Smale complexes is presented. The topology of the Morse-Smale complex (actually of its 1-skeleton) is encoded in a data structure equivalent to the Morse incidence graph. The geometry is referred to from the elements of the graph, arcs and nodes. We illustrate this data structure in Fig. 1.8.

The data structure encodes the nodes and arcs of the incidence graph in two arrays. Each element in the list of nodes has four fields, G_0, TAG, G_2/G_3 and A. The geometry (coordinates) of the corresponding critical point is stored in field G_0. The index of the critical point is stored in field TAG. A reference to the geometry of the associated Morse cell (depending on the index of p) is stored in field G_2/G_3: a descending 3-cell is associated with a maximum; an ascending 3-cell is associated with a minimum; a descending 2-cell is associated with a 2-saddle; an ascending 2-cell is associated with a 1-saddle. A pointer to an arc incident in the node (the first one in the list of such arcs) is stored in field A. Thus, the geometry of 0-, 2-, and 3-cells in the Morse complexes is referenced from the nodes.

Each element in the list of arc has five fields, G_1, CP_1, CP_2, A_1 and A_2. The geometry of the integral line (corresponding to a 1-cell in the Morse-Smale complex) encoded by the arc is stored in field G_1. The pointers to the nodes connected by the arc are stored in fields CP_1 and CP_2. Fields A_1 and A_2 contain pointers to the next arcs incident in nodes pointed at by CP_1 and CP_2, respectively.

The data structure in [25] is dimension-specific, because it represents 0-, 2- and 3-cells of the Morse complexes in the nodes, and 1-cells of the Morse-Smale complexes in the arcs of the incidence graph. The descending 1-cells in the Morse complex can be obtained as union of (the geometry associated with) two arcs incident in a node corresponding to a 1-saddle, and ascending 1-cells can be obtained as union of two arcs incident in a 2-saddle.

1.4.3 Comparison

The data structure in [25] encodes the combinatorial representation of the 1-skeleton of the Morse-Smale complex, which is equivalent to the encoding of the Morse incidence graph in [11].

Let us denote as n the number of nodes, and as a the number of arcs in the incidence graph. Both data structures encode the nodes and arcs of G in lists. Thus, the cost of maintaining those lists in both data structures is $n + a$. In the incidence-based representation in [11], for each arc there are two pointers pointing to it (one from each of its endpoints) and there are two pointers from the arc to its two endpoints. Thus, storing the connectivity information of the Morse incidence graph requires $4a$ pointers in [11]. In the data structure in [25], for each node there is a pointer to one arc incident in it, and for each arc there are four pointers, two pointing to its endpoints, and two pointing to the next arcs incident in the endpoints. This gives a total cost of $n + 4a$ pointers for storing the connectivity information of the graph in [25].

The difference between the two representations is how geometry is encoded. In the 3D instance of the incidence-based data structure, only the list of tetrahedra forming the ascending and descending 3-cells are encoded. This leads to a cost of twice the number of tetrahedra in the simplicial mesh Σ since each tetrahedron belongs to exactly one ascending and one descending 3-cell. The data structure in [25] encodes the geometry of the arcs (i.e., the 1-cells in the Morse-Smale complex), the geometry of the ascending and descending 3-cells in the Morse complexes, associated with the nodes encoding the extrema, and the geometry of the ascending and descending 2-cells in the Morse complexes associated with the nodes encoding the saddles. We cannot evaluate precisely the storage cost of this latter data structure, since in [25] it is not specified how the underlying geometry is encoded. However, the combinatorial part of the two data structures has almost the same cost. Thus, it is clear that the incidence-based representation is more compact since it encodes fewer geometric information.

1.5 Algorithms for Building 3D Morse and Morse-Smale Complexes

In this section, we describe and compare algorithms for extracting Morse and Morse-Smale complexes from a scalar field f defined on the vertices of a manifold simplicial mesh Σ in 3D. Similarly to the 2D case, extraction and classification of critical points is a usual preprocessing step. An algorithm performing this task is proposed in [21]. For each vertex p of Σ, the lower link $Lk^-(p)$ of p is considered. It consists of the vertices q in the link $Lk(p)$ of p such that $f(q) < f(p)$, and of the simplexes of $Lk(p)$ defined by these vertices. Vertex p is classified as a minimum if its lower link is empty. It is classified as a maximum if its lower link is the same as $Lk(p)$. Otherwise, p is classified based on the Betti numbers of $Lk^-(p)$ as a critical point composed of multiple 1- and 2-saddles. Intuitively, the Betti numbers β_0 and β_1 of $Lk^-(p)$ count the number of connected components and holes in $Lk^-(p)$, respectively.

The algorithms presented here can be classified, according to the approach they use, as *region-based* [11, 26], *boundary-based* [19, 21], or *based on Forman theory* [27, 31, 38]. Region-based algorithms extract only the minima and maxima of f, and do not explicitly extract saddle points. Boundary-based algorithms [19, 21] first extract and classify critical points of f (minima, maxima, and multiple 1- and 2-saddles) in the preprocessing step (using the method in [21]), and then compute the ascending and descending 1- and 2-cells associated with saddles. The algorithms in [27, 31, 38] construct a Forman gradient vector field V and its critical cells starting from a scalar field f.

The output of the algorithm in [26] is a decomposition of the vertices of Σ into 0-, 1-, 2- and 3-cells of the Morse complexes of f. Algorithms in [19, 21] produce 3-, 2-, 1- and 0-cells of the Morse and Morse-Smale complexes composed of tetrahedra, triangles, edges and vertices of Σ, respectively. The output of the algorithms based on Forman theory [27,31,38] (Forman gradient vector field V) can be used to obtain also the decomposition of the underlying mesh K into descending cells associated with critical cells of V. Each descending cell of a critical i-cell σ is composed of all i-cells of K that are reachable by tracing gradients paths of V starting from the boundary of σ. The algorithms in [11, 27] produce the graph encoding the connectivity of Morse and Morse-Smale complexes. In [27], an algorithm based on Forman theory has been developed to obtain the nodes and arcs of the graph. The algorithm in [11] obtains the graph starting from any segmentation of the tetrahedra of Σ in descending and ascending 3-cells of the Morse complexes of f.

1.5.1 A Watershed-Based Approach for Building the Morse Incidence Graph

In [11], a two-step algorithm is described for the construction of the Morse incidence graph of a scalar field f, defined on the vertices of a simplicial complex Σ with a manifold carrier. The first step is the decomposition of Σ in descending and ascending n-cells of the Morse complexes. In [11], this decomposition is obtained by extending the well-known watershed algorithm based on simulated immersion from image processing to n-dimensional manifold simplicial meshes [40]. The first step of the algorithm is, thus, dimension-independent. The second step of the algorithm, developed for the 2D and 3D cases, consists of the construction of the Morse incidence graph.

The watershed algorithm by simulated immersion has been introduced in [40] for segmenting a 2D image into regions of influence of minima, which correspond to ascending 2-cells. We describe the extension of this algorithm from images to scalar fields defined at the vertices of a simplicial mesh in arbitrary dimension. The vertices of the simplicial mesh Σ are sorted in increasing order with respect to the values of the scalar field f, and are processed level by level in increasing order of function values. For each minimum m, an ascending region $A(m)$ is iteratively constructed through a breadth-first traversal of the 1-skeleton of the simplicial mesh Σ (formed by its vertices and edges). For each vertex p, its adjacent, and already processed, vertices in the mesh are examined. If they all belong to the same ascending region $A(m)$, or some of them are watershed points, then p is marked as belonging to $A(m)$. If they belong to two or more ascending regions, then p is marked as a watershed point. Vertices that are not connected to any previously processed vertex are new minima and they start a new ascending region.

Each maximal simplex σ (an n-simplex if we consider an n-dimensional simplicial mesh) is assigned to an ascending region based on the labels of its vertices. If all vertices of σ, that are not watershed points, belong to the same region $A(m)$, then σ is assigned to $A(m)$. If the vertices belong to different ascending regions $A(m_i)$, then σ is assigned to the region corresponding to the lowest minimum.

Descending regions associated with maxima are computed in a completely similar fashion.

The algorithm proposed in [11] for the construction of the Morse incidence graph of f works on a segmentation produced by the watershed algorithm, although any other segmentation algorithm can be used. In the (dimension-independent) preprocessing step, for each descending region in Γ_d, a maximum node in the incidence graph is created, and for each ascending region in Γ_a, a minimum node is created. The algorithm for the construction of saddle nodes is based on inspecting the adjacencies between the regions corresponding to maxima and minima, and is developed for the 2D and the 3D case.

In the 2D case, after the preprocessing step, two steps are performed: (i) creation of the nodes corresponding to saddles, and (ii) creation of the arcs of the incidence graph. To create the saddle nodes, 1-cells of the ascending (or of the descending) complex need to be created. Each 1-cell is a chain of edges of the triangle mesh. Each edge e of Σ is inspected, and is classified with respect to such chain of edges based on the labels of the ascending regions to which the two triangles separated by e belong. Each connected component of edges separating two ascending regions is subdivided into topological 1-cells. Thus, if necessary, new saddle nodes are created. Each saddle node (1-cell) p is connected to the two minima it separates. The arcs

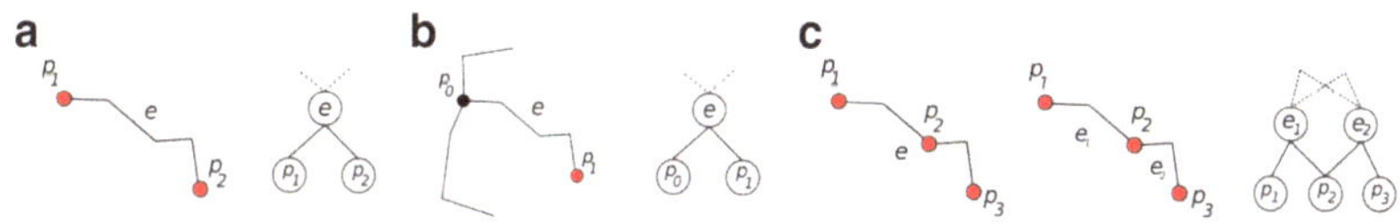

Fig. 1.9 Connection of the saddle nodes with maxima in 2D: (**a**) both endpoints of the 1-cell are maxima; (**b**) one of the endpoints is not a maximum; (**c**) there is a maximum in the interior of the 1-cell

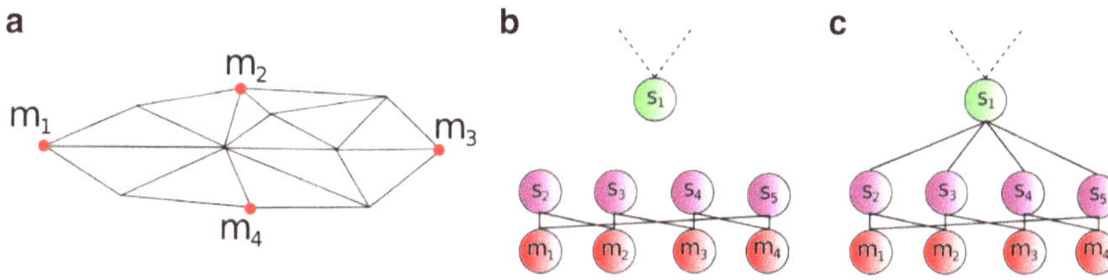

Fig. 1.10 Connection of the 1- and 2-saddle nodes in 3D: (**a**) maxima on the boundary of ascending 2-cell of 1-saddle s_1; (**b**) 2-saddles s_2, s_3, s_4 and s_5 connected to two maxima on the boundary of the 2-cell; (**c**) these 2-saddles are connected to 1-saddle s_1

between saddle nodes and nodes corresponding to maxima are created by inspecting the endpoints of the 1-cells. Three cases are distinguished, illustrated in Fig. 1.9: if the endpoints of 1-cell p are two maxima, the saddle node corresponding to p is connected to those maxima. If one of the endpoints is not a maximum, a new maximum is created and connected to the saddle. If there is a maximum inside 1-cell p, p is split in two 1-cells, each of which with that maximum as endpoint. If there is some maximum q not connected to any saddle, then that maximum must be inside some 2-cell in Γ_a. In this case, a 1-saddle is created by looking at the 2-cells corresponding to q and at its adjacent 2-cells in Γ_d.

The construction of the Morse incidence graph in 3D requires, after the preprocessing, other three steps, namely, (i) generation of the nodes corresponding to 1- and 2-saddles, (ii) generation of the arcs between 1-saddles and minima, and between 2-saddles and maxima, and (iii) generation of the arcs joining 1- and 2-saddles. The first two steps directly generalize the 2D algorithm. The third step consists of generating the arcs connecting the nodes corresponding to 1-saddles to those corresponding to 2-saddles. For each 2-cell s_1 in Γ_a (which corresponds to a 1-saddle), the set M_s of maxima connected to s_1 is considered, which correspond to the vertices of 2-cell s_1. For each pair of maxima m_1 and m_2 in M_s, it is verified if there exists a 2-cell s_2 (i.e., a 2-saddle) in the descending complex Γ_d between the 3-cells corresponding to m_1 and m_2. If s_2 exists, then the two nodes corresponding to 1-saddle s_1 and 2-saddle s_2 are connected in the MIG. The third step of the algorithm is illustrated in Fig. 1.10. A technique for processing 2-cells which are on the boundary of Σ has been also developed; it is not described here for the sake of brevity.

In summary, the algorithm in [11] is organized in two steps: segmentation of the simplicial mesh Σ into Morse complexes, and extraction of the incidence graph. The first step is dimension-independent. It is based on the extension of a watershed algorithm for intensity images to scalar fields defined on simplicial complexes in arbitrary dimension. The second step, developed for the 2D and 3D cases, constructs the nodes and arcs of the *MIG* encoding the Morse complexes generated at the first step.

1.5.2 A Boundary-Based Algorithm

The algorithm proposed in [21] builds a quasi-Morse-Smale complex (see Sect. 1.2), a complex that reflects the combinatorial structure of the Morse-Smale complex, but in which the arcs and quadrangles (1- and 2-cells) may not be those of maximal ascent and descent. The quasi-Morse-Smale complex is constructed during two sweeps over a simplicial complex Σ triangulating a 3-manifold M. The first sweep (in the direction of decreasing function value) computes the descending 1- and 2-cells and the second sweep (in the direction of increasing function value) the ascending 1- and 2-cells of the Morse complexes. The algorithm is boundary-based, as it computes the 1- and 2-cells which bound the 3-cells in the Morse complexes. During the first sweep, the descending 1- and 2-cells are computed simultaneously. A 1-cell is built as follows:

- If a current vertex p in the sweep is a 1-saddle, a descending 1-cell is started. The two arcs of the corresponding 1-cell are initialized by edges from p to the lowest vertex in each connected component of the lower link of p, as illustrated in Fig. 1.11a.
- If there is a descending arc ending at a current vertex p, it is expanded by adding an edge from p to the lowest vertex in its lower link. If p is a 1-saddle, later an ascending 2-cell will start at p and each descending arc is extended to the lowest vertex in the specific connected component of the lower link of p that is not separated from the arc by the ascending 2-cell.
- If p is a minimum, it becomes a node of the Morse-Smale complex, and the descending arcs end at p.

A 2-cell is built as follows:

- If a current vertex p in the sweep is a 2-saddle, a descending 2-cell is started. A cycle of edges in the lower link is constructed, which contains the lowest vertex in the lower link of p. Triangles determined by p and edges of the cycle form the initial descending 2-cell of p, as illustrated in Fig. 1.11b. Initially, the entire boundary of the descending 2-cell is unfrozen.
- A 2-cell is expanded by constructing a shortest-path tree in the lower link of the current (highest) vertex q on the unfrozen boundary of the 2-cell associated with 2-saddle p, and connecting q to the edges of this tree. If q is a critical point (a 1-saddle or a minimum), it is declared frozen together with its two incident edges on the boundary.
- When the complete boundary of a 2-cell is frozen the 2-cell is completed.

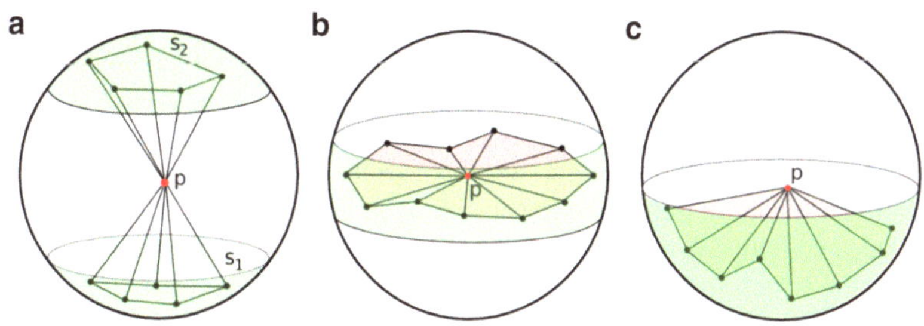

Fig. 1.11 (**a**) The 1-cell associated with 1-saddle p is initialized by connecting p to the two lowest vertices s_1 and s_2 in its lower link in [21]. (**b**) The 2-cell associated with a 2-saddle p is initialized by the triangles determined by p and a cycle of edges in the lower link of p in [21]. (**c**) Expanding a separating 2-cell at a regular vertex p in [19]

The next step consist of building the intersections between descending and ascending 2-cells by tracing ascending paths inside a descending 2-cell, starting from 1-saddles on the boundary of the descending 2-cell and ending at the 2-saddle that started the descending 2-cell. These intersections are used to guarantee the structural correctness of the extracted quasi-Morse-Smale complex. Each 2-saddle starts two arcs of an ascending 1-cell, which must not cross any already established descending 2-cells. The intersection curves between descending and ascending 2-cells, and the ascending 1-cells decompose each ascending 2-cell into quadrangles. The ascending cells are built one quadrangle at a time, similarly to descending 2-cells.

In summary, the algorithm in [20] extracts the boundaries of the 3-cells in the Morse-Smale complex. The extracted complex has the correct combinatorial structure described by a quasi-Morse-Smale complex. Each 3-cell in the extracted complex has quadrangular faces.

1.5.3 A Watershed-Based Labeling Algorithm

In [19], an algorithm is proposed that extracts 3-cells in the descending Morse complex starting from the values of a scalar field f defined over a triangulation Σ of a manifold M. To this aim the algorithm generates two functions on the simplexes of Σ: the *marking function* marks the simplexes of Σ that form the boundaries between descending 3-cells by 1, and the other simplexes of Σ are marked by 0; the *labeling function* labels each simplex σ of Σ marked by 0 by the label of the maximum whose descending 3-cell contains σ. The vertices are inspected in decreasing order of function value. Depending on the type of criticality of a current vertex p, the lower star of p (defined by p and simplexes in the lower link $Lk^{-}(p)$ of p) is processed.

- If p is a maximum, its lower link is equal to its link. Vertex p starts a new descending 3-cell. All simplexes in the lower star of p are labeled by the label of this 3-cell.
- If p is a regular point (see Fig. 1.11c), its lower link is a deformation retract of a disk. If there is a separating 2-cell that reached p, it is extended across p by creating a spanning tree in the lower link of p. The spanning tree is constructed so that it contains all vertices that belong to an already marked simplex (i.e., to a simplex which is part of the boundary between two descending 3-cells). All triangles and edges connecting p to this spanning tree are marked (extending a descending 2-cell between two 3-cells). Other non-labeled (and non-marked) simplexes in the star of p are labeled by copying from the neighbors. Otherwise (if there is no separating 2-manifold containing p), non-labeled simplexes in the star of p are labeled by copying from neighbors.
- If p is a 1-saddle, its lower link has two components, each a deformation retract of a disk. Each component of the lower link of p is processed in the same way as in the case of a regular point.
- If p is a 2-saddle, its lower link is a deformation retract of an annulus. Vertex p starts a new separating 2-cell. A cycle that encircles the whole lower link of p is created. All triangles and edges connecting p to this cycle are marked. They form the initial separating 2-cell associated with p. Other non-labeled simplexes in the star of p are labeled by copying from neighbors.
- If p is a minimum, its lower link is empty, and p is marked.

The descending 3-cells of maxima produced by the algorithm in [19] are topological cells.

1.5.4 A Region-Growing Algorithm

The algorithm proposed in [26] computes the Morse-Smale complex of a function f defined over the vertices of a simplicial complex Σ triangulating a manifold M. The ascending cells are computed through region growing, in the order of decreasing cell dimension. Descending cells are computed inside the ascending 3-cells, using the same region-growing approach. The ascending and descending cells of all dimensions are composed of vertices (0-simplexes) of Σ.

The computation of the ascending 3-cells consists of two steps. First, the set of minima of f are identified. Each minimum will be the origin for a set of vertices representing an ascending 3-cell. Then, each vertex p of Σ is classified as an internal vertex of an ascending cell, or as a boundary vertex. This depends on the number of connected components of the set of internal vertices in the lower link of p which are already classified as interior to some ascending 3-cell (see Fig. 1.12a). The classification is performed by sweeping Σ in the order of ascending function values.

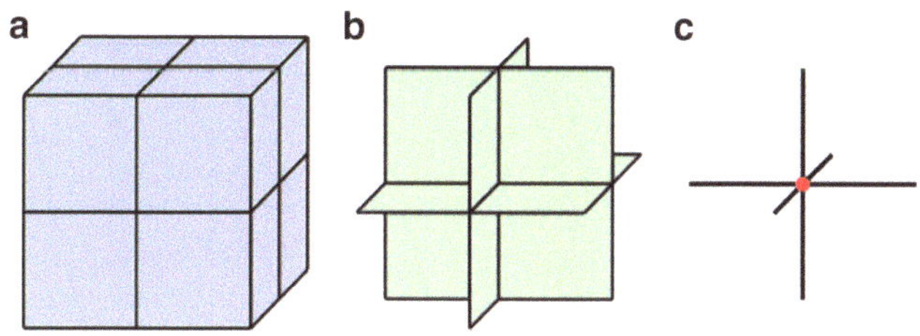

Fig. 1.12 Classification of the vertices of Σ as internal or boundary. (**a**) All the vertices are classified as internal to a 3-cell (*blue*) with the exception of vertices on the boundary of two or more 3-cells, which are classified as boundary. (**b**) The boundary vertices in (**a**) are classified again as internal (*green*) or boundary for 2-cells. (**c**) The boundary vertices of the 1-cells in (**b**) is classified as maxima (*red*)

Vertices classified as boundary in the first step of the algorithm are the input for the algorithm which builds the ascending 2-cells. An ascending 2-cell is created for each pair of adjacent 3-cells. The vertices of the 2-cells are classified as interior or boundary, based on local neighborhood information, similarly to the classification with respect to the 3-cells (see Fig. 1.12b). A 1-cell is created in every place where ascending 2-cells meet. Each 1-cell is composed of vertices classified as boundary in the previous step. Finally, each vertex p of an ascending 1-cell is classified as interior or boundary. Maxima are created at the boundaries between ascending 1-cells (see Fig. 1.12c). They form a small disjoint clusters of vertices.

For each ascending n-cell, the descending cells are computed in their interior. The region-growing steps are the same. Again here, iteration is performed in the order of decreasing dimension.

The main characteristics of the algorithm in [26] is that all the cells in the computed Morse complexes are composed of vertices of the simplicial mesh Σ. These cells are computed iteratively in order of decreasing dimension.

1.5.5 An Algorithm Based on Forman Theory

The algorithm proposed in [27] computes the Morse-Smale complex starting from a regular n-dimensional CW-complex K with scalar field f defined at the vertices of K. Intuitively, a (finite) CW complex is a finite collection of pairwise disjoint cells, in which the boundary of each cell is the disjoint union of cells of lower dimension. Function f is extended to a Forman function F, defined on all cells of K, such that $F(\sigma)$ is slightly larger than $F(\tau)$ for each cell σ and each face τ of σ. For the defined Forman function F, all cells of K are critical. A discrete gradient vector field is computed by assigning gradient arrows in a greedy manner in ordered sweeps over the cells of K according to increasing dimension and increasing F value. Each current non-paired and non-critical cell in the sweep is paired with its co-facet with only one facet not marked (as critical or as already paired). If there

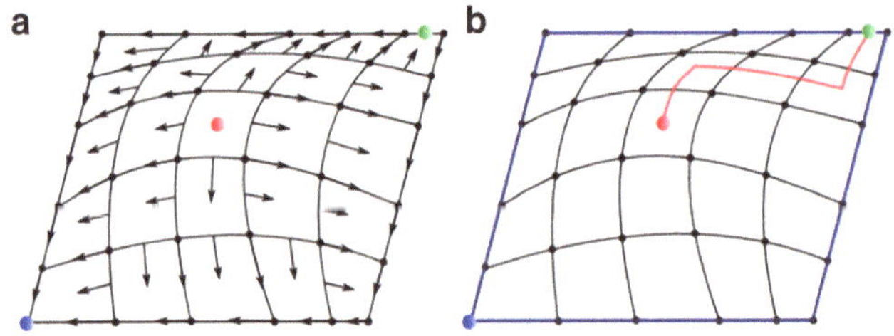

Fig. 1.13 (**a**) Construction of the Forman gradient vector field, and (**b**) of the incidence graph

are several of such co-facets the lowest is taken. If there is no such co-facet, a cell cannot be paired, and it is critical. This pairing defines a discrete gradient vector field, as illustrated in Fig. 1.13a.

The 1-skeleton of the Morse-Smale complex is computed starting from this gradient vector field. Critical cells of F (and not critical points of f) and the discrete gradient paths connecting them determine the nodes and arcs in the 1-skeleton of the Morse-Smale complex (incidence graph), as illustrated in Fig. 1.13b. In [23], this algorithm has been extended to extract topological 2- and 3-cells from a regular hexahedral 3D mesh.

The order in which the cells in K are processed by the algorithm is not completely deterministic, since there could be many different i-cells in K with the same value of function F. As a consequence, some unnecessary critical cells may be produced by the algorithm.

1.5.6 A Forman-Based Approach for Cubical Complexes

In [38], a dimension-independent algorithm is proposed for constructing a Forman gradient vector field on a cubical complex K with scalar field values given at the vertices, and applications to the 2D and 3D images are presented.

The algorithm processes the lower star of each vertex v in K independently. For each cell σ in the lower star, the value $\max_{p\in\sigma} f(p) = fmax(\sigma)$ is considered. An ascending order Δ is generated based on the values $fmax(\sigma)$ and the dimension of σ, such that each cell σ comes after its faces in the order. If the lower star of vertex v is v itself, then v is a local minimum and it is added to the set C of critical cells. Otherwise, the first edge e in Δ is chosen and vertex v is paired with edge e (the vector field V at v is defined as $V(v) = e$).

The star of v is processed using two queues, *PQone* and *PQzero*, corresponding to i-cells with one and zero unpaired faces, respectively. All edges in the star of v different from e are added to *PQzero*. All cofaces of e are added to *PQone* if the number of unpaired faces is equal to one.

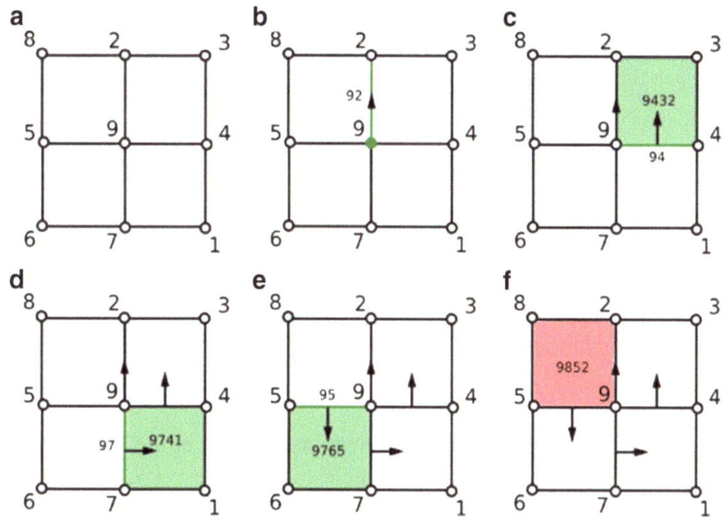

Fig. 1.14 Processing the lower star of vertex 9 using the algorithm in [38]

If queue *PQone* is not empty, the first cell α is removed from the queue. If the number of unpaired faces of α has become zero, α is added to *PQzero*. Otherwise, the vector field at the unique unpaired face $pair(\alpha)$ of α is defined as $V(pair(\alpha)) = \alpha$, $pair(\alpha)$ is removed from *PQzero* and all the co-faces of either α or $pair(\alpha)$ and with number of unpaired faces equal to one are added to *PQone*.

If *PQone* is empty and *PQzero* is not empty, one cell β is taken from *PQzero*. Cell β is added to the set C of critical points and all the co-faces of β with number of unpaired faces equal to one are added to *PQone*.

If both *PQzero* and *PQone* are empty, then the next vertex is processed. Result of the algorithm is the set C of critical cells and the pairing of non-critical cells, which define the Forman gradient vector field V.

In Fig. 1.14 we show the main steps of the algorithm in [38] when processing the lower star of vertex 9 (see Fig. 1.14a). Each vertex is labeled by its scalar field value. Other cells are labeled by the lexicographic order Δ. The lower star of 9 is not 9 itself, and thus 9 is not a minimum. The lowest edge starting from 9 (edge 92), is chosen to be paired with 9. All the other edges are inserted in *PQzero* and the cofaces of 92 with a single unpaired face (faces 9432 and 9852) are inserted in *PQone* (Fig. 1.14b). The first face is taken from *PQone* (face 9432) and coupled with its single unpaired face (edge 94). The face 9741, which is a coface of 94 with exactly one unpaired face, is inserted in *PQone* and edge 94 is removed from *PQzero* (Fig. 1.14c). Face 9741 is taken from *PQone* and paired with edge 97, which is removed from *PQzero*. Face 9765 is inserted in *PQone* and successively removed to be paired with edge 95 (Fig. 1.14d, e). Face 9852 is removed from *PQone* and declared critical, as it has no unpaired faces (Fig. 1.14f).

In the 3D case, the algorithm in [38] does not create spurious critical cells. The extracted critical cells are in a one-to-one correspondence with the changes in topology in the lower level cuts of cubical complex K.

1.5.7 A Forman-Based Approach for Simplicial Complexes

The algorithm proposed in [31] takes as input a scalar field f defined over the vertices of a 3D simplicial complex Σ and a persistence value $p \geq 0$. It computes a Forman gradient vector field V by subdividing the simplexes of Σ into three lists, denoted as A, B and C, such that lists A and B are of the same length, and for each i-simplex $\sigma_j \in A$, $V(\sigma_j) = \tau_j$, τ_j is the $(i+1)$-simplex in B, and C is the set of critical simplexes.

The algorithm builds the Forman gradient vector field in the lower link $Lk^-(v)$ of each vertex v in Σ, and extends this field to the cone $v * Lk^-(v)$. Lists A, B and C are initialized as empty. For each vertex v in Σ, if $L(v)$ is empty, then v is a minimum and it is added to C. Otherwise, v is added to A and the algorithm is recursively called on the lower link $L^-(v)$, producing lists A', B', C' that define the Forman gradient vector field V' on $Lk^-(v)$. The lowest critical vertex w is chosen from C' and edge $[v,w]$ is added to B. Thus, $V(v) = [v,w]$. For each i-simplex σ (different from w) in C' the $(i+1)$-simplex (cone) $v * \sigma$ is added to C. For each i-simplex σ in A' the $(i+1)$-simplex $v * \sigma$ is added to A and the $(i+2)$-simplex $v * V'(\sigma)$ is added to B. Thus, $V(v * \sigma) = v * V'(\sigma)$.

Once all the lower links of vertices in Σ have been processed, a *persistence canceling step* is performed in increasing order of dimension i. For each critical i-simplex σ, all the gradient paths to critical $(i-1)$-simplexes are found. A critical i-simplex σ can be cancelled with critical $(i-1)$-simplex γ if and only if there is only one gradient path from σ to γ. The effect of a cancellation is to reverse the gradient path connecting σ and γ. Cancellations are applied in the order of increasing persistence. The function that extends the scalar field f to the simplexes of Σ, and whose values are considered in the definition of persistence, is given by $fmax(\sigma) = \max_{p \in \sigma} f(p)$.

We illustrate the algorithm in [31] in Fig. 1.15. The star of vertex 9 is shown in Fig. 1.15a. The application of the algorithm to the lower link $Lk^-(9)$ of vertex 9 produces the following lists:

A′ = 3; 4;6; 7; 8
B′ = [3,2]; [4,1]; [6,5]; [7,1]; [8,2]
C′ = 1; 2; [4,3]; 5; [7,6]; [8,5]

The corresponding Forman gradient vector field V' on $Lk^-(9)$, and V' after the cancellation of vertex 2 and edge [3,4], and cancellation of vertex 5 and edge [6,7], are shown in Fig. 1.15b, c, respectively. The extension of V' to the cone $9 * Lk^-(9)$ (the lower star of vertex 9) is shown in Fig. 1.15d. Descending and ascending regions of critical cells of Forman vector field V constructed in [31] are computed in [29].

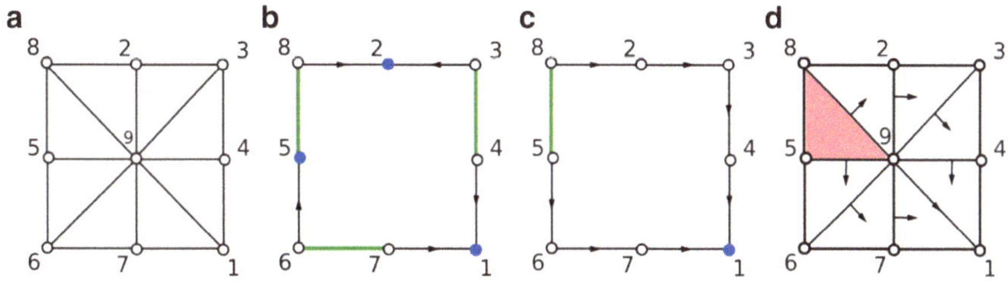

Fig. 1.15 (**a**) The lower star of vertex 9. The Forman gradient vector field V' on the link of 9 (**b**) before and (**c**) after the cancellation of critical edge [4,3] and vertex 2, and edge [7,6] and vertex 5. The critical vertices are *blue*, and the critical edges are *green*. (**d**) The Forman gradient vector field V in the lower star of vertex 9

Table 1.1 Reviewed algorithms classified according to the type of the input complex (simplicial or cell), dimension of the complex, approach used (region-based, boundary-based or based on Forman theory), type of extracted critical entities (critical points of the given scalar field f or critical cells of the constructed Forman gradient vector field V) and type of entities that form the extracted cells in the Morse complexes (cells or vertices of the input complex)

Algorithm	Input	Dimension	Approach	Critical points	Morse cells
[11]	Simplicial	3D	Region	Extrema	Simplexes of Σ
[21]	Simplicial	3D	Boundary	All	Simplexes of Σ
[19]	Simplicial	3D	Boundary	All	Simplexes of Σ
[26]	Simplicial	3D	Region	Extrema	Vertices of Σ
[27]	CW	nD	Forman	Critical cells	Cells of K
[38]	Cubical	nD	Forman	Critical cells	Cells of K
[31]	Simplicial	3D	Forman	Critical cells	Cells of Σ

If Σ is a triangulation of a 3D manifold and the scalar field f defined at the vertices of Σ has no multiple saddles, then there is a critical i-cell of the computed Forman gradient vector field V for each i-saddle of the scalar field f.

1.5.8 Analysis and Comparison

We have described seven algorithms for extracting Morse or Morse-Smale complexes of a scalar field f in the discrete case. We summarize the properties of the reviewed algorithms in Table 1.1.

We can classify the reviewed algorithms based on different criteria. All algorithms work on a 3D manifold simplicial complex Σ except the ones in [27, 38], which work on an arbitrary-dimensional CW-complex and cubical complex, respectively.

With respect to the approach used, some of the algorithms we reviewed [11, 26] can be classified as region-based, as they extract 3-cells corresponding to extrema. Others [19, 21] are boundary based, as they extract the boundaries between the

3-cells corresponding to extrema. The algorithms in [27, 31, 38] compute a Forman gradient vector field V starting from scalar field f.

The algorithms differ also in the type of critical points they extract before producing the segmentation of the input mesh: some algorithms [19,21] first classify all critical points of scalar field f (extrema and saddles); others [11, 26] extract only extrema of f, and obtain the other nodes in the Morse incidence graph from the adjacency relation between 3-cells associated with extrema. The algorithms in [27,31,38] extract the critical cells of a Forman gradient vector field V (and not the critical points of scalar field f) defined through f.

Finally, another difference among the algorithms is given by the entities used in the segmentation process: the algorithms in [11, 19, 21] assign the simplexes of Σ to cells in the Morse complexes; the algorithm in [26] assigns the vertices of Σ to cells in the Morse complexes; the algorithms in [27, 31, 38] assign the cells of the cell complex K to cells in the descending Morse complex.

The algorithm in [20] computes the segmentation of the 3D simplicial mesh with the correct combinatorial structure described by the quasi-Morse-Smale complex. The algorithm in [19] produces 3-cells in the descending Morse complex, which are topological cells. In the 3D case, the algorithm in [38] computes the critical cells of the Forman gradient vector field V that are in a one-to-one correspondence with the changes in topology in the lower level cuts of K. If Σ is a triangulation of a manifold M, and scalar field f has no multiple saddles, then the algorithm in [31] produces a critical i-cell of the Forman gradient vector field for each i-saddle of scalar field f. There are no formal claims about the critical cells of Forman gradient vector field computed by the algorithm in [26].

1.6 Simplification of 3D Morse and Morse-Smale Complexes

Although Morse and Morse-Smale complexes encode compactly the behavior of a scalar field f and the topology of its domain, simplification of these complexes is an important issue for two major reasons. The first is the over-segmentation (the presence of a large number of small and insignificant regions) produced by Morse and Morse-Smale complexes which is due to the presence of noise in the data sets, both in case they are obtained through measurements or as a result of a simulation algorithm. Simplification of the complexes through elimination of pairs of critical points can be used to eliminate noise. Each simplification is performed guided by *persistence*, which measures the importance of the pair of eliminated critical points, and is equal to the absolute difference in function values between them. Usually by using a threshold on persistence equal to 5–10 % of the maximum persistence value, a reduction of the storage cost in the representation of the Morse or Morse-Smale complexes can be obtained which amounts to 10–20 % for 2D data sets, and 5–10 % for 3D ones.

Even after a simplification which removes insignificant regions considered to be noise, and leaves regions that correspond to actual features of the scalar field, the size of Morse and Morse-Smale complexes can still be large, due to the huge size and amount of available scientific data. Thus, the second requirement is to reduce the size of the complexes at different levels of resolution, while retaining information on important structural features of the field and guaranteeing the topological correctness of the simplified representations. The size of a Morse complex is about 0.4–3 % the size of the underlying simplicial mesh in 2D and about 2–8 % in 3D. For large 3D data sets (which have 15M tetrahedra), the size of a Morse complex can be up to 50 MBytes.

We describe here two approaches to the simplification of Morse and Morse-Smale complexes in 3D proposed in the literature. The first approach [24] applies the *cancellation* operator of critical points of a Morse function f [33] on the Morse-Smale complexes of f. The second approach [10] applies a new set of simplification operators, called *removal* and *contraction*, which, together with their inverse refinement ones, form a minimally complete basis of operators for performing any topologically consistent simplification on Morse and Morse-Smale complexes.

Both cancellation and removal/contraction operators eliminate a pair of critical points of scalar field f, i.e., a pair of cells in the Morse complexes and a pair of vertices from the Morse-Smale complex. The difference between the two approaches to simplification is that cancellation often introduces a large number of cells (of dimension higher than zero) in the Morse-Smale complex, while this never happens with removal/contraction. Cancellation operator applied on large data sets can create complexes that exceed practical memory capabilities [28]. Removal/contraction operator, on the other hand, reduces the size of the complexes at each step of the simplification process.

1.6.1 Cancellation in 3D

In this section, we review the cancellation operator, which simplifies a Morse function f defined on a manifold M by eliminating its critical points in pairs [33]. Two critical points p and q can be cancelled if

1. p is an i-saddle and q is an $(i + 1)$-saddle, and
2. p and q are connected through a unique integral line of f.

After the cancellation of p and q, each critical point t of index at least $i + 1$, which was connected through integral line of f to i-saddle p becomes connected to each critical point r of index at most i, which was connected to $(i+1)$-saddle q before the cancellation. Equivalently, in a descending (and, symmetrically, ascending) Morse complex, an i-cell p and an $(i + 1)$-cell q can be simultaneously cancelled if cell p appears exactly once on the boundary of cell q. After the cancellation, each cell r which was on the boundary of $(i + 1)$-cell q becomes part of the boundary of each cell t which was in the co-boundary of i-cell p. In the Morse-Smale complex, there is a new k-cell for each two cells t and r that become incident to each other in the Morse complexes after the cancellation and that differ in dimension by k.

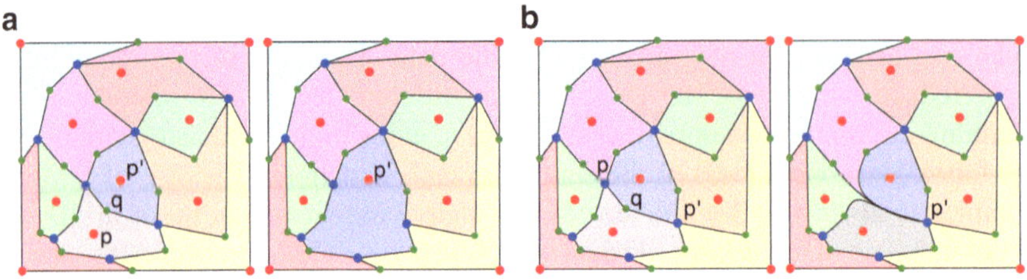

Fig. 1.16 (**a**) Cancellation of a maximum p and saddle q, and (**b**) cancellation of a minimum p and a saddle q, on the 2D descending Morse complex illustrated in Fig. 1.2a

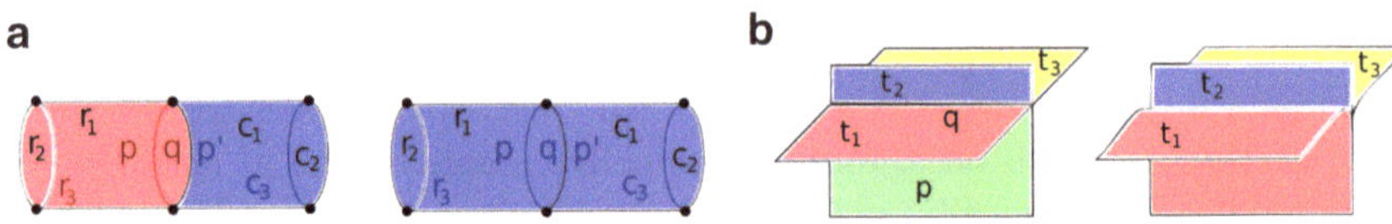

Fig. 1.17 Portion of a 3D descending Morse complex before and after (**a**) a cancellation of maximum p and 2-saddle q, and (**b**) a cancellation of a 1-saddle p and 2-saddle q

In 2D, there are two cancellation operators: cancellation of a maximum and a saddle, and cancellation of a minimum and a saddle. A cancellation of a maximum p and a saddle q is illustrated in Fig. 1.16a. It is feasible if 1-cell q is shared by exactly two 2-cells p and p'. After the cancellation, 1-cell q (corresponding to saddle) is deleted, and 2-cell p (corresponding to maximum) is merged into 2-cell p'. A cancellation of a minimum p and a saddle q is illustrated in Fig. 1.16b. It is feasible if 1-cell q is bounded by two different 0-cells p and p'. After the cancellation, 1-cell q is deleted, and 0-cell p is collapsed onto 0-cell p'.

In 3D, there are two instances of a cancellation: one cancels an extremum and a saddle (a maximum and a 2-saddle, or a minimum and a 1-saddle), the other cancels two saddle points. Cancellation of a maximum p and a 2-saddle q is feasible if 2-cell q is shared by exactly two different 3-cells p and p'. In the descending Morse complex, it removes 2-cell q, thus merging 3-cell p into 3-cell p', as illustrated in Fig. 1.17a. Cancellation of a minimum p and a 1-saddle q is feasible if 1-cell q is bounded by exactly two different 0-cells p and p'. In the descending complex Γ_d, it contracts 1-cell q with the effect of collapsing 0-cell p on 0-cell p'.

Cancellations that do not involve an extremum are more complex. The problem is that the number of cells in the Morse complexes that become incident to each other (and thus, the number of cells in the Morse-Smale complex) may increase after a cancellation. Let p and q be a 1-saddle and a 2-saddle, respectively. Let $R = \{r_j, j = 1, \ldots, j_{max}\}$ be the set of 2-saddles connected to p and different from q, and let $T = \{t_k, k = 1, \ldots, k_{max}\}$ be the set of 1-saddles connected to q different from p. The effect of the cancellation of 1-saddle p and 2-saddle q on a 3D descending Morse complex is illustrated in Fig. 1.17b. One-cell p and two-cell q are deleted, and the boundary of each 2-cell in R incident in p is extended to include 1-cells in T on the boundary of 2-cell q. Each 1-cell and each 0-cell that was on the boundary of 2-cell q (with the exception of 1-cell p) becomes part of the boundary of each 2-cell and each 3-cell incident in p (with the exception of 2-cell q), thus increasing the incidence relation on the descending Morse complex. The effect of the cancellation on the Morse-Smale complex consists of adding one arc for each pair (r_j, t_k) of critical points, where r_j belongs to R and t_k belongs to T, and deleting p and q, as well as all the arcs incident in them. Thus, a cancellation of p and q increases the number of arcs connecting 1- to 2-saddles in the complex by deleting $|R| + |T| + 1$ such arcs, but adding $|R| \cdot |T|$ arcs. Similarly, the number of 2- and 3-cells in the Morse-Smale complex may increase after the cancellation.

In [25], a macro-operator is defined, which consists of a 1-saddle-2-saddle cancellation, followed by a sequence of cancellation involving extrema. These latter cancellations eliminate the new incidences in the Morse complexes, the new cells in the Morse-Smale complex, and the new arcs in the incidence graph.

1.6.2 Removal and Contraction Operators

Motivated by the fact that cancellation operator is not a real simplification operator, two new basic dimension-independent simplification operators are introduced [10], called *removal* and *contraction*. They are defined by imposing additional constraints on the feasibility of a cancellation, and can be seen as merging of cells in the Morse complexes. There are two types of both a removal and contraction operator. For simplicity, we describe only the 3D instances of the operators of the first type.

A removal $rem(p, q, p')$ of index i of the first type of $(i+1)$-cell p and i-cell q is feasible if i-cell q appears once on the boundary of exactly two different $(i+1)$-cells p and p'. Intuitively, a removal $rem(p, q, p')$ removes i-cell q and merges $(i+1)$-cell p into $(i+1)$-cell p' in the descending Morse complex Γ_d. In the dual ascending Morse complex Γ_a, it contracts $(n-i)$-cell q and collapses $(n-i-1)$-cell p onto $(n-i-1)$-cell p'.

In 2D, there is one removal operator (of index 1). It is the same as a cancellation of a maximum and a saddle, illustrated in Fig. 1.16a.

In 3D, there are two removal operators: a removal of index 1 of 1-saddle q and 2-saddle p, and a removal of index 2 of 2-saddle q and maximum p. This latter is the same as the maximum-2-saddle cancellation illustrated in Fig. 1.17a.

A removal $rem(p, q, p')$ of index 1 in 3D is different from a cancellation, since it requires that 1-cell q bounds exactly two 2-cells p and p' in the descending complex. An example of the effect of a removal $rem(p, q, p')$ of index 1 on a 3D descending Morse complex is illustrated in Fig. 1.18a. After the removal, in the simplified descending Morse complex Γ_d', 1-cell q is deleted, and 2-cell p is merged with the unique 2-cell p' in the co-boundary of q and different from p. The boundary of p becomes part of the boundary of p'. Figure 1.18b illustrates the effect of removal $rem(p, q, p')$ on the dual ascending complex Γ_a. In Γ_a, q is a 2-cell bounded by exactly two different 1-cells p and p'. After the removal, 2-cell q is contracted, 1-cell p is collapsed onto 1-cell p'. All cells in the co-boundary of p become part of the co-boundary of p'.

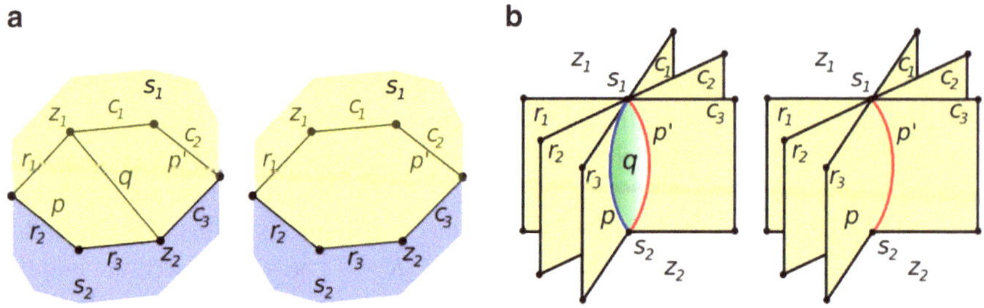

Fig. 1.18 Portion of a 3D descending (**a**) and ascending (**b**) Morse complex before and after a removal $rem(p, q, p')$ of index 1. The boundary of 2-cell p, consisting of 1-cells r_1, r_2 and r_3, is merged into the boundary of 2-cell p' in Γ_d. The co-boundary of 1-cell p, consisting of 2-cells r_1, r_2 and r_3, is merged into the co-boundary of 1-cell p' in Γ_a

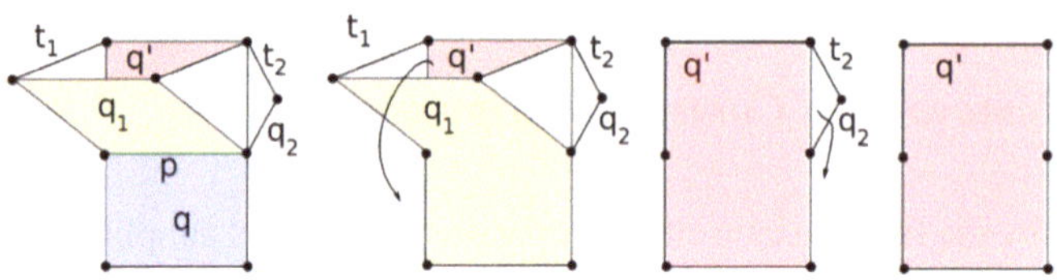

Fig. 1.19 A sequence consisting of a cancellation of 1-saddle p and 2-saddle q, followed by removals, which eliminate 2- and 3-saddles connected to p, on a 3D descending Morse complex

Contraction operators are dual to removal operators. The effect of a contraction of index i on a descending complex Γ_d is the same as the effect of a removal of index $(n-i)$ on an ascending complex Γ_a. Figure 1.18a, b illustrates the effect of a contraction $con(p, q, p')$ of index 2 on a descending and ascending Morse complex, respectively, and thus also the duality between removal and contraction operators.

In [10], it has been shown that removal and contraction simplification operators, together with their inverse ones, form a basis for the set of topologically consistent operators on Morse and Morse-Smale complexes on a manifold M. In particular, the macro-operator defined in [25], illustrated in Fig. 1.19, which cancels 1-cell p and 2-cell q and eliminates the cells created by this cancellation in the Morse-Smale complex, can be expressed as a sequence of removal and contraction operators, illustrated in Fig. 1.20. One-cell p is incident to four 2-cells, and 2-cell q is incident to four 1-cells. To be able to apply one of our operators (e.g. a removal of 1-cell p), which will eliminate 1-cell p and 2-cell q, we need to have only two 2-cells q and q' incident to 1-cell p. We can reduce the complex Γ_d to this situation by applying two removals of index 2, until all 2-cells incident to 1-cell p, with the exception of 2-cell q and one other 2-cell q' are eliminated. Now, we can apply a removal $rem(q, p, q')$, which eliminates 1-cell p and 2-cell q. Such sequence of removals consists of the same number of operators as a macro-operator consisting of a sequence of cancellations (macro-1-saddle-2-saddle operator), and it maintains simpler Morse and Morse-Smale complexes at each step.

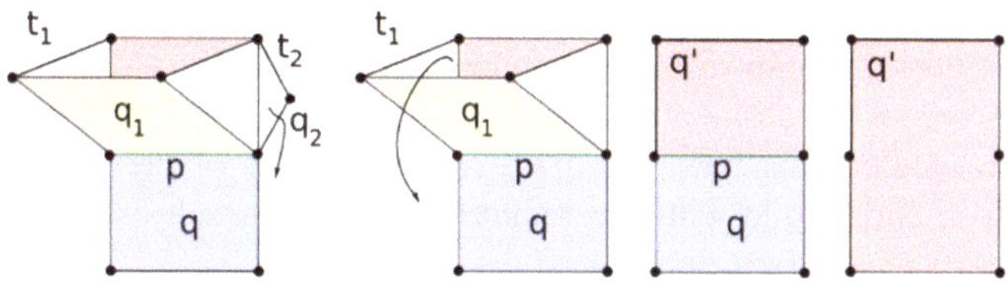

Fig. 1.20 A sequence consisting of removals, which eliminate 2- and 3-saddles connected to p, followed by a removal of index 1 that eliminates 1-saddle p and 2-saddle q on a 3D descending Morse complex

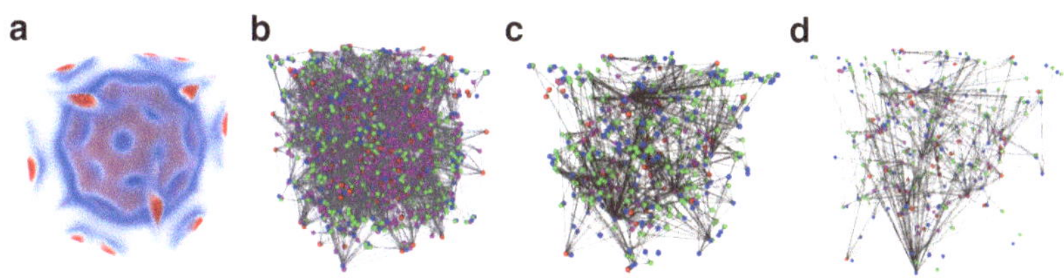

Fig. 1.21 (**a**) Field behavior for the Bucky Ball data set. (**b**) The incidence graph at full resolution, and the incidence graph after (**c**) 200 and (**d**) 400 simplifications

We have developed a simplification algorithm on the Morse complexes based on the removal and contraction simplification operators. Simplifications are applied in increasing order of persistence [20]. Our simplification algorithm can be applied not only to scalar fields representing the elevation, but to any scalar field, such as for example a discrete distortion [42], which generalizes the notion of curvature.

In Fig. 1.21, we illustrate the result of our simplification algorithm on a 3D Bucky Ball data set, depicted in Fig. 1.21a, which represents a carbon molecule having 60 atoms arranged as a truncated icosahedron. The full-resolution graph is shown in Fig. 1.21b. The incidence graphs after 200 and 400 simplifications are shown in Fig. 1.21c, d, respectively.

1.7 Concluding Remarks

The problem of modeling and simplifying Morse and Morse-Smale complexes in 2D has been extensively studied in the literature. Here, we have reviewed some recent work which extends these results to three and higher dimensions. We have described and compared data structures for representing Morse and Morse-Smale complexes. We have described and compared algorithms for extracting these complexes starting from the values of a scalar field f given at the vertices of a simplicial or a cell complex triangulating a manifold M. Finally, we have described and compared existing simplification operators on Morse and Morse-Smale complexes.

Simplification operators, together with their inverse refinement ones, form a basis for the definition of a multi-resolution model of Morse and Morse-Smale complexes [18].

The next challenge is how to extract representations of the geometry of the field which is compatible with the reduced incidence graph extracted from the multiresolution model [6, 16, 41].

Acknowledgements This work has been partially supported by the Italian Ministry of Education and Research under the PRIN 2009 program, and by the National Science Foundation under grant number IIS-1116747.

References

1. Bajaj, C.L., Shikore, D.R.: Topology preserving data simplification with error bounds. Comput. Graph. **22**(1), 3–12 (1998)
2. Banchoff, T.: Critical points and curvature for embedded polyhedral surfaces. Am. Math. Mon. **77**(5), 475–485 (1970)
3. Beucher, S.: Watershed, hierarchical segmentation and waterfall algorithm. In: Proceedings of the Mathematical Morphology and Its Applications to Image Processing, Fontainebleau, France, pp. 69–76 (1994)
4. Biasotti, S., De Floriani, L., Falcidieno, B., Frosini, P., Giorgi, D., Landi, C., Papaleo, L., Spagnuolo, M.: Describing shapes by geometrical-topological properties of real functions. ACM Comput. Surv. **40**, Article 12 (2008)
5. Bremer, P.-T., Edelsbrunner, H., Hamann, B., Pascucci, V.: A multi-resolution data structure for two-dimensional Morse functions. In: Proceedings IEEE Visualization 2003, Seattle, pp. 139–146. IEEE Computer Society (2003)
6. Bremer, P.-T., Edelsbrunner, H., Hamann, B., Pascucci, V.: A topological hierarchy for functions on triangulated surfaces. Trans. Vis. Comput. Graph. **10**(4), 385–396 (2004)
7. Bremer, P.-T., Pascucci, V., Hamann, B.: Maximizing adaptivity in hierarchical topological models. In: Belyaev, A.G., Pasko, A.A., Spagnuolo, M. (eds.) Proceedings of the International Conference on Shape Modeling and Applications 2005 (SMI '05), Cambridge, MA, pp. 300–309. IEEE Computer Society, Los Alamitos (2005)
8. Bremer, P.-T., Weber, G.H., Pascucci, V., Day, M.S., Bell, J.B.: Analyzing and tracking burning structures in lean premixed hydrogen flames. IEEE Trans. Vis. Comput. Graph. **16**(2), 248–260 (2010)
9. Cazals, F., Chazal, F., Lewiner, T.: Molecular shape analysis based upon the Morse-Smale complex and the connolly function. In: Proceedings of the Nineteenth Annual Symposium on Computational Geometry, San Diego, pp. 351–360. ACM, New York (2003)
10. Čomić, L., De Floriani, L.: Dimension-independent simplification and refinement of Morse complexes. Graph. Models **73**(5), 261–285 (2011)
11. Čomić, L., De Floriani, L., Iuricich, F.: Building morphological representations for 2D and 3D scalar fields. In: Puppo, E., Brogni, A., De Floriani, L. (eds.) Eurographics Italian Chapter Conference, Genova, pp. 103–110. Eurographics (2010)
12. Čomić, L., Mesmoudi, M.M., De Floriani, L.: Smale-like decomposition and Forman theory for discrete scalar fields. In: DGCI, Nancy, pp. 477–488 (2011)
13. Danovaro, E., De Floriani, L., Mesmoudi, M.M.: Topological analysis and characterization of discrete scalar fields. In: Asano, T., Klette, R., Ronse, C. (eds.) Geometry, Morphology, and Computational Imaging. Lecture Notes in Computer Science, vol. 2616, pp. 386–402. Springer, Berlin Heidelberg (2003)

14. Danovaro, E., De Floriani, L., Magillo, P., Mesmoudi, M.M., Puppo, E.: Morphology-driven simplification and multiresolution modeling of terrains. In: Hoel, E., Rigaux, P. (eds.) Proceedings ACM GIS 2003 – The 11th International Symposium on Advances in Geographic Information Systems, New Orleans, pp. 63–70. ACM (2003)
15. Danovaro, E., De Floriani, L., Vitali, M., Magillo, P.: Multi-scale dual Morse complexes for representing terrain morphology. In: GIS '07: Proceedings of the 15th Annual ACM International Symposium on Advances in Geographic Information Systems, Seattle, Washington, WA, pp. 1–8. ACM, New York (2007)
16. Danovaro, E., De Floriani, L., Magillo, P., Vitali, M.: Multiresolution Morse triangulations. In: Elber, G., Fischer, A., Keyser, J., Kim, M.-S. (eds.) Symposium on Solid and Physical Modeling, Haifa, pp. 183–188. ACM (2010)
17. De Floriani, L., Hui, A.: Shape representations based on cell and simplicial complexes. In: Eurographics 2007, Prague, Czech Republic, State-of-the-Art Report, Sept 2007
18. Čomić, L., De Floriani, L., Iuricich, F.: Dimension-independent multi-resolution Morse complexes. Comput. Graph. **36**(5), 541–547 (2012)
19. Edelsbrunner, H., Harer, J.: The persistent Morse complex segmentation of a 3-manifold. In: Magnenat-Thalmann, N. (ed.) 3DPH, Zermatt. Lecture Notes in Computer Science, vol. 5903, pp. 36–50. Springer (2009)
20. Edelsbrunner, H., Harer, J., Zomorodian, A.: Hierarchical Morse complexes for piecewise linear 2-manifolds. In: Proceedings of the 17th ACM Symposium on Computational Geometry, Medford, pp. 70–79 (2001)
21. Edelsbrunner, H., Harer, J., Natarajan, V., Pascucci, V.: Morse-Smale complexes for piecewise linear 3-manifolds. In: Proceedings of the 19th ACM Symposium on Computational Geometry, San Diego, pp. 361–370 (2003)
22. Forman, R.: Morse theory for cell complexes. Adv. Math. **134**, 90–145 (1998)
23. Gyulassy, A., Pascucci, V.: Computing simply-connected cells in three-dimensional Morse-Smale complexes. In: Peikert, R., Hauser, H., Carr, H., Fuchs, R. (eds.) Topological Methods in Data Analysis and Visualization: Theory, Algorithms, and Applications. Mathematics and Visualization, pp. 31–46. Springer, Heidelberg (2012)
24. Gyulassy, A., Natarajan, V., Pascucci, V., Bremer, P.-T., Hamann, B.: Topology-based simplification for feature extraction from 3D scalar fields. In: Proceedings IEEE Visualization'05, Minneapolis, Minnesota, MN, pp. 275–280. ACM (2005)
25. Gyulassy, A., Natarajan, V., Pascucci, V., Bremer, P.-T., Hamann, B.: A topological approach to simplification of three-dimensional scalar functions. IEEE Trans. Vis. Comput. Graph. **12**(4), 474–484 (2006)
26. Gyulassy, A., Natarajan, V., Pascucci, V., Hamann, B.: Efficient computation of Morse-Smale complexes for three-dimensional scalar functions. IEEE Trans. Vis. Comput. Graph. **13**(6), 1440–1447 (2007)
27. Gyulassy, A., Bremer, P.-T., Hamann, B., Pascucci, V.: A practical approach to Morse-Smale complex computation: scalability and generality. IEEE Trans. Vis. Comput. Graph. **14**(6), 1619–1626 (2008)
28. Gyulassy, A., Bremer, P.-T., Hamann, B., Pascucci, V.: Practical considerations in Morse-Smale complex computation. In: Pascucci, V., Tricoche, X., Hagen, H., Tierny, J. (eds.) Topological Methods in Data Analysis and Visualization: Theory, Algorithms, and Applications. Mathematics and Visualization, pp. 67–78. Springer, Heidelberg (2011)
29. Jerše, G., Mramor Kosta, N.: Ascending and descending regions of a discrete Morse function. Comput. Geom. Theory Appl. **42**(6–7), 639–651 (2009)
30. Kelley, J.L.: General Topology. Van Nostrand, Princeton (1955)
31. King, H., Knudson, K., Mramor, N.: Generating discrete Morse functions from point data. Exp. Math. **14**(4), 435–444 (2005)
32. Lewiner, T., Lopes, H., Tavares, G.: Applications of Forman's discrete Morse theory to topology visualization and mesh compression. Trans. Vis. Comput. Graph. **10**(5), 499–508 (2004)

33. Matsumoto, Y.: An Introduction to Morse Theory. Translations of Mathematical Monographs, vol. 208. American Mathematical Society, Providence (2002)
34. Milnor, J.: Morse Theory. Princeton University Press, Princeton (1963)
35. Natarajan, V., Wang, Y., Bremer, P.-T., Pascucci, V., Hamann, B.: Segmenting molecular surfaces. Comput. Aided Geom. Des. **23**(6), 495–509 (2006)
36. Ni, X., Garland, M., Hart, J.C.: Fair Morse functions for extracting the topological structure of a surface mesh. In: International Conference on Computer Graphics and Interactive Techniques ACM SIGGRAPH, Los Angeles, pp. 613–622 (2004)
37. Pascucci, V.: Topology diagrams of scalar fields in scientific visualization. In: Rana, S. (ed.) Topological Data Structures for Surfaces, pp. 121–129. Wiley, Chichester/Hoboken (2004)
38. Robins, V., Wood, P.J., Sheppard, A.P.: Theory and algorithms for constructing discrete Morse complexes from grayscale digital images. IEEE Trans. Pattern Anal. Mach. Intell. **33**(8), 1646–1658 (2011)
39. Takahashi, S., Ikeda, T., Kunii, T.L., Ueda, M.: Algorithms for extracting correct critical points and constructing topological graphs from discrete geographic elevation data. Comput. Graph. Forum **14**, 181–192 (1995)
40. Vincent, L., Soille, P.: Watershed in digital spaces: an efficient algorithm based on immersion simulation. IEEE Trans. Pattern Anal. Mach. Intell. **13**(6), 583–598 (1991)
41. Weinkauf, T., Gingold, Y.I., Sorkine, O.: Topology-based smoothing of 2D scalar fields with ^{1}-continuity. Comput. Graph. Forum **29**(3), 1221–1230 (2010)
42. Weiss, K., De Floriani, L., Mesmoudi, M.M.: Multiresolution analysis of 3D images based on discrete distortion. In: 20th International Conference on Pattern Recognition (ICPR), Istanbul, pp. 4093–4096 (2010)
43. Wolf, G.W.: Topographic surfaces and surface networks. In: Rana, S. (ed.) Topological Data Structures for Surfaces, pp. 15–29. Wiley, Chichester/Hoboken (2004)

Chapter 2
Geodesic Regression and Its Application to Shape Analysis

P. Thomas Fletcher

Abstract In this chapter, I present a regression method for modeling the relationship between a manifold-valued random variable and a real-valued independent parameter. The principle is to fit a geodesic curve, parameterized by the independent parameter, that best fits the data. Error in the model is evaluated as the sum-of-squared geodesic distances from the model to the data, and this provides an intrinsic least squares criterion. Geodesic regression is, in some sense, the simplest parametric model that one could choose, and it provides a direct generalization of linear regression to the manifold setting. A generalization of the coefficient of determination and a resulting hypothesis test for determining the significance of the estimated trend is developed. Also, a diagnostic test for the quality of the fit of the estimated geodesic is demonstrated. While the method can be generally applied to data on any manifold, specific examples are given for a set of synthetically generated rotation data and an application to analyzing shape changes in the corpus callosum due to age.

2.1 Introduction

Regression analysis is a fundamental statistical tool for determining how a measured variable is related to one or more potential explanatory variables. The most widely used regression model is linear regression, due to its simplicity, ease of interpretation, and ability to model many phenomena. However, if the response variable takes values on a nonlinear manifold, a linear model is not applicable. Such manifold-valued measurements arise in many applications, including those involving directional data, transformations, tensors, and shape. For example, in biology and medicine it is often critical to understand processes that change

P.T. Fletcher (✉)
University of Utah, Salt Lake City, 84112 UT, USA
e-mail: fletcher@cs.utah.edu

M. Breuß et al. (eds.), *Innovations for Shape Analysis*, Mathematics and Visualization,
DOI 10.1007/978-3-642-34141-0_2,

the shape of anatomy. The difficulty is that shape variability is inherently high-dimensional and nonlinear. An effective approach to capturing this variability has been to parameterize shape as a manifold, or shape space.

Statistical analysis of manifold data has been developed by several authors. The seminal work of Fréchet [10] generalized the concept of expectation from linear spaces to general metric spaces. This opened up the possibility of computing a sample mean statistic from a set of data on a manifold using the geodesic distance as metric. The Fréchet mean of a set of points, $y_1, \ldots, y_N$, in a Riemannian manifold M is given by

$$\mu = \arg\min_{y \in M} \sum_{i=1}^{N} d(y, y_i)^2,$$

where d is the geodesic distance between points on M. This equation generalizes the principle of least squares to the metric space setting. Karcher [12] provided conditions guaranteeing the existence and uniqueness of the Fréchet mean, which were later improved by Kendall [14]. Second-order statistics such as generalizations of principal components analysis [8] and Gaussian covariances [21] have also been developed and applied in the domain of image analysis. Related work includes statistical analysis of directional data (e.g., spheres) [16] and analysis on shape manifolds [5], where statistics are derived from probability distributions on specific manifolds (for example, the Fisher-von Mises distribution on spheres).

Several works have studied the regression problem on manifolds. Jupp and Kent [11] propose an unrolling method on shape spaces. Regression analysis on the group of diffeomorphisms has been proposed as growth models by Miller [18], nonparametric regression by Davis et al. [2], and second-order splines by Trouvé and Vialard [23]. Durrleman et al. [6] construct spatiotemporal image atlases from longitudinal data. Finally, Shi et al. [22] proposed a semiparametric model with multiple covariates for manifold response data. None of these methods provide a direct generalization of linear regression to manifolds. The purpose of this work is to develop such a generalization, called geodesic regression, which models the relationship between an independent scalar variable with a dependent manifold-valued random variable as a geodesic curve. Like linear regression, the advantages of this model are its simplicity and ease of interpretation. As will be shown, the geodesic regression model also leads to a straightforward generalization of the R^2 statistic and a hypothesis test for significance of the estimated geodesic trend. This chapter is an expanded exposition of the geodesic regression method first introduced in [7]. Niethammer et al. [20] independently proposed geodesic regression for the case of diffeomorphic transformations of image time series.

2.2 Multiple Linear Regression

Before formulating geodesic regression on general manifolds, we begin by reviewing multiple linear regression in $\mathbb{R}^n$. Here we are interested in the relationship between a non-random *independent* variable $X \in \mathbb{R}$ and a random *dependent* variable Y taking values in $\mathbb{R}^n$. A multiple linear model of this relationship is given by

$$Y = \alpha + X\beta + \epsilon, \tag{2.1}$$

where $\alpha \in \mathbb{R}^n$ is an unobservable *intercept* parameter, $\beta \in \mathbb{R}^n$ is an unobservable *slope* parameter, and ϵ is an $\mathbb{R}^n$-valued, unobservable random variable representing the error. Geometrically, this is the equation of a one-dimensional line through $\mathbb{R}^n$ (plus noise), parameterized by the scalar variable X. For the purposes of generalizing to the manifold case, it is useful to think of α as the starting point of the line and β as a velocity vector.

Given realizations of the above model, i.e., data $(x_i, y_i) \in \mathbb{R} \times \mathbb{R}^n$, for $i = 1, \ldots, N$, the least squares estimates, $\hat{\alpha}, \hat{\beta}$, for the intercept and slope are computed by solving the minimization problem

$$(\hat{\alpha}, \hat{\beta}) = \underset{(\alpha,\beta)}{\arg\min} \sum_{i=1}^{N} \|y_i - \alpha - x_i \beta\|^2. \tag{2.2}$$

This equation can be solved analytically, yielding

$$\hat{\beta} = \frac{\frac{1}{N}\sum x_i\, y_i - \bar{x}\,\bar{y}}{\sum x_i^2 - \bar{x}^2},$$
$$\hat{\alpha} = \bar{y} - \bar{x}\,\hat{\beta},$$

where $\bar{x}$ and $\bar{y}$ are the sample means of the x_i and y_i, respectively. If the errors in the model are drawn from distributions with zero mean and finite variance, then these estimators are unbiased and consistent. Furthermore, if the errors are homoscedastic (equal variance) and uncorrelated, then the Gauss-Markov theorem states that they will have minimal mean-squared error amongst all unbiased linear estimators.

2.3 Geodesic Regression

Let $y_1, \ldots, y_N$ be points on a smooth Riemannian manifold M, with associated scalar values $x_1, \ldots, x_N \in \mathbb{R}$. The goal of geodesic regression is to find a geodesic curve γ on M that best models the relationship between the x_i and the y_i. Just as in linear regression, the speed of the geodesic will be proportional to the independent parameter corresponding to the x_i. Estimation will be set up as a least-squares problem, where we want to minimize the sum-of-squared Riemannian distances between the model and the data. A schematic of the geodesic regression model is shown in Fig. 2.1.

Before formulating the model, we review a few basic concepts of Riemannian geometry. We will write an element of the tangent bundle as the pair $(p, v) \in TM$,

Fig. 2.1 Schematic of the geodesic regression model

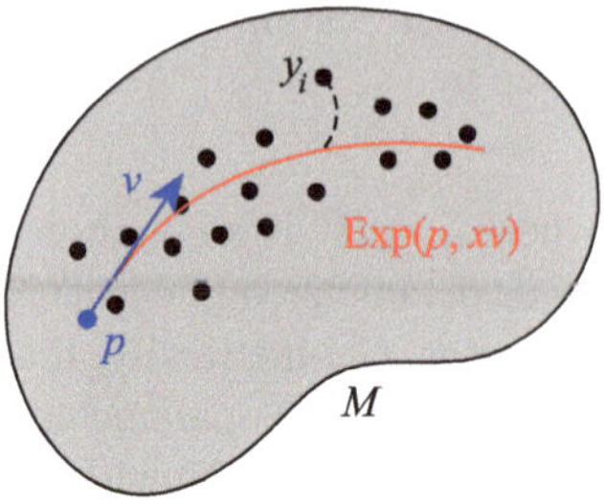

where p is a point in M and $v \in T_pM$ is a tangent vector at p. Recall that for any $(p, v) \in TM$ there is a unique geodesic curve γ, with initial conditions $\gamma(0) = p$ and $\gamma'(0) = v$. This geodesic is only guaranteed to exist locally. When γ is defined over the interval $[0, 1]$, the exponential map at p is defined as $\text{Exp}_p(v) = \gamma(1)$. In other words, the exponential map takes a position and velocity as input and returns the point at time 1 along the geodesic with these initial conditions. The exponential map is locally diffeomorphic onto a neighborhood of p. Let $V(p)$ be the largest such neighborhood. Then within $V(p)$ the exponential map has an inverse, the Riemannian log map, $\text{Log}_p : V(p) \to T_pM$. For any point $q \in V(p)$ the Riemannian distance function is given by $d(p, q) = \|\text{Log}_p(q)\|$. It will be convenient to include the point p as a parameter in the exponential and log maps, i.e., define $\text{Exp}(p, v) = \text{Exp}_p(v)$ and $\text{Log}(p, q) = \text{Log}_p(q)$.

Notice that the tangent bundle TM serves as a convenient parameterization of the set of possible geodesics on M. An element $(p, v) \in TM$ provides an intercept p and a slope v, analogous to the α and β parameters in the multiple linear regression model (2.1). In fact, β is a vector in the tangent space $T_\alpha\mathbb{R}^n \cong \mathbb{R}^n$, and thus (α, β) is an element of the tangent bundle $T\mathbb{R}^n$. Now consider an M-valued random variable Y and a non-random variable $X \in \mathbb{R}$. The generalization of the multiple linear model to the manifold setting is the *geodesic model*,

$$Y = \text{Exp}(\text{Exp}(p, Xv), \epsilon), \tag{2.3}$$

where ϵ is a random variable taking values in the tangent space at $\text{Exp}(p, Xv)$. Notice that for Euclidean space, the exponential map is simply addition, i.e., $\text{Exp}(p, v) = p + v$. Thus, the geodesic model coincides with (2.1) when $M = \mathbb{R}^n$.

2.3.1 Least Squares Estimation

Consider a realization of the model (2.3): $(x_i, y_i) \in \mathbb{R} \times M$, for $i = 1, \ldots, N$. Given this data, we wish to find estimates of the parameters $(p, v) \in TM$. First, define the sum-of-squared error of the data from the geodesic given by (p, v) as

$$E(p, v) = \frac{1}{2} \sum_{i=1}^{N} d(\text{Exp}(p, x_i v), y_i)^2. \tag{2.4}$$

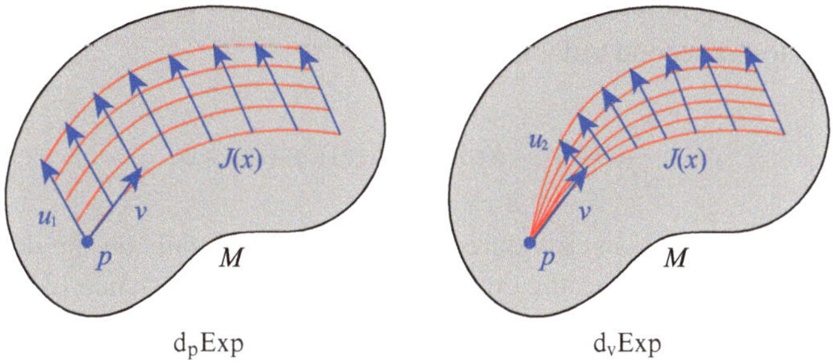

Fig. 2.2 Jacobi fields as derivatives of the exponential map

Following the ordinary least squares minimization problem given by (2.2), we formulate a least squares estimator of the geodesic model as a minimizer of the above sum-of-squares energy, i.e.,

$$(\hat{p}, \hat{v}) = \arg\min_{(p,v)} E(p, v). \tag{2.5}$$

Again, notice that this problem coincides with the ordinary least squares problem when $M = \mathbb{R}^n$.

Unlike the linear setting, the least squares problem in (2.5) for a general manifold M will typically not yield an analytic solution. Instead we derive a gradient descent algorithm. Computation of the gradient of (2.4) will require two parts: the derivative of the Riemannian distance function and the derivative of the exponential map. Fixing a point $p \in M$, the gradient of the squared distance function is $\nabla_x d(p, x)^2 = -2\mathrm{Log}_x(p)$ for $x \in V(p)$.

The derivative of the exponential map $\mathrm{Exp}(p, v)$ can be separated into a derivative with respect to the initial point p and a derivative with respect to the initial velocity v. To do this, first consider a variation of geodesics given by $c_1(s,t) = \mathrm{Exp}(\mathrm{Exp}(p, su_1), tv(s))$, where $u_1 \in T_pM$ defines a variation of the initial point along the geodesic $\eta(s) = \mathrm{Exp}(p, su_1)$. Here we have also extended $v \in T_pM$ to a vector field $v(s)$ along η via parallel translation. This variation is illustrated on the left side of Fig. 2.2. Next consider a variation of geodesics $c_2(s,t) = \mathrm{Exp}(p, su_2 + tv)$, where $u_2 \in T_pM$. (Technically, u_2 is a tangent to the tangent space, i.e., an element of $T_v(T_pM)$, but there is a natural isomorphism $T_v(T_pM) \cong T_pM$.) The variation c_2 produces a "fan" of geodesics as seen on the right side of Fig. 2.2.

Now the derivatives of $\mathrm{Exp}(p, v)$ with respect to p and v are given by

$$d_p\mathrm{Exp}(p,v) \cdot u_1 = \frac{d}{ds} c_1(s,t)\Big|_{s=0} = J_1(1)$$

$$d_v\mathrm{Exp}(p,v) \cdot u_2 = \frac{d}{ds} c_2(s,t)\Big|_{s=0} = J_2(1),$$

where $J_i(t)$ are *Jacobi fields* along the geodesic $\gamma(t) = \text{Exp}(p, tv)$. Jacobi fields are solutions to the second order equation

$$\frac{D^2}{dt^2} J(t) + R(J(t), \gamma'(t))\, \gamma'(t) = 0, \tag{2.6}$$

where R is the Riemannian curvature tensor. For more details on the derivation of the Jacobi field equation and the curvature tensor, see for instance [3]. The initial conditions for the two Jacobi fields above are $J_1(0) = u_1$, $J_1'(0) = 0$ and $J_2(0) = 0$, $J_2'(0) = u_2$, respectively. If we decompose the Jacobi field into a component tangential to γ and a component orthogonal, i.e., $J = J^\top + J^\perp$, the tangential component is linear: $J^\top(t) = u_1^\top + t u_2^\top$. Therefore, the only challenge is to solve for the orthogonal component.

Finally, the gradient of the sum-of-squares energy in (2.4) is given by

$$\nabla_p\, E(p, v) = -\sum_{i=1}^{N} d_p\text{Exp}(p, x_i v)^\dagger \text{Log}(\text{Exp}(p, x_i v), y_i),$$

$$\nabla_v\, E(p, v) = -\sum_{i=1}^{N} x_i\; d_v\text{Exp}(p, x_i v)^\dagger \text{Log}(\text{Exp}(p, x_i v), y_i),$$

where we have taken the adjoint of the exponential map derivative, e.g., defined by $\langle d_p\text{Exp}(p, v)u, w\rangle = \langle u, d_p\text{Exp}(p, v)^\dagger w\rangle$. As we will see in the next section, formulas for Jacobi fields and their respective adjoint operators can often be derived analytically for many useful manifolds.

2.3.2 R^2 Statistics and Hypothesis Testing

In regression analysis the most basic question one would like to answer is whether the relationship between the independent and dependent variables is significant. A common way to test this is to see if the amount of variance explained by the model is high. For geodesic regression we will measure the amount of explained variance using a generalization of the R^2 statistic, or coefficient of determination, to the manifold setting. To do this, we first define predicted values of y_i and the errors ϵ_i as

$$\hat{y}_i = \text{Exp}(\hat{p}, x_i \hat{v}),$$

$$\hat{\epsilon}_i = \text{Log}(\hat{y}_i, y_i),$$

where $(\hat{p}, \hat{v})$ are the least squares estimates of the geodesic parameters defined above. Note that the $\hat{y}_i$ are points along the estimated geodesic that are the best predictions of the y_i given only the x_i. The $\hat{\epsilon}_i$ are the residuals from the model predictions to the true data.

Now to define the total variance of data, $y_1, \ldots, y_N \in M$, we use the Fréchet variance, intrinsically defined by

$$\mathrm{var}(y_i) = \min_{y \in M} \frac{1}{N} \sum_{i=1}^{N} d(y, y_i)^2.$$

The unexplained variance is the variance of the residuals, $\mathrm{var}(\hat{\epsilon}_i) = \frac{1}{N} \sum \|\hat{\epsilon}_i\|^2$. From the definition of the residuals, it can be seen that the unexplained variance is the mean squared distance of the data to the model, i.e., $\mathrm{var}(\hat{\epsilon}_i) = \frac{1}{N} \sum d(\hat{y}_i, y_i)^2$. Using these two variance definitions, the generalization of the R^2 statistic is then given by

$$R^2 = 1 - \frac{\text{unexplained variance}}{\text{total variance}} = 1 - \frac{\mathrm{var}(\hat{\epsilon}_i)}{\mathrm{var}(y_i)}. \tag{2.7}$$

Fréchet variance coincides with the standard definition of variance when $M = \mathbb{R}^n$. Therefore, it follows that the definition of R^2 in (2.7) coincides with the R^2 for linear regression when $M = \mathbb{R}^n$. Also, because Fréchet variance is always nonnegative, we see that $R^2 \leq 1$, and that $R^2 = 1$ if and only if the residuals to the model are exactly zero, i.e., the model perfectly fits the data. Finally, it is clear that the residual variance is always smaller than the total variance, i.e., $\mathrm{var}(\hat{\epsilon}_i) \leq \mathrm{var}(y_i)$. This is because we could always choose $\hat{p}$ to be the Fréchet mean and $v = 0$ to achieve $\mathrm{var}(\hat{\epsilon}_i) = \mathrm{var}(y_i)$. Therefore, $R^2 \geq 0$, and it must lie in the interval [0, 1], as is the case for linear models.

We now describe a permutation test for testing the significance of the estimated slope term, $\hat{v}$. Notice that if we constrain v to be zero in (2.5), then the resulting least squares estimate of the intercept, $\hat{p}$, will be the Fréchet mean of the y_i. The desired hypothesis test is whether the fraction of unexplained variance is significantly decreased by also estimating v. The null hypothesis is $H_0 : R^2 = 0$, which is the case if the unexplained variance in the geodesic model is equal to the total variance. Under the null hypothesis, there is no relationship between the X variable and the Y variable. Therefore, the x_i are exchangeable under the null hypothesis, and a permutation test may randomly reorder the x_i data, keeping the y_i fixed. Estimating the geodesic regression parameters for each random permutation of the x_i, we can calculate a sequence of R^2 values, $R_1^2, \ldots, R_m^2$, which approximate the sampling distribution of the R^2 statistic under the null hypothesis. Computing the fraction of the R_k^2 that are greater than the R^2 estimated from the unpermuted data gives us a p-value.

2.4 Testing the Geodesic Fit

In any type of regression analysis, a choice is made as to the type of model that is fit to the data, whether it be linear, polynomial, or perhaps nonparametric. An important step in the analysis is to verify that the selected model is in fact appropriate. In linear regression, for example, one would want to test several assumptions: (1) that the trend in the data is truly linear, (2) that the error is homoscedastic, (3) that the model fit is not led astray by outliers, (4) that the errors are Gaussian distributed, etc. Several graphical and quantitative heuristic tests have been developed to test these assumptions. For a detailed treatment of these methods, see [9].

In this section we develop a diagnostic test of the model assumptions for geodesic regression. We focus on the following question: is a geodesic curve an appropriate model for the relationship between the independent and dependent variables? A geodesic curve is, in some sense, the "straightest" path one can take on a manifold. This begs the question if a more flexible model would do a better job at fitting the data. This is analogous to the model selection problem for real-valued data when one is making the choice between a linear model and something more flexible, such as a higher-order polynomial model. Of course, if a model is made more flexible there is a danger that the data will be overfit. One way to test if a model is "flexible enough" is to plot the residuals of the model versus the independent variable. If the model has captured the relationship between the independent and dependent variables, then the residuals should show no obvious trend. If they do show a trend, then a more flexible model is needed to capture the relationship between the data. However, for regression on manifolds, this is a difficult test to apply because the residuals are high-dimensional tangent vectors and are thus difficult to plot versus the independent variable. One solution might be to plot the magnitude of the residuals instead, but this loses most of the information contained in the residual vectors.

Instead, we will use nonparametric regression as a comparison model to test if a geodesic is sufficient to capture the relationships in the data. Nonparametric regression models, such as the kernel regression method described below, are highly flexible. Because there is no parametric model assumed for the functional relationship between the independent and dependent variables, these models can adapt to highly complex functions given enough data. Given a method to visualize the results of a manifold-valued regression, the diagnostic test is as follows. First, compute both a geodesic regression and a nonparametric regression of the data. Second, visualize the results of both regression methods. If the nonparametric regression trend is similar to the estimated geodesic, then this provides strong evidence that the geodesic model is sufficient. If the nonparametric trend deviates significantly from the estimated geodesic, then this indicates that the geodesic model is too inflexible to capture the relationship between the two variables.

An example of this procedure is given for synthesized univariate data in Fig. 2.3. The left figure shows data generated from a noisy linear trend. In this case the linear model and the nonparametric model give similar answers. The right figure shows data generated from a noisy nonlinear (sinusoidal) trend. Here the nonparametric regression adapts to the nonlinearities in the data, and the inadequacy of the linear trend can be seen as a difference between the two regression models. Of course, in the univariate case we can easily see that a linear trend is inadequate just by plotting the data even without comparing it to a nonparametric regression.

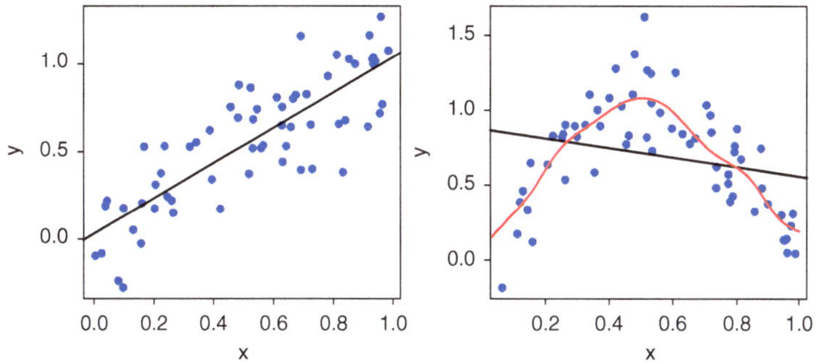

Fig. 2.3 Comparison of linear (*black*) and nonparametric (*red*) regressions as a test of fit. When the data is generated from a linear model (*left*), the two regression methods produce similar results. When the data is generated from a nonlinear model (*right*), the difference in the two models helps detect that a linear model is insufficient

However, for high-dimensional manifolds this type of plot is not available. This is where a comparison to a nonparametric trend is highly useful. In the results below (Sect. 2.6.2) we give an example of how this comparison to nonparametric regression can be used as a diagnostic of model fit in shape analysis applications. The nonparametric regression method that we use for comparison is the one given by Davis et al. [2], which we review now.

2.4.1 *Review of Univariate Kernel Regression*

Before reviewing the manifold version, we give a quick overview of univariate kernel regression as developed by Nadaraya [19] and Watson [25]. As in the linear regression setting, we are interested in finding a relationship between data $x_1, \dots, x_N \in \mathbb{R}$, coming from an independent variable X, and data $y_1, \dots, y_N \in \mathbb{R}$, representing a dependent variable Y. The model of their relationship is given by

$$Y = f(X) + \epsilon,$$

where f is an arbitrary function, and ϵ is a random variable representing the error. Contrary to linear regression, the function f is not assumed to have any particular parametric form.

Instead, the function f is estimated from the data by local weighted averaging.

$$\hat{f}_h(x) = \frac{\sum_{i=1}^{N} K_h(x - x_i) y_i}{\sum_{i=1}^{N} K_h(x - x_i)}.$$

In this equation, K is a function that satisfies $\int K(t)\,dt = 1$ and $K_h(t) = \frac{1}{h} K(\frac{t}{h})$, with bandwidth parameter $h > 0$. This is the estimation procedure shown in Fig. 2.3 (red curves).

2.4.2 Nonparametric Kernel Regression on Manifolds

The regression method of Davis et al. [2] generalizes the Nadaraya-Watson kernel regression method to the case where the dependent variable lives on a Riemannian manifold, i.e., $y_i \in M$. Here the model is given by

$$Y = \mathrm{Exp}(f(X), \epsilon),$$

where $f : \mathbb{R} \to M$ defines a curve on M, and $\epsilon \in T_{f(X)}M$ is an error term. As in the univariate case, there are no assumptions on the parametric form of the curve f.

Motivated by the definition of the Nadaraya-Watson estimator as a weighted averaging, the *manifold kernel regression estimator* is defined using a weighted Fréchet sample mean as

$$\hat{f}_h(x) = \arg\min_{y \in M} \frac{\sum_{i=1}^{N} K_h(x - x_i) d(y, y_i)^2}{\sum_{i=1}^{N} K_h(x - x_i)}.$$

Notice that when the manifold under study is a Euclidean vector space, equipped with the standard Euclidean norm, the above minimization results in the Nadaraya-Watson estimator.

2.4.3 Bandwidth Selection

It is well known within the kernel regression literature that kernel width plays a crucial role in determining regression results [24]. In particular, it is important to select a bandwidth that captures relevant population-wide changes without either oversmoothing and missing relevant changes or undersmoothing and biasing the results based on individual noisy data points. The 'Goldie Locks' method of tuning the bandwidth until the results are most pleasing is a common subjective method for bandwidth selection. However, non-subjective methods may be required, for example, when kernel regression is part of a larger statistical study. A number of automatic kernel bandwidth selection techniques have been proposed for this purpose [24].

One classic method for automatic bandwidth selection is based on least squares cross-validation. This method is easily extended to the manifold regression setting in the following way. The least squares cross-validation estimate for the optimal bandwidth h is defined as

$$\hat{h}_{\mathrm{LSCV}} = \arg\min_{h \in \mathbb{R}+} \frac{1}{N} \sum_{i=1}^{N} d\left(\hat{f}_h^{(i)}(x_i), y_i\right)^2,$$

where

$$\hat{f}_h^{(i)}(t) = \arg\min_{y \in M} \left(\frac{\sum_{j=1, j\neq i}^{N} K_h(x - x_j) d(y, y_j)^2}{\sum_{j=1, j\neq i}^{N} K_h(x - x_j)} \right)$$

is the manifold kernel regression estimator with the i-th observation left out. This cross-validation method was used to select the bandwidth for the kernel regression example in Fig. 2.3.

2.5 Results: Regression of 3D Rotations

2.5.1 *Overview of Unit Quaternions*

We represent 3D rotations as the unit quaternions, $\mathbb{Q}_1$. A quaternion is denoted as $q = (a, v)$, where a is the "real" component and $v = bi + cj + dk$. Geodesics in the rotation group are given simply by constant speed rotations about a fixed axis. Let $e = (1, 0)$ be the identity quaternion. The tangent space $T_e\mathbb{Q}_1$ is the vector space of quaternions of the form $(0, v)$. The tangent space at an arbitrary point $q \in \mathbb{Q}_1$ is given by right multiplication of $T_e\mathbb{Q}_1$ by q. The Riemannian exponential map is $\text{Exp}_q((0, v) \cdot q) = (\cos(\theta/2), 2v \cdot \sin(\theta/2)/\theta) \cdot q$, where $\theta = 2\|v\|$. The log map is given by $\text{Log}_q((a, v) \cdot q) = (0, \theta v/\|v\|) \cdot q$, where $\theta = \arccos(a)$.

Being a unit sphere, $\mathbb{Q}_1$ has constant sectional curvature $K = 1$. In this case the orthogonal component of the Jacobi field equation (2.6) along a geodesic $\gamma(t)$ has the analytic solution

$$J(t)^{\perp} = u_1(t) \cos(Lt) + u_2(t) \frac{\sin(Lt)}{L},$$

where u_1, u_2 are parallel vector fields along γ, with initial conditions $u_1(0) = J(0)^{\perp}$ and $u_2(0) = J'(0)^{\perp}$, and $L = \|\gamma'\|$. While the Jacobi field equation gives us the differential of the exponential map, we really need the adjoint of this operator for geodesic regression. However, from the above equation it is clear that $d_p\text{Exp}$ and $d_v\text{Exp}$ are both self-adjoint operators. That is, the above Jacobi field equation provides us both the differential and its adjoint.

2.5.2 *Geodesic Regression of Simulated Rotation Data*

To test the geodesic regression least squares estimation on $\mathbb{Q}_1$, synthetic rotation data was simulated according to the geodesic model (2.3). The intercept was the identity rotation: $p = (1, 0, 0, 0)$, and the slope was a rotation about the z-axis: $v = (0, 0, 0, \pi/4)$. The x_i data were drawn from a uniform distribution on $[0, 1]$.

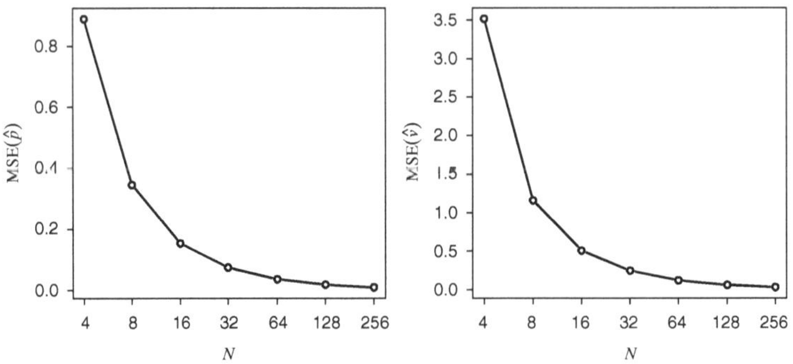

Fig. 2.4 Results for simulated rotation data: MSE of the geodesic regression estimates for the intercept (*left*) and slope (*right*) as a function of sample size

The errors in the model were generated from an isotropic Gaussian distribution in the tangent space, with $\sigma = \pi/8$. The resulting data (x_i, y_i) were used to compute estimates of the parameters $(\hat{p}, \hat{v})$. This experiment was repeated 1,000 times each for sample sizes $N = 2^k, k = 2, 3, \ldots, 8$. We would expect that as the sample size increases, the mean squared error (MSE) in the estimates $(\hat{p}, \hat{v})$, relative to the true parameters, would approach zero. The MSE is defined as

$$\mathrm{MSE}(\hat{p}) = \frac{1}{M} \sum_{i=1}^{M} d(\hat{p}_i, p)^2, \qquad \mathrm{MSE}(\hat{v}) = \frac{1}{M} \sum_{i=1}^{M} \|\hat{v}_i \cdot (\hat{p}_i^{-1} p) - v\|^2,$$

where $M = 1{,}000$ is the number of repeated trials, and $(\hat{p}_i, \hat{v}_i)$ is the estimate from the ith trial. Notice the multiplication by $(\hat{p}_i^{-1} p)$ in the second equation is a right-translation of $\hat{v}_i$ to the tangent space of p. Figure 2.4 shows plots of the resulting MSE for the slope and intercept estimates. As expected, the MSE approaches zero as sample size increases, indicating at least empirically that the least squares estimates are consistent.

2.6 Results: Regression in Shape Spaces

One area of medical image analysis and computer vision that finds the most widespread use of Riemannian geometry is the analysis of shape. Dating back to the groundbreaking work of Kendall [13] and Bookstein [1], modern shape analysis is concerned with the geometry of objects that is invariant to rotation, translation, and scale. This typically results in representing an object's shape as a point in a nonlinear Riemannian manifold, or *shape space*. Recently, there has been a great amount of interest in Riemannian shape analysis, and several shape spaces for 2D and 3D objects have been proposed [8, 15, 17, 26]. We choose here to use Kendall's shape space, but geodesic regression is applicable to other shape spaces as well. It could also be applied to spaces of diffeomorphisms, using the Jacobi field calculations given by Younes [27]. In fact, Niethammer et al. [20] recently independently developed geodesic regression for diffeomorphic transformations of image time series. They solve the gradient descent problem with an elegant control theory approach, constraining the regression curve to be a geodesic using Lagrange multipliers. The resulting update to the geodesic's initial conditions involves a numerical integration of an adjoint equation backwards along the geodesic with jump conditions at the data points.

2.6.1 Overview of Kendall's Shape Space

We begin with derivations of the necessary computations for geodesic regression on Kendall's shape space. A configuration of k points in the 2D plane is considered as a complex k-vector, $z \in \mathbb{C}^k$. Removing translation, by requiring the centroid to be zero, projects this point to the linear complex subspace $V = \{z \in \mathbb{C}^k : \sum z_i = 0\}$, which is equivalent to the space $\mathbb{C}^{k-1}$. Next, points in this subspace are deemed equivalent if they are a rotation and scaling of each other, which can be represented as multiplication by a complex number, $\rho e^{i\theta}$, where ρ is the scaling factor and θ is the rotation angle. The set of such equivalence classes forms the complex projective space, $\mathbb{C}P^{k-2}$. As Kendall points out, there is no unique way to identify a shape with a specific point in complex projective space. However, if we consider that the geodesic regression problem only requires computation of exponential/log maps and Jacobi fields, we can formulate these computations without making an explicit identification of shapes with points in $\mathbb{C}P^{k-2}$.

Thus, we think of a centered shape $x \in V$ as representing the complex line $L_x = \{z \cdot x : z \in \mathbb{C}\backslash\{0\}\}$, i.e., L_x consists of all point configurations with the same shape as x. A tangent vector at $L_x \in V$ is a complex vector, $v \in V$, such that $\langle x, v\rangle = 0$. The exponential map is given by rotating (within V) the complex line L_x by the initial velocity v, that is,

$$\mathrm{Exp}_x(v) = \cos\theta \cdot x + \frac{\|x\| \sin\theta}{\theta} \cdot v, \quad \theta = \|v\|. \tag{2.8}$$

Likewise, the log map between two shapes $x, y \in V$ is given by finding the initial velocity of the rotation between the two complex lines L_x and L_y. Let $\pi_x(y) = x \cdot \langle x, y\rangle / \|x\|^2$ denote the projection of the vector y onto x. Then the log map is given by

$$\mathrm{Log}_x(y) = \frac{\theta \cdot (y - \pi_x(y))}{\|y - \pi_x(y)\|}, \quad \theta = \arccos \frac{|\langle x, y\rangle|}{\|x\|\,\|y\|}. \tag{2.9}$$

Notice that we never explicitly project a shape onto $\mathbb{C}P^{k-2}$. This has the effect that shapes computed via the exponential map (2.8) will have the same orientation and scale as the base point x. Also, tangent vectors computed via the log map (2.9) are valid only at the particular representation x (and not at a rotated or scaled version of x). This works nicely for our purposes and implies that shapes along the estimated geodesic will have the same orientation and scale as the intercept shape, $\hat{p}$.

The sectional curvature of $\mathbb{C}P^{k-2}$ can be computed as follows. Let u, w be orthonormal vectors at a point $p \in \mathbb{C}P^{k-2}$. These vectors may be thought of as vectors in $\mathbb{C}^{k-1} \cong \mathbb{R}^{2k-2}$. Writing the vector w as $w = (w_1, \ldots, w_{2k-2})$, define the operator

$$j(w) = (-w_k, \ldots, -w_{2k-2}, w_1, \ldots, w_{k-1}).$$

(This is just multiplication by $i = \sqrt{-1}$ if we take w as a complex vector with the $k-1$ real coordinates listed first.) Using this operator, the sectional curvature is given by

$$K(u, w) = 1 + 3\langle u, j(w)\rangle^2.$$

When $k = 3$, $\mathbb{C}P^1$ is the space of triangle shapes and is isomorphic to the sphere, S^2, and thus has constant sectional curvature, $K = 1$. For $k > 3$, $\mathbb{C}P^{k-2}$ has sectional curvature in the interval $K \in [1, 4]$. Furthermore, let $u \in T_p\mathbb{C}P^{k-2}$ be any unit length vector. If we decompose the tangent space into an orthonormal basis $e_1, \ldots, e_{2k-2}$, such that $e_1 = j(u)$, then we have $K(u, e_1) = 4$ and $K(u, e_i) = 1$ for $i > 1$. This leads to the following procedure for computing the Jacobi field equation on $\mathbb{C}P^{k-2}$ along a geodesic γ. Given initial conditions for $J(0)^\perp$ and $J'(0)^\perp$, decompose $J(0)^\perp = u_1 + w_1$, so that u_1 is orthogonal to $j(\gamma')$ and w_1 is tangential to $j(\gamma')$. Do the same for $J'(0)^\perp = u_2 + w_2$. As before, extend these vectors to parallel fields, $u_i(t), w_i(t)$, along γ. Then the orthogonal component of the Jacobi field along γ is given by

$$J(t)^\perp = u_1(t)\cos(Lt) + u_2(t)\frac{\sin(Lt)}{L} + w_1(t)\cos(2Lt) + w_2(t)\frac{\sin(2Lt)}{2L}.$$

As was the case for rotations, both $d_p\mathrm{Exp}$ and $d_v\mathrm{Exp}$ are self-adjoint operators.

2.6.2 Application to Corpus Callosum Aging

The corpus callosum is the major white matter bundle connecting the two hemispheres of the brain. A midsagittal slice from a magnetic resonance image (MRI) with segmented corpus callosum is shown in Fig. 2.5. Several studies have shown that the volume of the corpus callosum decreases with normal aging [4]. However, less is known about how the *shape* of the corpus callosum changes with age. Understanding shape changes may provide a deeper understanding of the anatomical and biological processes underlying aging. For example, does the corpus callosum shrink uniformly in size, or do certain regions deteriorate faster than others? This type of question can be answered by geodesic regression in shape spaces.

To understand age-related changes in the shape of the corpus callosum, geodesic regression was applied to corpus callosum shape data derived from the OASIS brain database (www.oasis-brains.org). The data consisted of MRI from 32 subjects

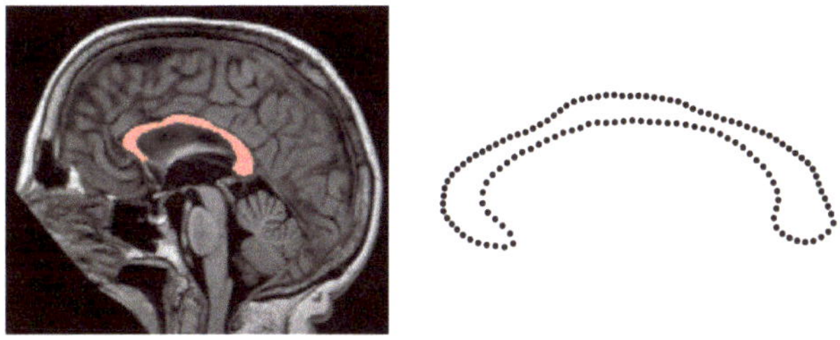

Fig. 2.5 Corpus callosum segmentation and boundary point model for one subject

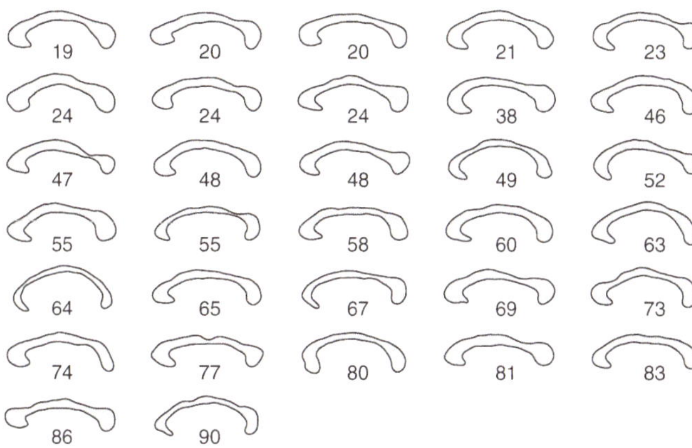

Fig. 2.6 The input corpus callosum shape data and corresponding subject ages in years

with ages ranging from 19 to 90 years old. The corpus callosum was segmented in a midsagittal slice using the ITK SNAP program (www.itksnap.org). These boundaries of these segmentations were sampled with 128 points using ShapeWorks (www.sci.utah.edu/software.html). This algorithm generates a sampling of a set of shape boundaries while enforcing correspondences between different point models within the population. An example of a segmented corpus callosum and the resulting boundary point model is shown in Fig. 2.5. The entire collection of input shapes and their ages is shown in Fig. 2.6 (boundary points have been connected into a boundary curve for visualization purposes). Each of these preprocessing steps were done without consideration of the subject age, to avoid any bias in the data generation.

Geodesic regression was applied to the data (x_i, y_i), where x_i was the ith subject's age, and y_i was the ith subject's corpus callosum, generated as above and represented as a point in Kendall's shape space. First, the average age of the group, $\bar{x}$, was subtracted from each x_i, which was done to make the intercept term correspond to the shape at the mean age, rather than the shape at age 0, which would be far outside the data range. Least squares estimates $(\hat{p}, \hat{v})$ were generated according to (2.5), and using the above calculations for $\mathbb{C}P^{k-2}$. The resulting estimated geodesic is shown in Fig. 2.7 as a sequence of shapes: $\hat{\gamma}(t_k) = \mathrm{Exp}(\hat{p}, (t_k - \bar{x})\hat{v})$, for

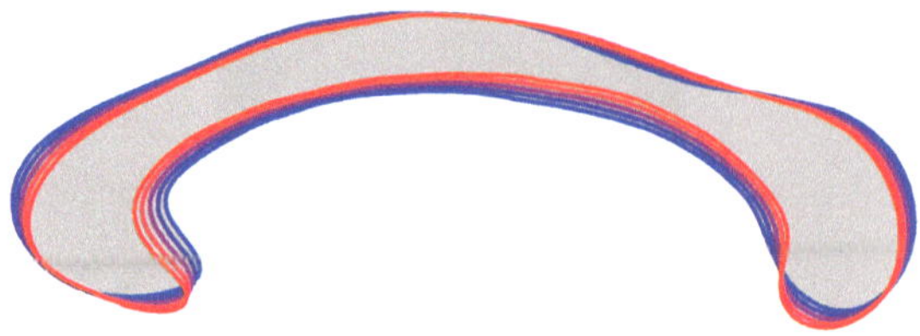

Fig. 2.7 Geodesic regression of the corpus callosum. The estimated geodesic is shown as a sequence of shapes from age 19 (*blue*) to age 90 (*red*)

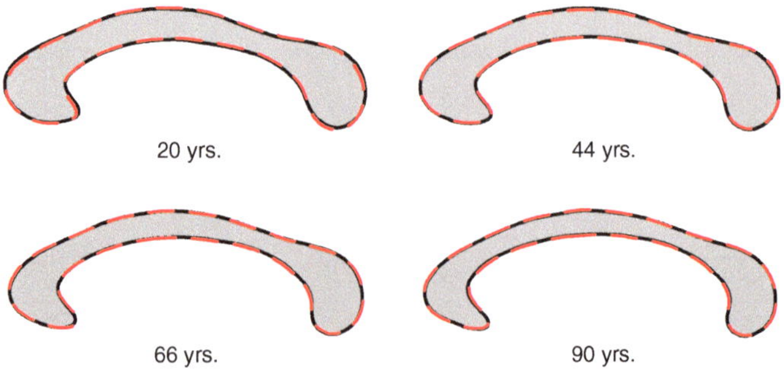

Fig. 2.8 Comparison of geodesic regression (*solid black*) and nonparametric kernel regression (*dashed red*) of the corpus callosum shape versus age

$t_k = 19, 36, 54, 72, 90$. The shape trend shows a very clear thinning of the corpus callosum, with the largest effects in the posterior part of the body and in the genu (anterior end).

The statistical significance of the estimated trend was tested using the permutation test described in Sect. 2.3.2, using 10,000 permutations. The p-value for the significance of the slope estimate, $\hat{v}$, was $p = 0.009$. The coefficient of determination (for the unpermuted data) was $R^2 = 0.12$. The low R^2 value must be interpreted carefully. It says that age only describes a small fraction of the shape variability in the corpus callosum. This is not surprising: we would expect the intersubject variability in corpus callosum shape to be difficult to fully describe with a single variable (age). However, this does not mean that the age effects are not important. In fact, the low p-value says that the estimated age changes are highly unlikely to have been found by random chance.

Finally, the appropriateness of the resulting geodesic model fit was tested using a comparison to nonparametric regression, as outlined in Sect. 2.4. First, a nonparametric kernel regression of the corpus callosum data versus age was computed using the method developed by Davis et al. [2] and reviewed in Sect. 2.4.2. The kernel regression was performed on the same Kendall shape space manifold and the bandwidth was chosen automatically using the cross-validation procedure described in Sect. 2.4.3. Next, the resulting corpus callosum shape trend generated by the kernel regression method was compared to the result of the geodesic regression. This was done by again generating shapes from the geodesic model $\hat{\gamma}(t_k)$ at a sequence of ages, t_k, and overlaying the corresponding generated shapes from the kernel regression model at the same ages. The results are plotted for ages

$t_k = 20, 44, 66$, and 90 (Fig. 2.8). Both regression methods give strikingly similar results. The two regression models at other values of ages, not shown, are also close to identical. This indicates that a geodesic curve does capture the relationship between age and corpus callosum shape, and that the additional flexibility offered by the nonparametric regression does not change the estimated trend. However, even though both methods provide a similar estimate of the trend, the geodesic regression has the advantage that it is simpler to compute and easier to interpret, from the standpoint of the R^2 statistic and hypothesis test demonstrated above.

2.7 Conclusion

We introduced a geodesic regression analysis method for Riemannian manifolds. The geodesic regression model is the natural generalization of linear regression and is parameterized by an intercept and slope term. We also developed a generalization of the R^2 statistic and a permutation test for the significance of the estimated geodesic trend. There are several avenues for future work. First, the hypothesis test presented here could be extended to test for group differences, for example, to test if age-related anatomical changes are different in a disease population compared to controls. Second, theoretical properties of geodesic regression, such as unbiasedness and consistency, would be of interest. Finally, regression diagnostics and model selection procedures need to be developed to assess the appropriateness of a geodesic model for a particular data set.

References

1. Bookstein, F.L.: Size and shape spaces for landmark data in two dimensions (with discussion). Stat. Sci. **1**(2), 181–242 (1986)
2. Davis, B., Fletcher, P.T., Bullitt, E., Joshi, S.: Population shape regression from random design data. In: Proceedings of IEEE International Conference on Computer Vision (2007)
3. do Carmo, M.: Riemannian Geometry. Birkhäuser, Boston (1992)
4. Driesen, N., Raz, N.: The influence of sex, age, and handedness on corpus callosum morphology: a meta-analysis. Psychobiology **23**(3), 240–247 (1995)
5. Dryden, I., Mardia, K.: Statistical Shape Analysis. Wiley, Chichester (1998)
6. Durrleman, S., Pennec, X., Trouvé, A., Gerig, G., Ayache, N.: Spatiotemporal atlas estimation for developmental delay detection in longitudinal datasets. In: Medical Image Computing and Computer-Assisted Intervention, pp. 297–304 (2009)
7. Fletcher, P.T.: Geodesic regression on Riemannian manifolds. In: MICCAI Workshop on Mathematical Foundations of Computational Anatomy, pp. 75–86 (2011)
8. Fletcher, P.T., Lu, C., Joshi, S.: Statistics of shape via principal geodesic analysis on Lie groups. In: IEEE CVPR, pp. 95–101 (2003)
9. Fox, J.: Applied Regression Analysis, Linear Models, and Related Methods. Sage, Thousand Oaks (1997)
10. Fréchet, M.: Les éléments aléatoires de nature quelconque dans un espace distancié. Ann. Inst. H. Poincaré **10**(3), 215–310 (1948)

11. Jupp, P.E., Kent, J.T.: Fitting smooth paths to spherical data. Appl. Stat. **36**(1), 34–46 (1987)
12. Karcher, H.: Riemannian center of mass and mollifier smoothing. Commun. Pure Appl. Math. **30**, 509–541 (1977)
13. Kendall, D.G.: Shape manifolds, Procrustean metrics, and complex projective spaces. Bull. Lond. Math. Soc. **16**, 18–121 (1984)
14. Kendall, W.S.: Probability, convexity, and harmonic maps with small image I: uniqueness and fine existence. Proc. Lond. Math. Soc. **3**(61), 371–406 (1990)
15. Klassen, E., Srivastava, A., Mio, W., Joshi, S.: Analysis of planar shapes using geodesic paths on shape spaces. IEEE PAMI **26**(3), 372–383 (2004)
16. Mardia, K.V.: Directional Statistics. Wiley, Chichester (1999)
17. Michor, P.W., Mumford, D.: Riemannian geometries on spaces of plane curves. J. Eur. Math. Soc. **8**, 1–48 (2006)
18. Miller, M.: Computational anatomy: shape, growth, and atrophy comparison via diffeomorphisms. NeuroImage **23**, S19–S33 (2004)
19. Nadaraya, E.A.: On estimating regression. Theory Probab. Appl. **10**, 186–190 (1964)
20. Niethammer, M., Huang, Y., Viallard, F.-X.: Geodesic regression for image time-series. In: Proceedings of Medical Image Computing and Computer Assisted Intervention (2011)
21. Pennec, X.: Intrinsic statistics on Riemannian manifolds: basic tools for geometric measurements. J. Math. Imaging Vis. **25**(1), 127–154 (2006)
22. Shi, X., Styner, M., Lieberman, J., Ibrahim, J., Lin, W., Zhu, H.: Intrinsic regression models for manifold-valued data. J. Am. Stat. Assoc. **5762**, 192–199 (2009)
23. Trouvé, A., Vialard, F.-X.: A second-order model for time-dependent data interpolation: splines on shape spaces. In: MICCAI STIA Workshop (2010)
24. Wand, M.P., Jones, M.C.: Kernel Smoothing. Number 60 in Monographs on Statistics and Applied Probabilitiy. Chapman & Hall/CRC, London/New York (1995)
25. Watson, G.S.: Smooth regression analysis. Sankhya **26**, 101–116 (1964)
26. Younes, L.: Computable elastic distances between shapes. SIAM J. Appl. Math. **58**, 565–586 (1998)
27. Younes, L.: Jacobi fields in groups of diffeomorphisms and applications. Q. Appl. Math. **65**, 113–113 (2006)

Chapter 3
Segmentation and Skeletonization on Arbitrary Graphs Using Multiscale Morphology and Active Contours

Petros Maragos and Kimon Drakopoulos

Abstract In this chapter we focus on formulating and implementing on abstract domains such as arbitrary graphs popular methods and techniques developed for image analysis, in particular multiscale morphology and active contours. To this goal we extend existing work on graph morphology to multiscale dilation and erosion and implement them recursively using level sets of functions defined on the graph's nodes. We propose approximations to the calculation of the gradient and the divergence of vector functions defined on graphs and use these approximations to apply the technique of geodesic active contours for object detection on graphs via segmentation. Finally, using these novel ideas, we propose a method for multiscale shape skeletonization on arbitrary graphs.

3.1 Introduction

Graph-theoretic approaches have become commonplace in computer vision. Examples include the graph-cut approaches to segmentation [7,8,16,24] and the statistical inference on discrete-space visual data with graphical models [42]. In most of these cases, the image graphs are regular grids that result from uniform sampling of continuous space. In addition, in nowadays science and technology there exist both low-level and high-level visual data as well as many other types of data defined on *arbitrary graphs* with irregular spacing among their vertices. Examples

---

P. Maragos (✉)
National Technical University of Athens, School of Electrical and Computer Engineering, Athens 15773, Greece
e-mail: maragos@cs.ntua.gr

K. Drakopoulos
Massachusetts Institute of Technology, 77 Massachusetts Avenue, Cambridge, MA 02139-4307, USA
e-mail: kimondr@mit.edu

M. Breuß et al. (eds.), *Innovations for Shape Analysis*, Mathematics and Visualization, DOI 10.1007/978-3-642-34141-0_3,

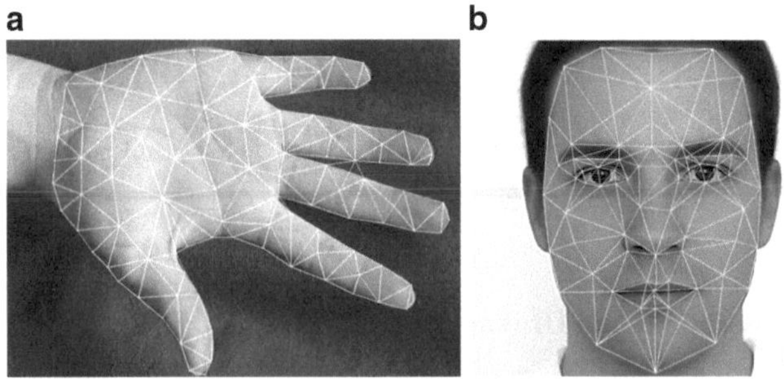

Fig. 3.1 Representing image or more general visual information on graphs. (**a**) Hand on a graph. (**b**) Face on a graph

from the vision area include region-based or part-based object representations, cluster analysis in pattern recognition, and graph-based deformable models for representing and recognizing shapes such as faces and gestures [15]. Two such examples from the last area are shown in Fig. 3.1. Examples from non-vision areas include network problems modeled with graphs, such as social nets, geographical information systems, and communications networks.

In this chapter we explore theoretically and algorithmically three topics related to shape morphology on arbitrary graphs: multiscale morphology on graphs, geodesic active contours on graphs, and multiscale skeletonization on graphs.

An important part in our work is how to define multiscale morphological operators on arbitrary graphs. We begin to approach this problem algebraically by extending the lattice definitions of morphological operators on arbitrary graphs which have been introduced in [20, 41] with some recent work in [17]. Then we focus on our major approach which is based on discretizing the PDEs generating continuous-scale morphological operators [1, 11] and the PDEs moving geodesic active contours [13] on arbitrary graphs. In this latter direction, a first approach to approximate morphological operators on graphs through mimicking the corresponding PDEs has been studied in Ta et al. [39]. Our approach is slightly different in our translation of the continuous gradient operator on arbitrary graph structures and in our usage of multiscale neighborhoods. (In the general field of approximating PDE-type problems on weighted graphs, a systematic analysis has been performed in [4, 14] by introducing discrete gradients and Laplacian and by studying Dirichlet and Neumann boundary value problems on graphs.) In the rest of our work, we propose approximations for computing the differential terms required in applying the technique of geodesic active contours to object detection on graphs. Finally, the modeling of multiscale morphology on graphs allows us to develop a method for multiscale skeletonization of shapes on arbitrary graphs.

3.2 Multiscale Morphology on Graphs

In this section we first review some basic concepts from lattice-based morphology. Then, we focus our review on (1) multiscale morphological image operators on a Euclidean domain, either defined algebraically or generated by nonlinear PDEs, and (2) on defining morphological operators on arbitrary graphs. Finally, we connect these two areas and define multiscale morphological operators on graphs.

3.2.1 *Background on Lattice and Multiscale Morphology*

A general formalization [19,37] of morphological operators views them as operators on complete lattices. A *complete lattice* is a set $\mathscr{L}$ equipped with a partial ordering $\leq$ such that $(\mathscr{L}, \leq)$ has the algebraic structure of a *partially ordered set* where the supremum and infimum of any of its subsets exist in $\mathscr{L}$. For any subset $\mathscr{K} \subseteq \mathscr{L}$, its *supremum* $\bigvee \mathscr{K}$ and *infimum* $\bigwedge \mathscr{K}$ are defined as the lowest (with respect to $\leq$) upper bound and greatest lower bound of $\mathscr{K}$, respectively. The two main examples of complete lattices used respectively in morphological shape and image analysis are: (i) the power set $\mathscr{P}(E) = \{X : X \subseteq E\}$ of all binary images or shapes represented by subsets X of some domain E where the $\bigvee / \bigwedge$ lattice operations are the set union/intersection, and (ii) the space of all graylevel image signals $f : E \rightarrow \mathscr{T}$ where $\mathscr{T}$ is a continuous or quantized sublattice of $\overline{\mathbb{R}} = \mathbb{R} \cup \{-\infty, \infty\}$ and the $\bigvee / \bigwedge$ lattice operations are the supremum/infimum of sets of real numbers. An operator ψ on $\mathscr{L}$ is called **increasing** if $f \leq g$ implies $\psi(f) \leq \psi(g)$. Increasing operators are of great importance; among them four fundamental examples are:

$$\delta \text{ is } \textbf{dilation} \Longleftrightarrow \delta(\bigvee_{i \in I} f_i) = \bigvee_{i \in I} \delta(f_i) \tag{3.1}$$

$$\varepsilon \text{ is } \textbf{erosion} \Longleftrightarrow \varepsilon(\bigwedge_{i \in I} f_i) = \bigwedge_{i \in I} \varepsilon(f_i) \tag{3.2}$$

$$\alpha \text{ is } \textbf{opening} \Longleftrightarrow \alpha \text{ is increasing, idempotent, and anti-extensive} \tag{3.3}$$

$$\beta \text{ is } \textbf{closing} \Longleftrightarrow \beta \text{ is increasing, idempotent, and extensive} \tag{3.4}$$

where I is an arbitrary index set, *idempotence* of an operator ψ means that $\psi^2 = \psi$, and *anti-extensivity* and *extensivity* of operators α and β means that $\alpha(f) \leq f \leq \beta(f)$ for all f. Operator products mean composition: $\phi\psi(f) = \phi(\psi(f))$. The notation ψ^r means r-fold composition.
Dilations and erosions come in pairs (δ, ε) called **adjunctions** if

$$\delta(f) \leq g \Longleftrightarrow f \leq \varepsilon(g) \tag{3.5}$$

Such pairs are useful for constructing openings $\alpha = \delta\varepsilon$ and closings $\beta = \varepsilon\delta$. The above definitions allow broad classes of signal operators to be studied under the unifying lattice framework.

In *Euclidean morphology*, the domain E becomes the d-dimensional Euclidean space $\mathbb{E}^d$ where $\mathbb{E} = \mathbb{R}$ or $\mathbb{E} = \mathbb{Z}$. In this case, the most well-known morphological operators are the translation-invariant Minkowski dilations $\oplus$, erosions $\ominus$, openings $\circ$, and closings $\bullet$, which are simple special cases of their lattice counterparts. If $X_{+b} = \{x + b : x \in X\}$ denotes the translation of a set/shape $X \subseteq \mathbb{E}^d$ by $b \in \mathbb{E}^d$, the simple Minkowski set operators are $X \oplus B = \bigcup_{b \in B} X_{+b}$, $X \ominus B = \bigcap_{b \in B} X_{-b}$, and $X \circ B = (X \ominus B) \oplus B$. The set B usually has a simple shape and small size, in which case it is called a *structuring element*. By denoting with $rB = \{rb : b \in B\}$ the r-scaled homothetic of B, where $r \geq 0$, we can define *multiscale* translation-invariant morphological set operators on $\mathbb{R}^d$:

$$\delta_{rB}(X) \triangleq X \oplus rB, \quad \varepsilon_{rB}(X) \triangleq X \ominus rB, \quad \alpha_{rB}(X) \triangleq X \circ rB \tag{3.6}$$

Similarly, if $(f \oplus B)(x) = \bigvee_{b \in B} f(x - b)$, $(f \ominus B)(x) = \bigwedge_{b \in B} f(x + b)$, and $(f \circ B)(x) = (f \ominus B) \oplus B$ are the unit-scale Minkowski translation-invariant flat (i.e. unweighted) function operators, their multiscale counterparts are

$$\delta_{rB}(f) \triangleq f \oplus rB, \quad \varepsilon_{rB}(f) \triangleq f \ominus rB, \quad \alpha_{rB}(f) \triangleq f \circ rB \tag{3.7}$$

If B is *convex*, then [30]

$$rB = \underbrace{B \oplus B \oplus \cdots \oplus B}_{r \text{ times}}, \quad r = 0, 1, 2, \ldots \tag{3.8}$$

This endows the above multiscale dilations and erosions with a semigroup property, which allows them to be generated recursively:

$$\delta_{(r+1)B} = \delta_B \delta_B^r, \quad \varepsilon_{(r+1)B} = \varepsilon_B \varepsilon_B^r, \quad r = 0, 1, 2, \ldots \tag{3.9}$$

and create the simplest case of a *morphological scale-space* [11, 28].

For *digital shapes and images*, the above translation-invariant morphological operators can be extended to multiple scales by using two alternative approaches. The first is an algebraic approach where if $B \subseteq \mathbb{Z}^d$ is a unit-scale discrete structuring graph, we define its scaled version rB for integer scales as in (3.8) and use (3.9) for producing multiscale morphological operators that agree with their continuous versions in (3.6) and (3.7) if B is convex. The second approach [1, 11] models the dilation and erosion scale-space functions $u(x,t) = f \oplus tB$ and $v(x,t) = f \ominus tB$ as generated by the nonlinear partial differential equations (PDEs)

$$\partial_t u = \|\nabla u\|_B, \quad \partial_t v = -\|\nabla v\|_B \tag{3.10}$$

where for a convex $B \subseteq \mathbb{R}^2$, $\|(x_1, x_2)\|_B = \sup_{(a_1,a_2)\in B} a_1x_1 + a_2x_2$. These PDEs can be implemented using the numerical algorithms of [32], as explored in [35]. In case of a shape X, the above PDEs can still be used to generate its multiscale morphological evolutions by treating u as the level function whose zero level set contains the evolving shape. Such PDE-based shape evolutions have been studied in detail by Kimia et al. [23]. Modern numerical algorithms for morphological PDEs can be found in [9, 10].

3.2.2 Background on Graph Morphology

We consider an undirected **graph** $G = (V, E)$ without loops and multiple edges, where $V = V(G)$ and $E = E(G)$ are the sets of its vertices (also called nodes) and edges, respectively. We denote edges by pairs (v, w) of vertices; these are symmetric, i.e. $(v, w) = (w, v)$, since the graph is undirected. If $V' \subseteq V$ and $E' \subseteq E$, the pair $G' = (V', E')$ is called a *subgraph* of G. A graph vertex mapping $\theta : V \to V'$ is called a *graph homomorphism* from G to G' if θ is one-to-one and preserves edges, i.e. $(v, w) \in E$ implies $(\theta(v), \theta(w)) \in E'$. If θ is a bijection, then it is called a *graph isomorphism*; if in addition $G' = G$, then it is called a *graph automorphism* or *symmetry* of G. The set of all such symmetries forms under composition the *symmetry group* $\mathrm{Sym}(G)$ of a graph. Symmetries play the role of 'generalized translations' on a graph.

Shapes $X \subseteq V$ and image functions $f : V \to \mathscr{T}$ defined on a graph G with values in a complete lattice $\mathscr{T}$ will be denoted by $(X|G)$ and $(f|G)$, respectively, and may be referred to as *binary graphs* and *multilevel graphs*. In case of multilevel graphs, the values of the functions $(f|G)$ may be discrete, e.g. $\mathscr{T} = \{0, 1, \ldots, m-1\}$, or continuous, e.g. $\mathscr{T} = \overline{\mathbb{R}}$. Similarly a graph operator for shapes or functions will be denoted by $\psi(\cdot|G)$. The argument G will be omitted if there is no risk of confusion. A graph operator ψ is called increasing if it is increasing in its first argument (shape or function), i.e. $X \subseteq Y$ implies $\psi(X|G) \subseteq \psi(Y|G)$, and G-*increasing* if it increases in G, i.e., $G' \subseteq G$ implies $\psi(f|G') \leq \psi(f|G)$ for all graph functions $(f|G)$. A graph operator ψ is called *invariant* under graph symmetries $\tau \in \mathrm{Sym}(G)$ if $\tau\psi = \psi\tau$.

Henceforth and until mentioned otherwise, we shall focus our discussion on binary graph operators. Given a graph $G = (V, E)$, the binary graph dilations and erosions on $\mathscr{P}(V)$ can be defined via a **graph neighborhood function** $N : V \to \mathscr{P}(V)$ which assigns at each vertex v a neighborhood $N(v)$. Taking the union of all such neighborhoods for the vertices of a shape $X \subseteq V$ creates a graph dilation of X; then, by using (3.5) we also find its adjunct erosion:

$$\delta_N(X|G) \triangleq \bigcup_{v\in X} N(v), \quad \varepsilon_N(X|G) \triangleq \{v \in V : N(v) \subseteq X\} \tag{3.11}$$

At each vertex v, the shape of $N(v)$ may vary according to the local graph structure and this inherently makes the above morphological graph operators *adaptive*. At each v, the *reflected neighborhood* is defined by

$$\check{N}(v) \triangleq \{w \in V : v \in N(w)\} \tag{3.12}$$

This is related to *operator duality* as follows. The dual (or negative) of a binary graph operator is defined by $\psi^*(X|G) = (\psi(X^*|G))^*$, where $X^* = V \setminus X$. Then, the dual graph dilation and erosion w.r.t. a neighborhood function coincide with the erosion and dilation, respectively, w.r.t. the reflected neighborhood function:

$$\delta_N^* = \varepsilon_{\check{N}}, \quad \varepsilon_N^* = \delta_{\check{N}} \tag{3.13}$$

If $N(v) = \check{N}(v)$ for each v, we have a *symmetric neighborhood* function. Such an example is Vincent's *unit-scale graph neighborhood* function [41]

$$N_1(v) \triangleq \{w \in V : (v, w) \in E\} \cup \{v\} \tag{3.14}$$

which, when centered at a vertex v, includes this vertex and all others that form an edge with it. If we use it in (3.11), this leads to the simplest unit-scale graph dilation $\delta_1(X|G)$ and erosion $\varepsilon_1(X|G)$. Since $(\delta_1, \varepsilon_1)$ is an adjunction, the composition $\alpha_1 = \delta_1\varepsilon_1$ and $\beta_1 = \varepsilon_1\delta_1$ is a graph opening and closing, respectively. See Fig. 3.2 for an example. All four of these operators inherit the standard increasing property from their lattice definition and are invariant under graph symmetries. However, δ_1 is G-increasing, ε_1 is G-decreasing, and α_1 and β_1 are neither of these.

Heijmans et al. [20,21] have generalized the above (symmetric neighborhood N_1) approach by introducing the concept of a **structuring graph *(s-graph)***. This is a graph $\mathscr{A} = (V_{\mathscr{A}}, E_{\mathscr{A}})$ of a relatively small size and has as additional structure two nonempty and possibly overlapping subsets: the *buds* $B_{\mathscr{A}} \subseteq V_{\mathscr{A}}$ and the *roots* $R_{\mathscr{A}} \subseteq V_{\mathscr{A}}$. It may not be connected and plays the role of a locally adaptive graph template, where (compared to Euclidean morphology) the buds correspond to the points of a structuring graph and the roots correspond to its origin. See Fig. 3.3 for examples. An s-graph $\mathscr{A}$ corresponds to the following neighborhood function

$$N_{\mathscr{A}}(v|G) = \bigcup \{\theta(B_{\mathscr{A}}) : \ \theta \text{ embeds } \mathscr{A} \text{ into } G \text{ at } v\} \tag{3.15}$$

where we say that θ *embeds* $\mathscr{A}$ into G at v if θ is a group homomorphism of $\mathscr{A}$ into G and $v \in \theta(R_{\mathscr{A}})$. Such an embedding matches the s-graph $\mathscr{A}$ with the local structure of the graph G. The simple neighborhood N_1 of (3.14) corresponds to the s-graph of Fig. 3.3a, with two vertices which both are buds and one of them is a root. Replacing (3.15) in (3.11) creates an adjunction of graph dilation and erosion $(\delta_{\mathscr{A}}, \varepsilon_{\mathscr{A}})$ by structuring graphs. These are *symmetry-invariant* operators, i.e. they commute with group symmetries τ, because the neighborhood function of their s-graph is invariant under group symmetries: i.e., $N_{\mathscr{A}}(\tau(v)|G) = \tau N_{\mathscr{A}}(v|G)$, where $\tau X = \{\tau(v) : v \in X\}$.

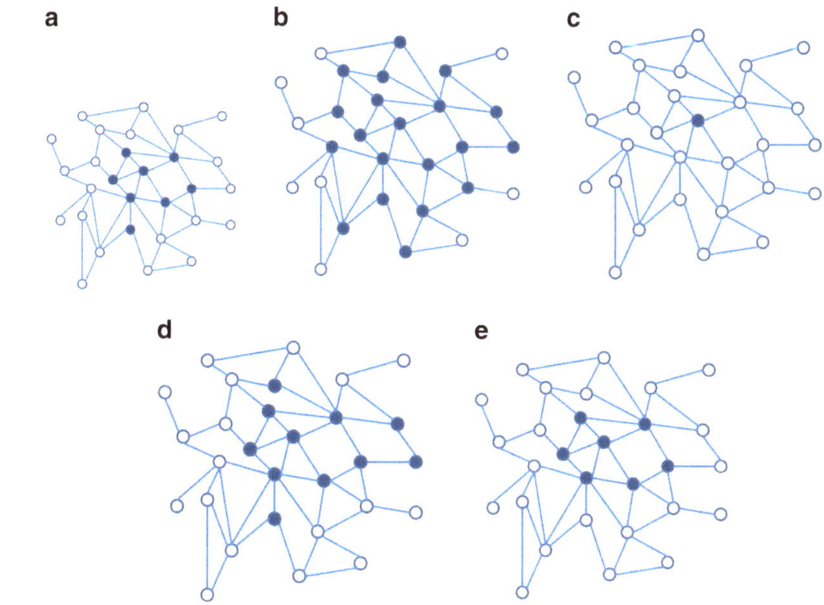

Fig. 3.2 Binary graph operators using a unit-scale symmetric neighborhood function. **(a)** The vertex set on which we apply morphological operators. **(b)** Dilation. **(c)** Erosion. **(d)** Closing. **(e)** Opening

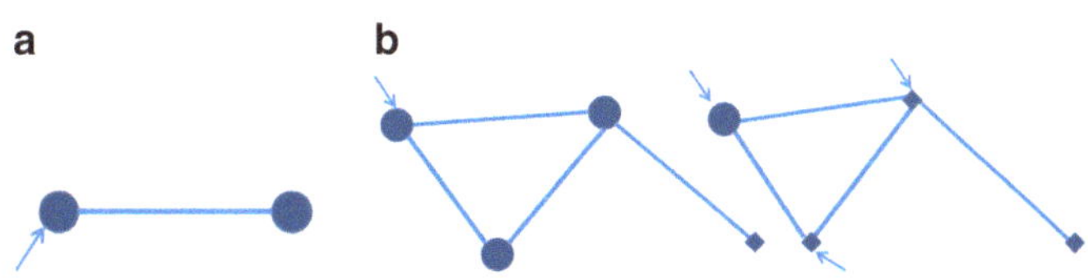

Fig. 3.3 Examples of structuring graphs. *Arrows* indicate roots. *Large circular nodes* denote buds. **(a)** The s-graph that corresponds to the simple neighborhood. Specifically, using this s-graph as a structuring element, the neighborhood of a node is the set of nodes that are adjacent to it. **(b)** A structuring graph and its reflection. The reflected s-graph has the same vertices and edges as the original s-graph but their bud and root sets are interchanged

Finally, the reflection of the neighborhood of an s-graph equals the neighborhood of another s-graph $\check{\mathscr{A}}$, called the *reflection* of $\mathscr{A}$:

$$\check{N}_{\mathscr{A}}(v|G) = N_{\check{\mathscr{A}}}(v|G) \tag{3.16}$$

The reflected s-graph $\check{\mathscr{A}}$ has the same vertices and edges as the original s-graph $\mathscr{A}$ but their bud and root sets are interchanged: $B_{\check{\mathscr{A}}} = R_{\mathscr{A}}$ and $R_{\check{\mathscr{A}}} = B_{\mathscr{A}}$ (see Fig. 3.3). The dual operator of a dilation by an s-graph is the erosion by its reflected s-graph, and vice-versa, as prescribed by (3.13).

All the previously defined binary graph operators are increasing and can be extended to multilevel graphs. Specifically, a multilevel graph $(f|G)$ can also be represented by its *level sets* $X_h(f|G) = \{v \in V : f(v) \geq h\}, h \in \mathscr{T}$:

$$(f|G)(v) = \sup\{h \in \mathscr{T} : v \in X_h(f|G)\} \quad (3.17)$$

By applying an increasing binary graph operator ψ to all level sets and using threshold superposition, we can extend ψ to a *flat operator* on multilevel graphs:

$$\psi(f|G)(v) = \sup\{h \in \mathscr{T} : v \in \psi(X_h(f)|G)\} \quad (3.18)$$

For example, if $\psi(X|G)$ is a set dilation by the s-graph $\mathscr{A}$, the corresponding function operator is

$$\delta_{\mathscr{A}}(f|G)(v) = \max_{w \in N_{\mathscr{A}}(v|G)} f(w) \quad (3.19)$$

Two useful choices for the function values are either discrete with $\mathscr{T} = \{0, 1, \ldots, m-1\}$, or continuous with $\mathscr{T} = \overline{\mathbb{R}}$.

3.2.3 Multiscale Morphology on Graphs

We need to discuss the notion of scale in graph morphology in order to obtain the graph counterparts of multiscale dilation and erosion defined in Sect. 3.2.1. Consider a graph $G = (V, E)$ and a nonempty subset $X \subseteq V$ of its vertices. Let $\mathscr{A}$ be an s-graph. One approach could be to define the dilation at scale $r = 1, 2, \ldots$ of a vertex subset X w.r.t. the s-graph $\mathscr{A}$ by $\delta_{r\mathscr{A}}(X|G)$ where $r\mathscr{A}$ denotes the r-fold graph dilation of the s-graph with itself. This approach would encounter the problem presented in Fig. 3.4. Specifically the scaled versions of the s-graph have complicated structure and in general, it would be highly unlikely to find an appropriate embedding in the graph at every node of X to calculate the dilation of the set.

Thus, we propose the following alternative new definition of the scaled versions of graph dilation and erosion in order to overcome the issues mentioned. We define recursively the graph dilation of X at integer scale $r = 1, 2, \ldots$ with respect to the s-graph $\mathscr{A}$ by

$$\delta^r_{\mathscr{A}}(X \mid G) = \delta_{\mathscr{A}}(\delta^{r-1}_{\mathscr{A}}(X \mid G) \mid G) \quad (3.20)$$

Essentially, we interchange the order with which we apply the dilation operators; in the classic framework in order to get the r-scale dilation we first find the r-scaling of the structuring graph and then perform the dilation with the set, whereas in our definition we dilate the set X with the structuring graph r times. Generalizing this notion of scale to multilevel dilation of a function $f : V \to \mathscr{T}$ we get the following definition. The dilation of f at integer scales r will be given at each $v \in V$ by

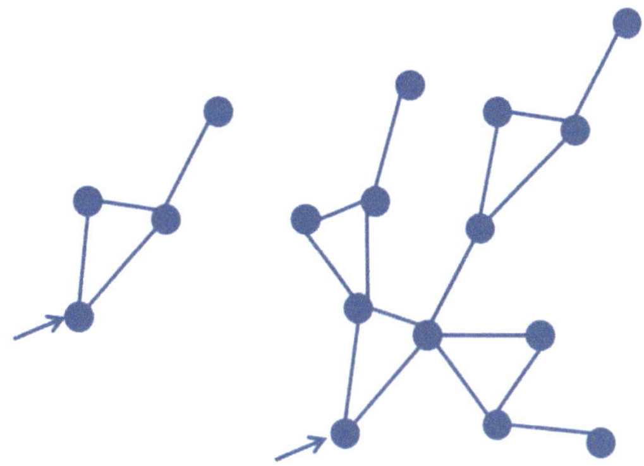

Fig. 3.4 *Left*: a structuring graph. *Right*: The scaled by $r = 2$ version of the s-graph. A scaling of a simple s-graph has increasingly complicated structure and therefore, for larger scales it is difficult or impossible to find an embedding to an arbitrary graph at each node. This fact necessitates an alternative definition of scale on graphs

$$\phi_r(v) \triangleq \delta_{\mathscr{A}}(\phi_{r-1}(f \mid G) \mid G)(v), \quad \phi_1(v) = \delta_{\mathscr{A}}(f \mid G)(v) \tag{3.21}$$

This provides a recursive computation of the multiscale dilations of a function $f : V \to \mathscr{T}$ and leads us to the following Proposition which offers an alternative recursive computation that involves a simple morphological gradient on a graph.

Proposition 3.1. *Given a graph $G = (V, E)$, the evolution of the multiscale dilation of a function $f : V \to \mathscr{T}$ by an s-graph $\mathscr{A}$ is described by the following difference equation at each vertex $v \in V$:*

$$\phi_{r+1}(v) - \phi_r(v) = \max_{w \in N_{\mathscr{A}}(v|G)} \{\phi_r(w) - \phi_r(v)\}. \tag{3.22}$$

Proof. By combining (3.21) with (3.19) we get

$$\phi_{r+1}(v) = \delta_{\mathscr{A}}(\phi_r(f \mid G))(v) = \max_{w \in N_{\mathscr{A}}(v|G)} \{\phi_r(w) - \phi_r(v)\} + \phi_r(v)$$

□

3.3 Geodesic Active Contours on Graphs

Kass et al. in [22] introduced the concept of energy minimizing snakes" driven by forces that pull it towards special features in the image like edges or lines. Specifically, the goal is to find in an image areas that are naturally distinguished from their background. The classical approach consists of a gradual deformation of an original curve towards the edges of those objects through the minimization, between successive time steps, of energy functionals which depend on the shape of the curve itself, its distance from the salient image features and finally terms that stabilize the snakes near local minima.

The main disadvantage of this initial approach is that the curve dynamics incurred do not allow changing the topology of the original contour; for example if the original curve contains two distinct objects the original snake will not be separated in two independent snakes. Heuristic solutions have been proposed in [31] but a topology-free approach has been given independently by Caselles et al. [12] and Malladi et al. [27]. These models are based on the theory of the curve evolution and geometric flows and the curve is propagating by means of a velocity that contains two terms, one related to the regularity of the curve and the other shrinks or expands towards the boundary. Finally, the curve dynamics take the form of a geometric flow (PDE) and can be implemented conveniently using the level set methods proposed by Osher and Sethian [32] that can accommodate changes in the topology between successive curves.

In particular let $\mathbf{C}(q) : [0,1] \to \mathbb{R}^2$ be parameterized planar curve and let $I : [0,1]^2 \to \mathbb{R}_+$ be the image in which one needs to detect the objects' boundaries. Note that we denote the curve by $\mathbf{C}(\cdot)$ when we interpret it as a vector-valued function and by C when we interpret it as a set of points. The energy functional associated with C can be written as follows:

$$E(C) = \alpha \int_0^1 \|\mathbf{C}'(q)\|^2 \, dq - \lambda \int_0^1 g(\|\nabla I(\mathbf{C}(q))\|) \, dq \qquad a, \lambda \geq 0 \tag{3.23}$$

Let $g : [0, +\infty) \to \mathbb{R}_+$ be a strictly decreasing function such that $g(r) \to 0$ as $r \to \infty$. Caselles et al. in [13] show that the problem of finding the minimum energy curve as defined in (3.23) is equivalent to finding the minimum length curve in a Riemannian space induced from image I, whose length is given by

$$L_R = \int_0^1 g(\|\nabla I(\mathbf{C}(q))\|)\|\mathbf{C}'(q)\| \, dq = \int_0^{L(\mathbf{C})} g(\|\nabla I(\mathbf{C}(q))\|) \, ds, \tag{3.24}$$

where $L(C)$ is the Euclidean length of the curve C. Furthermore, it is shown that a curve which is governed from the dynamics

$$\frac{\partial \mathbf{C}(t)}{\partial t} = g(I) \cdot \kappa \cdot \mathbf{N} - (\nabla g \cdot \mathbf{N}) \cdot \mathbf{N} \tag{3.25}$$

where κ is the Euclidean curvature and $\mathbf{N}$ is the unit inward normal, moves in the direction of the gradient of the length L_R.

Assume now that the curve $C(t)$ is a level set of a function $u : \mathbb{R}^2 \times \mathbb{R}_+ \to \mathbb{R}$. Namely, $C(t)$ is the set of points $\mathbf{x}$ for which $u(\mathbf{x}, t)$ is equal to a constant (for example $u = 0$). It is shown that if the function $u(\mathbf{x}, t)$ satisfies

$$\frac{\partial u}{\partial t} = g(I)\|\nabla u\|(c + \kappa) + \nabla g(I) \cdot \nabla u, \tag{3.26}$$

then the corresponding level set satisfies (3.25).

Our goal is to approximate all terms in the right hand side of (3.26) on graphs and finally construct a difference equation which would approximate the active contour dynamics for edge detection. The next subsection is devoted to the analysis of the simplest case of curve evolution on graphs, that is the constant velocity motion introducing useful ideas from graph morphology operators. Observe that this case corresponds to approximating the first term of the RHS of (3.26). Subsequently, we will approximate the rest of the terms participating in the curve dynamics to end up with a geodesic active contour model on graphs.

3.3.1 Constant-Velocity Active Contours on Graphs

We derive the difference equation that describes the evolution of the contour of a set that expands with constant velocity on a graph. In the continuous case, a contour undergoing such an evolution corresponds to the boundary of the multiscale dilation of the set by a unit disk. If this set is given as a level set of a graylevel function $u : \mathbb{R}^2 \times \mathbb{R}_+ \rightarrow \mathbb{R}$, then the evolution of u is described by

$$\frac{\partial u}{\partial t} = \|\nabla u\|. \tag{3.27}$$

Consider a subset X of V. Let $\mathscr{A}$ be a structuring graph. Imitating the continuous case, the constant velocity expansion of X corresponds to its r-scale dilation, denoted by X_r. If X is given as the level set of an original graylevel function $u_0 : X \rightarrow \mathbb{R}$ and X_r is the level set of the r-scale graylevel function $u_r : X \rightarrow \mathbb{R}$, then the difference equation that governs u_r, by using Proposition 3.1, is

$$u_{r+1}(v) - u_r(v) = \max_{w \in N_{\mathscr{A}}(v|G)} \{u_r(w) - u_r(v)\}. \tag{3.28}$$

The above expression is a discrete equivalent of the gradient magnitude on graphs. Similar expressions are being used in the literature, [4, 14, 39]. Our work extends previous results to more general structuring elements and exploits the revealed insight to approximate other geometric properties of differential operators on graphs in the next sections. In order to account for topological inhomogeneities of the graph one could calculate the gradient as the maximum rate of increase and its direction as the direction of the edge along which the rate of increase is larger. Therefore, (3.28) is the graph counterpart of (3.27), which implies the approximation of $\|\nabla u\|$ at node v by $\max_{w \in N_{\mathscr{A}}(v|G)}\{u(w) - u(v)\}$.

Summarizing, let X be a set of nodes whose contour expands with constant velocity c. Then, to implement its evolution on a graph we proceed as follows:

1. Let u_0 be the signed distance function from X, defined by

$$u_0(v) = \begin{cases} \min_{w \in G \setminus X} d_E(w, v) \text{ if } v \subset X, \\ -\min_{w \in X} d_E(w, v) \text{ if } v \notin X, \end{cases}$$

   where X is the zero level set of u_0, and $d_E(w, v)$ corresponds to the Euclidean distance between the nodes w and v.
2. Evolve u_r according to the following, at scales $r = 1, 2, \ldots$

$$u_{r+1}(v) - u_r(v) = c \cdot \max_{w \in N_{\mathcal{A}}(v|G)} \{u_r(w) - u_r(v)\}. \tag{3.29}$$

3. The set $X_r = \{v \in X : u_r(v) \geq 0\}$ corresponds to the r-scale dilation of X.

Figure 3.5 illustrates the results for the constant velocity expansion of a circular contour. Throughout this chapter for all our simulations we are using the simple structuring graph of Fig. 3.3a that generates the simple neighborhood. Moreover we embed our shapes on a *geometric random graph* on the unit square. The geometric random graph is characterized by two parameters; the number of nodes N and a radius ρ. N nodes are being placed uniformly at random on the unit square independently from one another. If the Euclidean distance between those two nodes is less than ρ then there is an edge between them. Typical values for N is 6,000–10,000 while ρ ranges from 0.015 to 0.04. Given the number of nodes, the parameter ρ affects the expected degree of a node and is proportional to the square root of its value.

3.3.2 Direction of the Gradient on Graphs

Observing Eq. (3.26), in order to obtain a model for active contours on graphs, we need to define, except from the magnitude of the gradient vector which we have already done in the previous section, its direction and also a characterization for the curvature of the corresponding curve.

Beginning from the first, trusting our intuition from real analysis, it would make sense to choose as the direction of the gradient on graphs the direction of the edge that corresponds to the maximum difference of the values of the function u. In other words, let a function u be defined on the set of nodes $v \in V$ of the graph $G = (V, E)$. Then

Having determined a meaning for the gradient of a function on a graph the only term remaining to be given a meaning on our more abstract graph structure is the curvature of the contour of each level set of the function u.

3.3.3 Curvature Calculation on Graphs

In the continuous case the curvature of a curve given as the contour of a level set of a function u can be computed using

$$\kappa = \operatorname{div}\left(\frac{\nabla u}{\|\nabla u\|}\right). \tag{3.33}$$

On the other hand we have derived expressions for the term $\frac{\nabla u}{\|\nabla u\|}$ on a graph. Therefore, the remaining step is to propose an expression for the computation of the divergence of a function on a graph. Consider a vector function $\mathbf{F} : \mathbb{R}^2 \to \mathbb{R}^2$. The *divergence* of $\mathbf{F}$ at a point $\mathbf{x}$ is defined as

$$\operatorname{div}\mathbf{F}(\mathbf{x}) = \lim_{S \to \{\mathbf{x}\}} \frac{\oint_{\Gamma(S)} \mathbf{F} \cdot \mathbf{n} d\ell}{|S|}, \tag{3.34}$$

where S is a two dimensional region, $\Gamma(S)$ its boundary, $\mathbf{n}$ the outward unit normal to that boundary, and $|S|$ its enclosed area.

To study the graph case consider Fig. 3.7. We can conclude that a good approximation for computing the divergence of $\mathbf{F}$ on a graph is the following

$$\operatorname{div}\mathbf{F}(v) = \frac{\sum_{w \in N_{\mathscr{A}}(v|G)} L_w \mathbf{F}(w) \cdot \mathbf{e}_{vw}}{S(v)} \tag{3.35}$$

where

- L_w corresponds to the length of the perpendicular to edge $\mathbf{e}_{vw}$,
- $S(v)$ corresponds to the area between the perpendicular to the edges lines,

as illustrated in Fig. 3.7.

At this point we can perform all the necessary calculations to compute the curvature of the contour of the level set of a graylevel function u on a graph. To illustrate the behavior of the expression proposed consider a circular shaped contour as in Fig. 3.8. We would expect the curvature for all points on the circle to be a positive number, if we were working in the continuous setting. On a graph, the curvature cannot be expected to be constant but the average value should be positive and the curvature at each point should oscillate around the average. This behavior is captured in Fig. 3.8.

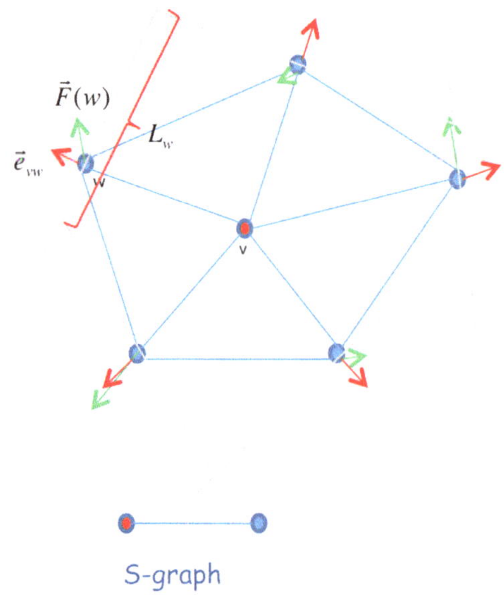

Fig. 3.7 Computing divergence on a graph: Let the *green* vectors denote the values of function **F**, the *red* vectors be the unit vectors corresponding to each edge and let the *gray* lines be perpendicular to the corresponding edge

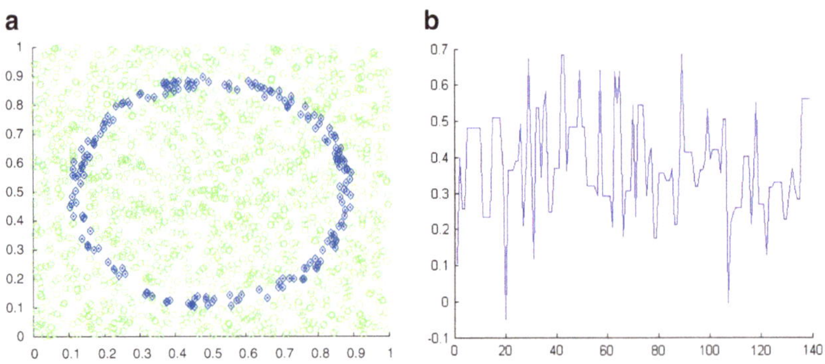

Fig. 3.8 The curvature on a circle calculated with the method proposed in Sect. 3.3.3. We omit the edges for illustration purposes. **(a)** Circle on geometric graph. **(b)** The curvature on the circle

3.3.4 *Convolution on Graphs*

The external image-dependent force is given by the edge-stopping function $g(I)$. The main goal of $g(I)$ is actually to stop the evolving curve when it arrives to the objects boundaries. Among the many choices proposed in the literature, we use the following taken from [12, 27]:

$$g(\|\nabla I_\sigma\|) = \frac{1}{1 + \frac{\|\nabla I_\sigma\|^2}{\lambda^2}} \tag{3.36}$$

where,

$$I_\sigma = I * G_\sigma, \quad G_\sigma(x,y) = \frac{1}{2\pi\sigma^2}\exp\left(-\frac{x^2+y^2}{2\sigma^2}\right). \tag{3.37}$$

In order to compute the smoothed version I_σ of I we need to define the convolution operation on graphs. Let $G = (V, E)$ denote the underlying graph. Let $d_E(v, w)$ denote the Euclidean distance between vertices v and w. For each $v, w \in V$ define the function $G_\sigma^w(v)$ as follows:

$$G_\sigma^w(v) = \frac{1}{\sqrt{2\pi\sigma^2}}\exp\left(-\frac{d_E(v,w)^2}{2\sigma^2}\right). \tag{3.38}$$

The smoothed version I_σ can be computed by mimicking the continuous convolution operation as follows:

$$I_\sigma(v) = \sum_{w\in V} I(v)G_\sigma^w(v). \tag{3.39}$$

3.3.5 *Active Contours on Graphs: The Algorithm*

Here we combine all the previous approximations to the PDE for geodesic active contours and summarize the algorithm for automatic graph segmentation.

Consider a graph $G = (V, E)$ and let a function $I : V \to \mathbb{R}$ assign a real value to each of the graph nodes.

Algorithm-Active Contour on Graphs

1. Compute the smoothed version I_σ of I as described in Sect. 3.3.4.
2. Compute the magnitude of ∇I_σ as described in Sect. 3.3.1 and then compute the function $g(\|\nabla I_\sigma\|)$.
3. Initiate the algorithm with a set of nodes that contains the objects to be found and let ϕ_o denote the signed distance function from the contour of the determined set.
4. For each $r \in \mathbb{N}$ compute $\nabla\phi_{r-1}$, $\|\nabla\phi_{r-1}\|$ and the curvature κ at each node v as described in Sects. 3.3.1 and 3.3.3. Iterate according to the following difference equation:

$$\phi_{r+1} - \phi_r = g(\|\nabla I_\sigma\|)\|\nabla\phi_{r-1}\|(c+\kappa) + g(\|\nabla I_\sigma\|)\cdot\nabla\phi_{r-1}, \quad c \geq 0 \tag{3.40}$$

Figure 3.9 illustrates the algorithm for the case of finding the boundaries of three disjoint objects (connected clusters of graph nodes).

3.4 Multiscale Skeletonization on Graphs

Since Blum's introduction of the skeleton or medial axis transform [5], it has received voluminous attention and has become one of the main tools for shape analysis and representation. The main process to find the skeleton is a distance wavefront propagation. In Euclidean spaces ($\mathbb{R}^d$, $d = 2, 3$) this can be modeled either using a continuous distance formulation [5] or via continuous-space morphology [26,36] or via PDEs that simulate these evolution operations [3, 18, 38, 40]. In the discrete 2D or 3D space $\mathbb{Z}^d$, the above approaches are replaced by discrete distance transforms and discrete morphology; for surveys and references see [34, 36]. The Chamfer distance transform is not always equivalent to the discrete morphology approach, unless the Chamfer ball is used as structuring element. Recent extensions of discrete distance transforms in 3D for skeletonization can be found in [2, 6]. One main disadvantage of the skeleton is its sensitivity on perturbations of the boundary. This can be partially addressed by using multiscale skeletons [29, 33], which provide a flexible framework of keeping only the skeleton parts that correspond to a smoothing of the original shape.

In this chapter we focus on the discrete skeleton transform obtained via multiscale morphology. For a discrete shape $X \subseteq \mathbb{Z}^2$ on regular grid, the morphological algorithm [29] computes the skeleton $S(X)$ of X, w.r.t. a disk-like unit-scale symmetric shape B,

$$S(X) = \bigcup_{n=0}^{N} S_n(X), \tag{3.41}$$

as a union of disjoint skeleton subsets $S_n(X)$, where

$$S_n(X) = (X \oplus nB) \setminus [(X \ominus nB) \circ B], \tag{3.42}$$

indexed by the discrete scale $n = 0, 1, \ldots, N$, with $N = \max\{n : X \ominus nB \neq \emptyset\}$, where nB denotes the n-fold dilation of B with itself. Reconstruction of the original shape or its opening-smoothed versions requires the morphological skeleton transform $(S_0, S_1, \ldots, S_N)$, or equivalently the skeleton $S(X)$ and the quench function (the restriction of the distance transform onto the skeleton set):

$$X \circ kB = \bigcup_{n \geq k} S_n(X) \oplus nB, \quad k = 0, 1, 2, \ldots \tag{3.43}$$

Some generalizations of the above algorithm can be found in [25].

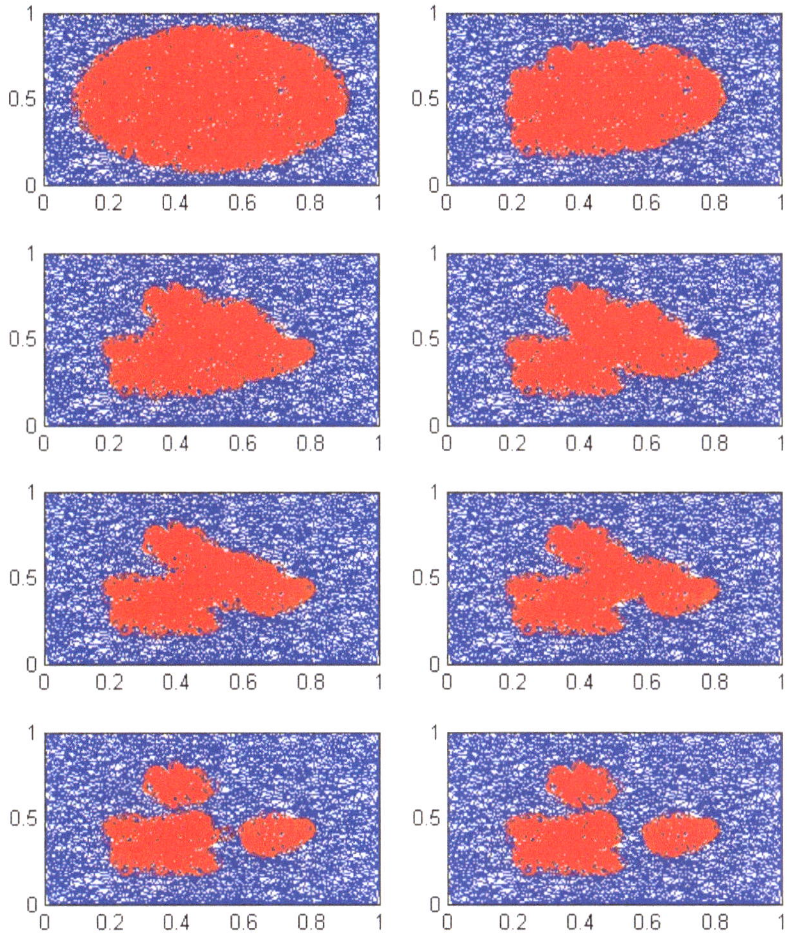

Fig. 3.9 Illustration of the active contour algorithms on graphs for finding three distinct objects on a graph. Note the change in the contour's topology. Time evolves from *top left* to *bottom right*

To extend multiscale skeletonization to shapes defined on arbitrary graphs $G = (V, E)$, we first provide a *lattice formulation* of the above discrete skeletonization algorithm adjusted for graph shapes. Let the shape be represented by a subset X of vertices of the graph and let $\mathscr{A}$ be a structuring graph. Then the skeleton of X can be obtained as follows:

$$S(X|G) \triangleq \bigcup_{n=0}^{N} S_n(X|G) \tag{3.44}$$

where

$$S_n(X|G) \triangleq \varepsilon^n_{\mathscr{A}}(X|G) \setminus \delta_{\mathscr{A}} \varepsilon^{n+1}_{\mathscr{A}}(X|G) \tag{3.45}$$

Taking the union of all or some of the skeleton subsets after dilating them in proportion to their scale yields respectively an exact or partial reconstruction of the graph shape:

$$\alpha_{k\mathscr{A}}(X|G) = \delta^k_{\mathscr{A}} \varepsilon^k_{\mathscr{A}}(X|G) = \bigcup_{n=k}^{N} \delta^n_{\mathscr{A}}[S_n(X|G)] \tag{3.46}$$

Namely, by not using the first k subsets, the above algorithm reconstructs the k-scale opening of the original graph shape.

Next we explore the application of the difference equation based techniques that we developed in order to calculate the skeleton of a shape defined on graphs. Specifically, we propose a calculation of the multiscale graph dilations and erosions involved using the active contour and level set approximations that we introduced in Sect. 3.3. The main idea can be summarized in the following:

Algorithm – Skeleton Calculation

Initialization:
$u(\cdot) = d_{sgn}(\cdot \mid X)$ (X is the shape whose skeleton we are computing)
$S \to \emptyset$
execution:
while $\max_{v \in G} u(v) > 0$ **do**
 $u'(v) = \min_{w \in N_{\mathscr{A}}(v)} u(w)$ (erosion-$\varepsilon^n_{\mathscr{A}}(X)$)
 $r(v) = \min_{w \in N_{\mathscr{A}}(v)} u'(w)$ (erosion- $\varepsilon^{n+1}_{\mathscr{A}}(X)$)
 $o(v) = \max_{w \in N_{\mathscr{A}}(v)} r(w)$ (dilation- $\delta_{\mathscr{A}} \varepsilon^{n+1}_{\mathscr{A}}(X)$)
 $S \leftarrow S \cup \{v : u'(v) \geq 0\} \setminus \{v : o(v) \geq 0\}$ (set difference)
 $u(v) = u'(v)$
end while

The results of our graph skeleton algorithm, simulated on a handshape object represented by a geometric random graph, are presented in Fig. 3.10.

3.5 Conclusions

In this chapter we have proposed an approximation to level set implementation of morphological operators, skeleton transforms and geodesic active contours on arbitrary graphs. Our motivation comes from the importance and the success of

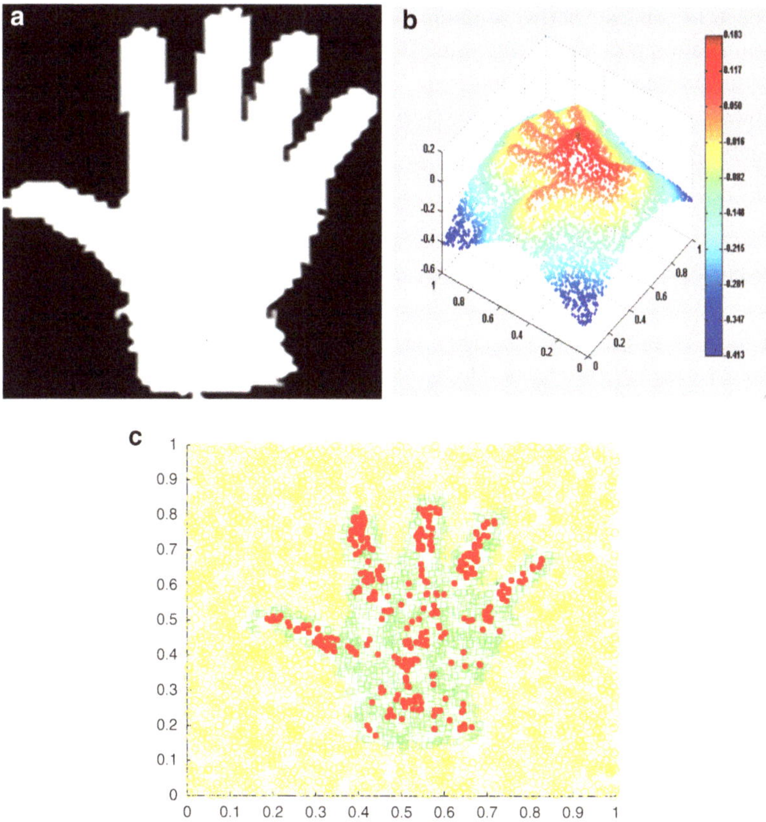

Fig. 3.10 The skeleton of a handshape image calculated using our algorithm. The *green* squares correspond to the embedding of the handshape image on the geometric random graph. The *red* dots correspond to the skeleton transform. The *yellow* circles are irrelevant nodes of the underlying graph. **(a)** Original handshape image. **(b)** The signed distance function on the graph. **(c)** The calculated skeleton

such concepts and techniques in image analysis as well as the existence of a strong theoretical background on graph morphology.

In our simulations we have mainly assumed geometric random graphs and simple structuring graphs. The choice of the simple s-graph is reasonable for any underlying graph structure with no prior information on the graph's characteristics. It is of great interest to other applications to correlate information on the underlying graph with the choice of the structuring graph.

Moreover, we are proposing approximations to concepts from calculus and there is space for better heuristics and modifications to this end. Keeping in mind the original energy minimization approach for geodesic active contours instead of the analytic solution, by properly defining and calculating an energy that corresponds to each contour and applying a step by step minimization procedure we obtain another approach to geodesic active contours. On the other hand, since the latter involves a minimization problem at each time step it is computationally less efficient but it may yield more accurate results.

Finally, regarding the application of the above ideas that we have introduced to multiscale shape skeletonization on arbitrary graphs, one research direction of interest is how to analyze or guarantee the connectedness of the resulting skeleton.

References

1. Alvarez, L., Guichard, F., Lions, P.L., Morel, J.M.: Axioms and fundamental equations of image processing. Arch. Ration. Mech. **123**(2), 199–257 (1993)
2. Arcelli, C., Sanniti di Baja, G., Serino, L.: Distance-driven skeletonization in voxel images. IEEE Trans. Pattern Anal. Mach. Intell. **33**(4), 709–720 (2011)
3. Aslan, C., Erdem, A., Erdem, E., Tari, S.: Disconnected skeleton: shape at its absolute scale. IEEE Trans. Pattern Anal. Mach. Intell. **30**(12), 2188–2203 (2008)
4. Bensoussan, A., Menaldi, J.: Difference equations on weighted graphs. J. Convex Anal. **12**(1), 13–44 (2005)
5. Blum, H.: A transformation for extracting new descriptors of shape. In: Proceedings of a Symposium on Models for the Perception of Speech and Visual Forms, Boston, Nov. 1964. MIT, Cambridge MA (1967)
6. Borgefors, G., Nyström, I., Sanniti di Baja, G.: Discrete skeletons from distance transforms in 2D and 3D. In: Siddiqi, K., Pizer, S. (eds.) Medial Representations: Mathematics, Algorithms and Applications. Springer, Dordrecht (2008)
7. Boykov, Y., Kolmogorov, V.: An experimental comparison of Min-Cut/Max-Flow Algorithms for energy minimization in vision. IEEE Trans. Pattern Anal. Mach. Intell. **26**(9), 1124–1137 (2004)
8. Boykov, Y., Veksler, O., Zabih, R.: Fast approximate energy minimization via graph cuts. IEEE Trans. Pattern Anal. Mach. Intell. **23**(11), 1222–1239 (2001)
9. Breuß, M., Weickert, J.: A shock-capturing algorithm for the differential equations of Dilation and Erosion. J. Math. Imaging Vis. **25**, 187–201 (2006)
10. Breuß, M., Weickert, J.: Highly accurate schemes for PDE-based morphology with general convex structuring elements. Int. J. Comput. Vis. **92**(2), 132–145 (2011)
11. Brockett, R.W., Maragos, P.: Evolution equations for continuous-scale morphological filtering. IEEE Trans. Signal Process. **42**(12), 3377–3386 (1994)
12. Caselles, V., Catte, F., Coll, T., Dibos, F.: A geometric model for active contours in image processing. Numerische Mathematik **66**(1), 1–31 (1993)
13. Caselles, V., Kimmel, R., Sapiro, G.: Geodesic active contours. Int. J. Comput. Vis. **22**(1), 61–79 (1997)
14. Chung, S.-Y., Berenstein, C.A.: Omega-Harmonic functions and inverse conductivity problems on networks. SIAM J. Appl. Math. **65**, 1200–1226 (2005)
15. Cootes, T.F., Edwards, G.J., Taylor, C.J.: Active appearance models. IEEE Trans. Pattern Anal. Mach. Intell. **23**(6), 681–685 (2001)
16. Couprie, C., Grady, L., Najman, L., Talbot, H.: Power watershed: a unifying graph-based optimization framework. IEEE Trans. Pattern Anal. Mach. Intell. **33**(7), 1384–1399 (2011)
17. Cousty, J., Najman, L., Serra, J.: Some morphological operators in graph spaces. In: Proceedings of the International Symposium on Mathematical Morphology. Lecture Notes in Computer Science, 5720. Springer, Berlin/New York (2009)
18. Hassouna, M.S., Farag, A.A.: Variational curve skeletons using gradient vector flow. IEEE Trans. Pattern Anal. Mach. Intell. **31**(12), 2257–2274 (2009)
19. Heijmans, H.J.A.M.: Morphological Image Operators. Academic, Boston (1994)

20. Heijmans, H.J.A.M., Nacken, P., Toet, A., Vincent, L.: Graph morphology. J. Vis. Commun. Image Represent. **3**(1), 24–38 (1992)
21. Heijmans, H.J.A.M., Vincent, L.: Graph morphology in image analysis. In: Dougherty, E.R. (ed.) Mathematical Morphology in Image Processing. Marcel Dekker, New York (1993)
22. Kass, M., Witkin, A., Terzopoulos, D.: Snakes: active contour models. Int. J. Comput. Vis. **1**(4), 321–331 (1988)
23. Kimia, B., Tannenbaum, A., Zucker, S.: On the evolution of curves via a function of curvature. I. The classical case. J. Math. Anal. Appl. **163**, 438–458 (1992)
24. Komodakis, N., Paragios, N., Tziritas, G.: MRF energy minimization and beyond via dual decomposition. IEEE Trans. Pattern Anal. Mach. Intell. **33**, 531–552 (2011)
25. Kresch, R., Malah, D.: Skeleton-based morphological coding of binary images. IEEE Trans. Image Process. **7**(10), 1387–1399 (1998)
26. Lantuejoul, C.: Skeletonization in quantitative metallography. In: Haralick, R.M., Simon, J.C. (eds.) Issues of Digital Image Processing. Sijthoff and Noordhoff, Groningen (1980)
27. Malladi, R., Sethian, J.A., Vemuri, B.C.: A Fast Level Set based Algorithm for Topology-Independent Shape Modeling. J. Math. Imaging Vis. **6**, 269–289 (1996)
28. Maragos, P.: Pattern spectrum and multiscale shape representation. IEEE Trans. Pattern Anal. Mach. Intell. **11**, 701–716 (1989)
29. Maragos, P., Schafer, R.W.: Morphological skeleton representation and coding of binary images. IEEE Trans. Acoust. Speech Signal Process. **34**, 1228–1244 (1986)
30. Matheron, G.: Random Sets and Integral Geometry. Wiley, New York (1975)
31. McInerney, T., Terzopoulos, D.: Topologically adaptable snakes. In: *Proceedings of the Int'l Conference on Computer Vision*, pp. 840–845 (1995)
32. Osher, S., Sethian, J.A.: Fronts propagating with curvature dependent speed: algorithms based on Hamilton-Jacobi formulations. J. Comput. Phys. **79**, 12–49 (1988)
33. Pizer, S.M., Siddiqi, K., Székely, G., Damon, J.N., Zucker, S.W.: Multiscale medial loci and their properties. Int. J. Comput. Vis. **55**(2/3), 155–179 (2003)
34. Rosenfeld, A., Kak, A.C.: Digital Picture Processing: Vol I and II. Academic, New York/London/Paris (1982)
35. Sapiro, G., Kimmel, R., Shaked, D., Kimia, B., Bruckstein, A.: Implementing continuous-scale morphology via curve evolution. Pattern Recognit. **26**(9), 1363–1372 (1993)
36. Serra, J.: Image Analysis and Mathematical Morphology. Academic, London/New York (1982)
37. Serra, J. (ed.): Image Analysis and Mathematical Morphology. Volume 2: Theoretical Advances. Academic, London/New York (1988)
38. Siddiqi, K., Buix, S., Tannenbaum, A., Zucker, W.: Hamilton-Jacobi skeletons. Int. J. Comput. Vis. **48**(3), 215–231 (2002)
39. Ta, V.-T., Elmoataz, A., Lézoray, O.: Nonlocal PDEs-based morphology on weighted graphs for image and data processing. IEEE Trans. Image Process. **20**(6), 1504–1516 (2011)
40. Tari, S.: Hierarchical shape decomposition via level sets. In: Proceedings of the International Symposium on Mathematical Morphology. Lecture Notes in Computer Science, 5720. Springer, Berlin/New York (2009)
41. Vincent, L.: Graphs and mathematical morphology. Signal Process. **16**(4), 365–388 (1989)
42. Yedidia, J.S., Freeman, W.T., Weiss, Y.: Constructing free energy approximations and generalized belief propagation algorithms. IEEE Trans. Inf. Theory **51**(7), 2282–2312 (2005)

Chapter 4
Refined Homotopic Thinning Algorithms and Quality Measures for Skeletonisation Methods

Pascal Peter and Michael Breuß

Abstract Topological skeletons are shape descriptors that have been applied successfully in practical applications. However, many skeletonisation methods lack accessibility, mainly due to the need for manual parameter adjustment and the shortage of tools for comparative analysis.
In this paper we address these problems. We propose two new homotopy-preserving thinning algorithms: Flux-ordered adaptive thinning (FOAT) extends existing flux-based thinning methods by a robust automatic parameter adjustment, maximal disc thinning (MDT) combines maximal disc detection on Euclidean distance maps with homotopic thinning. Moreover, we propose distinct quality measures that allow to analyse the properties of skeletonisation algorithms. Tests of the new algorithms and quality assessment tools are conducted on the widely used shape database CE-Shape-1.

4.1 Introduction

Shape analysis is a central problem for many applications in image processing and computer vision, such as object recognition or segmentation, see e.g. [27] for an overview. While boundary descriptors are a classic instrument for object representation, specific tasks in shape analysis demand alternative shape representations

P. Peter (✉)
Faculty of Mathematics and Computer Science, Mathematical Image Analysis Group, Saarland University, Campus E1.1, 66041 Saarbrücken, Germany
e-mail: peter@mia.uni-saarland.de

M. Breuß
Faculty of Mathematics and Computer Science, Mathematical Image Analysis Group, Saarland University, Campus E1.1, 66041 Saarbrücken, Germany

Applied Mathematics and Computer Vision Group, BTU Cottbus, Platz der Deutschen Einheit 1, HG 2.51, 03046 Cottbus, Germany
e-mail: breuss@mia.uni-saarland.de; breuss@tu-cottbus.de

M. Breuß et al. (eds.), *Innovations for Shape Analysis*, Mathematics and Visualization, DOI 10.1007/978-3-642-34141-0_4, © Springer-Verlag Berlin Heidelberg 2013

that unite geometrical and topological information. The widely-used *medial axis transform* (MAT) is such a shape descriptor. Initially, the MAT was introduced in 1967 by Blum [3] as a mathematical tool for modelling the biological problem of shape vision. It represents a shape by a thin set of lines or arcs that are centred in the shape, eliminating superfluous information of local symmetries. Intuitively, the MAT resembles bone structures, thus motivating the alternative term (topological) *skeleton*. Due to its useful properties, including equivalence to the shape boundary, homotopy to the original shape and invariance under Euclidean transformations, the MAT has been used for numerous practical applications; examples are object recognition [22, 28, 32] or medical imaging [31]. However, the results of existing skeletonisation algorithms largely depend on model parameters that often need to be manually adjusted. Moreover, the literature is lacking comprehensive quality analysis tools that allow to quantify important features of skeletonisation algorithms.

In this work, we deal with these problems. We propose here two new thinning algorithms for MAT computation, we call them FOAT and MDT. Both FOAT and MDT are robust w.r.t. the choice of parameters. In addition, MDT is particularly easy to understand and implement, and thus it is especially appealing for users. Moreover, we present new methods for comparative analysis of MAT algorithms.

Previous work. Algorithms for MAT computation are plentiful and diverse, both in their methods and their theoretical background. In general, three classes can be distinguished: Voronoi-based methods [10,18] that exploit the similarity of the MAT and Voronoi diagrams, approaches related to the distance map [15, 21], thinning algorithms [19,33] and morphological approaches [2, 12, 16, 24].

In our paper we focus on combinations of thinning algorithms and distance map methods with further enhancements by additional pruning steps. In particular, the Hamilton-Jacobi method by Siddiqi et al. [29] combines thinning and distance map methods with physical concepts of Hamiltonian mechanics, and it forms the basis for one of our algorithms. We also propose a new algorithm combining maximal disc detection using exact Euclidean distance maps with homotopic thinning. The method of Pudney [20] is similar to ours, but uses a different maximal disc detection method based on Chamfer distance maps. For mathematical definitions of thinning and pruning operators see [23].

Despite the large number of publications on MAT computation, systematic comparisons of existing methods are rare. Comparative studies of MAT algorithms are usually performed by evaluating them in the context of a particular application, e.g. for the purpose of object recognition. An example for such a comparison strategy is the matching approach by Siddiqi et al. [30], see also [1, 6, 9, 14, 32] for related approaches. One reason for this kind of indirect comparative studies is the lack of formal quality criteria for skeletonisation algorithms. However, from a more general point of view it is of interest to measure the qualities of skeletons independently from a specific application. In order to achieve this goal, we propose ways to assess in how far discrete algorithms mimic structural properties of the corresponding continuous-scale MAT. For this purpose, we also aim to quantify useful properties of the discrete skeletons. An additional problem, especially in the context of practical applications, is that many algorithms like the Hamilton-Jacobi method [29] require manual parameter adjustment for optimal results.

Our contribution. We address the aforementioned issues by proposing refined thinning algorithms that combine several state-of-the-art methods, and by introducing quality measures that allow to assess meaningful properties of MAT algorithms:

(i) *Flux-Ordered Adaptive Thinning* (FOAT) extends the Hamilton-Jacobi method [29] with a *secondary MAT detection* (SMD) step. This allows an automatic, robust adaption of input parameters to individual shapes.
(ii) *Maximal Disc Thinning (MDT)* is an extension of the maximal disc detection method of Remy and Thiel [21]. It combines the latter with the homotopy preserving thinning steps of the Hamilton-Jacobi method [29]. This method is homotopy preserving as well as largely independent of input parameters.
(iii) *Quality Criteria* are proposed that quantify useful properties of discrete skeletons in a new way, namely exactness of reconstruction, skeleton minimality and skeleton complexity. These new quality measures allow to distinguish important properties of MAT schemes.
(iv) *Comparative Skeleton Graph Matching* is employed as a means to test MAT algorithms for invariances. Skeletons are transformed into graphs that can be compared using a graph matching method [26] in a similar way as in object recognition approaches.

Tests of both the newly proposed algorithms and the quality measures were conducted on CE-Shape-1, a widely-used image database consisting of 1, 400 shapes, which was specifically designed for testing shape descriptors [13].

Paper organisation. In Sect. 4.2 we present our new skeletonisation algorithms along with general information about thinning methods. Quality assessment tools are discussed in Sect. 4.3. Section 4.4 contains the results of experiments conducted with the new algorithms and quality criteria. The paper is concluded by Sect. 4.5.

4.2 Algorithms

The algorithms described in the following paragraphs operate on a binary image $u : \Omega \subset \mathbf{R}^2 \to \{0, 1\}$. In the *image domain* Ω, the points of the shape form the object domain $\mathcal{O} = \{x \in \Omega \mid u(x) = 0\}$. The points of the skeleton $\Sigma \subset \mathcal{O}$ are exactly the centres of inscribed circles, that are tangent to the object boundary in at least two points. The *distance map* D of the *object boundary* $\partial\mathcal{O}$ is defined as

$$D : \Omega \to \mathbb{R}_0^+, \quad D(x) = \min_{y \in \partial\mathcal{O}} |y - x| \tag{4.1}$$

Homotopic thinning. Both of the new algorithms are thinning methods that share the same set of thinning rules. In each thinning step, the smallest point in terms of the *thinning order* $>_{\text{thin}}$ is deleted from the object domain $\mathcal{O}$, until only the skeleton

Σ remains. In addition to an appropriate thinning order, a termination criterion is needed to compute an approximation to the skeleton.

Current thinning methods enforce important MAT properties by applying additional thinning rules. In our work we use *homotopic thinning* [20, 29], which guarantees both homotopy to the original object and thinness of the skeleton. Homotopy to the original ensures that major topological features of the object are preserved in its skeleton [11]. In particular, the skeleton has the same number and configuration of connected components and holes. Thinness means for discrete skeletons that the maximal width of the medial axis is one pixel.

In practice, the homotopic thinning rules enforce that points are only removed from the object if they do not disconnect it nor introduce holes, i.e. if they do not change the discrete topology. Additionally, endpoints of thin lines are preserved.

Hamilton-Jacobi skeletons. The Hamilton-Jacobi method of Siddiqi et al. [29], also known as *Flux-Ordered Thinning* (FOT), is the basis for our FOAT algorithm. Exploiting properties from Hamiltonian mechanics, Siddiqi et al. [29] conclude that MAT points are exactly the sinks of the precomputed distance map D's gradient vector field ∇D, where $\nabla := \left(\frac{\partial}{\partial x}, \frac{\partial}{\partial y}\right)^\top$. Sinks of a vector field can be identified as points with large negative values of the *divergence* $\operatorname{div}\nabla D = \nabla \cdot \nabla D$ of the field.

In order to compute $\operatorname{div}\nabla D$ efficiently, the *average outward flux* $\mathscr{F} : \Omega \to \mathbb{R}$ is used, cf. [29] for a detailed discussion. Relating corresponding quantities via the divergence theorem [8], the average outward flux of a point $p \in \Omega$ can be computed as

$$\mathscr{F}(p) := \sum_{i=1}^{8} \frac{< \nabla D(n_i(p)), N(n_i(p)) >}{8} \tag{4.2}$$

Thereby, $n_1(p), \dots, n_8(p)$ are the direct neighbours of p on a discrete lattice, and $N(n_i(p))$ is the outward normal in $n_i(p)$ with respect to the corresponding circles around p. The resulting flux map $\mathscr{F}$ is used to define the thinning order $>_{\text{iflux}}$:

$$\forall p, q \in \Omega : \qquad p >_{\text{iflux}} q \quad \Leftrightarrow \quad \mathscr{F}(p) < \mathscr{F}(q) \tag{4.3}$$

In the thinning process of FOT, weak points with high flux values are removed first. As a termination criterion, thresholding of the flux map is applied to endpoints of thin lines resulting from the thinning process. Points with a flux value below the given threshold τ are marked as skeleton endpoints, other points are removed in accordance to homotopic thinning.

4.2.1 New Algorithm: Flux-Ordered Adaptive Thinning (FOAT)

The FOT method uses thresholding of the flux map to identify skeleton endpoints. Siddiqi et al. [29] propose choosing the threshold τ as a fixed value or as a quantile of the number of object points.

Both of these choices imply the need for manual adjustment in order to obtain good results, because geometric properties of the shapes are not taken into account for the definition of the threshold. While this is obvious w.r.t. the choice of a global fixed value for τ, let us note that quantiles guarantee that a chosen percentage of the image points is removed by applying thresholding to the flux map. However, quantiles do not take into account different relations of skeleton size to object size resulting from varying amounts of local symmetry in shapes. Extreme examples for this relation are a circle and a thin line of equal area. While the circle has exactly one valid skeleton point, its centre, the thin line is equal to the skeleton.

As a remedy, we propose to employ a *secondary MAT detection* (SMD) step. The underlying idea is to use a first approximation to the skeleton for performing an educated guess for the treshold parameter. The outward flux values of the preliminary skeleton $\hat{\Sigma}$ are used to determine the threshold by averaging:

$$\tau := (1-\lambda)\frac{\sum_{x\in\hat{\Sigma}} \mathscr{F}(x)}{|\hat{\Sigma}|} \tag{4.4}$$

where λ is an adjustment parameter that weights the removal of boundary artefacts against accuracy of reconstruction. While FOAT does not remove parameter dependence, the construction in (4.4) improves the robustness of the approach as we will show experimentally. In particular, the parameter λ does not need to be adjusted to the size and geometry of each shape. Instead, it influences the general balance of competing skeleton features such as size and exactness.

Any skeletonisation algorithm can be used for SMD, however, fast algorithms with less restrictive demands for skeleton properties like homotopy or thinness should be preferred in order to lessen the impact on overall performance. We choose the method of Remy and Thiel (RT) [21] for maximal disc detection which identifies MAT points as centres of maximal inscribed discs using lookup tables. In addition we use Meijster's algorithm [17] for distance map computation, which offers better performance and exactness than the D-Euclidean method [4, 5] proposed for this purpose in [29].

4.2.2 New Algorithm: Maximal Disc Thinning (MDT)

MDT is proposed as an extension of the RT maximal disc method [21] and acts as a largely parameter-independent alternative to FOT and FOAT. While MDT uses the same homotopic thinning process as FOT and FOAT, it entirely discards the flux components of the algorithm, thus removing the necessity for several computational steps, cf. Fig. 4.1.

The central idea of the MDT method is processing object points in order of their distance to the boundary, mimicking the wave propagation idea of the classical grass-fire model of Blum [3] and retaining only skeleton endpoints. In this way,

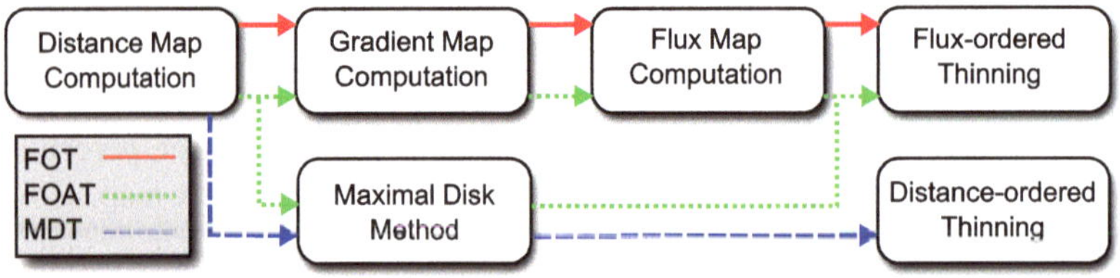

Fig. 4.1 Flow chart of steps used in the FOT, FOAT and MDT algorithms. Note that the maximal disc computation in the FOAT scheme can be performed completely in parallel to the gradient and flux map computation

whole level sets of the boundary's distance map are fully processed, before the algorithm moves on to the level set with the next higher distance value. This behaviour of the thinning process is achieved by defining the thinning order $>_{\text{dist}}$ via the distance map D from (4.1):

$$\forall p, q \in \mathcal{O}: \qquad p >_{\text{dist}} q \quad \Leftrightarrow \quad D(p) < D(q) \tag{4.5}$$

Object points are removed if they are (i) not identified as skeleton points by the RT scheme, and (ii) if they do not violate the homotopy. Endpoints of the preliminary skeleton branches after the RT step are removed, if they are not verified as endpoints during the thinning process.

In order to minimise the homotopy-induced occurence of large skeleton branches that connect scattered isolated skeleton points, we perform a pruning on the skeleton computed via the RT method. Isolated points correspond mainly to small boundary perturbations, or they are caused by discretisation artefacts. A pruning step is recommended in the context of skeletonisation methods, cf. [1,25].

Within our pruning step, we consider the number of points of the preliminary skeleton that are in a fixed neighbourhood of the pruning candidate. This candidate is removed from the preliminary skeleton, if the number of such neighbours does not exceed a predefined number. The latter is in fact the only parameter left in the MDT algorithm.

4.3 Analysis of MAT Algorithms

The performance of MAT algorithms is difficult to assess: Depending on the specific application, the desired priority of skeleton properties may vary significantly. For example, compression methods require skeletons of minimal size that produce reconstructions of maximal exactness. Other properties like homotopy, thinness or Euclidean invariances are not crucial for this application, but might be for others like object recognition or segmentation. Because of these requirements, we propose a novel modular approach to the analysis of MAT algorithms. This allows to express the quality of methods in terms of distinct properties that can be weighted as the application demands.

4.3.1 Quality Criteria

The aim of the quality criteria that we will introduce is to measure the difference between MAT algorithms (i.e. approximations of the MAT) and the exact MAT with respect to important properties of the latter.

Homotopy to the original shape and *thinness* are two binary criteria, that can be easily checked automatically. Within this paper, those criteria are not important for comparisons, since all of the proposed algorithms automatically guarantee both properties by their construction.

A natural quality criterion for discrete skeletons is the *exactness of reconstruction* that is automatically given in the continuous-scale MAT setting. In the discrete world, comparing the reconstructed shape to the original yields the set E of erroneous points, including false positives and false negatives. Normalising the number of errors in order to make the new classification number independent of the size of shapes yields

$$\hat{e}(u, \Sigma) := \frac{\min\{|E|, |\mathcal{O}|\}}{|\mathcal{O}|} \tag{4.6}$$

As we strive for a value of one (rather than zero) for exactness w.r.t. this property, we subtract $\hat{e}(u, \Sigma)$ from one, yielding finally as a quality measure

$$e(u, \Sigma) := 1 - \frac{\min\{|E|, |\mathcal{O}|\}}{|\mathcal{O}|} \quad \text{(exactness of reconstruction)} \tag{4.7}$$

Another central property of the discrete skeleton is its size. The exact MAT should ideally be the smallest, thin, connected subset of the object domain that yields an exact reconstruction of the original. Thus, the size of computed skeletons should be minimal. This defines the quality measure

$$m(u, \Sigma) := 1 - \frac{\min\{|\Sigma|, |\mathcal{O}|\}}{|\mathcal{O}|} \quad \text{(skeleton minimality)} \tag{4.8}$$

Additionally, noise and errors resulting from the discretisation of the boundary must be taken into account. Skeletons are quite sensitive to boundary perturbation, as small deviations from the exact boundary can lead to spurious skeleton branches. Thus, a robust algorithm should give a minimal number of branches, which can be identified via the set P of end and branching points of the skeleton. Consequently, the corresponding quality measure can be defined as

$$c(u, \Sigma) := 1 - \frac{\min\{|P|, |\Sigma|\}}{|\Sigma|} \quad \text{(skeleton complexity)} \tag{4.9}$$

Skeleton complexity is also important for applications that use the branches or their endpoints as graph nodes. The three quality measures above quantify major features of the skeleton and should be considered holistically.

4.3.2 Graph Matching for Invariance Validation

Graph matching is a standard tool in the context of using the MAT in object recognition, cf. [1, 7] for discussions. We investigate this approach in the context of desired invariances of the discrete skeleton, i.e. invariance under Euclidean transformation and noise. Rotational and translational invariance are natural properties of the continuous-scale MAT and should be preserved in the discrete setting. Invariance under noise is an obvious practical criterion while it is not included in a formal definition of the skeleton.

The quality of an algorithm's invariance behaviour can be assessed by a comparison of the transformed shape's skeleton to the skeleton of the original object. It is possible to use the quality criteria from the previous section for this task, although it is advisable to use unnormalised cardinality of the sets instead of the measures e, m and c, since in general, boundary perturbations change the cardinality of the object domain.

We propose here to use graph matching as an effective way of assessing quality w.r.t. invariances. A skeleton graph is composed of nodes that correspond to the end and branching points of the MAT. The graph's edges mirror the connections of the end and branching points in the skeleton, and branch lengths are used as edge weights. Similar set-ups for graph representations have been used before in the literature, and also refinements of it have already been proposed, see e.g. [6]. However, to our knowledge they have not been used before in order to investigate invariances. As for our purposes, we do not consider merge or cut operations on the skeleton graphs as recommended in [9, 14].

The graph matching method *graphdiff* by Shasha et al. [26] computes a one-to-one mapping of nodes of a query graph Q and a database graph D, using a scoring function that takes node types and edge weights into account. Each possible mapping has a total score, and the mapping with the highest score is considered as the best match. See Fig. 4.2 for an illustrative example.

If i, j are nodes in Q that are mapped to i' and j' in D, respectively, the score for the matches is determined by the matching edges of the nodes. The edge (i, j, w) with weight w matches (i', j', w') if i and i', as well as j and j' are of the same node type. Then, the score is computed as $\min(w/w', w'/w)$. These scores can be used directly as quality measures for invariances.

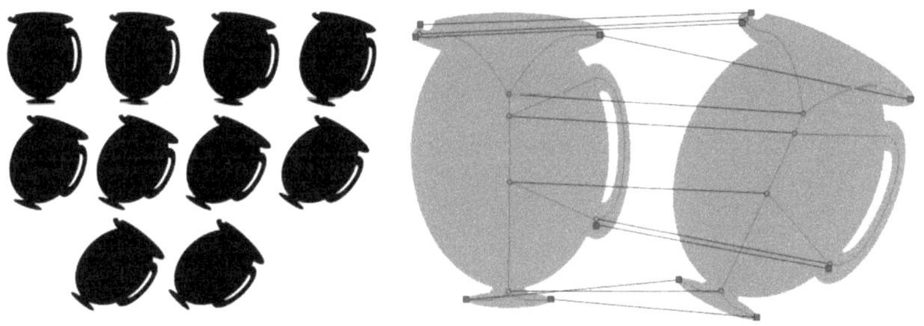

Fig. 4.2 Rotational invariances are tested with rotations of shapes from CE-Shape-1, varying from 5° to 45°. The image on the *right* displays the skeleton produced by FOAT and superimposed nodes of the corresponding skeleton graphs. *Square nodes* represent endpoints, *circles* stand for branch points. *Connection lines* between nodes mark the one-to-one matching produced by *graphdiff* [26]. The matching score of this example is 0.74 (perfect score: 1.0), indicating subtle differences in branch lengths

Table 4.1 Average matching scores for invariance tests using the CE-Shape-1 database. A score of 1.0 denotes a perfect match

	FOT	FOAT	MDT
Noise	0.60	0.64	0.57
Rotation	0.85	0.88	0.76

4.4 Experiments

The experiments conducted for this work serve two distinct purposes: (i) To investigate if the newly proposed quality assessment methods give new meaningful information; (ii) to compare the new algorithms to the FOT method. The general quality of the skeletonisation results is tested with the MPEG-7 core experiment database for shape descriptors CE-Shape-1.

Invariance tests are performed with images from CE-Shape-1 with added boundary noise, translation and rotation. Both the new quality criteria and the graph matching method are used to assess the invariance behaviour of the MAT algorithms along with visual observations.

Comparison of algorithms. The *invariance tests* reveal that the new algorithms either surpass FOT or yield comparable results. Table 4.1 shows that FOAT has consistently higher scores than FOT in respect to rotational and noise invariance.

The MDT method is often competitive to FOT, but in some tests a careful choice of parameters in FOT can give slightly better results. In particular, the behaviour of FOT/FOAT and MDT concerning rotational and noise invariance is different. FOT and FOAT tend to produce large spurious branches connecting the location of boundary perturbations with the exact skeleton, while MDT skeletons vary little in size, but show subtle structural changes, cf. Fig. 4.3.

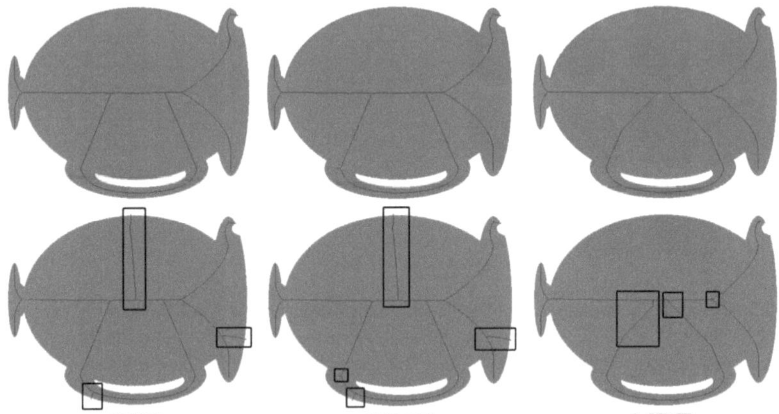

Fig. 4.3 Results of a *noise invariance test*. The shapes in the *upper row* contain 15 distinct boundary perturbations. For all algorithms, the corresponding skeletons show little or none deviations from the unmodified version. In the *lower row*, 30 perturbations were added to the boundary. FOT (*lower left*) and FOAT (*centre*) produce spurious branches of significant size, while for MDT (*lower right*), skeleton size varies only slightly but structural changes occur

Table 4.2 Quality averages for the CE-Shape-1 database. The abbreviations e, m, c denote the normalised quality measures as defined in (4.7)–(4.9). Ideally, scores should be close to 1

	RT	MDT	FOAT	FOT
$e(u, \Sigma)$	0.9622	0.9535	0.9532	0.9539
$m(u, \Sigma)$	0.9484	0.9726	0.9719	0.9722
$c(u, \Sigma)$	0.8803	0.9708	0.9715	0.9712

We do not compare against the RT method here, since it does often not result in connected skeletons. However, the construction of a graph-based representation useful for graph matching is beyond the scope of this paper.

Turning to the *quality averages* for the shape database, these reveal that FOAT produces skeletons that are of slightly simpler structure than the ones of FOT, while skeleton exactness and absolute size is in average slightly higher. The MDT scheme confirms its tendency to produce skeletons of minimal length, staying nevertheless quite accurate but introducing some small structures that result in a slightly higher complexity as FOT and FOAT, see Table 4.2. Let us note in this context that the parameters used for the FOT and FOAT results displayed in Table 4.2 are chosen as to give the best balance in terms of all three quality scores e, m and c, i.e. one could obtain e.g. better results w.r.t. e at the expense of weaker scores in c and m.

As an additional algorithm for our comparison, we employ here the accurate, yet not homotopy-preserving RT method which we also used within the SMD computation. We show these results in order to clarify the benefit of using a homotopy-based approach w.r.t. skeleton size and complexity.

Table 4.3 Comparison of methods for the CE-Shape-1 database. Parameters for FOT and FOAT are chosen as to match the accuracy $e(u, \Sigma) = 0.9535$ of the MDT method

	MDT	FOAT	FOT
$m(u, \Sigma)$	0.9726	0.9719	0.9723
$c(u, \Sigma)$	0.9708	0.9714	0.9716

Table 4.4 Parameter sensitivity for the CE-Shape-1 database. The table gives the maximal deviation in the measures e, m, c as by (4.7)–(4.9) when varying the parameter in an interval of length 0.1. Ideally, scores should be as small as possible

	$e(u, \Sigma)$	$m(u, \Sigma)$	$c(u, \Sigma)$
FOT	0.0095	0.0210	0.0470
FOAT	0.0079	0.0017	0.0010

All of the algorithms work on average with high precision. However, let us also point out that due to the size of 1,400 shapes in our test database, one may infer that the first three to four digits after the comma in the error bear significance: As computational results of the three algorithms are similar for many shapes leading to about the same error measure contributing in (4.7)–(4.9), deviations from this rule may be significant but do not show their influence in the first one or two digits.

Making algorithms even more comparable. With our new quality measures we have the tools at hand to study how the algorithms perform at a common level of accuracy. To this end, we have performed a systematic parameter search for FOT and FOAT that gives a uniform exactness score of $e(u, \Sigma) = 0.9535$, matching the exactness of the MDT method (which is parameter-free). The result is given in Table 4.3. We observe that MDT favors in comparison to the other schemes skeletons that are of more minimal size but of more complex structure. Moreover, a bit surprisingly, we observe that at the accuracy of the MDT method – or, more generally speaking, at a given average accuracy – FOT seems to slightly outperform FOAT.

Robustness vs. parameter choice. By the results documented in Table 4.3, one may come to the conclusion that the FOAT method may be not better than the already existing FOT scheme. However, having at hand the new rigorous measures for the quality of algorithms, we can now perform a systematic study of the sensitivity of the FOT and FOAT methods vs. the parameter choice. To this end, we have sampled a window of length 0.1 around the optimal parameter choice employed before and studied the deviation in the error measures. As by Table 4.4 we observe that the FOAT method is generally much less sensitive to parameter variations than the FOT scheme; w.r.t. skeleton minimality and complexity it turns out that this can be measured to be in the order of a considerable factor of 10 and 40, respectively. Moreover, for FOT deviations appear already in the second digit after the comma which indicates a considerable difference.

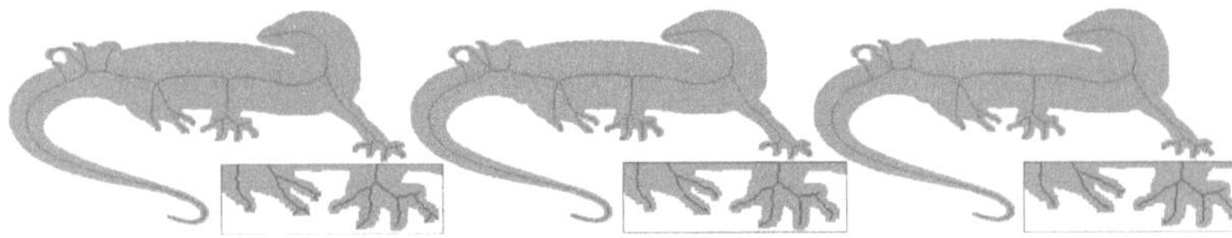

Fig. 4.4 The lizard shape is an example of skeletons that are similar in size and exactness of reconstruction, but vary in complexity. FOAT (*middle*) and MDT (*right*) feature less branches in the small details of the lizard such as in its toes. FOT (*left*) produces additional branches, in particular for jagged edges that result from the discretisation of the shape boundary, as it strives in this example for a very exact representation

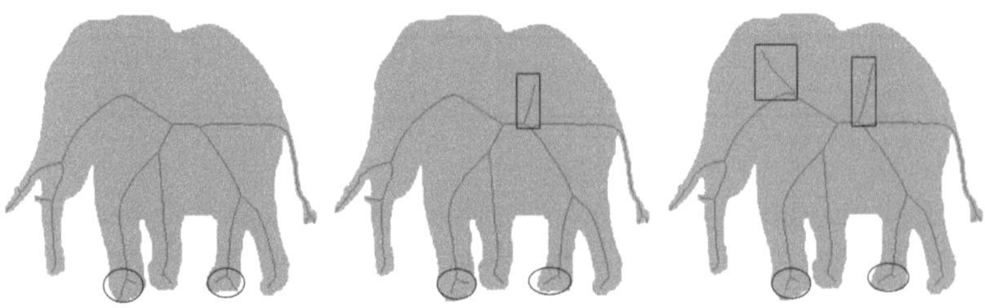

Fig. 4.5 The skeletons of the elephant shape reveal the improved adaption capabilities of FOAT (*middle*) and MDT (*right*) to individual shapes. Similar to Fig. 4.4, FOT (*left*) produces spurious branches at jagged edges (e.g. at the feet of the elephant), but omits branches that would give a refined reconstruction of the elephant's head and back

The score differences illustrate the distinctive behaviours of the algorithms. When attempting to fine tune the threshold parameter τ (see Sect. 4.2) or when making use of the same value of τ for different shapes, FOT easily produces more and larger branches at narrow regions of a shape. There, additional skeleton points increase size and complexity, but yield only slight differences in exactness of reconstruction, see Fig. 4.4. On some shapes however, FOT also omits large branches, thus losing boundary information. Both effects can occur simultaneously, as displayed in Fig. 4.5.

A refined analysis of the new MDT algorithm. We now illustrate that the new quality measures can be useful to infer information about the properties of an algorithm if a related method is at hand.

Let us recall that, by construction, the MDT algorithm relies on the RT method. The latter gives as observable via Table 4.2 in average considerably more accurate skeletons than the MDT scheme. While the RT method is missing thinness, homotopy and connectivity, one may wonder why the loss in the accuracy quality score is relatively large as in the second digit after the comma. The question arises, if there is a general loss of accuracy for the shapes in our database (which would indicate a shortcoming of qualities like thinness, homotopy and connectedness), or if it is due to a few outliers that diminish the average quality score.

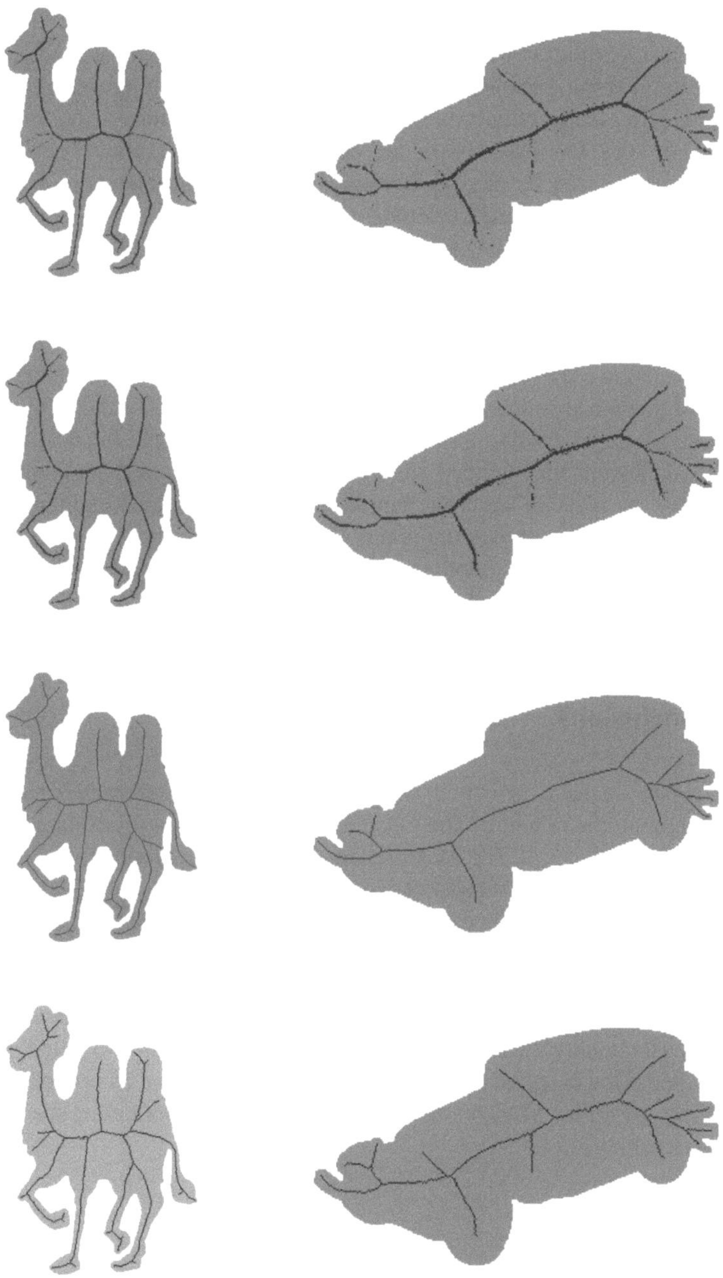

Fig. 4.6 Comparison of skeletonisation results illustrating the MDT algorithm. *Top row.* The algorithm of Remy and Thiel without pruning. *Second row.* The algorithm of Remy and Thiel with an additional pruning step. *Third row.* The new MDT scheme including pruning. Note that not all branches of the pruned RT skeleton are preserved during the MDT thinning phase due to the thinning order. For these selected shapes the problem appears that important branches are pruned. *Bottom row.* Result of modified MDT algorithm which will give a better accuracy for such shapes

We now investigate this issue. For an illustration of the evolution step introduced by the MDT algorithm in comparison to the RT method, we present some typical skeletonisation results, cf. Fig. 4.6. By adding the homotopy-preserving thinning component, we clearly obtain thin, connected skeletons with no spurious points. However, in very few cases as in the car or the camel example given here, a significant branch can be missing because of the general tendency of the MDT algorithm towards producing skeletons as minimal as possible. In such singular cases the accuracy is obviously highly diminished; in turn one can infer by the high average accuracy score in Table 4.2 that these cases are very rare.

As a proof-of-concept that one can easily give a remedy for such rare cases (that can in principle be identified e.g. by computing individually for each shape the reconstruction error e, or by a more local procedure) we propose to fine tune the basic MDT method as follows. In a first step, the modified MDT scheme uses the RT pilot points to construct a preliminary skeleton that is homotopic, but not necessarily thin. This skeleton is then thinned according to homotopy preserving thinning rules that do obey the distance ordering, but do not remove skeleton endpoints, i.e. no branches are deleted from the preliminary skeleton. Finally, branches of very small length are removed with a simple graph based pruning algorithm. See the last row of Fig. 4.6 for typical results of this procedure.

4.5 Conclusions

In this paper we have proposed two new MAT algorithms, FOAT and MDT, based on homotopic thinning. The FOAT scheme relies on a robustification of the thresholding step in FOT. The MDT scheme is with the exception of the pruning step parameter-free. As it is also relatively easy to implement and gives in general results of good quality, it is in our opinion an attractive algorithm for applications.

Additionally, we have proposed quality assessment techniques that give an insight into properties of MAT algorithms. Moreover, they allow to perform a systematic parameter sensitivity analysis, and to analyse construction steps in an algorithm. The quality measures are easy to evaluate and can be applied with any MAT method. However, the evaluation of the invariance tests as proposed here is not straightforward for MAT methods that do not preserve homotopy since it relies on graph matching.

References

1. Bai, X., Latecki, L.J.: Path similarity skeleton graph matching. IEEE Trans. Pattern Anal. Mach. Intell. **30**(7), 1282–1292 (2008)
2. Bloomberg, D.S.: Connectivity-preserving morphological image transformations. In: Proceedings of the SPIE Visual Communications and Image Processing '91: Image Processing Conference, Boston, MA, pp. 302–334 (1991)

3. Blum, H.: A transformation for extracting new descriptors of shape. In: Wathen-Dunn, W. (ed.) Models for the Perception of Speech and Visual Form, pp. 362–380. MIT, Cambridge, MA (1967)
4. Borgefors, G.: Distance transformations in arbitrary dimensions. Comput. Vis. Graph. Image Process. **27**(3), 321–345 (1984)
5. Danielsson, P.E.: Euclidean distance mapping. Comput. Vis. Graph. Image Process. **14**, 227–248 (1980)
6. Demirci, F., Shoukoufandeh, A., Keselman, Y., Bretzner, L., Dickinson, S.: Object recogniton as many-to-many feature matching. Int. J. Comput. Vis. **69**(2), 203–222 (2006)
7. Di Ruberto, C.: Recognition of shapes by attributed skeletal graphs. Pattern Recognit. **37**(1), 21–31 (2004)
8. Dimitrov, P., Damon, J., Siddiqi, K.: Flux invariants for shape. In: Proceedings of the IEEE Conference on Computer Vision and Pattern Recognition, Madison, WI, pp. 835–841 (2003)
9. Geiger, D., Liu, T., Kohn, R.V.: Representation and self-similarity of shapes. IEEE Trans. Pattern Anal. Mach. Intell. **25**(1), 86–99 (2003)
10. Kimmel, R., Shaked, D., Kiryati, N., Bruckstein, A.M.: Skeletonization via distance maps and level sets. Comput. Vis. Image Underst. **62**, 382–391 (1995)
11. Kong, T.Y., Rosenfeld, A.: Digital topology: introduction and survey. Comput. Vis. Graph. Image Process. **48**, 357–393 (1989)
12. Lantuejoul, C.: Skeletonization in quantitative metallography. In: Haralick, R.M., Simon, J.C. (eds.) Issues in Digital Image Processing, pp. 107–135. Sijthoff and Noordoff, Amsterdam (1980)
13. Latecki, L., Lakämper, R., Eckhardt, U.: Shape descriptors for non-rigid shapes with a single closed contour. In: Proceedings of the IEEE Conference on Computer Vision and Pattern Recognition, Hilton Head Island, South Carolina, pp. 424–429 (1967)
14. Liu, T., Geiger, D.: Approximate tree matching and shape similarity. In: Proceedings of the 7th IEEE Conference on Computer Vision, Kerkyra, Greece, pp. 456–462 (1999)
15. Malandain, G., Fernández-Vidal, S.: Euclidean skeletons. Image Vis. Comput. **16**(5), 317–327 (1998)
16. Maragos, P., Schafer, R.W.: Morphological skeleton representation and coding of binary images. IEEE Trans. Acoust. Speech Signal Process. **34**(5), 1228–1244 (1986)
17. Meijster, A., Roerdink, J.B.T.M., Hesselink, W.H.: A general algorithm for computing distance transforms in linear time. In: Goutsias, J., Vincent, L., Bloomberg, D.S. (eds.) Mathematical Morphology and Its Applications to Image and Signal Processing. Volume 18 of Computational Imaging and Vision, pp. 362–380. Springer, Dordrecht (2002)
18. Ogniewicz, R.L., Kübler, O.: Hierarchic Voronoi skeletons. Pattern Recognit. **28**, 343–359 (1995)
19. Palagyi, K., Nemeth, G.: Fully parallel 3D thinning algorithms based on sufficient conditions for topology preservation. In: Proceedings of the International Conference on Discrete Geometry for Computer Imagery. Volume 5810 of Lecture Notes in Computer Science, pp. 481–492. Springer, Berlin (2009)
20. Pudney, C.: Distance-ordered homotopic thinning: a skeletonization algorithm for 3D digital images. Comput. Vis. Image Underst. **72**, 404–413 (1998)
21. Rémy, E., Thiel, E.: Exact medial axis with euclidean distance. Image Vis. Comput. **23**, 167–175 (2005)
22. Sebastian, T.B., Klein, P.N., Kimia, B.B.: Recognition of shapes by editing their shock graphs. IEEE Trans. Pattern Anal. Mach. Intell. **26**, 550–571 (2004)
23. Serra, J. (ed.): Image Analysis and Mathematical Morphology, vol. 1. Academic, London (1982)
24. Serra, J. (ed.): Image Analysis and Mathematical Morphology, vol. 2. Academic, London (1988)
25. Shaked, D., Bruckstein, A.: Pruning medial axes. Comput. Vis. Image Underst. **69**(2), 156–169 (1998)

26. Shasha, D., Wang, J.: Graphdiff: approximate graph matcher and clusterer. http://cs.nyu.edu/shasha/papers/agm.html (2000). Accessed 29 December 2011
27. Siddiqi, K., Pizer, S.M. (eds.): Medial Representations: Mathematics, Algorithms and Applications. Volume 37 of Computational Imaging. Springer, Dordrecht (2008)
28. Siddiqi, K., Shokoufandeh, A., Dickinson, S.J., Zucker, S.W.: Shock graphs and shape matching. Int. J. Comput. Vis. **35**, 13–32 (1999)
29. Siddiqi, K., Bouix, S., Tannenbaum, A., Zucker, S.W.: Hamilton-Jacobi skeletons. Int. J. Comput. Vis. **48**, 215–231 (2002)
30. Siddiqi, K., Zhang, J., Macrini, D., Dickinson, S., Shokoufandeh, A.: 3D model retrieval using medial surfaces. In: Siddiqi, K., Pizer, S.M. (eds.) Medial Representations: Mathematics, Algorithms and Applications. Volume 37 of Computational Imaging, pp. 309–325. Springer, Dordrecht (2008)
31. Sorantin, E., Halmai, C., Erdohelyi, B., Palagyi, K., Nyul, L., Olle, K., Geiger, B., Lindbichler, F., Friedrich, G., Kiesler, K.: Spiral-CT-based assessment of tracheal stenoses using 3-D-skeletonization. Med. Imaging **21**, 263–273 (2002)
32. Zhu, S.C., Yuille, A.L.: Forms: a flexible object recognition and modeling system. Int. J. Comput. Vis. **20**(3), 187–212 (1996)
33. Zhu, Y., Seneviratne, L.D., Earles, S.W.E.: A fast boundary based thinning algorithm. In: Proceedings of IAPR Workshop on Machine Vision Applications, Kawasaki, Japan, pp. 548–551 (1994)

Chapter 5
Nested Sphere Statistics of Skeletal Models

Stephen M. Pizer, Sungkyu Jung, Dibyendusekhar Goswami, Jared Vicory, Xiaojie Zhao, Ritwik Chaudhuri, James N. Damon, Stephan Huckemann, and J.S. Marron

Abstract We seek a form of object model that exactly and completely captures the interior of most non-branching anatomic objects and simultaneously is well suited for probabilistic analysis on populations of such objects. We show that certain nearly medial, skeletal models satisfy these requirements. These models are first mathematically defined in continuous three-space, and then discrete representations formed by a tuple of spoke vectors are derived. We describe means of fitting these skeletal models into manual or automatic segmentations of objects in a way stable enough to support statistical analysis, and we sketch means of modifying these fits to provide good correspondences of spoke vectors across a training population of objects. Understanding will be developed that these discrete skeletal models live in an abstract space made of a Cartesian product of a Euclidean space and a collection of spherical spaces. Based on this understanding and the way objects change under various rigid and nonrigid transformations, a method analogous to principal component analysis called *composite principal nested spheres* will be seen to apply to learning a more efficient collection of modes of object variation

S.M. Pizer (✉) · D. Goswami · J. Vicory · X. Zhao · R. Chaudhuri · J.N. Damon
University of North Carolina, Chapel Hill, NC, USA
e-mail: pizer@cs.unc.edu

S. Jung
University of Pittsburgh, Pittsburgh, PA, USA
e-mail: sungkyu@pitt.edu

S. Huckemann
Institute for Mathematical Stochastics, University of Göttingen, Goldschmidtstrasse 7, 37077 Göttingen, Germany
e-mail: huckeman@math.uni-goettingen.de

J.S. Marron
Department of Statistics and Operations Research, University of North Carolina, Chapel Hill, NC 27599, USA
e-mail: marron@email.unc.edu

M. Breuß et al. (eds.), *Innovations for Shape Analysis*, Mathematics and Visualization, DOI 10.1007/978-3-642-34141-0_5,

about a new and more representative mean object than those provided by other representations and other statistical analysis methods. The methods are illustrated by application to hippocampi.

5.1 Object Models Suitable for Statistics

The objective of statistics on a population of 3D (or 2D) objects is to produce such entities as the mean object and a shape space spanned by principal modes of variation of the object. These statistical descriptors are derived from training cases, each typically given by a pile of object boundary contours. Statistics on objects has been applied to quite a variety of types of geometric model derived from the boundary cases:

1. Boundary point distribution models [4, 8, 17]. This popular form of model has been analyzed by principal component analysis (PCA), although Kendall [16] has shown this is not strictly appropriate because such models live in a space formed by a Cartesian product of a Euclidean space and a quotient of a high-dimensional sphere modulo the rotation group. However, even the correct analysis ignores the shape of the object interior and, partially as a result, has difficulty limiting its shape space to models whose boundaries do not cross themselves.
2. Deformation-of-atlas models, wherein the displacement of each voxel in the atlas is provided [1, 21] . These models have an enormous dimension, with the result that the statistical analysis is expensive and unstable with respect to the sampling into training cases and to noise in those training cases.
3. Implicit models, such as level surfaces of pseudo-signed-distance functions [18, 31] that do not live in Euclidean feature spaces but are often analyzed (by PCA) as if they did.
4. Skeletal models. While skeletal modeling is common, as summarized in the book by Siddiqi and Pizer [27], work on statistics of skeletal models has been the major topic of only one group, namely ours. That work emphasized that skeletal models, like those in the first two categories, live on abstract manifolds that are curved. Work on skeletal and other object models that live on abstract manifolds have been the subject of study of Fletcher [10]. Our previous work on skeletal model statistics, on which Fletcher was an early contributor, is laid out in Chaps. 8 and 9 of Siddiqi and Pizer [27]. The strength of skeletal models is that they richly model the object interior and boundary and that they yield an object-relative coordinate system for the object interiors. The difficulty with general skeletal models, and especially the medial models that are their most common form, is that even for objects that do not intuitively feel like a tree of figures, the model takes the form of the tree. However, the branching structure of the tree does not stay the same over the object population. This makes statistical analysis of these models very difficult. In our previous work we have solved this problem by using skeletal models whose boundaries are only approximately that of the training cases but which have no branches.

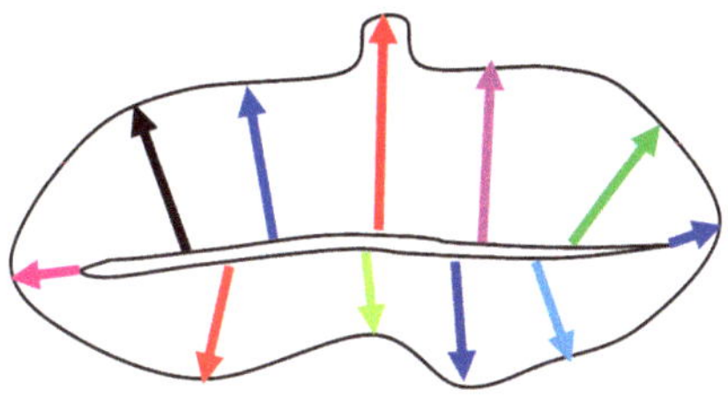

Fig. 5.1 An interior-filling skeletal model in 2D. The two sides of the skeletal locus are shown slightly separated, but in fact they are coincident. In the continuous form there is a spoke vector at every point on the folded curve forming the skeleton

From the previous discussion one can conclude that no form of model has been introduced that accurately and richly represents the training data and truly is fully appropriate for statistical modeling. In this paper, we define a form of skeletal model that precisely captures most non-branching anatomic objects and a form of statistical analysis more well suited to these models than ones that have been previously applied. The result of this statistical analysis, we show by illustration, produces more appropriate object means, lower-dimensional shape spaces, and more descriptive modes of variation than PCA-based statistical analysis of other skeletal or nonskeletal models. While our object modeling approach allows us to capture branching objects as well, the statistical techniques available to us or others is not yet capbable of handling structures with variable branching structure. Therefore, here we do not attempt here to model all object populations.

5.2 Skeletal Models of Non-branching Objects

Skeletal models capture the interior of objects (the egg), and as such they more stably and richly capture object shape than models that capture only the boundary of objects (the eggshell). We define a continuous interior-filling skeletal model of an object as a locus of spoke vectors $(\underline{p}, \mathbf{S})$ with tail at $\underline{p}$ and tip at $\underline{p} + \mathbf{S}$ such that

1. None of the vectors cross each other;
2. The union of the vectors forms the interior of the object and the union of the spoke tips forms the boundary of the object. The consequence of conditions 1 and 2 is that each point in the object is reached by precisely one spoke.
3. The union of the tails, which we call the "skeletal locus", forms a fully folded, multi-sided locus (double-sided for slab-shaped objects). That is, except at the fold, in slabs each position p appears twice in the set of spokes (see Fig. 5.1). The reader should not confuse this with the more traditional characterization of the skeletal locus as an unfolded point set that happens to have two spokes at every point but the end points.

We call such a representation an interior-filling s-rep, or for this paper just an *s-rep*. We will refer to the whole s-rep by the notation $\underline{\mathbf{m}}$. In this paper we restrict

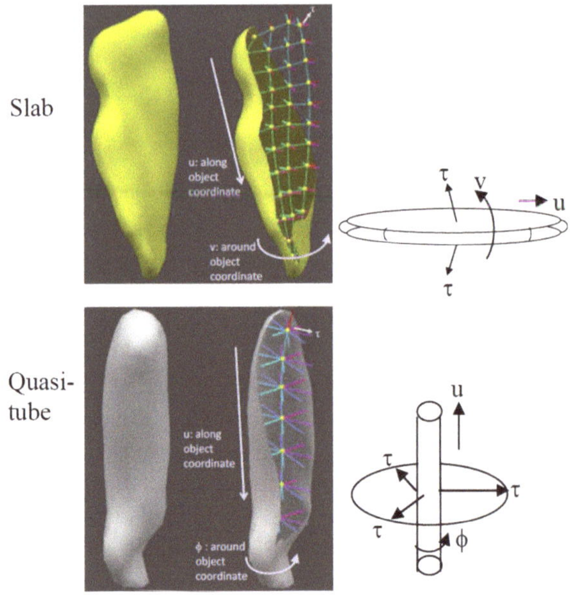

Fig. 5.2 Slabular and quasi-tubular s-reps and their parameter spaces. In the slab, $\underline{u} = (u, v)$ parametrize by mapping the long (u) and middle (v) dimensions of the object to the respective dimensions of an ellipsoid collapsed along its short axis, which in turn is mapped to the sphere by mapping its crest to the equator, its north side to the sphere's northern hemisphere, and its south side to the sphere's southern hemisphere. In the quasi-tube, $\underline{u} = (u, \phi)$ parametrize by mapping the along-tube dimension (u) to the axis of a cylinder with infinitesimal radius and mapping the around-tube dimension (ϕ) to the normal directions of the infinitesimal cylinder. In both forms of s-reps, τ parametrizes fractional distance along the spokes, where $\tau = 0$ at the skeletal locus and $\tau = 1$ at the object boundary. In both forms (u, τ) parametrize the interior of the object

ourselves to s-reps of 3D objects that have the topology of a sphere (have no interior cavities or through holes).

We define a non-branching 3D object as a region of space for which there exists an s-rep with a single cyclic fold curve, i.e., for which the skeleton does not branch. Examples of such objects are the kidney and the esophagus (into which we swallow).

Medial models, as defined by Blum [2] are a form of skeletal model, but they cannot exactly capture the entire interior of most non-branching anatomic objects without an arbitrary number of skeletal branches and variation of the number of branches and their branching structure. These render the model unsuitable for probabilistic analysis. Instead we seek a skeletal model [6] that allows overcoming this branching instability. In particular, our target is that all instances of the object have the simplest, stable branching pattern, i.e., without branches. In these s-reps the skeletal locus is *smooth* except at the fold points.

As illustrated in Fig. 5.2, there are two types of s-reps. In the first type, which we call *slabular*, the smooth points form two copies of a smooth sheet, with the two sheets pasted to each other, and the fold points form themselves into a single smooth cyclic curve, called the *end curve*. In the second type, which we call *quasi-tubular*, the smooth points take the form of a bent cylinder of infinitesimal radius, i.e., a cylinder whose surface is made up of copies of an axis curve with all the copy curves pasted to each other, and the fold points consist of two discrete points, called the *end points*. In both types a continuum of spokes, formed by vectors with tails on the skeletal locus and tips on the object boundary, cover the interior of the object.

Such s-reps capture all non-branching anatomic objects in the human body. This includes not only slabular objects such as the kidney, most muscles, the heart, and the bladder, and quasi-tubular objects, such as the esophagus and the rectum, but also folded slabular objects such as the cerebral cortex and folded quasi-tubular objects such as the intestines. These models capture objects with small pimples and dimples; the main objects not captured are objects with hook-shaped or curvy protrusions and indentations and those that branch multiple times.

In slabular s-reps the skeletal locus forms a two-sided sheet $\underline{p}(\underline{u})$ with a cyclic fold curve. The parametrization $\underline{u}$ has the topology of a sphere. In this work, for slabular s-reps we will parametrize by the sphere and let the equator of the parametrizing sphere map onto the fold of skeletal locus. Thus we will refer to the north and south sides of the sheet. For each $\underline{u}$ a spoke $\mathbf{S}(\underline{u})$ exists with its tail on $\underline{p}(\underline{u})$ and its tip on an object boundary point $\underline{b}(\underline{u})$. $\underline{u}$ parametrizes the skeletal locus along the object and across and around the object; it also parametrizes the spokes $\mathbf{S}(\underline{u})$, which pass within the object, as well as the object boundary $\underline{b}(\underline{u})$. The lengths of the spokes, $|\mathbf{S}(\underline{u})|$, which we will call $r(\underline{u})$, and the directions of the spokes $\mathbf{U}(\underline{u}) = \mathbf{S}(\underline{u})/r(\underline{u})$ are also defined. Thus $\underline{u}$ parametrizes the whole s-rep $\underline{\mathbf{m}}(\underline{u}) = (\underline{p}(\underline{u}), \mathbf{U}(\underline{u}), r(\underline{u}))$.

In quasi-tubular s-reps the skeletal locus forms a collapsed (infinitesimal radius) bent cylinder with hemispherical caps at both ends, such that the axis of the cylinder is a space curve and the spokes emanating from each axis-orthogonal cut of the cylinder end in a common plane. Here the parametrization $\underline{u}$ is typically on the unit-radius, possibly hemispherically capped right-circular cylinder (Fig. 5.2); it parametrizes the axis of (forming) the skeletal locus by the along-cylinder scalar variable u, and it parametrizes the angle around the circular tubular cross sections by another cyclic scalar variable ϕ. As with the slabular s-rep, $\underline{u}$ parametrizes the spokes $\mathbf{S}(\underline{u})$, their directions $\mathbf{U}(\underline{u})$, their lengths $r(\underline{u})$, and the object boundary $\underline{b}(\underline{u})$. Quasi-tubular s-reps are used to model roughly tubular objects such as the esophagus, but they can also be used to model objects like muscles whose cross-sections are star-shaped but not roughly tubular.

For both slabular and quasi-tubular s-reps the position along a spoke from its tip to its tail is parametrized by the proportion-of-spoke-length variable τ, so the union of the interior and the boundary of the object is parametrized by $(\underline{u}, \tau)$. In the remainder of this paper we restrict ourselves to slabular s-reps.

The mathematical theory of skeletal models is presented in Damon's Chap. 3 of the book *Medial Representations* by Siddiqi and Pizer [27] and more completely in Damon's papers: [6, 7]. There the geometry of our class of s-reps is shown to be neatly divisible into (a) the differential geometry of the two sides and fold of the skeletal locus $\underline{p}(\underline{u})$ and (b) a variant of differential geometry, which we call radial geometry, of the (tail-less) spokes $\mathbf{S}(\underline{u})$ with respect to motions on the tangent plane of the skeletal locus. The radial geometry of the spoke directions $\mathbf{U}(\underline{u})$ is of special interest; it can be divided into a *radial shape operator* $S_{rad}(\underline{u})$ applicable at non-fold medial points and an edge shape operator $S_E(\underline{u})$ (for $\underline{u}$ on the equator of the parametrizing sphere or the ends of the parametrizing cylinder) applicable at the fold curve. Each of these operators are represented by a 2×2 matrix. These shape operators describe $\mathbf{U}(\underline{u})$ motion in a way analogous to the way the ordinary shape operator from differential geometry describes the motion of boundary normals. There are special matters dealing with the appropriate coordinate system in which to describe $\mathbf{U}(\underline{u})$ motion and the need for projection of that motion onto the tangent plane to $\underline{p}(\underline{u})$, but we will not go into those in this paper.

Damon shows that an interior-filling locus $(\underline{p}(\underline{u}), \mathbf{U}(\underline{u}), r(\underline{u}))$ has no locally crossing spokes (is *geometrically legal*) if and only if for all $\underline{u}$ the positive eigenvalues of $S_{rad}(\underline{u}) < 1/r(\underline{u})$ and any positive generalized eigenvalue of $(S_E(\underline{u}), J) < 1/r(\underline{u})$, where J is the 2×2 matrix all of whose elements are 0 but for the upper left element, which is 1.

Discrete s-reps sample $\underline{\mathbf{m}}(\underline{u})$ into a network of spoke vectors. In our work with slabular s-reps the sampling is into a north $m \times n$ grid of spokes, a south $m \times n$ grid of spokes, for some m, n, and a cyclic chain of $2m + 2n - 4$ edge (crest) spokes (Fig. 5.2). These are organized into an $m \times n$ grid of "skeletal atoms", each consisting of two or three spokes with a common tail position. For interior grid points there are two spokes, one on the north side and one on the south side. For exterior grid points there are three spokes, one on the north side, one on the south side, and a crest spoke bisecting the other two.

Interpolation of a continuous s-rep from a discrete s-rep proceeds according to the method of Han [11], in which $\underline{p}(\underline{u})$, $r(\underline{u})$, and $\mathbf{U}(\underline{u})$ are interpolated in an intertwined way. The spokes are interpolated using the fact that $S_{rad}(\underline{u})$ or $S_E(\underline{u})$ times a small step vector on the tangent plane to the skeletal locus at $\underline{p}(\underline{u})$ allows the calculation of the spoke direction swing corresponding to that step. The $rS_{rad}(\underline{u})$ or $rS_E(\underline{u})$ matrix is first interpolated and then the implied spoke swing and length scaling for various small steps is integrated. The skeletal axis positions $\underline{p}(\underline{u})$ are interpolated using a Hermite-like interpolation using both the discrete positions and skeletal sheet normals there, where the normal is computed as the difference of the two spoke direction vectors with a common skeletal sheet position. This relation for the normal holds only for Blum medial sheets, but given the "as medial as possible" criterion for skeletal sheets discussed below, it is sensible to use it for the skeletal sheet interpolation.

5.3 Obtaining s-Reps Suitable for Probabilistic Analysis

There are three properties necessary for making a population of s-reps suitable for probabilistic analysis. The first is that the branching topology of all members of the population should be the same. There is nice, albeit not yet usable, work on statistics of trees with variable branching topology [28,30], but this is beyond the scope of this paper. In this paper we restrict the branching situation further to having no skeletal branches at all. This requires branching objects such as blood vessels to be treated by compositing single branch components, but this, too, is beyond the scope of this paper. This is the reason for the restriction mentioned in Sect. 5.1 to skeletal loci that are smooth except at the fold curve.

The second property needed to do probabilistic analysis is that small deformations in the objects yield small deformations in the skeletal model. This can be made to occur when the process of fitting s-reps into the data describing the object has certain behaviors. These will be discussed in Sect. 5.3.1.

The third property needed to do probabilistic analysis is that each spoke in one s-rep in the training population be at a position, orientation, and length that geometrically corresponds to that of the same spoke in the other training s-reps. Accomplishing this is discussed in Sect. 5.3.2.

5.3.1 Fitting Unbranched s-Reps to Object Description Data

Typically object data is provided in the form of a tiled object boundary, a pile of parallel planes each containing a boundary contour, or a binary discrete image. To fit s-reps to such data, we first transform the data into a signed distance image, where the distance is negative inside the object and positive outside and where the zero level curve of distance represents the object boundary. We have methods, using constrained flow controlled by the Laplacian of curvature, that yield a distance image representing a smooth boundary staying within a half-voxel of the object data [22].

Our objective is to study shape variability by statistical means, so the pair of distance images for many population samples of our object will be provided. For any object sample there are very many s-reps that can be fitted to this data. In particular, the skeletal locus could in principle be anywhere across the short dimension of the object. In the extreme it could be along one side of the object, such that the object was covered by lines going from one side of the object (e.g., the south side) to the other side of the object. Since our goal is to do statistical analysis of the fitted s-reps, we need to fit s-reps that have as little variability over the training cases as possible.

The closer the skeletal locus is to bisecting the top and bottom sides, the wider the range of pimples and dimples on the boundary that could be accommodated. More importantly, we can more stably constrain the geometry of the skeletal locus if we make it bisect the two sides. However, as indicated earlier, the Blum medial locus will not do, because it generates branches for pimples and dimples. So in brief, we want a skeletal locus smooth but at the folds that is space filling and, within those constraints, *as medial as possible*. The following section gives properties of truly medial fits that can be used in measuring how medial an s-rep is. The properties are described in the form of penalties that contribute to a weighted sum that is minimized in the fitting process.

5.3.1.1 As Medial as Possible

The following medial properties should be approximate targets for an as-medial-as-possible s-rep. Each property yields a penalty term in the aforementioned weighted sum. Some depend on the input signed distance image $D(\underline{x})$, and some involve the relation between the north (south) side spoke $\mathbf{S}(\underline{u})$ and $\mathbf{S}(\underline{v})$, its counterpart on the other side of the (double-sided) skeletal sheet with the same $\underline{p}(\underline{u})$ value.

1. Equal $r(\underline{u})$ values for the northside and southside spokes at the same $\underline{p}(\underline{u})$: in the fitting of a discrete s-rep we penalize according to $|r(\underline{u}) - r(\underline{v})|$).
2. Spokes are boundary normals: in the fitting we penalize according to the deviation from boundary normality, i.e., $cos^{-1}(\widehat{\nabla D(\underline{x})} \cdot \mathbf{U}(\underline{u}))$ at the spoke end, where, given a vector $\mathbf{W}$, $\widehat{\mathbf{W}}$ is the unit vector in the direction of $\mathbf{W}$.
3. The (interpolated) skeletal locus $\underline{p}(\underline{u})$ bisects $\mathbf{S}(\underline{u})$ and $\mathbf{S}(\underline{v})$. Equivalently, the normal to $\underline{p}(\underline{u})$ is in the same direction as $\mathbf{U}(\underline{u}) - \mathbf{U}(\underline{v})$: in the fitting, for each spoke we penalize according to the angle between the normal to $p(\underline{u})$ and $\mathbf{U}(\underline{u}) - \mathbf{U}(\underline{v})$. This penalty has not yet been implemented in our code to fit s-reps to objects.
4. End curves' spokes are at crests of the zero level curve of the distance image that forms the object boundary; this involves three properties, all involving the principal directions w_1 and w_2 and the associated principal curvature κ_1 of the level surface at the spoke end, where κ_1 is the lesser (more negative, i.e., more convex) principal curvature:

   (a) The spoke end is at an extremum of κ_1 along the integral curve of w_1, so the directional derivative of κ_1 in the w_1 direction should be zero. In the fitting we have attempted to penalize according to the magnitude of that directional derivative. However, the high order derivative of the distance function that this involves proved to be unstable, so for now this term is not included in the penalty list.
   (b) The plane of the spoke and its infinitesimally near neighbors along the principal curve of w_1 should be orthogonal to w_2. We penalize according to the angle between w_2 and $\mathbf{U}(\underline{u}) \times \mathbf{U}(\underline{v})$, the normal to the plane of the spokes at that end curve point.
   (c) The boundary implied by the s-rep at the spoke end and the level surface of the distance at the spoke end should agree in the first principal curvature. We have not yet implemented a penalty for deviating from this condition, but in a future implementation we plan also to penalize according to the magnitude of the difference between the two relevant radii of curvature.

This method of forcing spokes into the distance function's crest is good when there is a strong crest for the end curve to fit into, but in some objects this crest is weak or, even worse, the object has more than one crest and we wish the s-rep to fit between them. This situation is illustrated in Fig. 5.3, where the skeletal sheet bisects the twin crests on the right side of the object, instead of favoring one crest over the other.

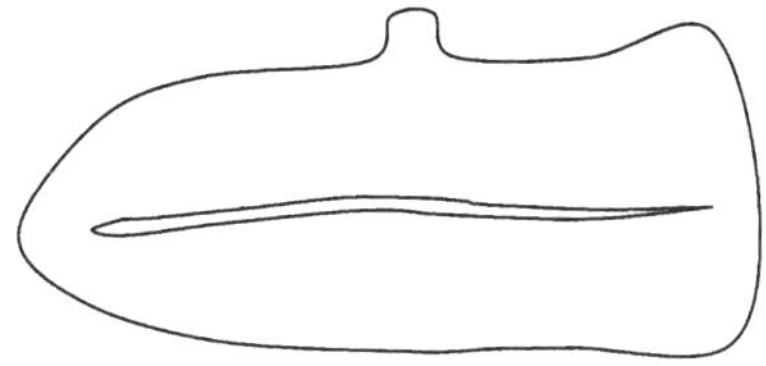

Fig. 5.3 A 2D object with multiple vertices (the 2D correspondent to a crest in 3D), together with the desired skeletal sheet

5.3.1.2 Fitting s-Reps to Distance Images

Fitting s-reps to such signed distance images is done in five stages.

In the first stage, to align the training cases, a reference s-rep is translated, rotated, and possibly uniformly scaled into the space of the distance image via matching of moments of boundary positions or via minimizing sum of squared distances between designated s-rep spoke ends and object landmarks provided by the user.

In the second stage we restrict the corresponding northside and southside spokes to having the same length and move each atom into place while maintaining a regular array that is geometrically legal. In the third (spoke) stage we adjust the spoke lengths and angles to match the object boundary quite closely. After the PCA-like statistical analysis described in Sect. 5.5 on the results of the third stage, the fourth stage obtains better correspondence among the fits to the various training cases. It accomplishes that by restricting the fits to the shape space resulting from the statistical analysis. In the fifth stage we tighten the fit to the distance function by allowing small deviations from the shape space; in particular, we again adjust the spoke lengths and angles to match the object boundary quite closely. The following describes these stages in more detail.

In the second stage, iteratively over the skeletal atoms (each consisting of a spoke duple or triple with a common hub) we optimize an objective function [19] that is a sum of two major penalty terms summed over the two or three spokes. Each major term is a sum of subsidiary terms. The first major term measures how well the candidate spokes are fit into the distance image; the second of these major terms measures the geometric fitness of the candidate s-rep. Both terms include penalties for deviation from being medial.

For the spoke $\mathbf{S}(\underline{u})$ the spoke-to-object-mismatch penalty sums terms penalizing misfit to $D(\underline{b}(\underline{u}))$ (the signed distance value at the spoke end) and its derivatives. The geometric fitness penalty sums terms penalizing general geometric misbehavior such as lack of adherence to the medial properties.

Spoke-to-object mismatch penalties: All of these penalties are summed over spokes interpolated from the original discrete set. In our application to hippocampi, each quad of 4 spokes is subdivided by 4 in each dimension, producing a quad of 25 interpolated spokes. No interpolation is done, at present, around the crest.

1. Zeroth order fit to the distance image: $|D(\underline{b}(\underline{u}))|$.
2. First order fit to the distance image: the angle between $\mathbf{S}(\underline{u})$ and $\nabla D(\underline{b}(\underline{u})))$, as described in Sect. 5.3.1.1, item 2.
3. The crest property penalties described in Sect. 5.3.1.1, item 4.

Geometric fitness penalties:

1. S-rep irregularity: To achieve roughly uniform coverage of the object interior and, given the crest-fitting of the end spokes, to provide approximate correspondence of spokes across cases, a penalty for each spoke is applied proportional to its difference from the average of its neighbors. At present, the average is calculated in a way consistent with the principal geodesic analysis approach described in Sect. 5.5.1.1.
2. Illegality of spoke crossings: a penalty is added for S_{rad} or S_E values for a spoke being greater than the reciprocal of their spoke lengths (see Sect. 5.2).
3. Deviation from medial properties, as described in items 1 and 3 of Sect. 5.3.1.1.
4. The distance between $\mathbf{S}(\underline{u})$ and the result of the first stage for that spoke. The distance measure reflects positional deviations at the spoke end.

In the third, i.e., spoke, stage the spoke tail $p(\underline{u})$ is fixed, and the optimization is only over the spoke direction $\mathbf{U}(\underline{u})$ and the spoke length $r(\underline{u})$. The spoke-to-object mismatch penalties are as above, but the geometric fitness penalty is only on the magnitude of the difference of both the spoke direction and the spoke length from the result at the first stage.

In the fourth stage the optimization is over the shape space determined by the statistical analysis of those third stage fits that are adequately good, and the intialization is always from the mean s-rep of those fits. In that stage the geometric atypicality penalty is the negative log prior density (up to a constant multiplier and additive constant) of being in a population of s-reps, given by the Mahalanobis distance function also determined from the statistical analysis. In Sects. 5.5.2 and 5.5.3 we describe how to compute a set of eigenmodes $\mathbf{v}^i$ of s-reps and associated principal variances σ_i^2, using the actual abstract space in which an s-rep falls. Any s-rep in that population is designated by a tuple of scores $\underline{\alpha}$ of the respective eigenmodes, and $\sum_i (\alpha_i^2/\sigma_i^2)$ forms the Mahalanobis distance of that s-rep.

The fifth, spoke, stage is just like the third stage, but on and relative to the results of the fourth stage.

In all of the stages conjugate gradient optimization of the objective function is applied.

Figure 5.4 shows some sample hippocampus fits by this method, as described in Sect. 5.6.

5.3.2 Achieving Correspondence of Spoke Vectors

The previous work on achieving correspondence in training populations of anatomic object models has involved shifting points on object boundaries in point-distribution models (*PDM*s) [3, 5, 20]. We find the most attractive method is that of Cates et al. [3] and Oguz et al. [20], which minimizes a sum of two entropy terms: $H(\underline{\mathbf{z}}) - \alpha \sum_i H(\underline{\mathbf{x}}_i)$. The first term measures the tightness of the probability distribution on the representation entities, here possibly including not only the boundary points but also derived values such as curvature. The second term sums

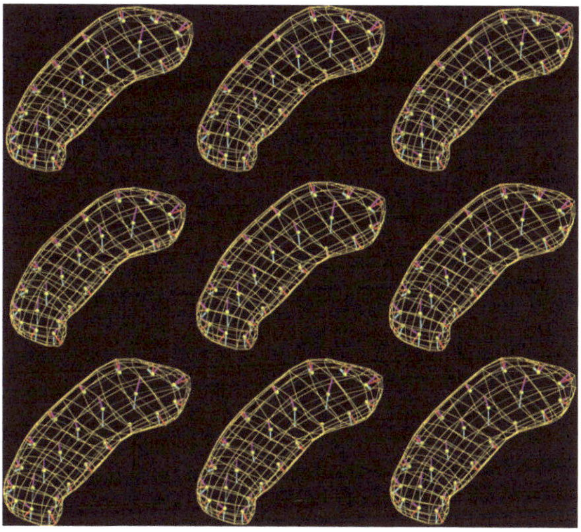

Fig. 5.4 S-reps fitted to 9 of the 62 hippocampi used for the statistical analysis (see Sect. 5.6)

negatives of entropies over the training cases, with each entropy measuring the uniformity of the distribution of the points on the respective object boundary. We are in the process of extending this method to spoke correspondences. In our case the representation entities are spokes, and the entropy possibly includes not only skeletal sheet curvatures (eigenvalues of the shape operator at spoke tails) but also spoke direction curvatures (eigenvalues of the S_{rad} or S_E operators of spokes). Also, the uniformity measurement must be of spokes within the object volume.

However, this method has not been fully implemented at the time this paper is written. Thus the following statistics are based on fits whose correspondence come from common shape-based penalties used for fitting the training cases and the common shape space in which they fall before refinement by the final spoke stage.

5.4 The Abstract Space of s-Reps and Common Configurations of s-Rep Families in that Space

5.4.1 The Abstract Space of s-Reps

Let each s-rep $\underline{\mathbf{m}}^k$ in a training population consist of n spoke vectors $\{(\underline{p}_i, r_i, \mathbf{U}_i) | i = 1, 2, \ldots, n\}$ that correspond across the population. The set $\{\underline{p}_i\}$ of points on each training discrete s-rep's skeletal locus form a PDM (point distribution model) that can be centered so that its center of gravity is the origin. Also, the result can be scaled by a factor making the sum of the squared distances of the points to the origins equal to unity. Thereby each PDM is describable by a scaling term represented by the log of its scale factor γ and a tuple of scaled, centered spoke tail points that abstractly live on the unit $3n - 4$ dimensional sphere S^{3n-4}.

Each spoke direction $\mathbf{U}_i$ abstractly lives on the unit two-sphere. Each $\log r_i$ value, as well as $\log \gamma$, abstractly live in Euclidean space. Thus a discrete s-rep lives in $\mathscr{R}^{n+1} \times S^{3n-4} \times (S^2)^n$, i.e., the Cartesian product of $n+2$ manifolds, one of which is Euclidean and all of the rest of which are spheres. If the number of training cases N is less than $3n-4$, the space of the scaled, centered spoke tails is restricted to the intersection of S^{3n-4} with the Euclidean space of dimension $N-1$ passing through the center of S^{3n-4}, i.e., in S^{N-1}. Also in that case, the n radii live in $\mathscr{R}^{N-1}$. In the experiment on hippocampi reported in Sect. 5.5, $N = 62$ and $n = 66$. Thus each hippocampus s-rep is a point in the composite space $\mathscr{R}^{62} \times S^{61} \times (S^2)^{66}$.

5.4.2 *Families of s-Rep Components on Their Spheres*

The statistical analysis method we will describe in Sect. 5.5 involves includes analysis on each of the here, 67 component spherical spaces, then compositing the analyzed information with one another and with the Euclidean space data, and analyzing the composited data. Therefore, it is of interest to consider how the s-rep data points distribute themselves on each sphere.

Certain basic transformations of the objects can be expected: global rigid rotations, rotational folding about an axis, and twisting about an axis. We have shown [24] the following behaviors of the spoke-direction points on S^2 under these transformations. For each transformation each spoke moves on a small circle on S^2 about the rotational axis of the transformation; different spokes move on different small circles, but all the circles share the same axis. In rotational folding and twisting different spoke-direction points rotate in opposite directions as the fold or twist progresses.

Also, for rigid rotation the points on S^{3n-4} or S^{N-1} (here S^{61}), each describing the tuple of skeletal points for a separate training s-rep, move on a small circle (1-sphere).

We have experimentally confirmed (see Fig. 5.5) the relevance of these ideas, by observing that real data s-rep spokes tend to arrange themselves near small circles.

We will now examine methods for statistically analyzing such data on a d-dimensional sphere in a PCA-like fashion, to describe the data by a suitable notion of mean and a shape space formed from a limited number of eigenmodes of variation.

5.5 Training Probability Distributions in Populations of Discrete s-Reps

5.5.1 *Previous Methods for Analyzing Data on a Sphere*

To understand the methods we will describe, it is useful to describe two approaches to PCA in a d-dimensional Euclidean space. The two approaches are called

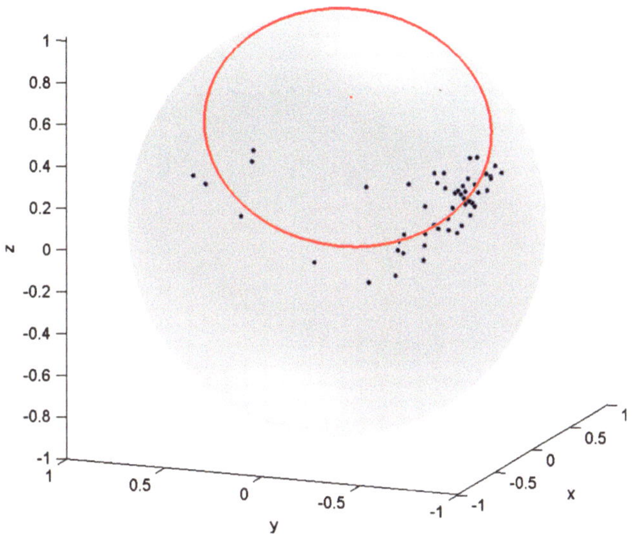

Fig. 5.5 Data points for an s-rep spoke in 62 hippocampi (see Sect. 5.6)

"forward" and "backward" according to the order in which the dimensions are analyzed. In a Euclidean space the two approaches are equivalent, but on spheres (and indeed any curved space) they can give far from the same results.

In the forward approach, the mean is computed as the 0-dimensional space (point) best fitting the d-dimensional data. Then the first principal component is computed as the 1-dimensional space (line) best fitting the d-dimensional data. Then the second principal component is computed by first finding the 2-dimensional space (plane) best fitting the d-dimensional data and then taking the direction in that plane orthogonal to the first component. And so on. The analysis proceeds from dimension 0 to 1 to 2 to ... to $d - 1$. At each step the fit is to the d-dimensional data.

The backward approach begins by fitting the best hyperplane of dimension $d - 1$ to the d-dimensional data. *Then the data is geodesically (orthogonally) projected onto this hyperplane*. Next the fit of a hyperplane of dimension $d - 2$ is made to this projected data. And then the data from the hyperplane of dimension $d - 1$ is projected onto this hyperplane of dimension $d - 2$. And so on, until we have a line fitted to the data that has been projected on a plane and we project the data from the plane onto the line. Finally, the mean is the 0-dimensional best fit to the data on the line. In this method the fit at the kth step is to the data projected onto a space of dimension $d - k + 1$.

In PCA each training point can be expressed as the mean plus a weighted sum of the d eigenmodes. The weights are called scores. The score of eigenmode $d - k + 1$, $k = 1, 2, \ldots, d$ for that training point is the signed projection distance of that data point from the space of dimension $d - k + 1$ to the space of dimension $d - k$.

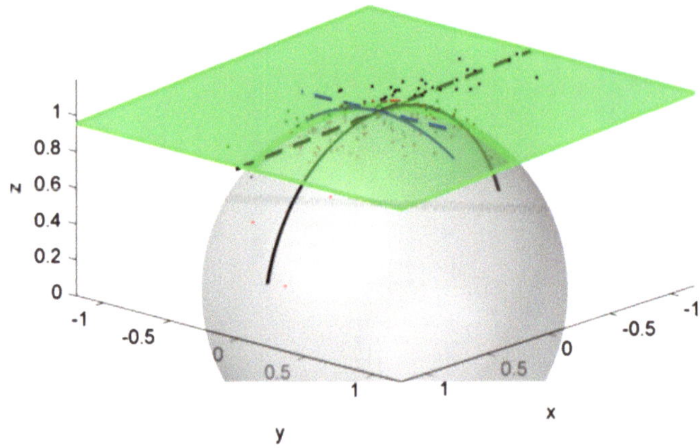

Fig. 5.6 Data clustered on a sphere

5.5.1.1 Principal Geodesic Analysis

The method we had previously used to analyze m-reps (s-reps with equal-length spokes at common spoke tails) was the Principal Geodesic Analysis (PGA) method of Fletcher [9]. This method is a forward approach. It begins with computing the intrinsic forward Fréchet mean on each component, spherical and Euclidean, in the composite space. The mean on a sphere is the point on the sphere whose sum of squared geodesic distances to the data points is minimum. The intrinsic Fréchet mean is a forward mean because the best zero-dimensional approximation to the data, the mean, is computed relative to the full d-dimensional sphere.

Given this mean, the form of PGA that we use moves the data from its manifold onto a tangent hyperplane at the mean with a Euclidean metric determined by keeping directions and the condition that every point on the sphere has the same Euclidean distance to the mean on the tangent hyperplane as its geodesic spherical distance to the mean on the sphere. On the data on this tangent plane the method carries out PCA, and its eigenvectors are mapped back to the original sphere, yielding geodesics through the mean. Strictly speaking, the method is applied to the composite space, not sphere by sphere, but it could be applied sphere by sphere.

This method works very well when the data cluster on each of the composite spheres (and the Euclidean space in the composite) because then data projected on the tangent plane represents the original data well (Fig. 5.6). However, in our case the data frequently live on small circles that are far from the pole, whereas the mean may be found at the pole. It is then not surprising that PGA will not be optimal.

5.5.1.2 GPCA: Geodesics with the Mean on the First Geodesic

Huckemann et al. [12, 13] proposed a method called geodesic principal component analysis (GPCA) which inspired our final method by having an important backward component. They realized that the Fréchet mean may be far from the best fitting great circle. Hence, if the data lives on or near a circle on a sphere, the mean should be computed on that circle after projection of the data onto it. Their method found the best fitting geodesic (great) circle on the sphere, projected the data onto that, and found the Fréchet mean on that great circle of the projected points. This is a backward mean, since it computed after projection from the original d-space onto a 1-dimensional subspace.

The method of Huckemann et al. on a d-sphere then went on to compute principal geodesic circles through the mean in a forward fashion, in a way quite similar to PCA, albeit on a sphere. Its weakness for our data comes from two problems. First, it fits great circles, whereas our data live near small circles. Second, it is backward only at the first step, whereas the fully backward method of Jung [15] described next appears superior.

5.5.1.3 Composite Principal Nested Spheres

A method better recognizing the loci along which the s-reps tend naturally to vary is the Composite Principal Nested Spheres (CPNS) method of Jung et al. [15]. It consists of (1) component by component estimation of principal modes of each of the spherical manifolds separately, each producing scores that are suitable for Euclidean analysis; then (2) compositing those analyses with one another and the Euclidean data, followed by PCA on the result.

5.5.2 Training Probability Distributions on s-Rep Components Living on Spheres: Principal Nested Spheres

We begin by describing the principal nested spheres (PNS) approach that is applied to the data on each sphere separately. This method has already been described in [15]. As suggested, the approach is backwards analysis, and at each step the best fitting subsphere of one dimension lower is fit, irrespective of whether that subsphere is a great or small subsphere.

The approach begins with the data on the d-dimensional sphere. The best subsphere of $d - 1$ dimensions is fit by an analytical computation. As illustrated in Fig. 5.7, the result is formed by (and recorded as) the pole (axis) position $\mathbf{w}^1$ of the subsphere on the d-sphere and its latitude (angle from the pole) ψ_1. Then it projects the data onto the subsphere of dimension $d - 1$. The process begins anew, finding the $(d - 2)$-dimensional subsphere best fitting the data that has been projected onto the $(d - 1)$-dimensional sphere. The process repeats itself until a 1-sphere (a not necessarily great circle) with points on it has been arrived at. The final step is to find the mean (best-fitting 0-sphere) as the best fitting point to (geodesic mean of) the data points that have been projected onto the circle.

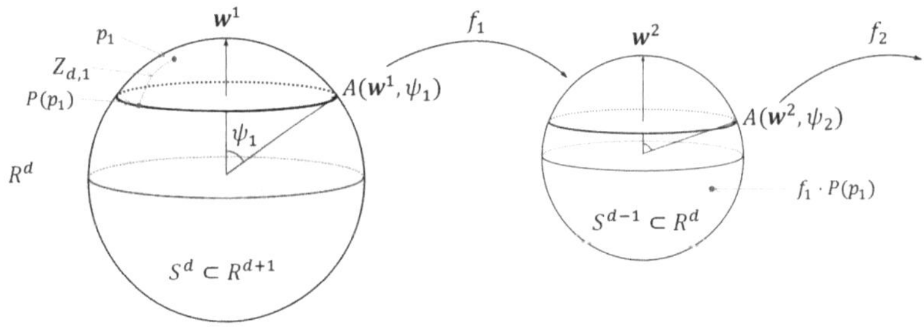

Fig. 5.7 Computation of subsphere, projection, and scores in principal nested spheres

The mean is successively projected forward through the dimensions $2, 3, \ldots, d$ using the subsphere-to-sphere transformations that have been recorded. The resulting point on the d-sphere gives the place in the data space which is the backwards mean of the shape component that was analyzed on this sphere.

As with PCA, we record the score of the jth data point relative to the projection of the subsphere of dimension $d - k + 1$, $k = 1, 2, \ldots, d$, to the subsphere of dimension $d - k$ as the signed projection distance $z_{d-k+1,j}$ of that data point between those two spheres (see Fig. 5.7). The random variable $z_{d-k+1,\cdot}$ is a 0-mean variable that can be understood as being in Euclidean space. Thus, the $d \times N$ array Z with elements k, j being $z_{d-k+1,j}$ represents the Euclideanized, 0-mean distribution of the respective shape component of the data on the d-sphere.

5.5.3 Compositing Component Distributions into an Overall Probability Distribution

In our method of analysis of CPNS the distribution of the 62-case s-rep data on each of the component spheres is analyzed by PNS. The compositing approach has already been described in [14]; here we detail it for analysis of s-reps. In that case, with 67 spheres, this yields 67 Z arrays. One of them (for the spoke tail points) is 61×62 and the other 66 (for the spoke directions) are 2×62. We must now account for the correlations between the various s-rep components, including both those that live on spheres and those that live in a Euclidean space. To prepare for that, for each s-rep the Euclidean components, $k = 1, 2, \ldots, 67$, each of the form log length or log scale factor, must each have their mean computed (in the ordinary way of summing the entries across the data items and dividing by the number of entries, 62) and then subtracting that mean from each data item's corresponding component. (Strictly speaking, that data lives in a 62-dimensional space, but the SVD analysis to follow is insensitive to using the original 67 dimensions.) Call the result of this the 67×62 array $Z^{Euclidean}$.

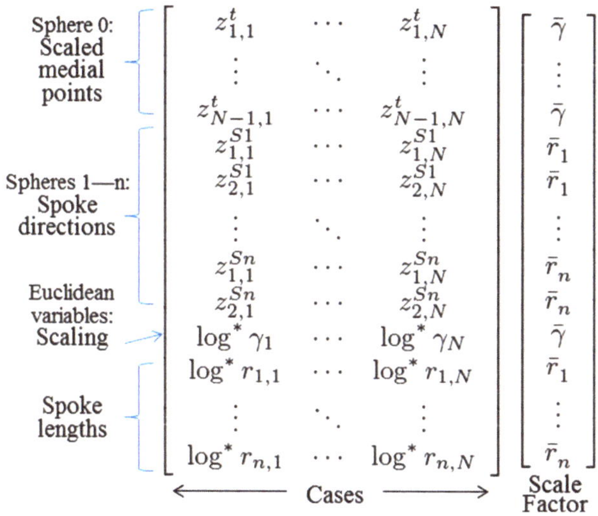

Fig. 5.8 *Left*: the composite array Z_{comp} before row scaling for N s-reps. * indicates that the variable's row mean has been subtracted. *Right*: the row scale factors

Next each Z array row must be made commensurate; we give each entry the units of the object boundary offset produced by any row entry. This is accomplished by

1. Multiplying the two rows giving z entries for each of the 66 spoke directions by the average of the length of that spoke over the training cases;
2. Multiplying the n (66) rows giving entries giving a mean-subtracted log spoke length by the average of the length of that spoke over the training cases;
3. Multiplying the $N - 1$ (61) rows giving z entries for the scaled and centered medial point tuple, as well as the row giving the log of the scale factor, by the geometric mean of the point tuple scale factors over the training cases.

We produce a 0-mean Euclidean representation of our data by piling all of these scaled Z arrays (67 from spherical analysis and 1 from Euclidean analysis) on top of one another, producing the 260×62 array Z_{comp} (Fig. 5.8), to which the same analysis as is done in conventional PCA can be applied. By this analysis, SVD (singular value decomposition) on Z_{comp} yields a list of eigenvectors (principal directions) $\mathbf{v}^i$ and eigenvalues (principal variances) σ_i^2. Together with the backward means for each of the spheres and the mean of the Euclidean variables, this is the PCA-like result reflecting our understanding that s-reps live on the Cartesian product of 67 spheres and a Euclidean space.

The CPNS mean is formed by compositing the backward means from each of the component spheres and the Euclidean mean. Each component mean is a point on its respective sphere or Euclidean space. The mean s-rep is achieved by compositing these points into spoke directions, spoke tail points, and spoke lengths. Unlike the Euclidean mean or the forward intrinsic mean, the spherical components of the

CPNS mean live not only on their spheres but on all the best fitting subspheres. It therefore can be expected that the CPNS mean is more representative of the training shapes than any of the forward means.

From the CPNS analysis, we can choose a number of eigenmodes that captures a desired fraction of the total variance. This forms a shape space for s-reps as implied by the training cases. Every s-rep in the shape space is given by a tuple $\underline{\alpha}$ of scores of the eigenmodes, with associated Mahalanobis distance $\Sigma_i(\alpha_i^2/\sigma_i^2)$. Given $\underline{\alpha}$, one can form the vector $\Sigma_i \alpha_i \mathbf{v}^i$ and then apply the row scale factors computed in preparing for the SVD. The entries in the resulting column vector can be divided into the tuples corresponding to each s-rep component, just as the columns in the array Z_{comp} were divided. Each component will have its value unscaled by the row scale factors previously discussed. The Euclidean components are then ready for adding their mean. Each group of components from a spherical space is ready for projection into its successive spherical spaces represented by the latitude angles ψ_i and polar directions $\mathbf{w}^i$, beginning with the backwards mean for that sphere. Compositing the results on each of the 68 spaces into spoke directions, spoke tails, and spoke lengths yields an s-rep that can be displayed or used to provide s-rep-relative coordinates for the points in the image space in which an s-rep was originally fit.

The results of a CPNS analysis on hippocampi, some of which were illustrated in Fig. 5.4, are described in the following section.

5.6 Analyses of Populations of Training Objects

We analyze 62 left hippocampi segmented from MR images. These are from control (normal) individuals in a study of schizophrenia [26]. Each had an initial model that had been fit years ago using an earlier version of the first three stages of the s-rep fitting program; these fits were neither tight with the object boundary nor did they satisfy the spoke-to-boundary orthogonality and crest fitting objectives. These initial models were each translationally, rotationally, and scale aligned to their distance functions, and the resulting models were fit using the geometric (atom-by-atom and spoke) stages described as the second and third stages in Sect. 5.3.1.2. Our objective was to compare fitting using CPNS followed by statistical analysis via CPNS with fitting using PGA followed by statistical analysis via PGA.

Fifty of the third-stage fits were chosen as satisfactory, and this set was dealt with in both ways. Fifteen eigennmodes were chosen from each set as the dominant eigenmodes. The result was two shape spaces, one with a mean and eigenmodes based on PGA on the 50 fits and one with a mean and eigenmodes based on CPNS of the same 50 fits. Following this analysis, two applications, one PGA-based and one CPNS-based, of the fourth (shape space) and fifth (spoke) stages (see Sect. 5.3.1.2) were applied to produce the two sets of fitted s-reps that we wished to compare. In each application, each of the 62 hippocampi was initialized by the mean appropriate for that set, and it was first translated, rotated, and scaled into the hippocampus distance image to be fit. That result was taken as the initialization of an optimization

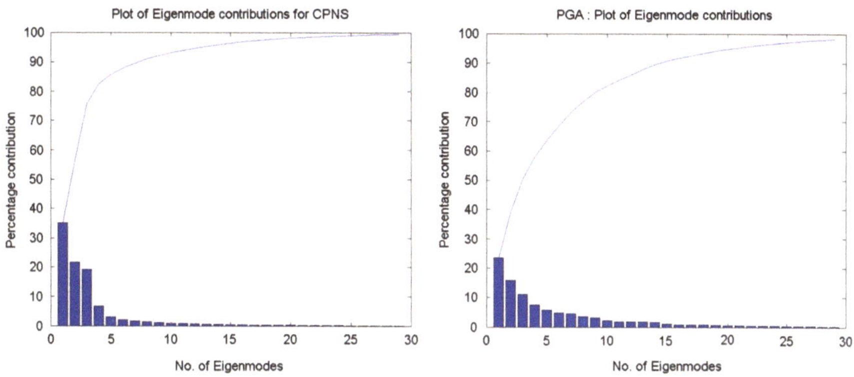

Fig. 5.9 Eigenvalues (*bars*) and cumulative eigenvalue (*curve*) plots as percentage of total variance for CPNS (*left*, requiring seven modes for 90 % of variance) and for PGA (*right*, requiring 14 modes for 90 % of variance) after removal of the first mode

over the corresponding shape space; the spoke optimization of the fifth stage was applied to the result. These fits seemed generally satisfactory for all 62 hippocampi. Let us refer to the results as the PGA-based fits and the CPNS-based fits.

CPNS analysis was applied to the CPNS-based fits, and PGA was applied to the PGA-based fits. In the following we compare the eigenvalue plots, the means, and the eigenmodes between these two analyses.

In both analyses the first mode of variation dominated, and it was largely uniform scaling. It formed 70 % of the total variance in the CPNS analysis and 90 % of the total variance in the PGA analysis. This is consistent with a longstanding approach in neuroimage analysis of focusing on hippocampus volume as a discriminative parameter. Removing this source of variation from both the eigenvalue list and the total variance in the corresponding analysis yielded the eigenvalue and cumulative eigenvalue plots shown in Fig. 5.9. There is an obvious compression to the left of the eigenvalues in the CPNS plot relative to those in the PGA plot. This is quantified by looking at how many eigenmodes were necessary to achieve 90 % of the remaining variance. The CPNS-based fitting and analysis requires only seven additional modes to achieve 90 % (thus 97 % of the total variance when the first mode is included), whereas the PGA-based fitting and analysis requires 14 additional modes to achieve 90 %. We also notice that, whereas the fitting after the first three stages, i.e., the geometry-based stages, produced analyses by either PGA or CPNS that required 20 modes each to reach 90 % of the variance in modes 1 on, the amount of fitting-based noise produced by the stage optimizing over the statistical shape space followed by spoke stage refinement is so lessened that only eight modes (the first, scaling mode and seven more) are now necessary to reach 97 % of the variance in modes 1 on.

Now we compare the CPNS-based backward mean s-rep to the PGA-based forward mean s-rep. Both means look like a hippocampus, but as seen in Fig. 5.10, the CPNS mean is longer, related to the fact that the PGA-based fits were noisier near the tail and head of the hippocampus.

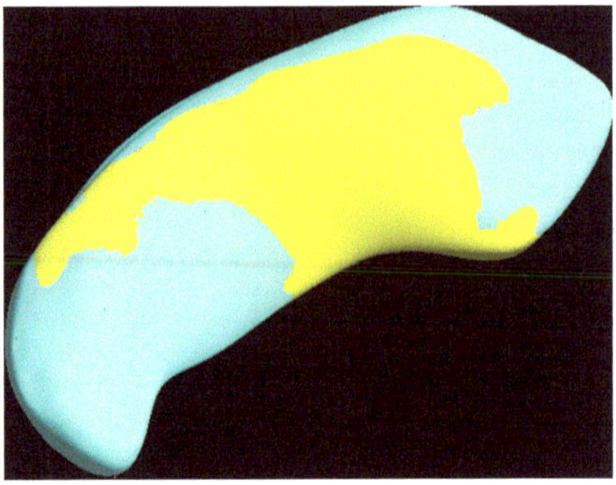

Fig. 5.10 Implied boundaries of CPNS (*cyan*) and PGA (*yellow*) means of the 62 left hippocampi of normal individuals

Finally, we compare the two analyses in terms of the shape changes shown by their eigenmodes. Because the contrast is too complex to show here, we summarize what we have observed. Beyond the first, largely scaling, eigenmode, none of the following four eigenmodes from the PGA set qualitatively match any of the following four eigenmodes from the CPNS set, and vice versa. Roughly, the CPNS eigenmodes seem to capture more in the way of bendings and twistings, and the PGA eigenmodes seem to capture more in the way of rotations and small variations in eccentricity. We suggest that the reason is that CPNS by its design is more sensitive to changes in local directions (e.g., of spokes), so, relative to PGA, its preliminary shape space enables more efficient modeling of variations of that type. Moreover, its use in the final statistical analysis senses those variations more efficiently.

5.7 Extensions and Discussion

The method of composite principal nested spheres applies to any data that are best analyzed as a composite of components lying on spheres and possibly including Euclidean components. The analyses of such data done so far have suggested that CPNS can provide more efficient analysis (fewer eigenmodes), stabler s-rep fitting, and possibly more information about twisting and bending changes than PGA and other forward methods. We have seen that fitting a geometric model, here an s-rep, to object segmentations is significantly improved by the use of a shape space based on CPNS-based analysis on preliminary fittings; the improvement is both in terms of quality of fit and in terms of making a second statistical analysis on the models informative.

Shape analysis via CPNS extends beyond s-reps to essentially every other representation of shape. Shape inherently involves directions and curvatures (derivatives on directions). As a result, it can be expected that CPNS will provide similar advantages when applied to other representations of shape. The following list some candidate representations suitable for CPNS analysis.

- Point distribution models (PDMs), made of a large tuple of boundary (and possibly other) points. On ten lung PDMs for ten phases of a patient's respiration, CPNS required only required one to two eigenmodes vs. three for PCA. As with s-reps, CPNS required fewer modes than the forward method, here PCA.
- Point and normal distribution models: i.e., boundary point PDMs with a boundary normal supplied with each point
- Quasi-tube s-reps
- Branching objects
- Multi-object complexes
- Warps, i.e., displacement vector images or time-indexed velocity vector images representing diffeomorphisms or other invertible deformations.
- Fiber conglomerates, such as those studied by Savadjiev et al. [23].
- Tensor images
- Covariance matrices

The method of s-rep fitting followed by correspondence improvement and CPNS analysis requires a variety of improvements, listed in the following, together with testing on many more shape populations. Once verified, it seems suitable for applications in segmentation by posterior optimization [27, Chap. 9], hypothesis testing [29], and classification [25].

Improvements in s-rep fitting needed include spoke interpolations at the crest using the S_{rad} and S_E matrices [11], fittings into multi-crest distance images, and generalization of entropy-based correspondence optimization [3] to spokes.

Acknowledgements We thank Jörn Schulz for studying the effects of global rotation, folding, and twisting on skeletal spoke vectors; Martin Styner for the hippocampus data; and Anna Snyder for help with the references.

References

1. Arsigny, V., Commowick, O., Pennec, X., Ayache, N.: A log-Euclidean framework for statistics on diffeomorphisms. In: Medical Image Computing and Computer-Assisted Intervention, vol. 9, pp. 924–931. Springer, Berlin/Heidelberg (2006)
2. Blum, H.: A transformation for extracting new descriptors of shape. In: Wathen-Dunn, W. (ed.) Models for the Perception of Speech and Visual Form. MIT, Cambridge, MA (1967)
3. Cates, J., Fletcher, P.T., Styner, M.E., Shenton, R.T.W.: Shape modeling and analysis with entropy-based particle systems. Inf. Process. Med. Imaging **20**, 333–345 (2007)
4. Cootes, T.F., Taylor, C., Cooper, D., Graham, J.: Training models of shape from sets of examples. In: Hogg, D., Boyle, R. (eds.) Proceedings of British Machine Vision Conference, pp. 9–18. Springer, Berlin (1992)
5. Cootes, T.F., Twining, C.J., Petrović, V.S., Babalola, K.O., Taylor, C.J.: Computing accurate correspondences across groups of images. IEEE Trans. Pattern Anal. Mach. Intell. **32**, 1994–2005 (2010)
6. Damon, J.: Smoothness and geometry of boundaries associated to skeletal structures I: sufficient conditions for smoothness. Ann. Inst. Fourier **53**, 1001–1045 (2003)

7. Damon, J.: Swept regions and surfaces: modeling and volumetric properties. Conf. Computational Alg. Geom. 2006, in honor of Andre Galligo. Spl. Issue Theor. Comp. Sci. **392**, 66–91 (2008)
8. Dryden, I.L., Mardia, K.V.: Statistical Shape Analysis. Wiley, Chichester (1998)
9. Fletcher, P.T., Lu, C., Pizer, S.M., Joshi, S.: Principal geodesic analysis for the study of nonlinear statistics of shape. IEEE Trans. Med. Imaging **23**, 995–1005 (2004)
10. Fletcher, P.T., Venkatasubramanian, S., Joshi, S.: The geometric median on Riemmanian manifolds with application to robust atlas estimation. NeuroImage **45**(1), S143–S152 (2009)
11. Han, Q., Pizer, S.M., Damon, J.N.: Interpolation in discrete single figure medial objects. In: Computer Vision and Pattern Recognition (CVPR) – Workshop on Mathematical Methods in Biomedical Image Analysis (MMBIA). IEEE Press (2006)
12. Huckemann, S., Ziezold, H.: Principal component analysis for Riemannian manifolds, with an application to triangular shape spaces. Adv. Appl. Probab. **38**(2), 299–319 (2006)
13. Huckemann, S., Hotz, T., Munk, A.: Intrinsic shape analysis: geodesic PCA for Riemannian manifolds modulo isometric lie group actions. Stat. Sin. **20**(1), 1–58 (2010)
14. Jung, S., Liu, X., Marron, J.S., Pizer, S.M.: Generalized PCA via the backward stepwise approach in image analysis. In: Angeles, J. (ed.) Brain, Body and Machine: Proceedings of an International Symposium on the 25th Anniversary of McGill University Centre for Intelligent Machines. Advances in Intelligent and Soft Computing, vol. 83, pp. 111–123. Springer, Berlin/Heidelberg (2010)
15. Jung, S., Dryden, I.L., Marron, J.S.: Analysis of principal nested spheres. Biometrika. **99**(3), 551–568 (2012)
16. Kendall, D.G., Barden, D., Carne, T.K., Le, H.: Shape and Shape Theory. Wiley, Chichester (1999)
17. Kurtek, S., Ding, Z., Klassen, E., Srivastava, A.: Parameterization-invariant shape statistics and probabilistic classification of anatomical surfaces. Inf. Process. Med. Imaging **22**, 147–158 (2011)
18. Leventon, M., Faugeras, O., Grimson, E., Kikinis, R., Wells, W.: Knowledge-based segmentation of medical images. In: Osher, S., Paragios, N. (eds.) Geometric Level Set Methods in Imaging, Vision, and Graphics. Springer, New York (2003)
19. Merck, D., Tracton, G., Saboo, R., Levy, J., Chaney, E., Pizer, S.M., Joshi, S.: Training models of anatomic shape variability. Med. Phys. **35**, 3584–3596 (2008)
20. Oguz, I., Cates, J., Fletcher, T., Whitaker, R., Cool, D., Aylward, S., Styner, M.: Entropy-based particle systems and local features for cortical correspondence optimization. In: Proceedings of the IEEE International Symposium on Biomedical Imaging (ISBI), pp. 1637–1641. IEEE Press (2008)
21. Pennec, X.: Statistical computing on manifolds: from Riemannian geometry to computational anatomy. Emerg. Trends Vis. Comput. **5416**, 347–386 (2008)
22. Saboo, R., Niethammer, M., North, M., Pizer, S.M.: Anti-aliasing discretely sampled object boundaries using fourth-order Laplacian of curvature flow. http://midag.cs.unc.edu/bibliography.html (2011). Accessed 3 Mar. 2012
23. Savadjiev, P., Campbell, J.S.W., Descoteaux, M., Deriche, R., Pike, G., Siddiqi, K.: Labeling of ambiguous sub-voxel fibre bundle configurations in high angular resolution diffusion MRI. NeuroImage **41**(1), 58–68 (2008)
24. Schulz, J., Jung, S., Huckemann, S.: A collection of internal reports submitted to UNC-Göttingen study group on s-rep change under rotational transformations. University of North Carolina at Chapel Hill (2011)
25. Sen, S.K., Foskey, M., Marron, J.S., Styner, M.A.: Support vector machine for data on manifolds: an application to image analysis. In: Proceedings of the IEEE International Symposium on Biomedical Imaging (ISBI), pp. 1195–1198. IEEE Press (2008)
26. Shi, X., Styner, M., Lieberman, J., Ibrahim, J.G., Lin, W., Zhu, H.: Intrinsic regression models for manifold-valued data. In: Medical Image Computing and Computer-Assisted Intervention, vol. 5762, pp. 192–199. Springer, Berlin/Heidelberg (2009)

27. Siddiqi, K., Pizer, S.: Medial Representations: Mathematics, Algorithms and Applications. Springer, Dordrecht (2008)
28. Sorensen, P., Lo, J., Petersen, A., Dirksen, A., de Bruijne, M.: Dissimilarity-based classification of anatomical tree structures. Inf. Process. Med. Imaging **22**, 475–485 (2011)
29. Terriberry, T., Joshi, S., Gerig, G.: Hypothesis testing with nonlinear shape models. Inf. Process. Med. Imaging **19**, 15–26 (2005)
30. Wang, H., Marron, J.S.: Object oriented data analysis: sets of trees. Ann. Stat. **35**(5), 1849–1873 (2007)
31. Yang, J., Staib, L., Duncan, J.: Neighbor-constrained segmentation with 3D deformable models. Inf. Process. Med. Imaging **18**, 198–209 (2003)

Chapter 6
3D Curve Skeleton Computation and Use for Discrete Shape Analysis

Gabriella Sanniti di Baja, Luca Serino, and Carlo Arcelli

Abstract A discrete 3D curve skeletonization algorithm is described, based on iterated voxel removal guided by the distance transform of the object and on anchor point selection. The criterion suggested for anchor point selection guarantees that the obtained curve skeleton reflects object's shape quite well since it consists of curves symmetrically placed within the regions intuitively perceived as composing the object. Then, the use of the curve skeleton for object decomposition in the framework of the structural approach to shape analysis is discussed. A suitable partition of the skeleton is presented that originates object decomposition in accordance with human intuition.

6.1 Introduction

The skeleton of an object is a subset of the object with lower dimensionality, symmetrically placed within the object, characterized by the same topology as the object, and reflecting the geometrical features of the object itself. In the case of 3D objects, two different kinds of skeleton can be considered, namely the surface skeleton, consisting of the union of surfaces and curves aligned along the principal symmetry planes and axes of the object, and the curve skeleton, exclusively consisting of curves aligned along the principal symmetry axes of the object. Obviously, the curve skeleton can be computed only for objects rid of cavities. In fact, to guarantee that the skeleton and the object have the same topology, a closed surface surrounding any cavity of the object should be present in the skeleton, which therefore would not be characterized by linear structure. Moreover, the curve skeleton should be used only to represent objects with the main symmetry

---

G. Sanniti di Baja (✉) · L. Serino · C. Arcelli
Institute of Cybernetics "E. Caianiello", CNR, Via Campi Flegrei 34, 80078 Pozzuoli (Naples), Italy
e-mail: g.sannitidibaja@cib.na.cnr.it; l.serino@cib.na.cnr.it; c.arcelli@cib.na.cnr.it

M. Breuß et al. (eds.), *Innovations for Shape Analysis*, Mathematics and Visualization,
DOI 10.1007/978-3-642-34141-0_6, © Springer-Verlag Berlin Heidelberg 2013

axes lying in the objects themselves. In fact, when the main symmetry axes are outside the object, the curve skeleton consists of curves that do not adequately represent the input object. For example, for a tubular object like a hollow cylinder, the curve skeleton is the ring placed in the middle of the cylindrical surface.

A wide literature exists presenting both continuous and discrete approaches to the computation of the skeleton of 2D and 3D objects [1–14] as well as to the use of the skeleton in a number of applications [15–22]. We follow the discrete approach and in this paper we deal with the computation of the curve skeleton of 3D objects in voxel images and with object decomposition via skeleton partition.

Skeletonization can be accomplished by means of iterated topology preserving removal. Each iteration of the process consists of two sub-iterations. During the first sub-iteration, the voxels constituting the current border of the object are identified; then, during the second sub-iteration, border voxels are sequentially checked for removal and are actually removed if their maintenance is not necessary for topology preservation or to account for relevant object's shape information. The process terminates when all object voxels are border voxels and these are all necessary for topology preservation or to account for shape properties.

One of the drawbacks affecting skeletonization via iterated removal is its computational cost, since the number of iterations, and hence the number of times the image has to be scanned, is proportional to object's thickness. This drawback can be solved by using methods where at each iteration only the neighbors of the points that have been modified during the previous iteration are inspected (see e.g., [23]).

Another crucial issue is the difficulty of finding a reliable and simple criterion to preserve from removal elements accounting for shape information. These elements, generally called end points, are often detected by setting a threshold on the maximum number of object's neighbors that a border voxel can have to be preserved from removal when inspected. Such a criterion does not guarantee an isotropic behavior since the neighborhood configurations of voxels, placed in object parts with the same structure but different orientation, depend on the order in which these voxels are visited. This drawback can be alleviated by identifying suitable anchor points, i.e., voxels whose removal is inhibited, before starting the removal process. For example, border voxels whose distance from the interior of the object overcome an a priori fixed threshold can be regarded as anchor points since they belong to significant protrusions (see, e.g. [24, 25]).

An alternative approach to skeletonization is based on the research done by Blum on the medial axis transform, MAT [26]. To define the MAT, initially for 2D objects and successively for objects in higher dimensions, the notion of symmetry points and a growth process were introduced. In the continuous space, for a 2D (3D) object, a symmetry point is a point that is center of a circle (sphere), bitangent two distinct sections of the boundary of the object and entirely contained in the object. A symmetry point can be associated with the radius of the corresponding circle (sphere). In turn, the circle (sphere) can be built via a growing process that, starting from the symmetry point, incorporates all object's points whose distance from the symmetry point does not overcome the radius associated to the symmetry point itself. The envelope of the circles (spheres) coincides with the object and the

MAT is the locus of the centers, associated with the corresponding radii. Blum also studied the geometric properties of the MAT and showed the usefulness of the MAT for shape analysis [27].

MAT computation is rather heavy, especially in 3D, so that it is of interest a skeletonization method based on the computation of a reasonably good approximation of the MAT. A possible solution in the discrete space is to resort to the use of the distance transform of the object X, $DT(X)$, and to the selection of the set of centers of maximal balls, $CMB(X)$. In fact, $DT(X)$ is a multi-valued replica of the object, where each voxel of the object is assigned the value of its distance from the boundary. Thus, each object voxel in $DT(X)$ can be interpreted as the center of a ball with radius equal to the distance value of the voxel itself and touching the object boundary. As the name suggests, a voxel of $CMB(X)$ is a voxel whose associated ball is maximal, i.e., is not included by any other single ball, touches the object boundary in at least two different parts, and is obviously totally included in the object. Moreover, the union of all the maximal balls coincides with the object. This indicates that $CMB(X)$ consists of symmetry points and constitutes a discrete approximation of the MAT.

As we will see in more detail in the following sections, the set $CMB(X)$ does not satisfy all the properties expected for the skeleton. We note that, except for the case of 3D objects having tubular shape, voxels of $CMB(X)$ do concentrate along symmetry planes rather than along symmetry axes, so that not all voxels of $CMB(X)$ are of interest for curve skeletonization. Moreover, $CMB(X)$ is not necessarily characterized by the same topology as the represented object. Thus, on one side skeletonization needs to accept as skeleton elements other voxels besides those of $CMB(X)$ in order to obtain a topologically correct result. On the other side, skeletonization should not accept all voxels of $CMB(X)$ as skeleton voxels so as to originate a skeleton with a linear structure easy to manage and still able to represent the object sufficiently well. Of course, accepting only a subset of $CMB(X)$ in the skeleton also means that the union of the balls associated with the voxels of the skeleton does not coincide with the object itself.

Actually, the voxels of $CMB(X)$ are generally not all equally important. For example, a number of voxels of $CMB(X)$ do not carry shape information, since they are due only to the discrete nature of the digital space. In this respect, let us consider a cylinder with its main axis oriented along one of the three axes in a Cartesian coordinate system. In the continuous space, sections of the cylinder perpendicular to such an axis are circles and, for sections sufficiently far apart from the bases of the cylinder, the distance transform appears as consisting of concentric circular wave-fronts, each including points with the same distance value. For each of these sections only one center of maximal ball exists, which is the symmetry point placed along the main symmetry axis of the cylinder. In the digital space, the circular sections of the cylinder are unavoidably transformed into polygonal shapes and the distance transform appears as due to the propagation of a polygonal wave-front. Since voxels of $CMB(X)$ are located where the wave-front folds upon itself, besides the centers of maximal balls along the main axis, also other centers of maximal balls are detected. These are due to the existence of weak convexities in correspondence with the vertices of the polygonal wave-front. This fact causes folding of the wave-front upon itself.

More in general, voxels of $CMB(X)$ can be found in correspondence of object parts characterized by different properties, not all equally important for the specific task. Thus, it is profitable to devise criteria to select an appropriate subset of $CMB(X)$, depending on the object properties of interest for the task at hand [28]. Once suitable criteria to filter out irrelevant centers of maximal balls are used, then the surviving voxels of $CMB(X)$ can be taken as anchor points and skeletonization can be accomplished by using iterated voxel removal to guarantee a topologically correct result.

In this paper, we describe a skeletonization method based on the selection of anchor points in the distance transform and on iterated voxel removal. Besides using the distance transform to identify the subset of $CMB(X)$ whose voxels are taken as anchor points, we also exploit it to strongly reduce the computational cost of iterated voxel removal. In fact, we avoid repeated inspections of the image by directly accessing in increasing distance value order the voxels that undergo removal. The curve skeleton *CS* of an object X is computed in two phases.

During the first phase, the Pseudo Surface Skeleton *PSS* is detected by following the strategy suggested in [14]. The set *PSS* is a subset of X and is used as input to the second phase. To obtain *PSS*, the distance transform of X is computed and suitable centers of maximal balls are taken as anchor points. Topology preserving removal operations are then applied to object voxels that are not anchor points to get an at most two-voxel thick set. Final thinning is performed to gain unit thickness and pruning of peripheral lines is done for trimming out negligible details.

During the second phase, a classification of the voxels of *PSS* into curve, internal, junction, extended junction, branching and edge voxels is done. The distance transform of *PSS* is then computed. Voxels that are classified as branching or junction voxels, and centers of maximal balls of *PSS* are taken as anchor points. Topology preserving removal operations are then applied to reduce *PSS* to an at most two-voxel thick set of curves. Final thinning and pruning complete the second curve skeletonization phase.

Though the union of the balls associated with the voxels of the curve skeleton generally provides only a rough version of the input object, the curve skeleton is of interest in the framework of the structural approach to object analysis to reduce the complexity of the recognition task. In particular, the skeleton can be used for object decomposition (see e.g., [29–32]).

As discussed in [33–36], the human visual system is likely to represent objects with complex shape in terms of simpler parts, by decomposing the objects into parts and by organizing object representation in terms of the parts and of their spatial relationships. Such a structured representation is characterized by a significantly greater robustness under changes in viewing conditions.

Decomposition guided by skeleton partition has been followed particularly for objects than can be perceived as articulated in parts with tubular shape, where a one-to-one correspondence exists between the individual curves composing the skeleton and the individual parts of the object. In this case, the object can be interpreted as consisting of a number of parts, called geons. The underlying theory is that human object understanding is based on recognition-by-component. Each individual curve of the skeleton can be used to analyze the part of the object it represents, and the spatial relations among the curves of the skeleton can be used to derive information on the organization of the parts constituting the object.

When the object also consists of parts that cannot be interpreted as geons, individual skeleton curves do not necessarily correspond to perceptually significant object parts. Thus, skeleton curves have to be suitably grouped to generate a skeleton partition where each partition component corresponds to a perceptually meaningful object part.

In this paper, we present an object decomposition method based on curve skeleton partition. We use the notion of zone of influence, i.e., the region union of the balls associated with a given subset of the skeleton. In particular, we consider the zones of influence associated with the branch points, i.e., the voxels where different curves of the skeleton meet. By using these zones of influence, the skeleton is partitioned into three different kinds of subsets. Then, the skeleton partition components are used to build the object's parts. Since pairs of adjacent parts may be separated by non-planar surfaces, voxels in each pair of adjacent parts are possibly redistributed between the two parts in such a way to obtain an almost planar separation. Finally, a merging process is accomplished to obtain a decomposition of the object into a smaller number of perceptually significant parts.

We have applied our algorithms for curve skeletonization and for object decomposition to a variety of images, taken from publicly available shape repositories [37, 38]. A few examples are shown in the paper.

6.2 Notions and Definitions

We consider a binary voxel image in a cubic grid, where the object X is the set of 1s and the background B is the set of 0s. To avoid topological paradoxes, we use the 26-connectedness and the 6-connectedness for X and B, respectively.

The $3 \times 3 \times 3$ neighborhood $N(p)$ of a voxel p is the set including the 26 neighbors of p. These are the 6 face-, the 12 edge- and the 8 vertex-neighbors of p, respectively denoted by n_f, n_e, n_v. We also consider the set $N^*(p)$ that includes only 18 neighbors of p. These are the 6 face- and the 12 edge-neighbors of p. In the following, the term neighbor or the notation n_i are used when a generic neighbor of p is considered.

We assume that the set B consists of a single six-connected component. This presupposition is necessary since we aim at the computation of the curve skeleton. This computation is not possible for an object including cavities, each of which should be surrounded by a closed surface in the resulting skeleton in order to have a topologically correct skeleton.

An object voxel p is a simple point if the two objects with and without p have the same topology. Voxel simplicity can be stated by checking that topology is locally preserved. Thus, if we denote by c_p the number of 26-connected object components in $N(p)$, and by c_p^* the number of six-connected background components having p as face-neighbor and computed in $N^*(p)$, a voxel p is simple if $c_p = 1$ and $c_p^* = 1$ [39,40].

A path linking two voxels p and q is a sequence of unit moves from p to q through voxels that are face-, edge- or vertex-neighbors of each other. To give a reasonably good approximation of the Euclidean length of the path, the unit moves towards face- edge- and vertex-neighbors encountered along the path should be measured by different weights, suitably taking into account the different Euclidean lengths of these moves. The integer weights $w_f = 3$, $w_e = 4$ and $w_v = 5$ are adequate to provide a good approximation of the length of the path [41].

Though efficient algorithms exist to compute the Euclidean distance between two voxels p and q (e.g., see [42]), when following discrete approaches, the distance between p and q is preferably defined as the length of a shortest path linking p to q [43]. In this work, we resort to the use of the $\langle 3, 4, 5 \rangle$ weighted distance that combines the simplicity of path-based distances with a reasonable approximation to the Euclidean distance. The main reason for preferring a path-based distance to the true Euclidean distance is that with the former metric strong constraints exist on the distance values that are possible for neighboring voxels, which will allow us to devise easily local operations for various tasks, e.g., for center of maximal ball detection.

The curve skeleton *CS* of a 3D object X is a set union of curves, symmetrically placed within X, with the same topology of X and reflecting the main geometrical features of X. *CS* may be two-voxel thick at the intersection of curves. A voxel p of the skeleton is an end point when it has only one neighboring skeleton voxel in $N(p)$. A voxel p of the skeleton is a branch point when it has more than two neighboring skeleton voxels in $N(p)$.

6.2.1 Distance Transform

The distance transform $DT(X)$ is a labeled replica of an object X, where each voxel p of X is labeled with the length of a shortest path, entirely consisting of voxels of X, connecting p to the boundary of X. In the following, we will denote by $d(p)$ the distance value of p in $DT(X)$.

$DT(X)$ can be interpreted as due to a local propagation process from the boundary of X towards the innermost part of X. During propagation, each internal voxel p of X receives distance information from those of its neighbors that are closer to the boundary than p, and propagates distance information to its object neighbors that are farther than p from the boundary.

The object boundary is composed by the object voxels delimiting the object itself. For an object consisting of three-dimensional manifolds, the boundary is a set union of closed surfaces that in the discrete space includes all and only the object's voxels with at least a face-neighbor in the background, while the internal voxels are the object's voxels having no face-neighbors in B. In turn, when X consists of two-dimensional manifolds, its boundary will be a set union of curves. To identify the boundary of X in this latter case, we need to classify the voxels of X. To this aim, we use a modified version of the classification scheme introduced in [14] and classify as follows the voxels of an object X consisting exclusively of intersecting surfaces and curves:

- An object voxel with at most two disjoint object neighbors is a *curve voxel.*
- A not yet classified object voxel with a neighboring curve voxel is a *branching voxel.*
- A not yet classified voxel p is a branching voxel if $c_p \neq 1$ and p has a neighboring branching voxel, or if the object neighbors of p are all branching voxels.
- An object voxel p with $c_p^* \neq 1$ is an *internal voxel.*
- An internal voxel p with more than two 6-connected components of background voxels in $N(p)$ face- or edge-adjacent to p, or being any of the eight internal voxels in a $2 \times 2 \times 2$ configuration is re-classified as *junction voxel.*
- An internal voxel having a face- or an edge-neighbor classified as junction voxel, is re-classified as *extended junction voxel.*
- Any other object voxel is an *edge voxel.*

The boundary of a set X consisting exclusively of intersecting surfaces and curves, is the set of edge voxels, branching voxels and curve voxels. Branching voxels and junction voxels are the voxels where curves and surfaces intersect and will act as barriers to the propagation of distance information. Extended junction voxels are internal voxels playing also the role of junction voxels in presence of intersecting surfaces, where voxels of a surface have edge-neighbors in another surface. Differently from the junction voxels, which will neither receive nor propagate distance information, the extended junction voxels are allowed to receive, but are not allowed to propagate, this information to their neighbors. Due to the presence of barriers, voxels internal in one of the surfaces constituting X will receive distance information exclusively from the edge voxels delimiting that surface.

The above classification allows us also to distinguish peripheral and internal surfaces. The first are surfaces at least partially delimited by edge voxels, the latter are completely surrounded by junction voxels.

For the computation of $DT(X)$, boundary voxels are assigned the minimal distance value, which is equal to 3 when using the $\langle 3, 4, 5\rangle$ weighted distance. Distance information is then propagated to the internal voxels of X. If X consists of three-dimensional manifolds, $DT(X)$ can be computed by means of the standard two-scan algorithm [41]. When X consists exclusively of surfaces and curves, an ordered propagation algorithm, e.g., [44], is more convenient due to the presence of junction voxels and branching voxels acting as barriers to the propagation.

For the sake of clarity, we point out that when X consists exclusively of surfaces and curves, more than one application of distance transformation may be necessary to obtain the complete $DT(X)$. In fact, the internal voxels belonging to internal surfaces cannot be reached by the distance information propagated from the boundary, due to the role of barriers played by the junction voxels and by the extended junction voxels. To assign distance values also to the internal voxels of internal surfaces, we change the status of some particular junction voxels into that of delimiting junction voxels. These particular voxels are the junction voxels having both neighboring voxels already reached by the propagated distance information, and neighboring extended junction voxels not yet labeled with distance value. The delimiting junction voxels are taken as constituting a new set of boundary voxels from which to compute the distance transform for internal surfaces. The extended junction voxels that are neighbors of delimiting junction voxels change their status into that of internal voxels. To guarantee some continuity in distance values, the delimiting junction voxels are assigned a value related to the maximum distance value, *max*, assigned during the previous application of distance transformation. Namely, delimiting junction voxels are set to $max + w_v$.

6.2.2 Centers of Maximal Balls and Anchor Points

Any voxel p in the distance transform can be interpreted as the center of a ball with radius $d(p)$. The ball associated to p can be constructed by applying to p the reverse distance transformation [45].

A center of maximal ball is any voxel p whose associated ball is included in the object, but is not included by any other single ball in the object. The set of centers of maximal balls, $CMB(X)$, is computed by comparing the distance value $d(p)$ of each p of X, with the distance values of the neighbors of p, by taking into account the weights w_f, w_e and w_v [46]. In detail, p is a center of maximal ball if for each of its neighbors n_i there results

$$d(n_i) - d(p) < w_i \tag{6.1}$$

We note that the distance value 3 must be replaced by the equivalent value 1 before detecting the centers of maximal balls. In fact, balls with radii 3 and 1 are identical and replacement of 3 by 1 is necessary to avoid the erroneous detection as center of maximal ball of a voxel whose associated ball is not maximal (see [1] Chap. 5, for details on the equivalent values).

We also note that when X exclusively consists of surfaces and curves, rather than talking of balls we should more correctly talk of geodesic discs. The geodesic disc associated with a voxel p of X can be obtained by considering the intersection between X and the ball associated to p by the reverse distance transformation. If the intersection consists of more than one connected component, the geodesic disc associated to p is the component that includes p.

In principle, at least during the first phase of skeletonization, we should take as anchor points all centers of maximal balls of X since they are symmetry points. However, we are aware that the curve skeleton of X could include almost all centers only when X is composed by parts with tubular shape. Moreover, if all centers are taken as anchor points, the structure of the set resulting at the end of the first

Fig. 6.1 From *left* to *right*, a 3D object X, the set of centers of maximal balls $CMB(X)$, and the set of relevant centers of maximal balls $RCMB(X)$

phase of skeletonization would be so complex to make the second phase of the process rather hard and heavy. Thus, we prefer to devise some criteria to filter out the less important centers of maximal balls. As described below, we use one of the criteria that we suggested in [28] to characterize the centers of maximal balls and will include in the curve skeleton only the relevant centers of maximal balls.

We note that even if some neighbors of p have distance values larger than $d(p)$, the voxel p can still satisfy the detection rule (6.1). Thus, we can use a stricter detection rule, where the difference $d(n_i) - d(p)$ is compared with a value smaller than the appropriate weight. Namely, we call relevant center of maximal ball any voxel p of X such that for each of its neighbors n_i there results

$$d(n_i) - d(p) < w_i - 1 \tag{6.2}$$

The detection rule (6.2) strongly reduces the number of centers of maximal balls with respect to those selected by rule (6.1). At the same time, the set obtained by applying the reverse distance transformation to the set of the relevant centers of maximal balls, $RCMB(X)$, does not significantly differ from the input object, showing that a skeleton based on the selection as anchor points of the relevant centers of maximal balls still satisfactorily represents the object. In the average, about 95.25 % of the input object X is recovered by applying the reverse distance transform to $RCMB(X)$.

A 3D object X, the set $CMB(X)$, and the set $RCMB(X)$ are shown in Fig. 6.1.

6.3 The Curve Skeleton

The curve skeleton CS of a 3D object X is obtained in two phases, the first of which is devoted to the computation of PSS, which is a subset of X consisting of surfaces and curves used as input to the second phase.

The first phase of skeletonization includes the following tasks: (1) computation of the $\langle 3,4,5 \rangle$ weighted distance transform of the object, $DT(X)$; (2) detection of the set of anchor points, i.e., of the set $RCMB(X)$ of the relevant centers of maximal balls; (3) application of topology preserving removal operations, based on the notion of simple point, to object voxels that are not anchor points. Object voxels are accessed in increasing distance value order. A currently inspected voxel p is removed, i.e., its value is set to the background value 0, if it is both $c_p = 1$ and $c_p^* = 1$; (4) final thinning applied to the obtained at most two-voxel thick set *PSS* so as to gain unit thickness; (5) pruning of peripheral curves, done for trimming out negligible details.

During the second phase, the following tasks are done: (1) the voxels of *PSS* are classified; (2) the distance transform $DT(PSS)$ is computed; (3) voxels of *PSS* that are classified as branching voxels or junction voxels and the relevant centers of maximal balls of *PSS* are taken as anchor points; (4) topology preserving removal operations are applied to the voxels of *PSS* that are not anchor points; (5) the two-voxel thick set *CS* is transformed into a linear set; (6) pruning of peripheral curves is finally accomplished.

Details for implementing most of the above tasks have already been provided in the previous sections. Thus, in the next section we give the details necessary for accomplishing the remaining tasks.

6.3.1 Final Thinning and Pruning

To perform final thinning and pruning during the first phase of skeletonization, we resort to the four processes P1–P4, respectively devoted to (1) Reduction of thickening in face-direction; (2) Reduction of thickening in edge-direction; (3) Removal of scarcely significant peripheral surfaces; (4) Removal of scarcely significant peripheral curves. The sequence P1, P2 and P3 is applied as far as voxel removal is possible. Then, P4 is accomplished.

To identify the voxels of *PSS* that undergo P1, P2 and P3, we partition the voxels of *PSS* in two classes: type1 voxels, including all voxels characterized by $c_p^* \neq 1$, and type2 voxels, including all other voxels of *PSS*. For sure, type1 voxels are not simple. Thus, only type2 voxels undergo P1, P2 and P3. Of course, not all type2 voxels can be removed. In fact, not all of them are simple and not all the simple type2 voxels are really placed where two-voxel thickening of *PSS* occurs.

To identify the type2 voxels in parts of *PSS* that are two-voxel thick in face- or edge-direction, we use directional templates. Each template consists of four voxels, aligned along one of the 6 face-directions or along one of the 12 edge-directions. The two central voxels of the template are requested to be type2 voxels, while the remaining two voxels are requested to be background voxels.

Each application of the process P1 (P2) actually consists in the application of 6 (12) directional processes. Type1 and type2 voxels are identified before each face-directional (edge-directional) process. Each directional process involves sequential removal in one of the 6 face-directions (12 edge-directions) and is concerned only with voxels where thickening in that face-direction (edge-direction) occurs.

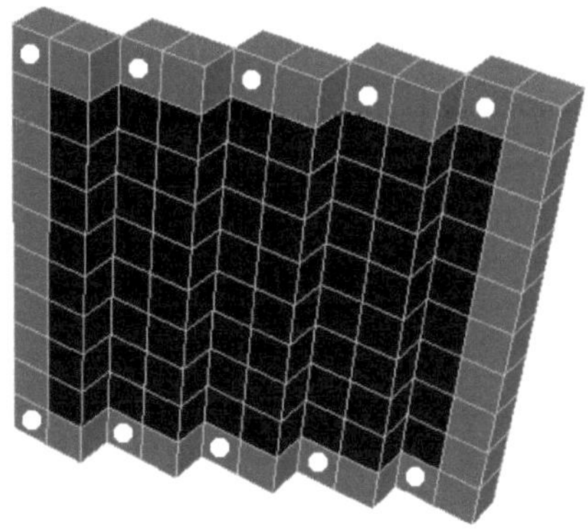

Fig. 6.2 Type1 and type2 voxels are shown in *black* and in *gray*, respectively. The same directional templatc is satisfied for all *dotted* voxels. However, sequential removal of the *dotted* voxels would cause undesirable zig-zags along the boundary of the object

As for the removal conditions, a voxel p, for which the current template occurs, is removed if the following three conditions are satisfied in $N(p)$:

C1. There exists exactly one 26-connected component of object voxels.
C2. There exists exactly one 26-connected component of type2 voxels.
C3. There exists more than one object voxel.

Note that C1 guarantees topology preservation. In fact, the template is satisfied for p, which means that p is a type2 voxel and, hence, is characterized by $c_p^* = 1$. Thus, once C1 is also satisfied for p, it results $c_p = 1$, i.e., p is simple. In turn, C2 and C3 are tailored to remove voxels in two-voxel thick parts of *PSS*, without affecting the parts of *PSS* that are already unit-wide. If C1 and C2 are not used, unwanted zig-zags along the boundary of unit-wide parts of *PSS* and shortening of curves of *PSS* would be created. C2 and C3 respectively prevent zig-zags and shortening. See Fig. 6.2, as far as the role of C2 is concerned.

Process P3 does not require the use of templates. Each application of P3 sequentially removes type2 voxels satisfying conditions C1 and C3. In this way, the size of all peripheral surfaces of *PSS* diminishes without altering topology and without shortening possibly existing curves, since only boundary voxels of peripheral surfaces are removed.

The process P4 is commonly known as pruning (e.g., [47, 48]) and is aimed at removing peripheral skeleton curves with limited significance. Our pruning involves two steps, during each of which a peripheral curve is either totally removed or remains untouched. To this aim, each peripheral curve is traced starting from its end point until a branch point is met. The total number L of voxels of the curve and the number N of anchor points along the curve are counted. The first pruning step removes peripheral curves for which $L \leq \theta_1$, where θ_1 is an a priori fixed threshold (the default value used in this paper is $\theta_1 = 4$). The second pruning step removes peripheral curves for which $N/L \leq \theta_2$, where θ_2 is a threshold to be fixed depending on problem domain (the default value used in this paper is $\theta_2 = 0.25$).

During the second phase of skeletonization, final thinning and pruning of *CS* are accomplished analogously to what has been done during the first phase. Obviously,

Fig. 6.3 A few objects, *top*, and their curve skeletons, *bottom*

P3 is not taken into account, since *CS* does not include surfaces. Moreover, some constraints in P1 and P2 can be relaxed: distinction between type1 and type2 voxels is no longer necessary; the middle voxels in the directional templates are simply object voxels; only removal conditions C1 and C3 are considered. P4 is done exactly as during the first phase and the same values are used for the two pruning thresholds ($\theta_1 = 4$ and $\theta_2 = 0.25$).

In Fig. 6.3, a few 3D objects and their curve skeletons are shown.

Actually, we have applied our skeletonization algorithm to a significantly larger number of 3D objects and have observed that in the average the curve skeleton consists of about 0.80 % of the input object. Thus, the curve skeleton is very efficient as concerns data reduction. In turn, if the reverse distance transformation is applied to the curve skeleton, where skeleton voxels are assigned the distance value pertaining their homologous voxels in the distance transform of the object, we note that the average object recovery from the curve skeleton is about 69 %. The limited recovery ability of the curve skeleton is clearly due to the fact that only a fraction of the centers of maximal balls of the object are included in the curve skeleton. This notwithstanding, the curve skeleton provides a stylized version of the object useful for a number of tasks, e.g., for object decomposition.

6.4 Object Decomposition

Though the curve skeleton, where skeleton voxels are assigned the distance value pertaining their homologous voxels in the distance transform of the object, has limited recovery ability, it can still be used to decompose the 3D object it represents.

To this aim, we first partition the curve skeleton into three types of subsets, that we call *simple curves*, *complex sets*, and *single points*. Then, we recover the object parts associated to the skeleton parts. The recovered object parts are respectively called *simple regions*, *bumps* and *kernels*. Kernels can be interpreted as a sort of main bodies of the object, from which simple regions and bumps protrude. Finally, we perform a merging process, aimed at reducing the number of object regions to the perceptually most relevant ones.

6.4.1 Skeleton Partition

To partition the curve skeleton, we apply the reverse distance transformation to the branch points. As a result, we obtain a number of regions, that we call *zones of influence*, which is at most equal to the number of branch points. Actually, the number of zones of influence is generally smaller than the number of branch points. In fact, the balls associated to branch points that are neighbors of each other or are closer to each other than the sum of the corresponding radii overlap and merge into a single connected component.

The simple curves of the partition are constituted by the connected components of skeletal voxels that are not included in any zone of influence. The single points of the partition, if any, are constituted by the end points of the skeleton that result to be included in zones of influence. When single points exist in the partition, the skeleton branches that can be traced from each of the corresponding end points until a branch point is encountered are called *linking branches*. Finally, the complex sets of the partition are constituted by the connected components of all the skeletal voxels included in the zones of influence, except for the end points.

Figure 6.4 shows the partition into simple curves, single points and complex sets for the skeletons of the test objects.

6.4.2 Simple Regions, Bumps and Kernels

The construction of the simple regions, bumps and kernels into which the object can be seen as decomposed requires four steps.

During the first step, the reverse distance transformation is applied to each individual skeleton partition component, by taking into account the following issues: (i) voxels that can be recovered from more than one individual partition component are assigned to the kernel they are closer to, so as to originate a decomposition into disjoint regions; (ii) linking branches do not take part in recovery, i.e., the reverse distance transformation is not applied to the linking branches. In fact, if a linking branch contributes, together with the other branches in a complex set, to the recovery of the corresponding kernel, the bump protruding from that kernel is characterized by a very small volume and the perceptual relevance of the bump is not enhanced.

Fig. 6.4 Skeleton partition into simple curves, *black*, single points, *red*, and complex sets, *light blue*

In turn, if the linking branch contributes to recovery of the bump, rather than to recovery of the kernel, the bump intrudes in the kernel.

Aim of the second step is to improve the planarity of the surface in between adjacent recovered regions. The desired effect is obtained by means of a local concavity filling process that assigns to bumps and simple regions the voxels placed in local concavities of the boundary of bumps and simple regions and belonging to adjacent kernels. Concavity filling is iterated as far as local concavities can be detected along the boundary of bumps and simple regions. Since the concavity filling method that we use, [49], employs $3 \times 3 \times 3$ operations but derives curvature information from a $5 \times 5 \times 5$ neighborhood, the separating surfaces obtained at the end of the process reasonably well approximate planar surfaces.

The third step guarantees that a one-to-one correspondence exists between skeleton partition components and object decomposition regions. In fact, the assignment to kernels of voxels at the same distance from more than one partition component and the re-assignment of boundary voxels during concavity filling may cause splitting of a region in sub-parts. When a simple region or a kernel results as split into sub-parts, we preserve the sub-part that includes at least one skeletal voxel, while assign to the background the other sub-parts. When a bump results to be split into sub-parts, we take as significant only the sub-part with the largest size and assign the other parts to the background. In fact, the criterion used for simple regions and kernels cannot be used also for bumps, since the end points responsible of recovery of the bumps are definitely not included in the bumps.

The fourth step completes the recovery of bumps, simple regions and kernels. To this purpose, the voxels of the input object that were not recovered when the reverse

Fig. 6.5 Object decomposition before merging

distance transformation was applied to the curve skeleton are assigned to the closest bumps, simple regions and kernels. Object voxels at the same distance from more than one object part are assigned to the part with which they have the largest number of neighbors.

In Fig. 6.5, the decompositions into simple regions, bumps and complex regions corresponding to the simple curves, single points and complex sets found in the skeletons of the test objects are shown.

6.4.3 Merging

Merging criteria to reduce the number of regions to the perceptually most relevant ones are of interest. Since we regard kernels as a sort of main bodies, we candidate the remaining regions, i.e., bumps and simple regions, to be merged to the kernels from which they protrude.

We distinguish simple regions into peripheral regions, i.e., those adjacent to one kernel only, and non-peripheral regions, i.e., those delimited by two kernels. Bumps are obviously all peripheral regions.

Let R_i be the region currently considered for merging with the adjacent kernel(s). The surface of R_i consists of the object voxels of R_i having at least one face-neighbor either in the background or in the adjacent kernel(s). The total area of the surface of R_i is given by the number of surface voxels of R_i. The area of the visible portion of the surface, $Av(R_i)$, is equal to the number of voxels of R_i that are face-adjacent to the background, while the area of the non-visible portion of the surface, $Anv(R_i)$, is given by the number of voxels of R_i that are face-adjacent to the kernel(s).

We regard a region R_i as scarcely visible if the ratio $Av(R_i)/Anv(R_i)$ is smaller than an a priori fixed threshold γ. The value of γ should be fixed based on the problem domain. To fix the default value of γ, we consider as scarcely visible a cylinder whose height, h, is smaller than the diameter, $2r$.

The first merging step is concerned only with non-peripheral scarcely visible simple regions. In this case, according to our model, we merge a region R_i for which it results $Av(R_i)/Anv(R_i) = 2\pi rh/2\pi r^2 < \gamma$. Since it is $h < 2r$, the default value for the threshold is $\gamma = 2$.

Fig. 6.6 Final decompositions

Let R_j be any such a region. The multiplicity degree $m(K_i)$ of each kernel K_i is computed as the number of non-peripheral scarcely visible simple regions adjacent to K_i.

If for both kernels K_i and K_h adjacent to R_j the multiplicity degree is equal to one, R_j is merged into a unique region with K_i and K_h.

If for R_j one of its delimiting kernels, K_i, is characterized by $m(K_i) = 1$, while for the other kernel K_h it is $m(K_h) > 1$, R_j is merged with K_i only. Once all possible merging cases involving kernels with multiplicity degree equal to one have been treated, merging with kernels with multiplicity degree larger than one is considered. Merging with a kernel K_h characterized by $m(K_h) > 1$ is done if all the non-peripheral scarcely visible regions adjacent to K_h have been actually merged to their other delimiting kernels. Merging involves K_h and the adjacent region with which K_h shares the largest portion of its surface.

The second merging step regards peripheral regions, which can be bumps or simple regions. In this case, according to our model, we merge a region R_i for which it results $Av(R_i)/Anv(R_i) = (2\pi rh + \pi r^2)/\pi r^2 < \gamma$. Since it is $h < 2r$, the default value for the threshold is $\gamma = 5$.

To merge a peripheral region R_i to the adjacent kernel K_i we require that, besides being scarcely visible, R_i is such that the ratio between the volume (measured as number of voxels) of the region union of R_i and K_i, and the volume of K_i is smaller than an a priori fixed threshold τ (set to 1.2 in this work). Using the visibility criterion also when dealing with peripheral regions is done to avoid merging elongated and narrow regions that satisfy the condition on volume.

The final results for the test patterns are shown in Fig. 6.6.

6.5 Discussion and Conclusion

We have applied our skeletonization and decomposition algorithms to a large number of 3D objects taken from publicly available shape repositories [37, 38]. We have generally obtained satisfactory results, some of which have been shown in this paper.

Any curve skeletonization method should be evaluated in terms of the properties expected for the curve skeleton. The curve skeleton should have lower

Fig. 6.7 Different poses of the object "horse", *top*, the skeletons before pruning, *middle*, and after pruning, *bottom*. For all poses the same threshold values have been used during pruning

dimensionality, should be placed symmetrically within the object, should be characterized by the same topology as the object, and should reflect the geometrical features of the object itself.

For our curve skeleton, topology is certainly preserved, since only simple voxels are sequentially removed. The skeleton is symmetrically placed within the object, since the selected anchor points are all symmetrically placed within the 3D object. Dimensionality reduction is also satisfied, since the curve skeleton is a set union of curves. Object's geometry is also reflected by the curve skeleton, since a correspondence exists between skeleton curves and parts intuitively perceived as composing the object.

Since we have used the $\langle 3,4,5\rangle$ weighted distance, which is a good approximation of the Euclidean distance, our skeletonization algorithm is almost stable under object rotation. As for the sensitivity to noise, we point out that skeletonization should be applied to reasonably clean objects. In fact, skeletonization preserves topology, and the geometrical structure of the skeleton is conditioned by the geometry of the object's boundary. When noise consists essentially in noisy protrusions of the object, pruning improves the robustness of skeletonization.

Pruning is particularly important when the curve skeleton is used for image analysis tasks, e.g., for object decomposition. In fact, whichever skeletonization algorithm is used, the skeleton of an object in different poses or scales is not guaranteed to have in all cases exactly the same structure before pruning. Thus, a different partition of the skeleton and, hence, a different decomposition of the object could be obtained. As an example, refer to Fig. 6.7 where different poses of the object "horse" are shown together with the skeletons before and after pruning. Note that these skeletons have been computed by using the same values for the two pruning thresholds θ_1 and θ_2 for all poses. However, to obtain skeletons having the

Fig. 6.8 Decomposition for different poses of "horse" before merging, *top*, and after merging, *bottom*

same structure after pruning a value for θ_1 larger than that suggested in this work as default value has been used. Namely, for the various poses of "horse" we have used the thresholds $\theta_1 = 6$ and $\theta_2 = 0.25$.

We point out that our partition of the skeleton has a key role in obtaining a decomposition of the object where the surfaces separating adjacent object parts cut the object in correspondence with negative minima of curvature along the boundary of the object. In particular, complex sets are crucial to obtain a perceptually significant decomposition.

For the different poses of "horse", the decompositions obtained before and after merging are shown in Fig. 6.8. The default values have been used for the thresholds γ and τ.

Both curve skeletonization and object decomposition algorithms are easy to implement, have limited computational cost and produce results in agreement with human intuition.

References

1. Siddiqi, K., Pizer, S.M. (eds.): Medial Representations: Mathematics, Algorithms and Applications. Springer, Berlin (2008)
2. Arcelli, C., Sanniti di Baja, G.: A width-independent fast thinning algorithm. IEEE Trans. PAMI **7**, 463–474 (1985)
3. Lam, L., Lee, S.W., Suen, C.Y.: Thinning methodologies-a comprehensive survey. IEEE Trans. PAMI **14**(9), 869–885 (1992)
4. Arcelli, C., Sanniti di Baja, G.: Euclidean skeleton via center-of-maximal-disc extraction. Image Vis. Comput. **11**, 163–173 (1993)
5. Kimmel, R., Shaked, D., Kiryati, N., Bruckstein, A.: Skeletonization via distance maps and level sets. CVIU **62**(3), 382–391 (1995)
6. Sanniti di Baja, G., Thiel, E.: Skeletonization algorithm running on path-based distance maps. Image Vis. Comput. **14**, 47–57 (1996)
7. Chuang, J., Tsai, C., Ko, M.-C.: Skeletonization of three-dimensional object using generalized potential field. IEEE Trans. PAMI **22**(11), 1241–1251 (2000)

8. Svensson, S., Nyström, I., Sanniti di Baja, G.: Curve skeletonization of surface-like objects in 3D images guided by voxel classification. Pattern Recognit. Lett. **23**(12), 1419–1426 (2002)
9. Xie, W., Thompson, R.P., Perucchio, R.: A topology-preserving parallel 3D thinning algorithm for extracting the curve skeleton. Pattern Recognit. **36**(7), 1529–1544 (2003)
10. Bouix, S., Siddiqi, K., Tannenbaum, A.: Flux driven automatic centerline extraction. Med. Image Anal. **9**, 209–221 (2005)
11. Lohou, Ch., Bertrand, G.: Two symmetrical thinning algorithms for 3D binary images, based on P-simple points. Pattern Recognit. **40**(8), 2301–2314 (2007)
12. Couprie, M., Coeurjolly, D., Zrour, R.: Discrete bisector function and Euclidean skeleton in 2D and 3D. Image Vis. Comput. **25**(10), 1543–1556 (2007)
13. Ju, T., Baker, M.L., Chiu, W.: Computing a family of skeletons of volumetric models for shape description. Comput.-Aided Des. **39**(5), 352–360 (2007)
14. Arcelli, C., Sanniti di Baja, G., Serino, L.: Distance driven skeletonization in voxel images. IEEE Trans. PAMI **33**(4), 709–720 (2011)
15. Thibault, D., Gold, C.M.: Terrain reconstruction from contours by skeleton generation. Geo Inf. **4**(4), 349–373 (2000)
16. He, T., Hong, L., Chen, D., Liang, Z.: Reliable path for virtual endoscopy: ensuring complete examination of human organs. IEEE Trans. Vis. Comput. Graph. **7**(4), 333–342 (2001)
17. Gagvani, N., Silver, D.: Animating volumetric models. Academic Press Professional. Graph. Models **63**(6), 443–458 (2001)
18. Wade, L., Parent, R.E.: Automated generation of control skeletons for use in animation. Vis. Comput. **18**(2), 97–110 (2002)
19. Fridman, Y., Pizer, S.M., Aylward, S., Bullitt, E.: Extracting branching tubular object geometry via cores. Med. Image Anal. **8**(3), 169–176 (2004)
20. Chaturvedi, A., Lee, Z.: Three-dimensional segmentation and skeletonization to build an airway tree data structure for small animals. Phys. Med. Biol. **50**(7), 1405–1419 (2005)
21. Sadleir, R.J.T., Whelan, P.F.: Fast colon centerline calculation using optimized 3D topological thinning. Comput. Med. Imaging Graph. **29**(4), 251–258 (2005)
22. Cornea, N.D., Silver, D.: Curve-skeleton properties, applications, and algorithms. IEEE Trans. Vis. Comput. Graph. **13**(3), 530–548 (2007)
23. Verwer, B.J.H., Van Vliet, L.J., Verbeek, P.W.: Binary and grey-value skeletons: metrics and algorithms. IJPRAI **7**(5), 1287–1308 (1993)
24. Arcelli, C., Sanniti di Baja, G.: A thinning algorithm based on prominence detection. Pattern Recognit. **13**, 225–235 (1981)
25. Svensson, S., Borgefors, G., Nystrom, I.: On reversible skeletonization using anchor-points from distance transforms. J. Vis. Commun. Image Represent. **10**, 379–397 (1999)
26. Blum, H.: A transformation for extracting new descriptors of shape. In: Wathen-Dunn, W. (ed.) Models for the Perception of Speech and Visual Form, pp. 362–380. MIT, Cambridge, MA (1967)
27. Blum, H., Nagel, R.: Shape description using weighted symmetric axis features. Pattern Recognit. **10**(3), 167–180 (1978)
28. Serino, L., Arcelli, C., Sanniti di Baja, G.: A characterization of centers of maximal balls in the $\langle 3,4,5 \rangle$ weighted distance transform of 3D digital objects. In: Proceeding of the Workshop on Applications of Discrete Geometry and Mathematical Morphology, pp. 12–16 (2010)
29. Sanniti di Baja, G., Thiel, E.: (3,4)-weighted skeleton decomposition for pattern representation and description. Pattern Recognit. **27**, 1039–1049 (1994)
30. Cornea, N.D., Silver, D., Yuan, X., Balasubramanian, R.: Computing hierarchical curve-skeletons of 3D objects. Vis. Comput. **21**(11), 945–955 (2005)
31. Lien, J.-M., Geyser, J., Amato, N.M.: Simultaneous shape decomposition and skeletonization. In: Proceedings of the 2006 ACM Symposium on Solid and Physical Modeling, pp. 219–228. ACM, New York (2006)
32. Serino, L., Sanniti di Baja, G., Arcelli, C.: Object decomposition via curvilinear skeleton partition. In: Proceedings of ICPR 2010, Istanbul, pp. 4081–4084. IEEE Computer Society, Washington, DC (2010)

33. Marr, D., Nishihara, H.K.: Representation and recognition of three-dimensional shapes. Proc. R. Soc. Lond. B **200**, 269–294 (1978)
34. Hoffman, D.D., Richards, W.A.: Parts of recognition. Cognition **18**, 65–96 (1984)
35. Biederman, I.: Recognition-by-components: a theory of human image understanding. Psychol. Rev. **94**, 115–147 (1987)
36. Singh, M., Hoffman, D.D.: Part-based representations of visual shape and implications for visual cognition. In: Shipley, T., Kellman, P. (eds.) From Fragments to Objects: Segmentation and Grouping in Vision, pp. 401–459. Elsevier, New York (2001)
37. AIM@SHAPE Shape Repository. http://shapes.aimatshape.net/viewmodels.php
38. Shilane, P., Min, P., Kazhdan, M., Funkhouser, T.: The Princeton Shape Benchmark. Shape Modeling International, Genova (2004)
39. Saha, P.K., Chaudhuri, B.B.: Detection of 3D simple points for topology preserving transformations with application to thinning. IEEE Trans. PAMI **16**(10), 1028–1032 (1994)
40. Bertrand, G., Malandain, G.: A new characterization of three-dimensional simple points. Pattern Recognit. Lett. **15**(2), 169–175 (1994)
41. Borgefors, G.: On digital distance transforms in three dimensions. CVIU **64**(3), 368–376 (1996)
42. Breu, H., Gil, J., Kirkpatrick, D., Werman, M.: Linear time Euclidean distance algorithms. IEEE Trans. PAMI **17**(5), 529–533 (1995)
43. Yamashita, M., Ibaraki, T.: Distances defined by neighborhood sequences. Pattern Recognit. **19**(3), 237–246 (1986)
44. Piper, J., Granum, E.: Computing distance transformations in convex and non-convex domains. Pattern Recognit. **20**(6), 599–615 (1987)
45. Nyström, I., Borgefors, G.: Synthesising Objects and Scenes Using the Reverse Distance Transformation in 2D and 3D. Lecture Notes in Computer Science, vol. 974, pp. 441–446. Springer, Berlin (1995)
46. Svensson, S., Sanniti di Baja, G.: Using distance transforms to decompose 3D discrete objects. Image Vis. Comput. **20**, 529–540 (2002)
47. Shaked, D., Bruckstein, A.M.: Pruning medial axes. CVIU **69**(2), 156–169 (1998)
48. Svensson, S., Sanniti di Baja, G.: Simplifying curve skeletons in volume images. CVIU **90**(3), 242–257 (2003)
49. Borgefors, G., Sanniti di Baja, G.: Analyzing non-convex 2D and 3D patterns. CVIU **63**(1), 145–157 (1996)

Chapter 7
Orientation and Anisotropy of Multi-component Shapes

Joviša Žunić and Paul L. Rosin

Abstract There are many situations in which several single objects are better considered as components of a multi-component shape (e.g. a shoal of fish), but there are also situations in which a single object is better segmented into natural components and considered as a multi-component shape (e.g. decomposition of cellular materials onto the corresponding cells). Interestingly, not much research has been done on multi-component shapes. Very recently, the orientation and anisotropy problems were considered and some solutions have been offered. Both problems have straightforward applications in different areas of research which are based on a use of image based technologies, from medicine to astrophysics.

The object orientation problem is a recurrent problem in image processing and computer vision. It is usually an initial step or a part of data pre-processing, implying that an unsuitable solution could lead to a large cumulative error at the end of the vision system's pipeline. An enormous amount of work has been done to develop different methods for a spectrum of applications. We review the new idea for the orientation of multi-component shapes, and also its relation to some of the methods for determining the orientation of single-component shapes. We also show how the anisotropy measure of multi-component shapes, as a quantity which indicates how consistently the shape components are oriented, can be obtained as a by-product of the approach used.

J. Žunić (✉)
Computer Science, University of Exeter, Exeter EX4 4QF, UK

Mathematical Institute Serbian Academy of Sciences and Arts, Belgrade, Serbia
e-mail: J.Zunic@ex.ac.uk

P.L. Rosin
School of Computer Science and Informatics, Cardiff University, Cardiff, CF24 3AA, UK
e-mail: Paul.Rosin@cs.cardiff.ac.uk

M. Breuß et al. (eds.), *Innovations for Shape Analysis*, Mathematics and Visualization, DOI 10.1007/978-3-642-34141-0_7, © Springer-Verlag Berlin Heidelberg 2013

7.1 Introduction

Shape is one of the object characteristics which enables many numerical characterizations suitable for computer supported manipulations. Because of that, different shape concepts are intensively used in object recognition, object identification or object classification tasks. Many approaches to analyse and characterise shapes have been developed. Some of them are very generic, like moment invariants or Fourier descriptors, while others relate to specific object characteristics, e.g. descriptors like convexity, compactness, etc. Another distinction among these approaches is based on which points of shapes are used for analysis. Some approaches use all shape points (area-based ones), other use boundary information only (boundary-based ones), but there are methods which use only specific shape points (convex hull vertices or boundary corners) or hybrid methods (shape compactness is often computed from the relation between shape perimeter and shape area).

In the most research and applications to date, shapes are treated as single objects, even if very often several objects form a group (vehicles on the road, group of people, etc.), and thus, it could be beneficial to consider them as multi-component shapes.

It is not difficult to imagine situations in which it is better to decompose a single shape into its naturally defined components and then, by treating it as a multi-component shape, take some advantage over those methods that treat it as a single shape. There are many more situations where a multi-component shape approach may be appropriate – e.g., when analysing video sequences, multiple instances of objects over time can be grouped together and analysed as a multi-component shape. Several situations where it can be useful to treat objects as multi-component ones are displayed in Fig. 7.1.

Here we consider the multi-component shape orientation problem. We overview recently introduced methods (area-based and boundary-based) for the computation of orientation of multi-component shapes and its relation to the most standard shape orientation method (based on the computation of the axis of the least second moment of inertia). As a by-product of the new method for the computation of orientation of compound shapes, an anisotropy measure of such shapes can be derived. This is a first shape measure defined for multi-component shapes and it indicates the consistency of a set of shape component orientations.

All discussions are illustrated with suitably selected experiments. Strict proofs are omitted and for them the readers are referred to the source references.

7.2 Shape Orientation

Computation of shape orientation is a common problem which appears in a large number of applications in both $2D$ and $3D$, and also in higher-dimensional spaces. Due to the variety of shapes as well as to the diversity of applications, there is

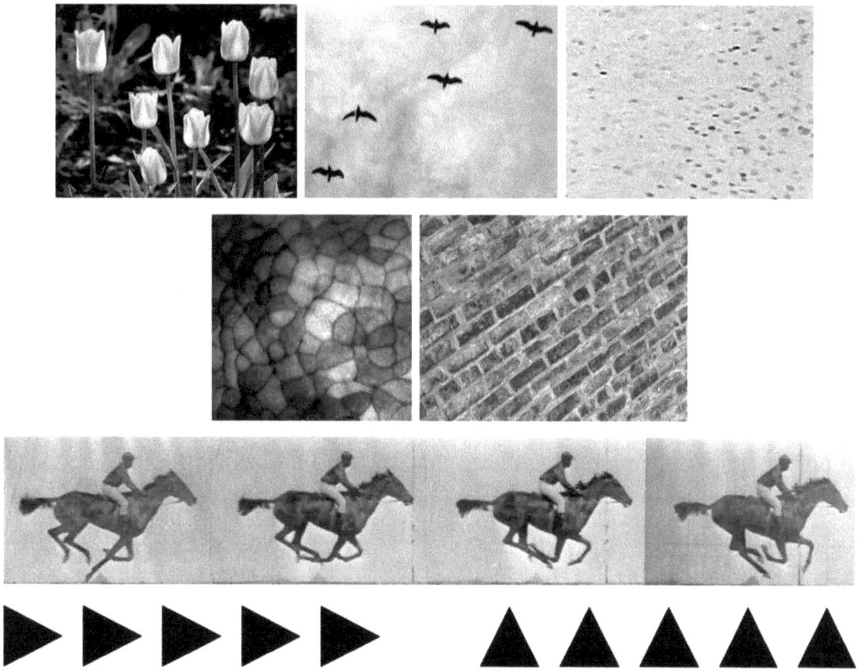

Fig. 7.1 *First row*: a group of static objects (*flowers*), a group of moving objects (*birds*), a group of different objects (*blood cell*) make multi-component objects. *Second row*: the tissue and texture displayed are sometimes better to be decomposed and analysed as multi-component objects. *Third row*: the appearance of a moving object, in a frame sequence, can be considered and analysed as a multi-component object. *Fourth row*: different arrangements of a simple object make two multi-component objects with perceptually different orientations [14]

no single method for the computation of shape orientation which outperforms the others in all situations. Therefore, many methods have been developed, and different techniques have been used, including those based on complex moments [20], Zernike moments [7], Fourier analysis [1], algebraic arguments [10], etc. The suitability of those methods strongly depends on the particular situation to which they are applied, as they each have their relative strengths and weaknesses. Due to new applications and increasing demands for better computational efficiency, in addition to the previously established methods, there are also many recent examples [3,5,11,12,16,24,27].

Several methods are defined by using a common approach: Consider a suitably chosen function $F(S,\alpha)$ which depends on a given shape S and rotation angle α, and define the orientation of S by the angle which optimizes $F(S,\alpha)$, i.e. by the angle α for which $F(S,\alpha)$ reaches its minimum or maximum. The most standard method for the computation of the shape orientation is such a method. More precisely, this method defines the orientation $O_{st}(S)$ of a given shape S by the, so called, axis of the least second moment of inertia, i.e. by the line which minimises

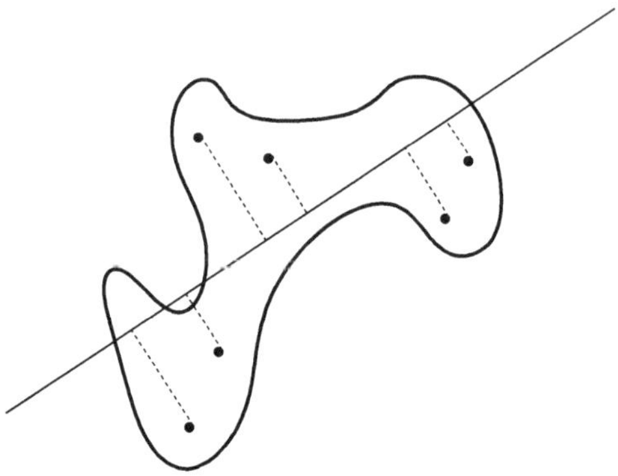

Fig. 7.2 The standard method defines the orientation of a given shape by the line which minimizes the integral of squared distances of the shape points to the line

the integral of the squared distances of the shape points to the line (see Fig. 7.2). Simple algebraic manipulation shows that such a line passes through the shape centroid. Note that the centroid of a given shape S is defined as

$$(x_S, y_S) = \left(\frac{\iint_S x dx dy}{\iint_S dx dy} \frac{\iint_S y dx dy}{\iint_S dx dy} \right). \tag{7.1}$$

So, in order to compute the orientation of a given shape S it is sufficient to find the minimum of the function

$$F_{st}(S, \alpha) = \iint_S r(x, y, \alpha)^2 dx dy \tag{7.2}$$

where $r(x, y, \alpha)^2$ is the perpendicular distance of the point (x, y) to the line which passes thought the centroid (x_S, y_S) of S and has a slope α. If we assume that the centroid of S coincides with the origin, i.e., $(x_S, y_S) = (0, 0)$, $r(x, y, \alpha)^2$ becomes $(x \cdot \sin \alpha - y \cdot \cos \alpha)^2$, and the optimizing function $F_{st}(S, \alpha)$ in (7.2) can be expressed as

$$\begin{aligned} F_{st}(S, \alpha) &= \iint_S r(x, y, \alpha)^2 dx dy \\ &= \mu_{2,0}(S) \cdot \sin^2 \alpha + \mu_{0,2}(S) \cdot \cos^2 \alpha - \mu_{1,1}(S) \cdot \sin(2\alpha), \end{aligned} \tag{7.3}$$

where $\mu_{p,q}(S)$ are the well-known centralised geometric moments [21] defined, for all $p, q \in \{0, 1, 2, \ldots\}$, as

$$\mu_{p,q}(S) = \iint_S (x - x_S)^p (y - y_S)^q dx dy. \tag{7.4}$$

Now, we come to the following definition of the orientation of a given shape.

Definition 7.1. The orientation of a given shape S is determined by the angle α where the function $F_{st}(S,\alpha)$ reaches its minimum.

This standard method defines the shape orientation in a natural way – by the line which minimizes the integral of the squared distances of shape points to this line. Such a definition matches our perception of what the shape orientation should be. Also, there is a simple formula for the computation of such orientation. It is easy to check [21] that the angle which minimizes $F_{st}(S,\alpha)$ satisfies the equation

$$\tan(2\cdot O(S)) = \frac{2\cdot\mu_{1,1}(S)}{\mu_{2,0}(S)-\mu_{0,2}(S)}. \tag{7.5}$$

These are desirable properties, but there are some drawbacks too. The main problem is that there are many situations where the method fails [23, 28] or does not perform satisfactorily. The situations were the method fails are easy to characterise. Indeed, considering the first derivative of $F_{st}(S,\alpha)$ (see (7.3))

$$\frac{dF_{st}(S,\alpha)}{d\alpha} = (\mu_{2,0}(S)-\mu_{0,2}(S))\cdot\sin(2\alpha) - 2\mu_{1,1}(S)\cdot\cos(2\alpha), \tag{7.6}$$

and looking for the conditions when $dF_{st}(S,\alpha)/d\alpha$ vanishes, it is easy to see that for all shapes S satisfying

$$\mu_{2,0}(S)-\mu_{0,2}(S) = 0 \qquad \text{and} \qquad \mu_{1,1}(S) = 0 \tag{7.7}$$

the function $F_{st}(S,\alpha)$ is constant and consequently does not tell which angle should be selected as the shape's orientation. N-fold rotationally symmetric shapes are shapes which satisfy (7.7) but there are many other (irregular) shapes which satisfy (7.7) and consequently could not be oriented by the standard method given by Definition 7.1.

In order to overcome such problems, [23] suggested a modification of the optimizing function $F_{st}(S,\alpha)$ by increasing the exponent in (7.2). The method from [23] defines the orientation of a given shape S whose centroid coincides with the origin, by the angle which minimizes

$$F_N(S,\alpha) = \iint_S r(x,y,\alpha)^{2N}\,dx\,dy \tag{7.8}$$

for a certain exponent $2N$. In such a way, the class of shapes whose orientation can be computed is expanded. On the other hand, there is not a closed formula (analogous to (7.5)) for the computation of the shape orientation by using $F_N(S,\alpha)$, for an arbitrary N.

Notice that difficulties in the computation of the shape orientation can be caused by the nature of certain shapes. While for many shapes their orientations are intuitively clear and can be computed relatively easily, the orientation of some other shapes may be ambiguous or ill defined. Problems related to the estimation of the degree to which a shape has a distinct orientation are considered in [29].

7.3 Orientation of Multi-component Shapes

As discussed before, there are many methods for the computation of the orientation of single-component shapes. On the other hand, as mentioned earlier, in many situations, several single objects usually appear as a group (e.g. the shoal of fish in Fig. 7.4, flock of birds in Fig. 7.1, vehicles on the road, etc.). Also, in many situations, it is suitable to consider a single object as a multi-component one, consisting of suitably defined components (as cells in embryonic tissue displayed in Fig. 7.1, or material micro structure elements, etc.). In addition, the appearances of the same object at different times can be also considered as components of a multi-component shape. Some examples where treating objects as multi-component shapes becomes very natural are in Fig. 7.1 and also in the forthcoming experiments.

In this section we consider the method for the computation of the orientation of multi-component shapes introduced by Žunić and Rosin [26]. Before that, note that most of the existing methods for the computation of the orientation of single component shapes do not have a (at least straightforward) extension which can be used to compute the orientation of compound shapes. The main reason for this is that most of the existing methods have a 180° ambiguity about the computed orientation. That is because they define the shape orientation by a line, not by a vector. Thus, the orientations of φ degrees and the orientation of $\varphi + 180°$ are considered to be the same. A consequence of such an ambiguity is that natural ideas how to compute the orientation of a multi-component shape from the orientations assigned to its components, do not work. For example, if $S_1, S_2, \ldots, S_n$ are components of a multi-component shape S, then most of the existing methods would compute their orientations as $\varphi_1 + a_1 \cdot 180°, \varphi_2 + a_2 \cdot 180°, \ldots, \varphi_n + a_n \cdot 180°$, where the numbers $a_1, a_2, \ldots, a_n$ are arbitrarily chosen from $\{0, 1\}$. Thus if, in the simplest variant, the orientation of multi-component shape $S = S_1 \cup S_2 \cup \ldots \cup S_n$ is computed as the average value of the orientations assigned to its components, then the orientation of S would be computed as

$$\frac{(\varphi_1 + a_1 \cdot 180°) + \ldots + (\varphi_n + a_n \cdot 180°)}{n} = \frac{\varphi_1 + \ldots + \varphi_n}{n} + \frac{(a_1 + \ldots + a_n) \cdot 180°}{n}$$

and obviously, for different choices of $a_1, a_2, \ldots, a_n$, the computed orientations are inconsistent (i.e. they could differ for an arbitrary multiple of the fraction $180°/n$). This is obviously unacceptable.

Now we consider the method for the computation of multi-component shapes described in [26]. The authors define the orientation of a multi-component shape by considering the integrals of the squared length of projections of all edges whose end points belong to a certain component. Before a formal definition, let us introduce the necessary denotations (see Fig. 7.3 for an illustration):

- Let $\overrightarrow{a} = (\cos\alpha, \sin\alpha)$ be the unit vector in the direction α;
- Let $|\mathbf{pr}_{\overrightarrow{a}}[AB]|$ denote the length of the projection of the straight line segment $[AB]$ onto a line having the slope α.

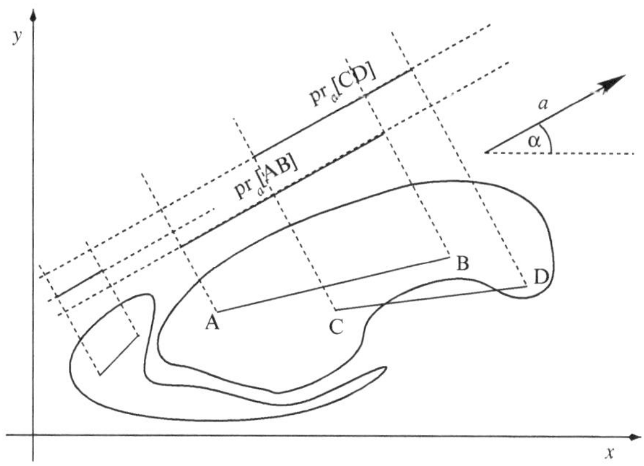

Fig. 7.3 The orientation of multi-component shapes is defined by the direction α which minimizes the integral of squared projections of line segments whose end points belong to a certain component

Definition 7.2. Let S be a multi-component shape which consists of m disjoint shapes $S_1, S_2, \ldots, S_m$. Then the orientation of S is defined by the angle that maximises the function $G_{comp}(S,\alpha)$ defined by

$$G_{comp}(S,\alpha) = \sum_{i=1}^{m} \iiiint\limits_{\substack{A=(x,y)\in S_i \\ B=(u,v)\in S_i}} |\mathbf{pr}_{\vec{a}}[AB]|^2 dx\, dy\, du\, dv. \tag{7.9}$$

The previous definition is naturally motivated, but also enables a closed formula for the computation of the orientation of multi-component shapes. This is the statement of the following theorem.

Theorem 7.1. *The angle α where the function $G_{comp}(S,\alpha)$ reaches its maximum satisfies the following equation*

$$\frac{\sin(2\alpha)}{\cos(2\alpha)} = \frac{2\cdot\sum\limits_{i=1}^{m} \mu_{1,1}(S_i)\cdot\mu_{0,0}(S_i)}{\sum\limits_{i=1}^{m} (\mu_{2,0}(S_i) - \mu_{0,2}(S_i))\cdot\mu_{0,0}(S_i)}. \tag{7.10}$$

To prove the theorem it is sufficient to enter the following two trivial equalities

$$|\mathbf{pr}_{\vec{a}}[AB]|^2 = ((x-u)\cdot\cos\alpha + (y-v)\cdot\sin\alpha)^2, \text{ for } A=(x,y), B=(u,v) \tag{7.11}$$

and

$$\iiiint\limits_{S\times S} x^p y^q u^r v^t\, dx\, dy\, du\, dv = \mu_{p,q}(S)\cdot\mu_{r,t}(S) \tag{7.12}$$

into the optimizing function $G_{comp}(S,\alpha)$. After that, simple calculus applied to the equation $\frac{dG_{comp}(S,\alpha)}{d\alpha} = 0$ establishes the proof. For more details we refer to [26].

The orientation of multi-component shapes computed by optimizing the function $G_{comp}(S,\alpha)$ is theoretically well founded and because of that it can be well understood. Some of properties are a direct consequence of the definition and can be proved by using basic calculus. We list some of them.

Property 7.1. If the method given by Definition 7.2 is applied to a single component shape, then the computed orientation is the same as the orientation computed by the standard definition, i.e. by the optimizing the function $F_{st}(S,\alpha)$ (see (7.2) and (7.5)). Note that the optimizing function $G(S,\alpha) = G_{comp}(S,\alpha)$, specified for single component shapes, and the optimizing function $F_{st}(S,\alpha)$ are different, but they are connected with the following, easily provable, equality

$$G(S,\alpha) + 2\cdot\mu_{0,0}(S)\cdot F_{st}(S,\alpha) = 2\cdot\mu_{0,0}(S)\cdot(\mu_{2,0}(S)+\mu_{0,2}(S)). \quad (7.13)$$

Since the right-hand side of (7.13) does not depend on α we deduce that the maximum of $G(S,\alpha)$ and the minimum of $F_{st}(S,\alpha)$ are reached at the same point. In other words, the direction α which defines the orientation of S by applying the standard method is the same as the direction which defines the orientation of S if the Definition 7.2 is applied to single component shapes.

Property 7.2. As it is expected, there are situations where the method given by Definition 7.2 fails. Due to the definition of the optimizing function $G_{comp}(S,\alpha)$, a simple characterization of such situations is possible. Indeed, by using (7.11) we deduce:

$$G_{comp}(S,\alpha) = \cos^2\alpha\cdot\sum_{i=1}^{m} 2\mu_{0,0}(S_i)\mu_{2,0}(S_i) + \sin^2\alpha\cdot\sum_{i=1}^{m} 2\mu_{0,0}(S_i)\mu_{0,2}(S_i)$$

$$+\sin(2\alpha)\cdot\sum_{i=1}^{m} 2\mu_{0,0}(S_i)\mu_{1,1}(S_i). \quad (7.14)$$

The last equality says immediately that the first derivative $\frac{dG_{comp}(S,\alpha)}{d\alpha}$ is identically equal to zero (i.e. $G_{comp}(S,\alpha)$ is constant) if and only if the following two conditions are satisfied

$$\sum_{i=1}^{m} \mu_{0,0}(S_i)\cdot\mu_{1,1}(S_i) = 0 \quad \text{and} \quad \sum_{i=1}^{m} \mu_{0,0}(S_i)\cdot(\mu_{2,0}(S_i)-\mu_{0,2}(S_i)) = 0. \quad (7.15)$$

So, under the conditions in (7.15) the optimizing function is constant and no direction can be selected as the shape orientation.

The Eq. (7.14) also says that the components S_i of a multi-component shape $S = S_1 \cup \ldots \cup S_m$ which satisfy

$$\mu_{1,1}(S_i) = 0 \quad \text{and} \quad \mu_{2,0}(S_i) - \mu_{0,2}(S_i) = 0,$$

(i.e. the shapes are not orientable by the standard method) do not contribute to the $G_{comp}(S,\alpha)$ and because of that, such components S_i, can be omitted when computing the orientation of S.

Notice that in case of $G_{comp}(S,\alpha) = constant$ we can increase the exponent in (7.9) and define the orientation of S by the direction which maximizes the following function

$$G_{comp,N}(S,\alpha) = \sum_{i=1}^{m} \iiiint\limits_{\substack{A=(x,y)\in S_i \\ B=(u,v)\in S_i}} |\mathbf{pr}_{\vec{a}}[AB]|^{2N} dx\, dy\, du\, dv.$$

In this way the class of multi-component shapes whose orientation is well defined would be extended. A drawback is that there is no closed formula (similar to (7.10)) which enables easy computation of such a defined multi-component shape orientation.

The next property seems to be a reasonable requirement for all methods for the computation of the orientation of multi-component shapes.

Property 7.3. If all components S_i of $S = S_1 \cup \ldots \cup S_m$ have an identical orientation α, then the orientation of S is also α.

To prove the above statement it is sufficient to notice that if all components S_i have the same orientation (see Property 7.1), then there would exist an angle α_0 such that all summands

$$\iiiint\limits_{\substack{A=(x,y)\in S_i \\ B=(u,v)\in S_i}} |\mathbf{pr}_{\vec{a}}[AB]|^{2} dx\, dy\, du\, dv. \tag{7.16}$$

in (7.9) reach their maximum for the angle $\alpha = \alpha_0$. An easy consequence is that $\alpha = \alpha_0$ optimizes $G_{comp}(S,\alpha) = \sum_{i=1}^{m} \iiiint\limits_{\substack{A=(x,y)\in S_i \\ B=(u,v)\in S_i}} |\mathbf{pr}_{\vec{a}}[AB]|^{2} dx\, dy\, du\, dv$, as well. This means that the orientation of S coincides with the orientation of its components S_i.

Property 7.4. The method established by Definition 7.2 is very flexible, in that the influence of the shape component's size (to the computed orientation) can vary. In the initial form in Definition 7.2 the moments $\mu_{1,1}(S_i)$, $\mu_{2,0}(S_i)$ and $\mu_{0,2}(S_i)$ are multiplied with the size/area of S_i, i.e. by $\mu_{0,0}(S_i)$. But the method allows $\mu_{0,0}(S_i)$ to be replaced with $\mu_{0,0}(S_i)^T$ for some suitable choice of T. In this way the influence of the shape components to the computed orientation can be controlled. The choice $T = -2$ is of a particular importance. In this case the size/area of the shape components does not have any influence to the computed orientation. This

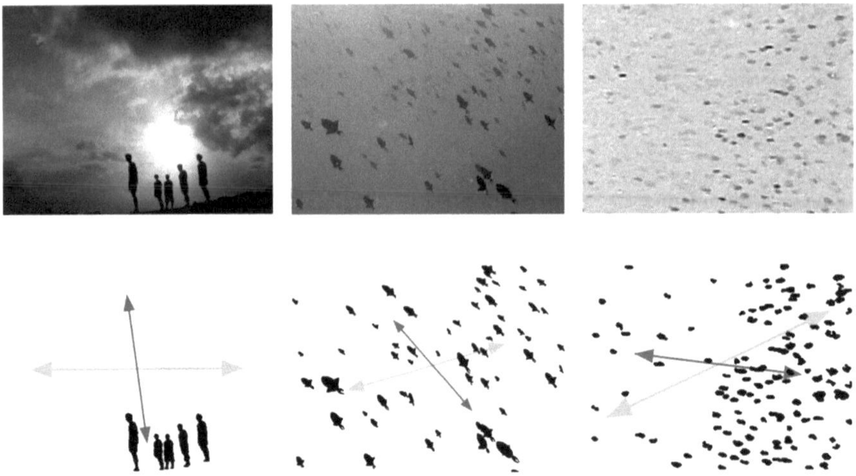

Fig. 7.4 Real images are in the first row. After thresholding, the image components are treated as components of a multi-component shape and then oriented by the method given by Definition 7.2 – the orientations computed are presented by *short dark arrows*. *Long grey arrows* represent the orientations computed by the standard method where all components are taken together to build a single shape

is very suitable since objects, which are of the same size in reality, often appear in images at varying scales since their size depends on their relative position with respect to the camera used to capture the image.

7.3.1 Experiments

This subsection includes several experiments which illustrate how the method for the computation of the orientation of multi-component shapes works in practice. Since this is the first method for the computation of the orientation of such shapes, there are no suitable methods for comparison. In the experiments presented, together with the orientations computed by the new method, the orientations computed by the standard method, which treats all the multi-component objects as a single object, are also displayed. However, orientations computed by the standard method are displayed just for illustrative purposes, not for qualitative comparison against the new method.

In the first example in Fig. 7.4 three images are displayed. In all three cases the objects that appear (humans, fish and blood cell components) are treated as components of a multi-component shape (a group of people, a shoal, a blood sample) and are then oriented. In the case of the group of people and the fish shoal the computed orientations (by the new method) are in accordance with our perception. As expected the computation by the standard method (i.e. treating the

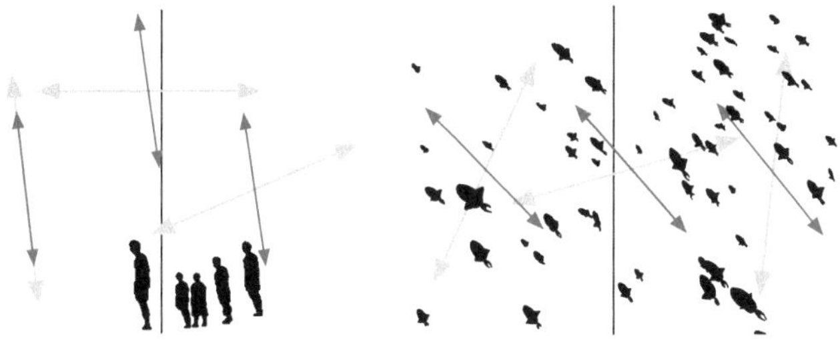

Fig. 7.5 The orientation computed by the method from Definition 7.2 for halves and the whole shape illustrate the "Frame independence" property of the new method for the orientation of multi-component shapes

multi-component shapes as single ones) does not lead to orientations which match our perception. The same could be said for the third example (blood cell) even though the our perception of what the orientation should be is not as strong as in the first two examples.

The next figure illustrates a very nice and useful property of the new method. It illustrates that for multi-component shapes whose components are relatively consistently oriented, the computed orientation of a subset of such shapes coincides with the orientation of the whole shape. Somehow it could be said that the computed orientations do not depend much on the frame used to capture a portion of the multi-component shape considered. The humans and fish shoal images (in Fig. 7.5) are split onto two halves. The halves are treated as multi-component shapes and oriented by the new method. The computed orientations are shown by the short dark arrows. As it can be seen such computed orientations are consistent – i.e. the orientations of the halves coincide with the orientation of the whole multi-component shape. As expected, the orientations computed by the standard method (long grey arrows) do not coincide.

The third example in this subsection illustrates a possible application of the method to the orientation of textures. Wood textures, displayed in the first row in Fig. 7.6, are not multi-component objects with clearly defined components. Nevertheless, after suitable thresholding the components become apparent, and the orientation of such obtained multi-component shapes can be computed. The results are in the third row and it could be said (in the absence of ground truth) that the orientations obtained are in accordance with our perception.

The last example in this subsection is somewhat different from the previous ones. In this case, a gait sequence is taken from NLPR Gait Database [25] and each appearance of a human silhouette in the sequence of the frames is considered as a component of the multi-component shape analysed. So, in this case the shape components are distributed temporally across the sequence (not spatially over the image, as in the previous examples). After segmentation many errors, which typically appear, have been removed using standard morphological techniques. However, several large errors remain and the task was to detect them. Due to the nature of the shapes it is expected that all components (i.e. silhouettes) are

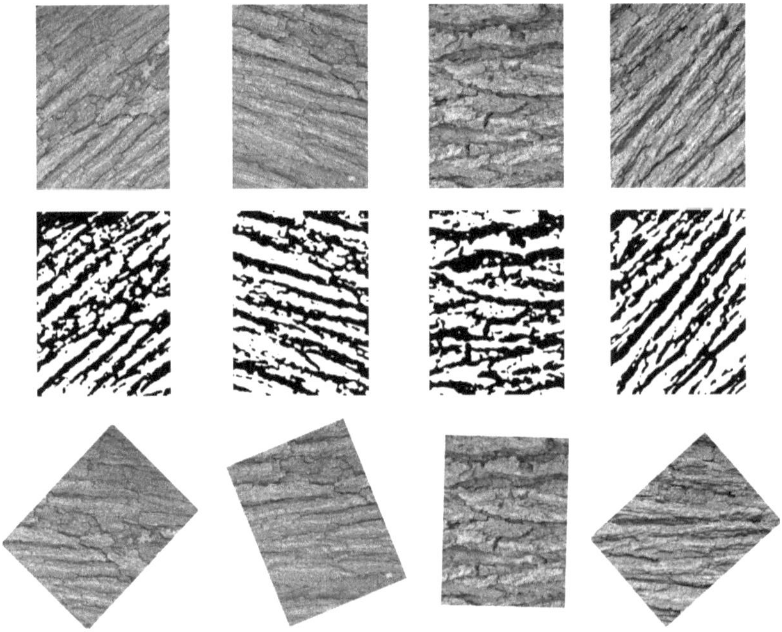

Fig. 7.6 Texture images, in the *first row*, are thresholded and their orientation is then computed as the orientation of multi-component shapes which correspond to the black-and-white images in the *second row*. The images in the *third row* are re-oriented in accordance with the computed orientations

Fig. 7.7 The extracted silhouettes from a gait sequence are displayed (the *first row*) and underneath is an intensity coding of each silhouette to show its degree of being an outlier (*dark* means high likelihood). A magnified view of the most outlying silhouette and its neighbours is in the *third row*

fairly consistently oriented if they are extracted properly. Thus, we make the hypothesis that the silhouettes with orientations inconsistent with the majority of silhouettes suffer from segmentation errors. The difference in multi-component shape orientation caused by removing the least consistent component has been used as a criterion to find possible outliers. In the example given, due to the errors in the processing chain that produced the sequence of binary images, the person's leading leg has been displaced and this silhouette/frame has been detected as an outlier, as shown in Fig. 7.7.

7.4 Boundary-Based Orientation

So far we have considered area-based methods only, i.e. the methods which use all the shape points for the computation of (in this particular case) shape orientation. But boundary-based methods for the computation of shape orientation are also considered in the literature. Apart from the fact that some area-based methods for the computation of the shape orientation have a straightforward extension to their boundary-based analogues, there are some boundary-based methods which cannot be derived in such a way – some examples are in [12, 27].

The methods considered in the previous section have an easy extension to boundary-based methods. For example, an analogue to the standard method for the computation of shape orientation is the method which orients a given shape by the line which minimizes the integral of the squared distance of the boundary points to this line. Since only the boundary points are used for the computation, this method is considered as a boundary-based one. The line which minimizes the optimizing integral can be obtained by following the same formalism as in the case of the standard method, but the appearing area integrals should be replaced by line integrals.

So, first we have to place S such that its boundary-based centroid coincides with the origin. The boundary-based centroid $(x_{\partial S}, y_{\partial S})$ is defined as the average of the boundary points. This means that

$$(x_{\partial S}, y_{\partial S}) = \left(\frac{\int_{\partial S} x(s)ds}{\int_{\partial S} ds} \frac{\int_{\partial S} y(s)ds}{\int_{\partial S} ds} \right). \qquad (7.17)$$

where the boundary ∂S of S is given in an arc-length parametrization: $x = x(s)$, $y = y(s)$, $s \in [0, perimeter\_of\_S]$. Note that such a choice of the boundary representation is suitable because it preserves rotational invariance – i.e. if the shape is rotated for a certain angle, then the computed orientation is changed by the same angle. In the rest of this chapter an arc-length parametrization of the appearing boundaries/curves will be always assumed, even not mentioned.

Next, in order to compute the boundary-based orientation of a given shape S, we have to find the minimum of the function

$$L_{bo}(\partial S, \alpha) = \int_{\partial S} r(x, y, \alpha)^2 ds \qquad (7.18)$$

where $r(x, y, \alpha)$ is, as in (7.2), the orthogonal distance of the point (x, y) to the line passing through the origin and having the slope α.

The optimizing function $L_{bo}(\partial S, \alpha)$ can be expressed as

$$L_{bo}(\partial S, \alpha) = \nu_{2,0}(\partial S) \cdot \sin^2 \alpha + \nu_{0,2}(\partial S) \cdot \cos^2 \alpha - \nu_{1,1}(\partial S) \cdot \sin(2\alpha), \qquad (7.19)$$

where $\nu_{p,q}(\partial S)$ are the normalised line moments [2] defined as

$$\nu_{p,q}(\partial S) = \int_{\partial S} (x - x_{\partial S})^p (y - y_{\partial S})^q ds, \qquad (7.20)$$

for all $p, q \in \{0, 1, 2, \ldots\}$.

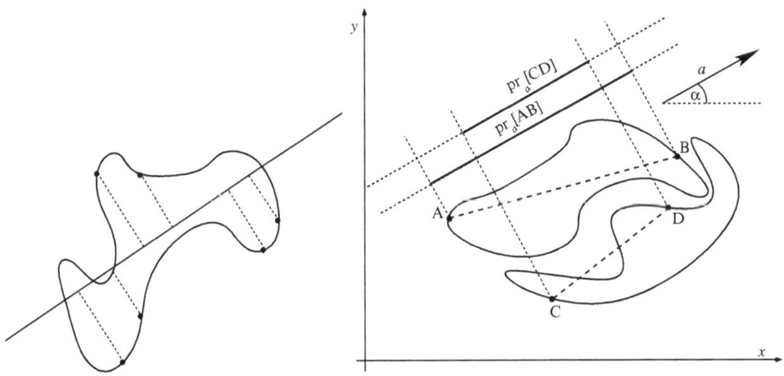

Fig. 7.8 On the *left*: The standard method defines the orientation of the shape by the line which minimizes the integral of squared distances of the shape boundary points to the line. On the *right*: the boundary-based method for computation of the orientation of multi-component shapes considers the projections of edges whose end points belong to the boundary of a certain component

Finally, the maximum of the optimizing function $L_{bo}(\partial S, \alpha)$ is reached for $\alpha = O_{bo}(\partial S)$ which satisfies the following equation:

$$\tan(2 \cdot O_{bo}(\partial S)) = \frac{2 \cdot \nu_{1,1}(\partial S)}{\nu_{2,0}(\partial S) - \nu_{0,2}(\partial S)}. \tag{7.21}$$

The equality (7.21) is an obvious analogue for the equality (7.5) related to the standard area-based method. This is as expected because the same formalism is used in both area-based and boundary-based cases.

The idea used in Sect. 7.3 to define the orientation of multi-component shapes also has a boundary-based analogue. The problem is studied in detail in [15]. Following the idea from [26], the authors consider the projections of the edges whose end points belong to the boundary of a certain component of a given multi-component shape and define the shape orientation by the line which maximises the integral of the squared values of the projections of such edges to the line. The formal definition follows.

Definition 7.3. Let $S = S_1 \cup \ldots \cup S_m$ be a multi-component shape and let the boundary of S be the union of the boundaries ∂S_i of the components of S: $\partial S = \partial S_1 \cup \ldots \cup \partial S_m$. The orientation of S (Fig. 7.8) is defined by the angle that maximises the function $L_{comp}(\partial S, \alpha)$ defined as follows

$$L_{comp}(\partial S, \alpha) = \sum_{i=1}^{m} \iint_{\substack{s \in [0, per(S_i)] \\ l \in [0, per(S_i)]}} |\mathbf{pr}_{\vec{a}}[A_i(s)B_i(l)]|^2 ds\, dl. \tag{7.22}$$

Definition 7.3 enables easy computation of the angle which defines the orientation of ∂S, as given by the following theorem.

Theorem 7.2. *Let S be a multi-component shape whose boundary is $\partial S = \partial S_1 \cup \ldots \cup \partial S_m$. The angle α where the function $L_{comp}(\partial S, \alpha)$ reaches its maximum satisfies the following equation*

$$\tan(2\alpha) = \frac{2 \cdot \sum_{i=1}^{m} \nu_{1,1}(\partial S_i) \cdot \nu_{0,0}(\partial S_i)}{\sum_{i=1}^{m} (\nu_{2,0}(\partial S_i) - \nu_{0,2}(\partial S_i)) \cdot \mu_{0,0}(\partial S_i)}. \tag{7.23}$$

Being derived in an analogous way as the area-based method for the computation of the multi-component shapes, the method given by Definition 7.3 also satisfies the Properties 7.1–7.4 listed in Sect. 7.3, before the experimental section. It is worth mentioning that the method also enables the control of the influence of the perimeter of the shape components to the computed orientation. If, for some applications, such an influence should be ignored then the moments $\nu_{1,1}(\partial S_i)$, $\nu_{2,0}(\partial S_i)$, and $\nu_{0,2}(\partial S_i)$ which appear in (7.23) should be multiplied by $(\nu_{0,0}(\partial S_i))^{-2}$.

To close this section, let us mention that (in general) boundary-based approaches to define the orientation of shapes allow some extra generalizations. One such generalisation was considered in [12]. Therein the shape orientation is computed based on the projections of the tangent vectors at the shape boundary points, weighted by the suitably chosen function of the boundary curvature at the corresponding points. The advantage of such an approach is not only that the boundary curvature, as an important shape feature, is involved in the computation of the shape orientation. This also provides an easy approach to overcome situations were the orientations are not computable. It is sufficient to modify the curvature based weighting function, and shapes which were not "orientable" by an initial choice of the weighting function can become orientable with another choice. The computation of the shape orientation remains possible by a closed form formula whether the weighting function is changed or not. For more details see [12].

7.4.1 Experiments

In this section we have two examples to illustrate how the boundary-based method (established by Definition 7.3) works.

First we consider the embryonic tissue displayed in Fig. 7.9. Cell boundaries are extracted and then the boundary-based method for the computation of the orientation of multi-component shapes is applied. The orientations are computed for the whole image and also separately for the upper and lower parts. The computed orientations are consistent (presented by short dark arrows) which actually implies that the tissue displayed has an inherent consistent orientation. The orientations computed by the boundary-based analogue (by optimizing $L_{bo}(\partial S, \alpha)$ from (7.18)), are different (long grey arrows), as it was expected.

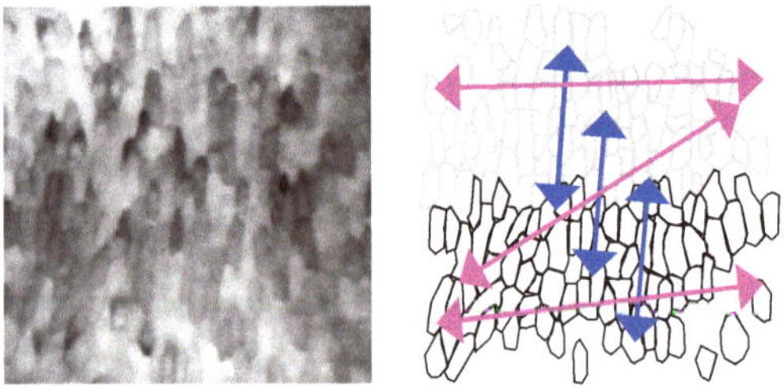

Fig. 7.9 Boundaries of cells of an embryonic tissue (on the *left*) are extracted, and then split onto an "upper" and "lower" part (on the *right*). Orientations computed by optimizing $L_{comp}(\partial S, \alpha)$ all coincide (*short dark arrows*). The orientations by the analogue of the standard method are shown by *long grey arrows*

Fig. 7.10 Figure: Signatures of subject s048 from Munich and Perona [13]. Signatures in the *top row* are re-oriented according to the standard method while the *bottom row* displays the same signatures re-oriented according to the multi-component method

In the second example the boundary-based method for multi-component shapes (Definition 7.3) has been used to compute the orientations of signatures. Better results were obtained than when the orientations were computed by the standard boundary-based method. Figure 7.10 displays signatures of subject s048 from Munich and Perona [13]. In the first row the signatures are oriented by applying the standard boundary-based method and the problems are obvious. Signatures are not oriented consistently, which is a problem because the similarity measure used in [13] was not rotationally invariant. In the second row are the same signatures but oriented by the boundary-based multi-component method. Prior to the computation of the orientation, the signatures were segmented at the vertices with small subtended angles. The obtained orientations are very consistent, as required.

7.5 Anisotropy of Multi-component Shapes

In all the methods presented here, shape orientation is computed by optimizing a suitably chosen function which depends on the orientation angle and the shape considered. Depending on the method selected, either the angle which maximizes the optimizing function or the angle which defines the minimum of the optimizing function is selected as the shape orientation. To have a reliable method for orienting the shape it is important to have distinct minima and maxima of the optimizing function. This is because in image processing and computer vision tasks we deal with discrete data and very often in the presence of noise. Thus, if the optima of the optimizing function are not distinct significant values then the computed orientations could arise due to noise or digitization errors, rather than from inherent properties of the shapes. Note that for a fixed shape the difference of maxima and minima of the optimizing function could dependent strongly on the method applied. That is why most of the methods only suit certain applications well. The question of whether a shape possess an inherent orientation or not is considered in [29]. The authors have introduced a *shape orientability measure* as a quantity which should indicate to which degree a shape has a distinct orientation.

The ratio $\mathscr{E}_{st}(S)$ of the maxima and minima of the optimizing function $F_{st}(S,\alpha)$, in the case of the standard method, is well studied and widely used in shape based image analysis. The quantity

$$\mathscr{E}_{st}(S) = \frac{\mu_{2,0}(S) + \mu_{0,2}(S) + \sqrt{4\cdot(\mu_{1,1}(S))^2 + (\mu_{2,0}(S) - \mu_{0,2}(S))^2}}{\mu_{2,0}(S) + \mu_{0,2}(S) - \sqrt{4\cdot(\mu_{1,1}(S))^2 + (\mu_{2,0}(S) - \mu_{0,2}(S))^2}}. \quad (7.24)$$

is well-known as the shape elongation measure. $\mathscr{E}_{st}(S)$ ranges over $[1,\infty)$ and takes the value 1 if S is a circle. A problem is that there are many other shapes whose elongation equals 1. In addition, $\mathscr{E}_{st}(S)$ is invariant with respect to translation, rotation and scaling transformations and is easily and accurately computable from the object images [8,9].

The elongation measure $\mathscr{E}_{st}(S)$ has its boundary-based analogue – the area moments in (7.24) have to be replaced with the corresponding line integrals along the shape boundaries. Another elongation measure is suggested by Stojmenović and Žunić [22].

Another related property is shape anisotropy. It has a natural meaning for single component shapes. For example, for a shape centred at the origin in both $2D$ and $3D$, it could be inversely related to the degree to which the shape points are equally distributed in all directions [18, 19]. It has been used as a useful feature in shape (object) classification tasks but also as a property which highly correlates with some mechanical characteristics of certain real objects and materials [4, 6]. It is also of interest when analyzing tracks of different species of animals [17].

The anisotropy measure of multi-component shapes has not been considered before, but it seems that it should be given different meaning than in the case of single component shapes. Our understanding is that an anisotropy measure for the multi-component shapes should indicate how consistently the shape's components are oriented. It has turned out that a quantity defined as the ratio between the maxima and minima of the function $L_{comp}(\partial S,\alpha)$ from Definition 7.3 provides such a measure. So, we give the following definition.

Definition 7.4. Let $S_1, \ldots, S_m$ be components of a compound shape S. Then the anisotropy $\mathscr{A}(S)$ of S is defined as

$$\mathscr{A}(S) = \frac{\max_{\alpha\in[0,2\pi)} L_{comp}(\partial S,\alpha)}{\min_{\alpha\in[0,2\pi)} L_{comp}(\partial S,\alpha)} = \frac{\max_{\alpha\in[0,2\pi)} L_{comp}(\partial S_1 \cup \ldots \cup S_m,\alpha)}{\min_{\alpha\in[0,2\pi)} L_{comp}(\partial S_1 \cup \ldots \cup S_m,\alpha)}. \tag{7.25}$$

The anisotropy measure $\mathscr{A}(S)$ for multi-component shapes ranges over $[1,\infty)$ and is invariant with respect to translations, rotations and scaling transformations. It also enables an explicit formula for its computation. By basic calculus it can be shown that the maxima and minima of $L_{comp}(\partial S,\alpha)$ are:

$$\max_{\alpha\in[0,2\pi)} L_{comp}(\partial S,\alpha) = C + \sqrt{A^2+B^2}, \qquad \min_{\alpha\in[0,2\pi)} L_{comp}(\partial S,\alpha) = C - \sqrt{A^2+B^2}$$

where the quantities A, B, and C are

$$A = \sum_{i=1}^{m} (\nu_{2,0}(\partial S_i) - \mu_{0,2}(\partial S_i)) \cdot \nu_{0,0}(\partial S_i),$$
$$B = \sum_{i=1}^{m} 2 \cdot \nu_{1,1}(\partial S_i) \cdot \nu_{0,0}(\partial S_i),$$
$$C = \sum_{i=1}^{m} (\nu_{2,0}(\partial S_i) + \nu_{0,2}(\partial S_i)) \cdot \nu_{0,0}(\partial S_i).$$

We illustrate how the anisotropy measure $\mathscr{A}(S)$ acts by two examples. Notice that the anisotropy measure, as defined here, also depends on the elongation of the shape components, not only on their orientations. This also seems acceptable, e.g. a stereotype for a multi-component shape with a high anisotropy $\mathscr{A}(S)$ is a shape whose components have high elongations and the same orientation.

The first example is in Fig. 7.11. The shapes in both rows are treated as a multiple component object (e.g. a herd of cattle and a group of cars). The anisotropy was first computed for just the cattle, giving a value of 3.49. The anisotropy of the cattle's shadows alone increases to 7.57 since the shadows are more consistently orientated, and are also slightly more elongated. Merging the cattle and their shadows produces even more elongated regions, which lead to an increase of the herd's anisotropy to 12.08. The anisotropy of the cars (in the second row) is 1.38, which is very small. This is to be expected since the orientations of the individual cars vary strongly.

The second example which indicates how the anisotropy measure $\mathscr{A}(S)$ acts is in Fig. 7.12, in which anisotropy is used to select appropriate elongated regions to enable skew correction of the document. The first image (on the left) in the first row is the original image. Its components are letters whose orientations vary strongly, but also many of the letters have a low elongation. This results in a very low anisotropy of this image, as it can be seen from the graph given in the second row.

After blurring is applied to the image the characters start to merge into words, which are both more consistently oriented and more elongated. This leads to an

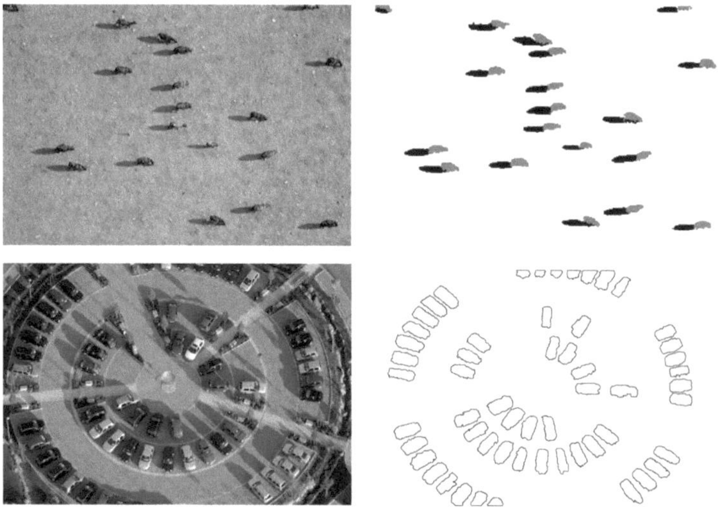

Fig. 7.11 Object boundaries are extracted from the original images and considered as components of the boundary of a multi-component shape. The highest anisotropy of 12.08 is measured if the cattle and their shadows are merged and considered as individual components of multi-component shapes. A low anisotropy measure of 1.38 is computed for the compound shape in the *second row*

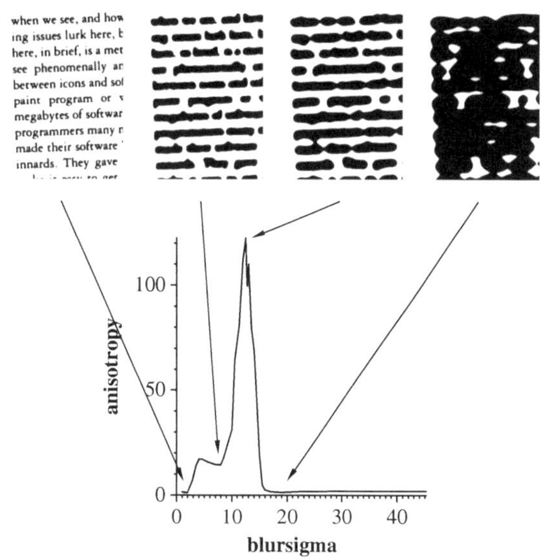

Fig. 7.12 The image on the *left* is blurred and thresholded, and the resulting component anisotropy is plotted. Three of the thresholded images are shown demonstrating that maximum anisotropy is achieved when many of the words are merged into lines

increase of the anisotropy (see the second image in the first row). If enough blurring is applied to merge characters/words into continuous lines the anisotropy increases dramatically (see the third image in the first row).

More blurring is counter productive to the task of skew correction, as sections of adjacent lines merge, and their anisotropy quickly drops (the last image in the first row).

7.6 Conclusion

Multi-component shapes have not been considered much in literature. This is somewhat surprising since there are many situations in which objects act as a part of a very compact group, or where single objects need to be decomposed onto components for analysis. There are also some less obvious situations were the multi-component approach can be useful.

We have focused on two problems related to multi-component shapes: computing orientation and anisotropy. Both problems have only recently been considered [15, 26] and some solution were offered. These problems do not have analogues in the existing single-component shape based methods. Thus, new ideas have to be developed.

The obtained results are promising, and have been justified with a number of experiments. The extension to the other shape based techniques will be investigated in the future. Due to the variety of ways that a multi-component shape can be defined, there are plenty of different demands which have to be satisfied by the methods developed. Moreover, there are some specific demands which do not exist when dealing with single-component shapes. To mention just two of them, which have been discussed in this chapter: the frame independence property and the tunable influence of the component size to the method performance.

References

1. Chandra, D.V.S.: Target orientation estimation using Fourier energy spectrum. IEEE Trans. Aerosp. Electron. Syst. **34**, 1009–1012 (1998)
2. Chen, C.-C.: Improved moment invariants for shape discrimination. Pattern Recognit. **26**, 683–6833 (1993)
3. Cortadellas, J., Amat, J., de la Torre, F.: Robust normalization of silhouettes for recognition application. Pattern Recognit. Lett. **25**, 591–601 (2004)
4. Ganesh, V.V., Chawla, N.: Effect of particle orientation anisotropy on the tensile behavior of metal matrix composites: experiments and microstructure-based simulation. Mat. Sci. Eng. A-Struct. **391**, 342–353 (2005)
5. Ha, V.H.S., Moura, J.M.F.: Afine-permutation invariance of 2-D shape. IEEE Trans. Image Process. **14**, 1687–1700 (2005)
6. Iles, P.J., Brodland, G.W., Clausi, D.A., Puddister, S.M.: Estimation of cellular fabric in embryonic epithelia. Comput. Method Biomech. **10**, 75–84 (2007)
7. Kim, W.-Y., Kim, Y.-S.: Robust rotation angle estimator. IEEE Trans. Pattern Anal. **21**, 768–773 (1999)

8. Klette, R., Žunić, J.: Digital approximation of moments of convex regions. Graph. Model Image Process. **61**, 274–298 (1999)
9. Klette, R., Žunić, J.: Multigrid convergence of calculated features in image analysis. J. Math. Imaging Vis. **13**, 173–191 (2000)
10. Lin, J.-C.: The family of universal axes. Pattern Recognit. **29**, 477–485 (1996)
11. Martinez-Ortiz, C., Žunić, J.: Points position weighted shape orientation. Electron. Lett. **44**, 623–624 (2008)
12. Martinez-Ortiz, C., Žunić, J.: Curvature weighted gradient based shape orientation. Pattern Recognit. **43**, 3035–3041 (2010)
13. Munich M.E., Perona. P.: Visual identification by signature tracking. IEEE Trans. Pattern Anal. **25**, 200–217 (2003)
14. Palmer, S.E.: What makes triangles point: local and global effects in configurations of ambiguous triangles. Cognit. Psychol. **12**, 285–305 (1980)
15. Rosin, P.L., Žunić, J.: Orientation and anisotropy of multi component shapes from boundary information. Pattern Recognit. **44**, 2147–2160 (2011)
16. Rosin, P.L., Žunić, J.: Measuring squareness and orientation of shapes. J. Math. Imaging Vis. **39**, 13–27 (2011)
17. Russell, J., Hasler, N., Klette, R., Rosenhahn, B.: Automatic track recognition of footprints for identifying cryptic species. Ecology **90**, 2007–2013 (2009)
18. Saha, P.K., Wehrli, F.W.: A robust method for measuring trabecular bone orientation anisotropy at in vivo resolution using tensor scale. Pattern Recognit. **37**, 1935–1944 (2004)
19. Scharcanski, J., Dodson, C.T.J.: Stochastic texture image estimators for local spatial anisotropy and its variability. IEEE Trans. Instrum. Meas. **49**, 971–979 (2000)
20. Shen, D., Ip, H.H.S.: Generalized affine invariant normalization. IEEE Trans. Pattern Anal. **19**, 431–440 (1997)
21. Sonka, M., Hlavac, V., Boyle, R.: Image Processing, Analysis, and Machine Vision. Thomson-Engineering, Toronto (2007)
22. Stojmenović, M., Žunić, J.: Measuring elongation from shape boundary. J. Math. Imaging Vis. **30**, 73–85 (2008)
23. Tsai, W.H., Chou, S.L.: Detection of generalized principal axes in rotationally symmetric shapes. Pattern Recognit. **24**, 95–104 (1991)
24. Tzimiropoulos, G., Mitianoudis, N., Stathaki, T.: A unifying approach to moment based shape orientation and symmetry classification. IEEE Trans. Image Process. **18**, 125–139 (2009)
25. Wang, L., Tan, T.N., Hu, W.M., Ning, H.Z.: Automatic gait recognition based on statistical shape analysis. IEEE Trans. Image Process. **12**, 1120–1131 (2003)
26. Žunić, J, Rosin, P.L.: An alternative approach to computing shape orientation with an application to compound shapes. Int. J. Comput. Vis. **81**, 138–154 (2009)
27. Žunić, J., Stojmenović, M.: Boundary based shape orientation. Pattern Recognit. **41**, 1785–1798 (2008)
28. Žunić, J., Kopanja, L., Fieldsend, J.E.: Notes on shape orientation where the standard method does not work. Pattern Recognit. **39**, 856–865 (2006)
29. Žunić, J., Rosin, P.L., Kopanja, L.: On the orientability of shapes. IEEE Trans. Image Process. **15**, 3478–3487 (2006)

Part II
Partial Differential Equations for Shape Analysis

Chapter 8
Stable Semi-local Features for Non-rigid Shapes

Roee Litman, Alexander M. Bronstein, and Michael M. Bronstein

Abstract Feature-based analysis is becoming a very popular approach for geometric shape analysis. Following the success of this approach in image analysis, there is a growing interest in finding analogous methods in the 3D world. Maximally stable component detection is a low computation cost and high repeatability method for feature detection in images.In this study, a diffusion-geometry based framework for stable component detection is presented, which can be used for geometric feature detection in deformable shapes.

The vast majority of studies of deformable 3D shapes models them as the two-dimensional boundary of the volume of the shape. Recent works have shown that a volumetric shape model is advantageous in numerous ways as it better captures the natural behavior of non-rigid deformations. We show that our framework easily adapts to this volumetric approach, and even demonstrates superior performance.

A quantitative evaluation of our methods on the SHREC'10 and SHREC'11 feature detection benchmarks as well as qualitative tests on the SCAPE dataset show its potential as a source of high-quality features. Examples demonstrating the drawbacks of surface stable components and the advantage of their volumetric counterparts are also presented.

R. Litman (✉) · A.M. Bronstein
School of Electrical Engineering, Tel Aviv University, Tel Aviv, Israel
e-mail: roeelitm@post.tau.ac.il; alex@eng.tau.ac.il

M.M. Bronstein
Institute of Computational Science, Faculty of Informatics, Universita della Svizzera Italiana, Lugano, Switzerland
e-mail: michael.bronstein@usi.ch

M. Breuß et al. (eds.), *Innovations for Shape Analysis*, Mathematics and Visualization, DOI 10.1007/978-3-642-34141-0_8,

8.1 Introduction

Following their success in image analysis, many feature-based methods have found their way into the world of 3D shape analysis. Feature descriptors play a major role in many applications of shape analysis, such as assembling fractured models [16] in computational archeology, or finding shape correspondence [36].

Some shape feature-works are inspired by and follow methods in image analysis, for example the histogram of intrinsic gradients used in [41] is similar in principle to the scale invariant feature transform (SIFT) [21] which has recently become extremely popular in image analysis. The concept of "bags of features" [32] was introduced as a way to construct global shape descriptors that can be efficiently used for large-scale shape retrieval [27, 37].

Other features were developed natively for 3D, as they rely on properties like the shape normal field, as in the popular spin image [17]. Another example is a family of methods based on the heat kernel [4, 35], describing the local heat propagation properties on a shape. These methods are deformation-invariant due to the fact that heat diffusion geometry is intrinsic, thus making descriptors based on it applicable in deformable shape analysis.

8.1.1 Related Work

The focus of this work is on another class of feature detection methods, one that finds stable components (or regions) in the analyzed image or shape. The origins of this approach are also in the image processing literature, this time in the form of the watershed transform [9, 39].

Matas et al. [22] introduced the stable component detection concept to the computer vision and image analysis community in the form of the maximally stable extremal regions (MSER) algorithm. This approach represents intensity level sets as a component tree and attempts to find level sets with the smallest area variation across intensity. The use of area ratio as the stability criterion makes this approach affine-invariant, which is an important property in image analysis, as it approximates viewpoint transformations. This algorithm can be made very efficient [25] in certain settings, and was shown to have superior performance in a benchmark done by Mikolajczyk et al. [24]. A deeper inspection of the notion of region stability was done by Kimmel et al. [18], also proposing an alternative stability criteria. The MSER algorithm was also expanded to gray-scale volumetric images in [12], though this approach was tested only in a qualitative way, and not evaluated as a feature detector.

Methods similar to MSER have been explored in the works on topological persistence [13]. Persistence-based clustering [7] was used by Skraba et al. [33] to perform shape segmentation. More recently, Dey et al. [10] researched the persistence of the Heat Kernel Signature [35] for detecting features from partial shape data. In Digne et al. [11], extended the notion of vertex-weighted component trees to meshes and proposed to detect MSER regions using the mean curvature. The approach was tested only in a qualitative way, and not evaluated as a feature detector.

A part of this study was published in the proceedings of the Shape Modeling International (SMI'11) conference [20].

8.1.2 Main Contribution

The contributions of our framework are three-fold:

First, in Sect. 8.3 we introduce a generic framework for stable component detection, which unifies vertex- and edge-weighted graph representations (as opposed to vertex-weighting used in image and shape maximally stable component detectors [11, 12, 22]). Our results (see Sect. 8.6) show that the edge-weighted formulation is more versatile and outperforms its vertex-weighted counterpart in terms of feature repeatability.

Second, in Sect. 8.2 we introduce diffusion geometric weighting functions suitable for both vertex- and edge-weighted component trees. We show that such functions are invariant under a large class of transformations, in particular, non-rigid inelastic deformations, making them especially attractive in non-rigid shape analysis. We also show several ways of constructing scale-invariant weighting functions. In addition, following Raviv et al. [29], we show that the suggested framework performs better on volumetric data removing the (sometimes) unwanted insensitivity to volume-changing transformations inherent to the surface model (see Fig. 8.8 for an illustration).

Third, in Sect. 8.6 we show a comprehensive evaluation of different settings of our method on a standard feature detection benchmark comprising shapes undergoing a variety of transformations. We also present a qualitative evaluation on the SCAPE dataset of scanned human bodies and demonstrate that our methods are capable of matching features across distinct data such as synthetic and scanned shapes.

8.2 Diffusion Geometry

Diffusion geometry is an umbrella term referring to geometric analysis of diffusion or random walk processes [8]. Let us consider the shape of a 3D physical object, modeled as a connected and compact region $X \subset \mathbb{R}^3$. The boundary of the region ∂X is a closed connected two-dimensional Riemannian manifold. In many application in graphics, geometry processing, and pattern recognition, one seeks geometric quantities that are invariant to inelastic deformations of the object X [19, 30, 35]. Traditionally in the computer graphics community, 3D shapes are modeled by considering their 2D boundary surface ∂X, and deformations as isometries of ∂X preserving its Riemannian metric structure. In the following, we refer to such deformations as *boundary isometries*, as opposed to a smaller class of *volume isometries* preserving the metric structure inside the volume X (volume isometries are necessarily boundary isometries, but not vice versa – see Fig. 8.8 for an illustration). Raviv et al. [29] argued that the latter are more suitable for modeling realistic shape deformations that boundary isometrics, which preserve the area of ∂X, but not necessarily the volume of X.

8.2.1 Diffusion on Surfaces

Recent line of works [4, 8, 19, 26, 30, 31, 35] studied intrinsic description of shapes by analyzing heat diffusion processes on ∂X, governed by the *heat equation*

$$\left(\frac{\partial}{\partial t} + \Delta_{\partial X}\right) u(t,x) = 0, \tag{8.1}$$

where $u(t,x) : [0,\infty) \times \partial X \rightarrow [0,\infty]$ is the heat value at a point x in time t, and $\Delta_{\partial X}$ is the positive-semidefinite Laplace-Beltrami operator associated with the Riemannian metric of ∂X. The solution of (8.1) is derived from

$$u(x,t) = \int_{\partial X} h_t(x,y) u_0(y) da(y) \tag{8.2}$$

and is unique given the initial condition $u(0,x) = u_0(x)$, and a boundary condition if the manifold ∂X has a boundary. The Green's function $h_t(x,y)$ is called the *heat kernel* and represents the amount of heat transferred on ∂X from x to y in time t due to the diffusion process. The heat kernel is the non-shift-invariant impulse response of (8.1), i.e $h_t(x,x_0)$ it is the solution to a point initial condition $u(0,x) = \delta(x,x_0)$. A probabilistic interpretation of the heat kernel $h_t(x,y)$ is the transition probability *density* of a random walk of length t from the point x to the point y. In particular, the diagonal of the heat kernel or the *auto-diffusivity function* $h_t(x,x)$ describes the amount of heat remaining at point x after time t. Its value is related to the Gaussian curvature $K(x)$ through

$$h_t(x,x) \approx \frac{1}{4\pi t}\left(1 + \frac{1}{6}K(x)t + \mathcal{O}(t^2)\right). \tag{8.3}$$

which describes the well-known fact that heat tends to diffuse slower at points with positive curvature, and faster at points with negative curvature. Due to this relation, the auto-diffusivity function was used by Sun et al. [35] as a local surface descriptor referred to as *heat kernel signature* (HKS). Being intrinsic, the HKS is invariant to boundary isometries of ∂X.

The heat kernel is easily computed using the spectral decomposition of the Laplace-Beltrami operator [19],

$$h_t(x, y) = \sum_{i \geq 0} e^{-\lambda_i t} \phi_i(x)\phi_i(y), \tag{8.4}$$

where $\phi_0 = \text{const}, \phi_1, \phi_2, \ldots$ and $\lambda_0 = 0 \leq \lambda_1 \leq \lambda_2 \ldots$ denote, respectively, the eigenfunctions and eigenvalues of $\Delta_{\partial X}$ operator satisfying $\Delta_{\partial X}\phi_i = \lambda_i \phi_i$.

The parameter t can be given the meaning of *scale*, and the family $\{h_t\}_t$ of heat kernels can be thought of as a scale-space of functions on ∂X. By integrating over all scales, a *scale-invariant* version of (8.4) is obtained,

$$\begin{aligned} c(x, y) &= \int_0^\infty h_t(x, y)dt = \sum_{i \geq 0} \left(\int_0^\infty e^{-\lambda_i t} dt \right) \phi_i(x)\phi_i(y) \\ &= \sum_{i \geq 1} \frac{1}{\lambda_i} \phi_i(x)\phi_i(y) \end{aligned} \tag{8.5}$$

This kernel is referred to as the *commute-time kernel* and can be interpreted as the transition probability density of a random walk of any length. Similarly, $c(x, x)$ express the probability density of remaining at a point x after any time

It is worthwhile noting that both the heat kernel and the commute time kernels constitute a family of low pass filters. In Aubry et al. [2], argued that for some shape analysis tasks kernels acting as band-pass filters might be advantageous. The proposed "wave kernel signature" is related to the physical model of a quantum particle on a manifold described by the Schrödinger equation. The study of this alternative model is beyond the scope of this paper.

8.2.2 *Volumetric Diffusion*

Instead of considering diffusion processes on the boundary surface ∂X, Raviv et al. [29] considered diffusion *inside* the volume X, arising from the Euclidean volumetric heat equation with Neumann boundary conditions,

$$\begin{aligned} \left(\frac{\partial}{\partial t} + \Delta \right) U(t, x) &= 0 \qquad x \in \text{int}(X); \\ \langle \nabla U(t, x), n(x) \rangle &= 0 \qquad x \in \partial X. \end{aligned} \tag{8.6}$$

Here, $U(t, x) : [0, \infty) \times \mathbb{R}^3 \to [0, \infty]$ is the volumetric heat distribution, Δ is the Euclidean positive-semidefinite Laplacian, and $n(x)$ is the normal to the surface ∂X at point x. The heat kernel of the volumetric heat equation (8.6) is given, similarly to (8.4) by

$$H_t(x, y) = \sum_{i \geq 0} e^{-\Lambda_i t} \Phi_i(x) \Phi_i(y), \tag{8.7}$$

where Φ_i and Λ_i are the eigenfunctions and eigenvalues of Δ satisfying $\Delta\Phi_i = \Lambda_i \phi_i$ and the boundary conditions $\langle \nabla \Phi_i(x), n(x) \rangle = 0$. A volumetric version of the commute-time kernel can be created in a similar manner by integration over all values of t, yielding $C(x, y) = \sum_{i \geq 1} \Lambda_i^{-3/2} \Phi_i(x) \Phi_i(y)$. The diagonal of the heat kernel $H_t(x, x)$ gives rise to the *volumetric HKS* (vHKS) descriptor [29], which is invariant to volume isometries of X.

8.2.3 *Computational Aspects*

Both the boundary of an object discretized as a mesh and the volume enclosed by it discretized on a regular Cartesian grid can be represented in the form of an undirected graph. In the former case, the vertices of the mesh form the vertex set V while the edges of the triangles constitute the edge set E. In the latter case, the vertices are the grid point belonging to the solid, and the edge set is constructed using the standard 6- or 26-neighbor connectivity of the grid (for points belonging to the boundary, some of the neighbors do not exist). With some abuse of notation, we will denote the graph by $X = (V, E)$ treating, whenever possible, both cases in the same way. Due to the possibility to express all the diffusion-geometric constructions in the spectral domain, their practical computation boils down to the ability to discretize the Laplacian.

8.2.3.1 Surface Laplace-Beltrami Operator

In the case of 2D surfaces, the discretization of the Laplace-Beltrami operator of the surface ∂X and can be written in the generic matrix-vector form as $\Delta_{\partial X} \mathbf{f} = \mathbf{A}^{-1} \mathbf{W} \mathbf{f}$, where $\mathbf{f} = (f(v_i))$ is a vector of values of a scalar function $f : \partial X \to \mathbb{R}$ sampled on $V = \{v_1, \ldots, v_N\} \subset \partial X$, $\mathbf{W} = \mathrm{diag}\left(\sum_{l \neq i} w_{il}\right) - (w_{ij})$ is a zero-mean $N \times N$ matrix of weights, and $\mathbf{A} = \mathrm{diag}(a_i)$ is a diagonal matrix of normalization coefficients [14, 40]. Very popular in computer graphics is the *cotangent weight* scheme [23, 28], where

$$w_{ij} = \begin{cases} (\cot \alpha_{ij} + \cot \beta_{ij})/2 & (v_i, v_j) \in E; \\ 0 & else, \end{cases} \tag{8.8}$$

where α_{ij} and β_{ij} are the two angles opposite to the edge between vertices v_i and v_j in the two triangles sharing the edge, and a_i are the discrete area elements. The eigenfunctions and eigenvalues of $\Delta_{\partial X}$ are found by solving the generalized eigendecomposition problem $\mathbf{W}\phi_i = \mathbf{A}\phi_i \lambda_i$ [19]. Heat kernels are approximated by taking a finite number of eigenpairs in the spectral expansion.

8.2.3.2 Volumetric Laplacian

In the 3D case, we used a ray shooting method to create rasterized volumetric shapes, i.e. every shape is represented as arrays of voxels on a regular Cartesian grid, allowing us to use the standard Euclidean Laplacian. The Laplacian was discretized using a 6-neighborhood stencil. We use the finite difference scheme to evaluate the second derivative in each direction in the volume, and enforced boundary conditions by zeroing the derivative outside the shape.

The construction of the Laplacian matrix under these conditions boils down to this element-wise formula (up to multiplicative factor):

$$(\Delta)_{ij} = \begin{cases} -1 & \text{if } i \neq j \text{ and } (v_i, v_j) \in E \\ -\sum_{k \neq j} (\Delta)_{kj} & \text{if } i = j \end{cases}$$

8.3 Maximally Stable Components

Let us now we go over some preliminary graph-theory terms, needed to cover the topic of the component-tree. As mentioned, we treat the discretization of a shape as an undirected graph $X = (V, E)$ with the vertex set V and edge set E. Two vertices v_1 and v_2 are said to be *adjacent* if $(v_1, v_2) \in E$. A *path* is an ordered sequence of vertices $\pi = \{v_1, \ldots, v_k\}$ such that for any $i = 1, \ldots, k-1$, v_i is adjacent to v_{i+1}. In this case, every pair of vertices on π are *linked* in X. If every pair of vertices in X is linked, the graph is said to be *connected.* A *subgraph* of X is every graph for which $Y = (V' \subseteq V, E' \subseteq E)$, and will be denoted by $Y \subseteq X$. Such Y will be a (connected) *component* of X if it is a maximal connected subgraph of X (i.e. for any connected subgraph Z, $Y \subseteq Z \subseteq X$ implies $Y = Z$). A subset of the graph edges $E' \subseteq E$, induces the graph $Y = (V', E')$ where $V' = \{v \in V : \exists v' \in V, (v, v') \in E'\}$, i.e. V' is a vertex set is made of all vertices belonging to an edge in E'.

A component tree can be built only on a weighted graph. A graph is called *vertex-weighted* if it is equipped with a scalar function $f : V \to \mathbb{R}$. Similarly, an *edge-weighted* graph is one that is equipped with a function $d : E \to \mathbb{R}$ defined on the edge set. In what follows, we will assume both types of weights to be non-negative.

In order to define the MSER algorithm on images, some regular connectivity (e.g., four-neighbor) was used. Gray-scale images may be represented as vertex-weighted graphs where the intensity of the pixels is used as weights. Using a function measuring dissimilarity of pairs of adjacent pixels one can obtain edge weights, as done by Forssen in [15]. Edge weighting is more general than vertex weighting, which is limited to scalar (gray-scale) images.

8.3.1 Component Trees

The *ℓ-cross-section* of a graph X is the sub-graph created by using only weights smaller or equal to ℓ (assuming $\ell \geq 0$). If the graph has a vertex-weight $f : V \rightarrow \mathbb{R}$, its ℓ-cross-section is the graph induced by $E_\ell = \{(v_1, v_2) \in E : f(v_1), f(v_2) \leq \ell\}$. Similarly, for a graph with an edge-weight $d : E \rightarrow \mathbb{R}$ the ℓ-cross-section is induced by the edge subset $E_\ell = \{e \in E : d(e) \leq \ell\}$. A connected component of the ℓ-cross-section is called the *ℓ-level-set* of the weighted graph.

The *altitude* of a component C, denoted by $\ell(C)$, is defined as the minimal ℓ for which C is a component of the ℓ-cross-section. Altitudes establish a partial order relation on the connected components of X as any component C is contained in a component with higher altitude.

The set of all such pairs $(\ell(C), C)$ therefore forms a tree called the *component tree*. The component tree is a data structure containing the nesting hierarchy the level-sets of a weighted graph. Note that the above definitions are valid for both vertex- and edge-weighted graphs.

8.3.2 Maximally Stable Components

Since we represent a discretized smooth manifold (or a compact volume) by an undirected graph, a measure of *area* (or *volume*) can associate with every subset of the vertex set, and therefore also with every component. Even though we are dealing with both surface and volume, w.l.o.g we will refer henceforth only to surface area as a measure of a component C, which will be denoted by $A(C)$. When dealing with regular sampling, like in the case images, the area of C can be thought of as its cardinality (i.e. the number of pixels in it). In the case of non-regular sampling a better discretization is needed, and a discrete area element $da(v)$ is associated with each vertex v in the graph. The area of a component in this case is defined as $A(C) = \sum_{v \in C} da(v)$.

The process of detection is done on a sequence of nested components $\{(\ell, C_\ell)\}$ forming a branch in the component tree. We define the *stability* of C_ℓ as a derivative along the latter branch:

$$s(\ell) = \frac{\text{Change in altitude}}{\text{\% Change in area}} = \frac{A(C_\ell)}{\frac{d}{d\ell} A(C_\ell)}. \tag{8.9}$$

In other words, the less the area of a component changes with the change of ℓ, the more stable it is. A component C_{ℓ^*} is called *maximally stable* if the stability function has a local maximum at ℓ^*. As mentioned before, maximally stable components are widely known in the computer vision literature under the name of *maximally stable extremal regions* or *MSERs* for short [22], with $s(\ell^*)$ usually referred to as the region *score*.

Pay attention that while in their original definition, both MSERs and their volumetric counterpart [12] were defined on a component tree of a vertex-weighted graph, while the latter definition allows for edge-weighted graphs as well and therefore more general. The importance of such an extension will become evident in the sequel. Also, the original MSER algorithm [22] assumes the vertex weights to be quantized, while our formulation is suitable for scalar fields whose dynamic range is unknown a priori (this has some disadvantages, as will be seen up next).

8.3.3 Computational Aspects

Najman et al. [25] introduced quasi-linear time algorithm for the construction of vertex-weighted component trees, and its adaptation to the edge-weighted case is quite straightforward. The algorithm is based on the observation that the vertex set V can be partitioned into disjoint sets which are merged together as one goes up in the tree. Maintaining and updating such a partition can be performed very efficiently using the *union-find* algorithm and related data structures. The resulting tree construction complexity is $O(N \log\log N)$. However, since the weights must be sorted prior to construction, the latter complexity is true only if the weights are quantized over a known range (which is not the case in weights we used), otherwise the complexity becomes $O(N \log N)$.

The stability function (8.9) contains a derivative along a branch of the component tree. Given a branch of the tree is a sequence of nested components $C_{\ell_1} \subseteq C_{\ell_2} \subseteq \cdots \subseteq C_{\ell_K}$, the derivative was approximated using finite differences scheme:

$$s(\ell_k) \approx \frac{A(C_{\ell_k})}{A(C_{\ell_{k+1}}) - A(C_{\ell_{k-1}})}(\ell_{k+1} - \ell_{k-1}). \tag{8.10}$$

Starting from the leaf nodes, a single pass over every branch of the component tree is in order to evaluate the function and detected its local maxima. Next, we filter out maxima with too low values of s (more details about this in the detector result section). Finally, we keep only the bigger of two nested maximally stable regions if they are overlapping by more that a predefined threshold.

8.4 Weighting Functions

Unlike images where pixel intensities are a natural vertex weight, 3D shapes generally do not have any such field. This could be a possible reason why MSER was shown to be extremely successful as a feature detector in images, but equivalent techniques for 3D shapes are quite rare. This method was recently proposed in [11], but due to the fact it uses mean curvature it is not deformation invariant and therefore not suitable for the analysis deformable shape. The diffusion geometric framework used in [33] is more appropriate for this task, as it allows a robust way to analyze deformable shapes. We follow this approach and show that it allows the construction both vertex and edge weights suitable for the definition of maximally stable components with many useful properties.

Note that even though all of the following weighting functions are defined on the surface of the shape (usually by using $h_t(x, y)$), they are easily adapted to the volumetric model (usually by using $H_t(x, y)$ instead).

The discrete auto-diffusivity function is a trivial vertex-weight when using diffusion geometry, and can be directly used given its value on a vertex v:

$$f(v) = h_t(v, v). \tag{8.11}$$

As will be with all diffusion geometric weights, the latter weights are intrinsic. Therefore, maximally stable components defined this way are invariant to non-rigid bending. In images, weighting functions are based on intensity values and therefore contain all the data needed abut the image. The above weighting function, however, does not capture the complete intrinsic geometry of the shape, and depends on the scale parameter t.

Every scalar field may also be used as an edge-weight simply by using $d(v_1, v_2) = |f(v_1) - f(v_2)|$. As mentioned, edge weights are a more flexible and allow us more freedom in selecting how to incorporate the geometric information. For example, a vector-valued field defined on the vertices of the graph can be used to define an edge weighting scheme by weighting an edge by the distance between its vertices' values, as done in [15].

There are ways to define edge weights without the usage of vector field defined on the vertices. Lets take the discrete heat kernel $h_t(v_1, v_2)$ as an example. Taking a weight function inversely proportional to its value is metric-like in essence, since it represents random-walk probability between v_1 and v_2. The resulting edge weight will be

$$d(v_1, v_2) = \frac{1}{h_t(v_1, v_2)}. \tag{8.12}$$

This function also contains fuller information about the shape's intrinsic geometry, for small values of t.

8.4.1 Scale Invariance

All of the three latter weighting functions are based on the heat kernel, and therefore are not scale invariant. If we globally scale the shape by a factor $a > 0$, both the time parameter and the kernel itself will be scaled by a^2, i.e. the scaled heat kernel will be $a^2 h_{a^2 t}(v_1, v_2)$. The volumetric heat kernel $H_t(v_1, v_2)$ will be scaled differently and will become $a^3 H_{a^2 t}(v_1, v_2)$ (this is why $C(v_1, v_2)$ is constructed using the eigenvalues with a power of $-3/2$).

The commute time kernel is scale invariant, and could be used as a replacement to the heat kernel. However, the numerical computation of the commute time kernel is more difficult as its coefficients decay polynomially (8.5), very slow compared to $h_t(v_1, v_2)$ whose coefficients decay exponentially (8.4). The slower decay translates to the need of many more eigenfunctions of the Laplacian for $c(v_1, v_2)$ to achieve the same accuracy as $h_t(v_1, v_2)$.

Let us point out that there is some invariance to scale, originating in the way the detector operates over the component tree (this is also notable in the detector result section). As noted by Matas et al. [22], the MSER detector is invariant to any monotonic transformation on the weights (originally pixel-intensities), a fact that can be harnessed to gain scale invariance in our detector. In practice, it is sufficient for us to limit the effect scaling on the weights to a monotonous transformation instead of completely undoing its effect. Such a weighting function will not be scale invariant by itself, nor the stability function (8.9) computed on such a component tree. However, the local maxima of (8.9), namely the stable components, will remain unaffected. An alternative to this approach would be designing a more sophisticated scale-invariant stability function. We intend to explore both of these options in follow up studies.

8.5 Descriptors

Usually, when using a feature-based approach, feature detection is followed by attaching a descriptor to each of the features. Once descriptors are manifested, we can measure similarity between a pair of features, which in turn, enables us to perform tasks like matching and retrieval. Since our detected feature are components of the shape, we first create a point-wise descriptor of the form $\alpha : V \to \mathbb{R}^q$ and then aggregate al the point-descriptors into a single region descriptor.

8.5.1 Point Descriptors

We consider a descriptor suitable for non-rigid shapes proposed in [35] – the *heat kernel signature* (HKS) HKS is computed by sampling the values of the discrete auto-diffusivity function at vertex v at multiple times, $\alpha(v) = (h_{t_1}(v, v), \ldots, h_{t_q}(v, v))$, where $t_1, \ldots, t_q$ are some fixed time values. The resulting descriptor is a vector of dimensionality q at each vertex. Since the heat kernel is an intrinsic quantity, the HKS is invariant to isometric transformations of the shape.

8.5.1.1 Scale-Invariant Heat Kernel Signature

The fact that HKS is based on $h_t(v, v)$, means it also inherits the drawback of its dependence on the scaling of the shape. As mentioned, scaling the shape globally by a factor $a > 0$ will result in the scaled heat kernel $a^2 h_{a^2 t}(v_1, v_2)$. A way of rendering $h_t(v, v)$ scale invariant was introduced in [4], by performing a sequence of transformations on it. First, we sample the heat kernel with logarithmical spacing in time. Then, we take the logarithm of the samples and preform a numerical derivative (with respect to time of the heat kernel) to undo the multiplicative constant. Finally, we perform the discrete Fourier transform followed by taking the absolute value, to undo the scaling of the time variable. Note that the latter sequence of transformations will also work on $H_t(v, v)$, as the effect of scaling differs only in the power of the multiplicative constant.

This yields the *modified heat kernel* of the form

$$\hat{h}_\omega(v, v) = \left| \mathscr{F} \left\{ \frac{\partial \log h_t(v, v)}{\partial \log t} \right\} (\omega) \right| \tag{8.13}$$

where ω denotes the frequency variable of the Fourier transform. The scale-invariant version of the HKS descriptor (SI-HKS) is obtained by replacing h_t with $\hat{h}_\omega$, yielding $\alpha(v) = (\hat{h}_{\omega_1}(v, v), \ldots, \hat{h}_{\omega_q}(v, v))$, where $\omega_1, \ldots, \omega_q$ are some fixed frequency values.

8.5.1.2 Vocabulary Based Descriptors

Another method for the descriptor construction is following the *bag of features* paradigm [32]. Ovsjanikov et al. [27] used this approach to create a global shape descriptor using point-wise descriptors. In the bag of features approach we perform off-line clustering of the descriptor space, resulting in a fixed "geometric vocabulary" $\alpha_1, \ldots, \alpha_p$. We then take any point descriptor $\alpha(v)$, and represent it using the vocabulary by means of vector quantization. This results in a new point-wise p-dimensional vector $\theta(v)$, where each of its elements follow a distribution of the form $[\theta(v)]_l \propto e^{-\|\alpha(v)-\alpha_l\|^2/2\sigma^2}$. The vector is then normalized in such a way that the elements $\theta(v)$ sum to one. Setting $\sigma = 0$ is a special case named "hard vector quantization" where the descriptor boils down to a query result for the nearest neighbor of $\alpha(v)$ in the vocabulary. In other words, we will get $[\theta(v)]_l = 1$ for α_l being the closest vocabulary element to $\alpha(v)$ in the descriptor space, and zero elsewhere.

8.5.2 *Region Descriptors*

After creating a descriptor $\alpha(v)$ at each vertex $v \in V$, we need to gather the information from a subset of vertices, i.e. component $C \subset V$. This will result in a *region descriptor*. The simplest way to do this is by computing the average of α in C, weighted by the discrete area elements:

$$\beta(C) = \sum_{v \in C} \alpha(v) da(v). \tag{8.14}$$

The resulting region descriptor $\beta(C)$ is a vector of the same dimensionality q as the point descriptor α.

Other methods to create region descriptors from point descriptors such as the region covariance descriptors [38], are beyond the scope of this text. The latter method may provide higher discriminativity due to the incorporation of spatial contexts.

8.6 Results

8.6.1 Datasets

The proposed approaches were tested both qualitatively and quantitatively. All datasets mentioned below were given as triangular meshes, i.e. as 2D manifolds. In the following experiments, meshes were down-sampled to at most $10,000$ vertices. For the computation of the volumetric regions, meshes were rasterized in a cube with variable number of voxels per dimension (usually around 100–130) in order for the resulting shapes to contain approximately 45,000 voxels.

8.6.1.1 Data for Quantitative Tests

While almost every data set can be used for a qualitative evaluation, only datasets with additional ground-truth data can be used to quantify the performance of the algorithm. We chose two such data sets: the SHREC'10 [6] and SHREC'11 [3] feature detection and description benchmarks.

SHREC'10[6] was used for evaluation only of the 2D version of the algorithm. The dataset consisted of three shape classes (human, dog, and horse), with simulated transformations applied to them. Shapes are represented as triangular meshes with approximately 10,000–50,000 vertices.

Each shape class is given in a so-called "native" form, coined *null*, and also in nine categories of transformations: *isometry* (non-rigid almost inelastic deformations), *topology* (welding of shape vertices resulting in different triangulation), *micro holes* and *big holes* simulating missing data and occlusions, *global* and *local scaling*, *additive Gaussian noise*, *shot noise*, and *downsampling* (less than 20 % of the original points). All mentioned transformations appeared in five different strengths, and are combined with isometric deformations. The total number of transformations per shape class is $45 + 1$ null shape, i.e. 138 shapes in total.

Vertex-wise correspondence between the transformed and the null shapes was given and used as the ground-truth in the evaluation of region detection repeatability. Since all shapes exhibit intrinsic bilateral symmetry, best results over the ground-truth correspondence and its symmetric counterpart were used.

SHREC'11 [3] was used for the comparison between the surface and volumetric approaches, due to the fact that results of both methods were too similar on SHREC'10. Having a wider and more challenging range and strength of transformations present in the SHREC'11 corpus was needed to emphasize the difference. The SHREC'11 dataset was constructed along the guidelines of its predecessor, SHREC'10. It contains one class of shapes (human) given in a null form, and also in 11 categories of transformations, in transformation appeared in five different strengths combined with isometric deformations. The strength of transformation is more challenging, in comparison to SHREC'10. The total number of shapes is $55 + 1$ null shape.

Most of the transformations appear also in SHREC'10: *isometry*, *micro holes* and *big holes*, *global*, *additive Gaussian noise*, *shot noise*, and *downsampling*. Two transformations were discarded (*topology* and *local scaling*) and some new ones were introduced: *affine transformation*, *partial* (missing parts), *rasterization* (simulating non-pointwise topological artifacts due to occlusions in 3D geometry acquisition), and *view* (simulating missing parts due to 3D acquisition artifacts). Some of the mentioned transformations are not compatible with our volumetric rasterization method. We did not include in our experiments the following transformations: big holes, partial and view.

Vertex-wise correspondence were given like in SHREC'10, including bilateral symmetry. Volumetric ground-truth had to be synthesized, however. For that purpose, the surface voxels of two shapes were first matched using the groundtruth correspondences; then, the interior voxels were matched using an onion-peel procedure.

8.6.1.2 Data for Qualitative Tests

In addition to the already mentioned sets, three more datasets without groundtruth correspondence were used to demonstrate the performance of the proposed method visually:

TOSCA [5] The dataset contains 80 high-resolution nonrigid shapes in a variety of poses, including cat, dog, wolves, horse, 6 centaur, gorilla, female figures, and two different male figures. Shapes have a varying number vertices, usually about 50,000. This dataset gives a good example of the potential of the detected features to be used for partial-matching of shapes (see Figs. 8.1–8.3).

SCAPE[1] The dataset contains watertight scanned human figures, containing around 12.5 K vertices, in various poses. This dataset gives a hint on performance of the detector on real life (not synthetic) data. Figure 8.7 shows that the detected components are consistent and remain invariant under pose variations.

Deformation Transfer for Triangle Meshes dataset by Sumner et al.[34] contains a few shapes, out of which we chose two animated sequences of a horse shape represented as a triangular mesh with approximately 8.5 K vertices.

Fig. 8.1 Maximally stable regions detected on different shapes from the TOSCA dataset. Note the invariance of the regions to strong non-rigid deformations. Also observe the similarity of the regions detected on the female shape and the upper half of the centaur (compare to the male shape from Fig. 8.2). Regions were detected using $h_t(v, v)$ as vertex weight function, with t =2,048

Fig. 8.2 Maximally stable regions detected on shapes from the SHREC'10 dataset using the vertex weight $h_t(v, v)$ with t =2,048. *First row*: different approximate isometries of the human shape. *Second row*: different transformations (*left*-to-*right*): holes, localscale, noise, shotnoise and scale

One sequence includes a series of boundary and volume isometries (gallop), while the other includes series of non-volume-preserving boundary isometries (collapsing). Figure 8.8 shows that while the surface MSERs are invariant to both types of transformations, the proposed volumetric MSERs remain invariant only under volume isometries, changing quite dramatically if the volume is not preserved – a behavior consistent with the physical intuition.

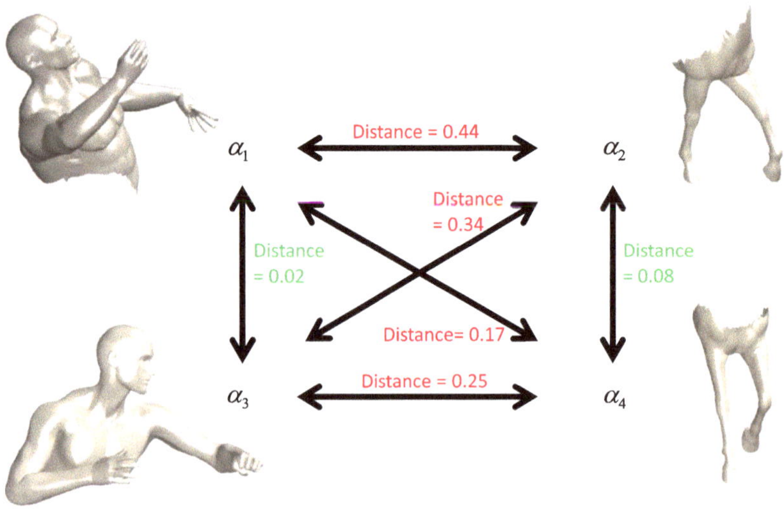

Fig. 8.3 A toy example showing the potential of the proposed method for partial shape matching. Shown are four maximally stable regions detected on the surface of three shapes from the TOSCA dataset using the edge weight $1/h_t(v_1, v_2)$ with t =2,048. The *top* two regions are the torso and the front legs of a centaur (marked with α_1 and α_2 respectively), and on the *bottom* are the torso of a human and the front legs of a horse (marked with α_3 and α_4 respectively). Each of the regions is equipped with a SI-HKS descriptor, and L_2 distance is shown between every pair

8.6.2 Detector Repeatability

8.6.2.1 Evaluation Methodology

We follow the spirit of Mikolajczyk et al. [24] in the evaluation of the proposed feature detector (and also of the descriptor, later on). The performance of a feature detector is measured mainly by its *repeatability*, defined as the percentage of regions that have corresponding counterpart in a transformed version of the shape. In order to measure this quantity we need a rule to decide if two regions are "corresponding". Every comparison is done between a transformed shape and its original null shape, coined Y and X respectively. We will denote the regions detected in X and Y as $X_1, \ldots, X_m$ and $Y_1, \ldots, Y_n$. In order to perform a comparison, we use the ground-truth correspondence to project a region Y_j onto X, and will denote it's projected version as X'_j. We define the *overlap* of two regions X_i and Y_j as the following area ratio:

$$O(X_i, X'_j) = \frac{A(X_i \cap X'_j)}{A(X_i \cup X'_j)} = \frac{A(X_i \cap X'_j)}{A(X_i) + A(X'_j) - A(X_i \cap X'_j)} \quad (8.15)$$

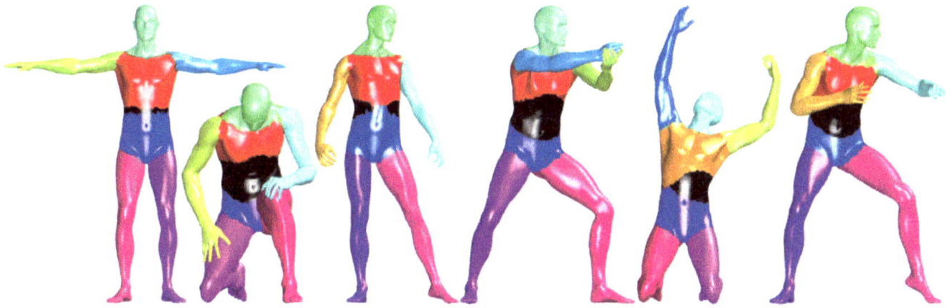

Fig. 8.4 Maximally stable regions detected on shapes from the SHREC'10 dataset using the edge weight $1/h_t(v_1, v_2)$ with t =2,048. Region coloring is arbitrary

The *repeatability at overlap o* is defined as the percentage of regions in Y that have corresponding counterparts in X with overlap greater than o. An ideal detector has the repeatability of 100 % even for $o \to 1$. Note that comparison was defined single-sidedly due to the fact that some of the transformed shapes had missing data compared to the null shape. Therefore, unmatched regions of the null shape did not decrease the repeatability score, while regions in the transformed shape that had no corresponding regions in the null counterpart incurred a penalty on the score.

8.6.2.2 Surface Detector

Two vertex weight functions were compared: discrete heat kernel (8.11) with $t = 2{,}048$ and the commute time kernel. These two scalar fields were also used to construct edge weights according to $d(v_1, v_2) = |f(v_1) - f(v_2)|$. In addition, we used the fact that these kernels are functions of a pair of vertices to define edge weights according to (8.12). Unless mentioned otherwise, $t = 2{,}048$ was used for the heat kernel, as this setting turned out to give best performance on the SHREC'10 dataset.

We start by presenting a qualitative evaluation on the SHREC'10 and the TOSCA datasets. Regions detected using the vertex weight $h_t(v, v)$ with $t = 2{,}048$ are shown for TOSCA in Fig. 8.1 and for SHREC'10 in Fig. 8.2. Figure 8.4 depicts the maximally stable components detected with the edge weighting function $1/h_t(v_1, v_2)$ on several shapes from the SHREC'10 dataset. These results show robustness and repeatability of the detected regions under transformation. Surprisingly, many of these regions have a clear semantic interpretation, like limbs or head. In Addition, the potential of the proposed feature detector for partial shape matching can be seen by the similarly looking regions that are detected on the male and female shapes, and the upper half of the centaur (see a toy example in Fig. 8.3).

Ideally, we would like a detector to have perfect repeatability, i.e. to produce a large quantity of regions with a corresponding counterpart in the original shape. This is unfeasible, and all detectors will produce some poorly repeatable regions. However, if the repeatability of the detected regions is highly correlated with their

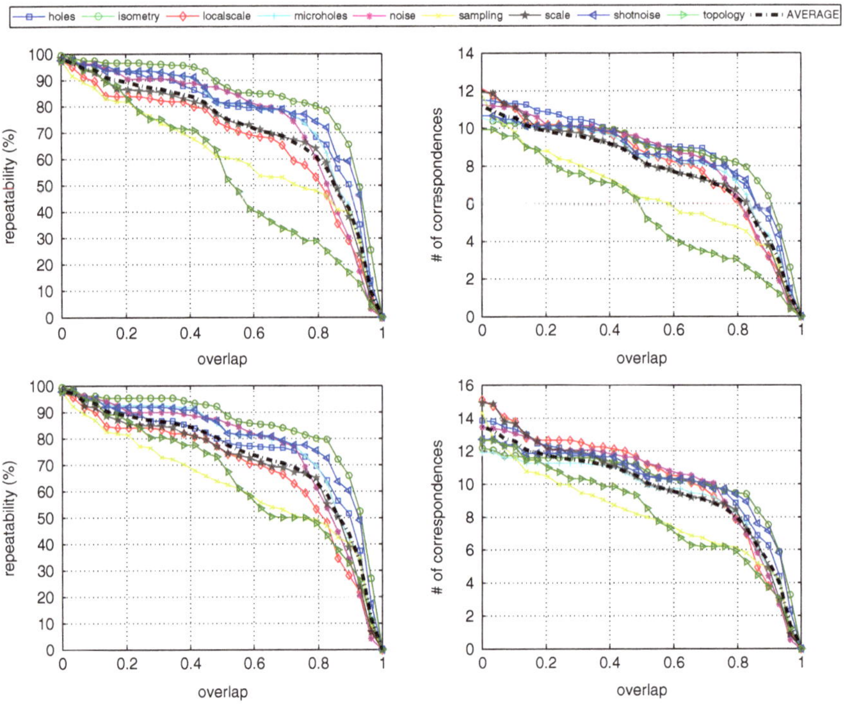

Fig. 8.5 Repeatability of maximally stable components with the vertex weight $h_t(v, v)$ (*first row*) and edge weight $1/h_t(v_1, v_2)$ (*second row*), t =2,048

stability scores, a poor detector can still be deemed good by selecting a cutoff threshold on the stability score. In other words, set a minimum region stability value that is accepted by the detector, such that the rejected regions are likely not to be repeatable. This cutoff value is estimated based on the empirical 2D distributions of detected regions as a function of the stability score and the overlap with the corresponding ground-truth regions. Of course, this can only be done given ground-truth correspondences. In some of the tested detectors, a threshold was selected for stability score to minimize the detection of low-overlap regions, in order to give an estimate for the theoretical limits on the performance of the weighting functions.

We now show the performance of the best four weighting functions in Figs. 8.5 and 8.6. These figures depict the repeatability and the number of correctly matching regions as the function of the overlap. One can infer that scale-dependent weighting generally outperform their scale-invariant counterparts in terms of repeatability. This could be explained by the fact that we have selected the best time value for our dataset's common scale, whereas scale-invariant methods suffer from its larger degree of freedom. The scalar fields corresponding to the auto-diffusivity functions perform well both when used as vertex and edge weights. Best repeatability is achieved by the edge weighting function $1/h_t(v_1, v_2)$. Best scale invariant weighting by far is the edge weight $1/c(v_1, v_2)$.

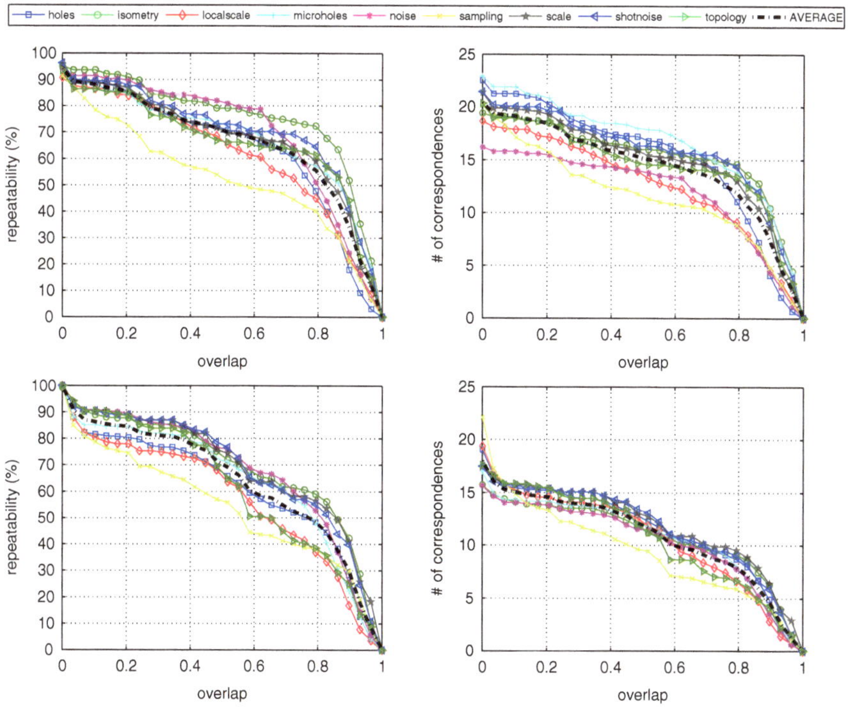

Fig. 8.6 Repeatability of maximally stable components with the edge weight $|h_t(v_1, v_1) - h_t(v_2, v_2)|$ (*first row*) and edge weight $1/c(v_1, v_2)$ (*second row*), t =2,048

8.6.2.3 Volume Detector

In order to asses the differences between the 2D and the 3D of the approach, we performed two experiments comparing between 3D MSER and 2D MSER: comparison of invariance of the two methods to boundary and volume isometric deformations, and a quantitative comparison evaluating the sensitivity of two methods to shape transformations and artifacts on the SHREC'11 benchmark. In addition, we performed one evaluation of volumetric (3D) MSER invariance on scanned human figures.

As mentioned before, all the datasets used in our experiments were originally represented as triangular meshes and were rasterized and represented as arrays of voxels on a regular Cartesian grid.

In the first experiment, we applied the proposed approach to the SCAPE dataset [1], containing a scanned human figure in various poses. Figure 8.7 shows that the detected components are consistent and remain invariant under pose variations. In the second experiment, we used the data from Sumner et al. [34]. The dataset contained an animated sequence of a horse shape and includes a series of boundary and volume isometries (gallop) and series of non-volume-preserving

Fig. 8.7 Stable volumetric regions detected on the SCAPE data [1]. Shown are volumetric regions (*first* and *third columns*) and their projections onto the boundary surface (*second* and *fourth columns*). Corresponding regions are denoted with like colors. The detected components are invariant to isometric deformations of the volume

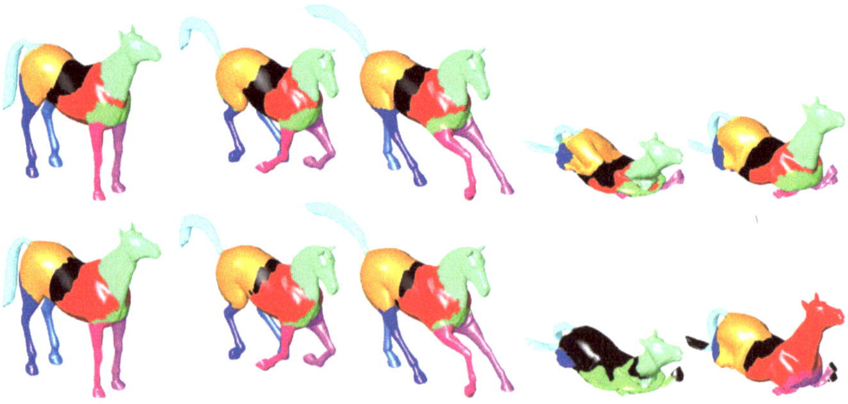

Fig. 8.8 Maximally stable components detected on two approximate volume isometries (*second* and *third columns*) and two volume-changing approximate boundary surface isometries (*two right-most columns*) of the horse shape (*left column*). Stable regions detected on the boundary surface (2D MSER, *first row*) remain invariant to all deformations, while the proposed volumetric stable regions (3D MSER, *second row*) maintain invariance to the volume-preserving deformations only. This better captures natural properties of physical objects. Corresponding regions are denoted with like colors. For easiness of comparison, volumetric regions are projected onto the boundary surface

boundary isometries (collapsing). Figure 8.8 shows that while the surface MSERs are invariant to both types of transformations, the proposed volumetric MSERs remain invariant only under volume isometries, changing quite dramatically if the volume is not preserved – a behavior consistent with the physical intuition.

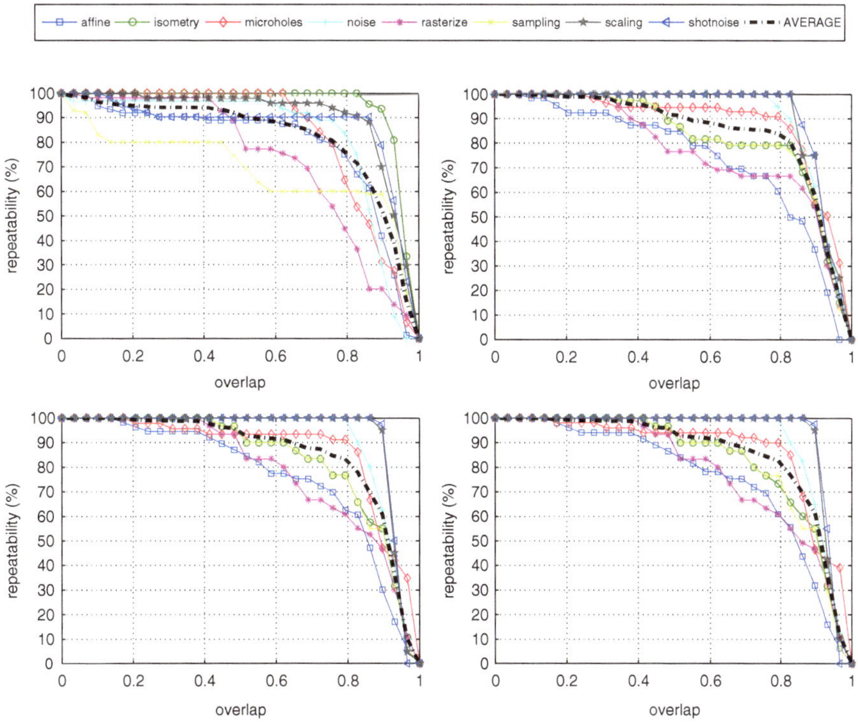

Fig. 8.9 Repeatability of region detectors on the SHREC'11 dataset. *Upper left*: 2D MSER using the edge weight $1/h_t(v_1, v_2)$, t =2,048. *Upper right*: 3D MSER using the commute-time vertex-weight. *Lower left*: 3D MSER using the edge weight $1/H_t(v_1, v_2)$, t =2,048. *Lower right*: 3D MSER using the vertex-weight $H_t(v, v)$, t =2,048

In the third experiment, we used the SHREC'11 feature detection and description benchmark [3] to evaluate the performance of the 2D and 3D region detectors and descriptors under synthetic transformations of different type and strength.[1] As mentioned, some of the transformations in SHREC'11 are not compatible with our volumetric rasterization method, so we did not include in our experiments the big-holes, partial, and view transformations.

Figure 8.9 shows the repeatability of the 3D and 2D MSERs. We conclude that volumetric regions exhibit similar or slightly superior repeatability compared to boundary regions, especially for large overlaps (above 80 %). We attribute the slightly lower repeatability in the presence of articulation transformations ("isometry") to the fact that these transformations are almost exact isometries of the

[1]In this evaluation we used SHREC11, rather than SHREC10 that was used previously in 2D. this is due to the fact that results of the 3D and 2D versions were too similar on SHREC10, and dataset with a wider, and more challenging range and strength of transformations was needed to emphasize the difference.

boundary, while being only approximate isometries of the volume. Another reason for the latter degradation may be local topology changes that were manifested in the rasterization of the shapes in certain isometries. These topology changes appear only in the volumetric detector, and they affected the quality of detected regions in their vicinity. Although the construction of the MSER feature detector is not affine-invariant, excellent repeatability under affine transformation is observed. We believe that this and other invariance properties are related to the properties of the component trees (which are stronger than those of the weighting functions) and intend to investigate this phenomenon in future studies.

8.6.3 Descriptor Informativity

8.6.3.1 Evaluation Methodology

In these experiments, we aim to evaluate the informativity of region descriptors. This is done by measuring the relation between the overlap of two regions and their distance in the descriptor space.

Keeping the notation from the previous section, we set a minimum overlap $\rho = 0.75$ deeming two regions Y_i and X_j *matched*, i.e. if $o_{ij} = O(X_i', X_j) \geq \rho$ (X_i' is the projection of Y_i on the null shape X). This threshold is needed to constitute the matching ground-truth.

Given a region descriptor β on each of the regions, we set a threshold τ on the distance between the descriptors of Y_i and X_j in order to classify them as *positives*, i.e. if $d_{ij} = \|\beta(Y_i) - \beta(X_j)\| \leq \tau$. For simplicity we assume the distance between the descriptors to be the standard Euclidean distance. We define the *true positive rate* (TPR) and *false positive rate* (FPR) as the ratios

$$\text{TPR} = \frac{|\{d_{ij} \leq \tau\}|}{|\{o_{ij} \geq \rho\}|} \quad ; \quad \text{FPR} = \frac{|\{d_{ij} > \tau\}|}{|\{o_{ij} < \rho\}|}$$

The *receiver operator characteristic* (ROC) curve is obtained a set of pairs (FPR, TPR), created by varying the threshold τ. The *false negative rate* defined as FNR $= 1 - $ TPR. The *equal error rate* (EER) is the point on the ROC curve for which the false positive and false negative rates coincide. EER is used as a scalar measure for the descriptor informativity, where ideal descriptors have EER $= 0$.

Another descriptor performance criterion is created by finding for each X_i its nearest neighbor in the descriptor space $Y_{j^*(i)}$, namely $j^*(i) = \arg\min_j d_{ij}$. We then define the *matching score*, as the ratio of correct first matches for a given overlap ρ, and m, the total number of regions in X:

$$\text{score}(\rho) = \frac{|\{o_{ij^*(i)} \geq \rho\}|}{m}. \tag{8.16}$$

Table 8.1 Equal error rate (EER) performance of different maximally stable component detectors and descriptors ($t = 2{,}048$ was used in all cases). p denotes the vocabulary size in the bag of features region descriptors

Weighting function	HKS avgerage	HKS BoF($p = 10$)	HKS BoF($p = 12$)	SI-HKS avgerage	SI-HKS BoF($p = 10$)	SI-HKS BoF($p = 12$)
$h_t(v, v)$	0.311	0.273	0.278	0.093	0.091	0.086
$1/h_t(v_1, v_2)$	0.304	0.275	0.281	0.104	0.093	0.090
$\|h_t(v_1, v_1) - h_t(v_2, v_2)\|$	0.213	0.212	0.222	0.085	0.091	0.094
$1/c(v_1, v_2)$	0.260	0.284	0.294	0.147	0.157	0.148

8.6.3.2 Surface Descriptor

Given the maximally stable components detected by a detector, region descriptors were calculated. We used two types of point descriptors: the heat kernel signature (HKS), and its scale invariant version SI-HKS. Each of these two point descriptors was also used as a basis to create vocabulary based descriptors. Region descriptors were created from every point descriptor using averaging (8.14). In the following experiments, the HKS was created based on the heat kernel signature $h_t(v, v)$, sampled at six time values $t = 16, 22.6, 32, 45.2, 64, 90.5, 128$. The SI-HKS was created by sampling the heat kernel time values $t = 2^1, 2^{1+1/16}, \dots, 2^{25}$ and taking the first six discrete frequencies of the Fourier transform, repeating the settings of [4]. Bags of features were tested on the two descriptors with vocabulary sizes $p = 10$ and 12, trained based on the SHREC'10 and TOSCA datasets. This sums up to a total of 6 descriptors – two in the "raw" form and four vocabulary based descriptors.

The four best weighting functions (shown in Figs. 8.5 and 8.6) were also selected to test region detectors. The performance in terms of EER of all the combinations of latter weighting functions and the six region descriptors is shown in Table 8.1.

Figures 8.10 and 8.11 show the number of correct first matches and the matching score as a function of the overlap for the two "raw" descriptors and the two best weighting functions: the vertex weight $h_t(v, v)$ and the edge weight $1/h_t(v_1, v_2)$. Figure 8.12 depicts the ROC curves of all of the descriptors based on maximally stable components of the same two weighting functions.

We conclude that the SI-HKS descriptor consistently exhibits higher performance in both the "raw" form and when using vocabulary, though the latter perform slightly worse. On the other hand, the bag of feature setting seems to improve the HKS descriptor in comparison to its "raw" form, though never reaching the scores of SI-HKS. Surprisingly, SI-HKS consistently performs better, even in transformations not including scaling, as can be seen from Figs. 8.10–8.11.

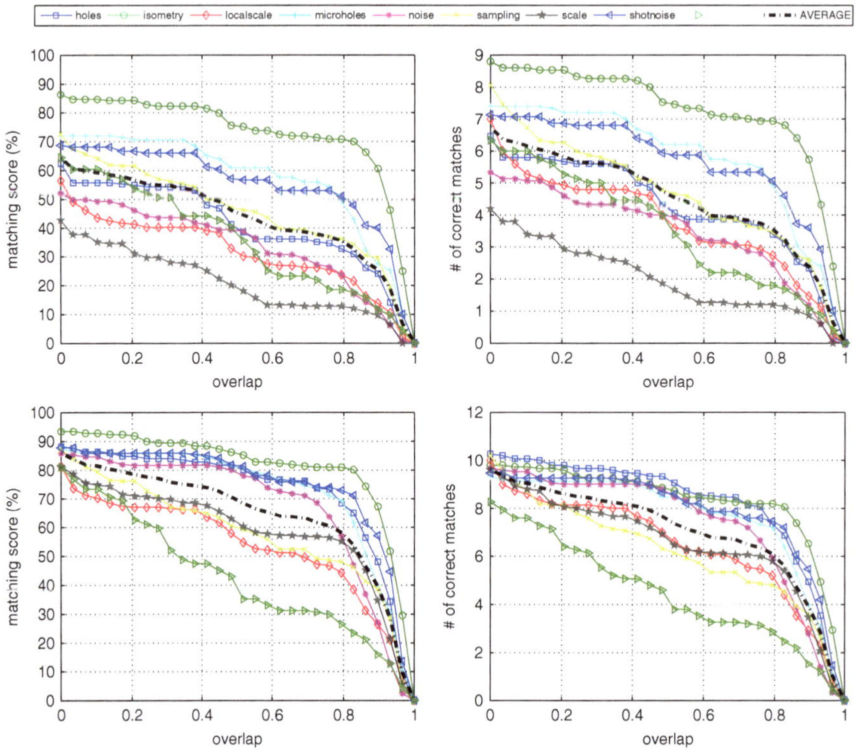

Fig. 8.10 Performance of region descriptors with regions detected using the vertex weight $h_t(v, v)$, t =2,048. Shown are the HKS descriptor (*first row*) and SI-HKS descriptor (*second row*)

8.6.3.3 Volume Descriptor

Scale invariant volume HKS descriptor (SI-vHKS) were calculated for every volumetric stable region detected in the previous section. When testing the volumetric setting we used only the SI-vHKS descriptor due to its superior performance in 2D.

As with the surface descriptors, the matching score was measured for each of the volumetric regions and is shown in Fig. 8.13. For comparison, we used the SI-HKS descriptors on the boundary for the detected 2D regions. The combination of volumetric regions with volumetric descriptors exhibited highest performance over the entire range of deformations.

A region matching experiment was performed on the volumetric regions, seeking the nearest neighbors of a selected query region. The query region was taken from the SCAPE dataset, and the nearest neighbors were taken from the TOSCA dataset which differs considerably from the former (SCAPE shapes are 3D scans of human figures, while TOSCA contains synthetic shapes). Despite the mentioned difference, correct matches were found consistently, as can be seen in Fig. 8.14

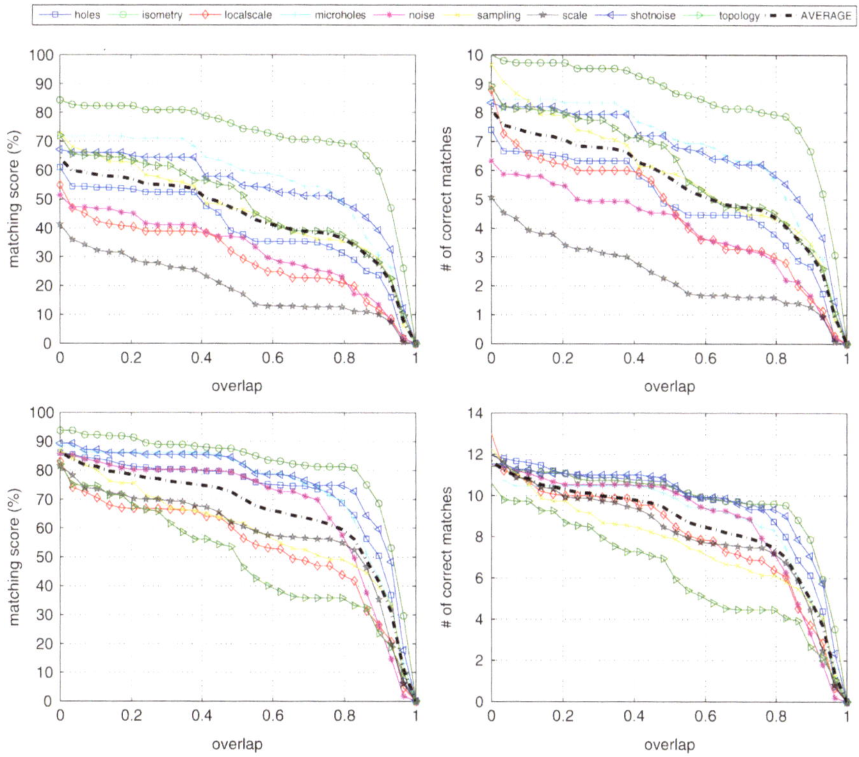

Fig. 8.11 Performance of region descriptors with regions detected using the edge weight $1/h_t(v_1, v_2)$, $t = 2{,}048$. Shown are the HKS descriptor (*first row*) and SI-HKS descriptor (*second row*)

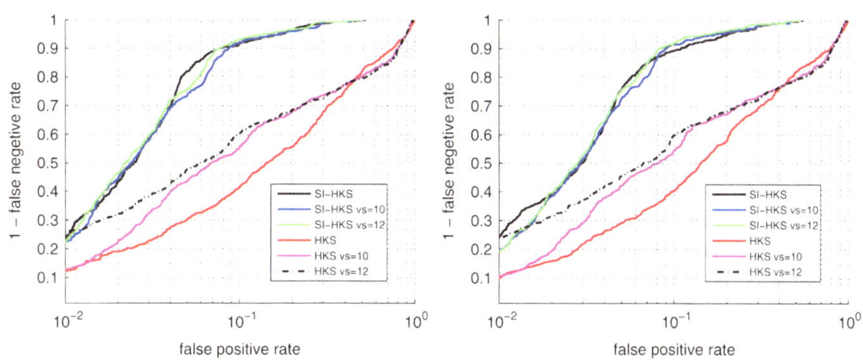

Fig. 8.12 ROC curves of different regions descriptors ("vs" stands for vocabulary size). The following detectors were used: vertex weight $h_t(v, v)$ (*left*), and edge weight $1/h_t(v_1, v_2)$ (*right*).

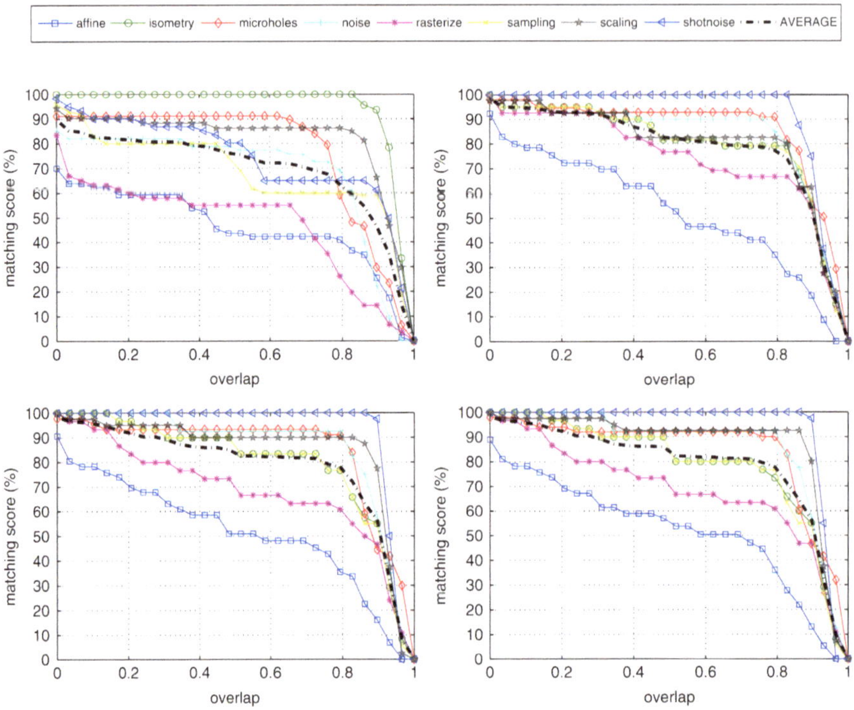

Fig. 8.13 Matching score of descriptors based on the 2D and 3D regions detected with the detectors shown in Fig. 8.9. Shown are the 2D SI-HKS (*upper left*) 3D SI-vHKS (*upper right* and two *bottom plots*) descriptors

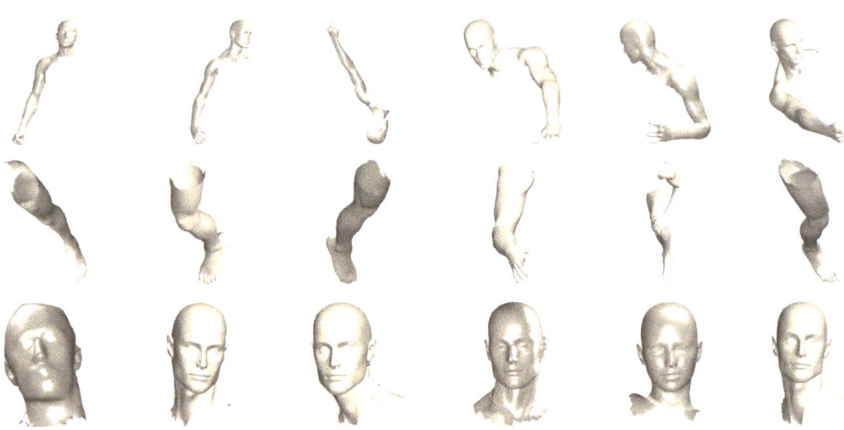

Fig. 8.14 Examples of closest matches found for different query regions from the SCAPE dataset on the TOSCA dataset. Shown from *left* to *right* are: query, 1st, 2nd, 4th, 10th, and 15th matches. Edge-weight $1/H_t(v_1, v_2)$ was used as the detector; average SI-vHKS was used as the descriptor

8.7 Conclusions

A generic framework for the detection of stable non-local features in deformable shapes is presented. This approach is based on a popular image analysis tool called MSER, where we maximize a stability criterion in a component tree representation of the shape. The framework is designed to unify the vertex or edge weights, unlike most of its predecessors. The use of diffusion geometry as the base of the weighting scheme make to detector invariant to non-rigid bending, global scaling and other shape transformations, a fact that makes this approach applicable in the challenging setting of deformable shape analysis.

The approach was shown to work with volumetric diffusion geometric analysis. In all experiments, our volumetric features exhibited higher robustness to deformation compared to similar features computed on the two-dimensional boundary of the shape. We also argue and exemplify that unlike features constructed from the boundary surface of the shape, our volumetric features are *not* invariant to volume-changing deformations of the solid object. We believe that this is the desired behavior in many applications, as volume isometries better model natural deformations of objects than boundary isometries.

We showed experimentally the high repeatability of the proposed features, which makes them a good candidate for a wide range of shape representation and retrieval tasks.

Acknowledgements We are grateful to Dan Raviv for providing us his volume rasterization and Laplacian disretization code. M. M. Bronstein is partially supported by the Swiss High-Performance and High-Productivity Computing (HP2C) grant. A. M. Bronstein is partially supported by the Israeli Science Foundation and the German-Israeli Foundation.

References

1. Anguelov, D., Srinivasan, P., Koller, D., Thrun, S., Rodgers, J., Davis, J.: SCAPE: shape completion and animation of people. TOG **24**(3), 408–416 (2005)
2. Aubry, M., Schlickewei, U., Cremers, D.: The wave kernel signature-a quantum mechanical approach to shape analyis. In: Proceedings of the CVPR, Colorado Springs (2011)
3. Boyer, E., Bronstein, A.M., Bronstein, M.M., Bustos, B., Darom, T., Horaud, R., Hotz, I., Keller, Y., Keustermans, J., Kovnatsky, A., Litman, R., Reininghaus, J., Sipiran, I., Smeets, D., Suetens, P., Vandermeulen, D., Zaharescu, A., Zobel, V.: Shrec '11: Robust feature detection and description benchmark. In: Proceedings of the 3DOR, Llandudno, pp. 71–78 (2011)
4. Bronstein, M.M., Kokkinos, I.: Scale-invariant heat kernel signatures for non-rigid shape recognition. In: Computer Vision and Pattern Recognition, San Francisco, pp. 1704–1711 (2010)
5. Bronstein, A., Bronstein, M.M., Bronstein, M., Kimmel, R.: Numerical Geometry of Non-rigid Shapes. Springer, New York (2008)
6. Bronstein, A., Bronstein, M.M., Bustos, B., Castellani, U., Crisani, M., Falcidieno, B., Guibas, L.J., Kokkinos, I., Murino, V., Ovsjanikov, M., et al.: SHREC 2010: robust feature detection and description benchmark. In: Eurographics Workshop on 3D Object Retrieval (2010)

7. Chazal, F., Guibas, L.J., Oudot, S.Y., Skraba, P.: Persistence-based clustering in riemannian manifolds. Research Report RR-6968, INRIA (2009)
8. Coifman, R.R., Lafon, S.: Diffusion maps. Appl. Comput. Harmonic Anal. **21**(1), 5–30 (2006)
9. Couprie, M., Bertrand, G.: Topological grayscale watershed transformation. In: SPIE Vision Geometry V Proceedings, San Diego, vol. 3168, pp. 136–146 (1997)
10. Dey, T.K., Li, K., Luo, C., Ranjan, P., Safa, I., Wang, Y.: Persistent heat signature for pose-oblivious matching of incomplete models. Comput. Graph. Forum **29**(5), 1545–1554 (2010)
11. Digne, J., Morel, J.-M., Audfray, N., Mehdi-Souzani, C.: The level set tree on meshes. In: Proceedings of the Fifth International Symposium on 3D Data Processing, Visualization and Transmission, Paris (2010)
12. Donoser, M., Bischof, H.: 3d segmentation by maximally stable volumes (msvs). In: Proceedings of the 18th International Conference on Pattern Recognition, vol. 1, pp. 63–66. IEEE Computer Society, Los Alamitos (2006)
13. Edelsbrunner, H., Letscher, D., Zomorodian, A.: Topological persistence and simplification. Discret.Comput. Geom. **28**(4), 511–533 (2002)
14. Floater, M.S., Hormann, K.: Surface parameterization: a tutorial and survey. In: Advances in Multiresolution for Geometric Modelling, vol. 1, pp. 157–186. Springer, Berlin (2005)
15. Forssen, P.E.: Maximally stable colour regions for recognition and matching. In: Proceedings of the CVPR, Minneapolis, pp. 1–8 (2007)
16. Huang, Q.X., Flöry, S., Gelfand, N., Hofer, M., Pottmann, H.: Reassembling fractured objects by geometric matching. ACM Trans. Graph. **25**(3), 569–578 (2006)
17. Johnson, A.E., Hebert, M.: Using spin images for efficient object recognition in cluttered 3D scenes. Trans. PAMI **21**(5), 433–449 (1999)
18. Kimmel, R., Zhang, C., Bronstein, A.M., Bronstein, M.M.: Are mser features really interesting? IEEE Trans. PAMI **33**(11), 2316–2320 (2011)
19. Levy, B.: Laplace-Beltrami eigenfunctions towards an algorithm that understands geometry. In: Proceedings of the IEEE International Conference on Shape Modeling and Applications 2006, pp. 13. IEEE Computer Society, Los Alamitos (2006)
20. Litman, R., Bronstein, A.M., Bronstein, M.M.: Diffusion-geometric maximally stable component detection in deformable shapes. Comput. Graph. **35**, 549–560 (2011)
21. Lowe, D.: Distinctive image features from scale-invariant keypoint. IJCV **60**(2), 91–110 (2004)
22. Matas, J., Chum, O., Urban, M., Pajdla, T.: Robust wide-baseline stereo from maximally stable extremal regions. Image Vis. Comput. **22**(10), 761–767 (2004)
23. Meyer, M., Desbrun, M., Schroder, P., Barr, A.H.: Discrete differential-geometry operators for triangulated 2-manifolds. In: Visualization and Mathematics III, pp. 35–57. Springer, Berlin (2003)
24. Mikolajczyk, K., Tuytelaars, T., Schmid, C., Zisserman, A., Matas, J., Schaffalitzky, F., Kadir, T., Gool, L.V.: A comparison of affine region detectors. IJCV **65**(1), 43–72 (2005)
25. Najman, L., Couprie, M.: Building the component tree in quasi-linear time. IEEE Trans. Image Proc. **15**(11), 3531–3539 (2006)
26. Ovsjanikov, M., Sun, J., Guibas, L.: Global intrinsic symmetries of shapes. Comput. Graph. Forum **27**(5), 1341–1348 (2008)
27. Ovsjanikov, M., Bronstein, A.M., Bronstein, M.M., Guibas, L.J.: Shape google: a computer vision approach to isometry invariant shape retrieval. In: Computer Vision Workshops (ICCV Workshops), Kyoto, pp. 320–327 (2009)
28. Pinkall, U., Polthier, K.: Computing discrete minimal surfaces and their conjugates. Exp. Math. **2**(1), 15–36 (1993)
29. Raviv, D., Bronstein, M.M., Bronstein, A.M., Kimmel, R.: Volumetric heat kernel signatures. In: Proceedings of the ACM Workshop on 3D Object Retrieval, Firenze, pp. 39–44 (2010)
30. Reuter, M., Wolter, F.-E., Peinecke, N.: Laplace-spectra as fingerprints for shape matching. In: Proceedings of the ACM Symposium on Solid and Physical Modeling, Cambridge, pp. 101–106 (2005)
31. Rustamov, R.M.: Laplace-Beltrami eigenfunctions for deformation invariant shape representation. In: Proceedings of the SGP, Barcelona, pp. 225–233 (2007)

32. Sivic, J., Zisserman, A.: Video google: A text retrieval approach to object matching in videos. In: Proceedings of the CVPR, Madison, vol. 2, pp. 1470–1477 (2003)
33. Skraba, P. Ovsjanikov, M., Chazal, F., Guibas, L.: Persistence-based segmentation of deformable shapes. In: Proceedings of the NORDIA, San Francisco, pp. 45–52 (2010)
34. Sumner, R.W., Popović, J.: Deformation transfer for triangle meshes. ACM Transactions on Graphics (Proceedings of the SIGGRAPH), Los Angeles, vol. 23, pp. 399–405 (2004)
35. Sun, J., Ovsjanikov, M., Guibas, L.: A concise and provably informative multi-scale signature based on heat diffusion. Comput. Graph. Forum **28**, 1383–1392 (2009)
36. Thorstensen, N., Keriven, R.: Non-rigid shape matching using geometry and photometry. In: Computer Vision – ACCV 2009, Xi'an, vol. 5996, pp. 644–654 (2010)
37. Toldo, R., Castellani, U., Fusiello, A.: Visual vocabulary signature for 3d object retrieval and partial matching. In: Proceedings of the 3DOR, Munich, pp. 21–28 (2009)
38. Tuzel, O., Porikli, F., Meer, P.: Region covariance: a fast descriptor for detection and classification. In: Computer Vision ECCV 2006. Lecture Notes in Computer Science, vol. 3952, pp. 589–600. Springer, Berlin/Heidelberg (2006)
39. Vincent, L., Soille, P.: Watersheds in digital spaces: an efficient algorithm based on immersion simulations. IEEE Trans. PAMI, **13**(6), 583–598 (2002)
40. Wardetzky, M., Mathur, S., Kaelberer, F., Grinspun, E.: Discrete laplace operators: no free lunch. In: Proceedings of the of Eurographics Symposium on Geometry Processing, Barcelona, pp. 33–37 (2007)
41. Zaharescu, A., Boyer, E., Varanasi, K., Horaud, R.: Surface feature detection and description with applications to mesh matching. In: Proceedings of the CVPR, Miami, pp. 373–380 (2009)

Chapter 9
A Brief Survey on Semi-Lagrangian Schemes for Image Processing

Elisabetta Carlini, Maurizio Falcone, and Adriano Festa

Abstract In this survey we present some semi-Lagrangian schemes for the approximation of weak solutions of first and second order differential problems related to image processing and computer vision. The general framework is given by the theory of viscosity solutions and, in some cases, of calculus of variations. The schemes proposed here have interesting stability properties for evolutive problems since they allow for large time steps, can deal with degenerate problems and are more accurate if compared to standard finite difference/element methods of the same order. Several examples on classical problems will illustrate these properties.

9.1 Introduction

Nonlinear Partial Differential Equations (PDEs) appear in many classical image processing problems and there is a need for accurate and efficient numerical methods to solve them. Usually we have to deal with several difficulties, including the fact that in most cases these are degenerate problems with non smooth solutions. Classical methods (Finite Elements, Finite Differences) must be modified and often stabilized to solve this kind of problems (the interested reader will find in the books by Strikwerda [46] and Brenner and Scott [6] a comprehensive introduction to these classical techniques and well as some recent results). The goal of this paper is to present semi-Lagrangian (SL in the sequel) approximation techniques and illustrate their properties through the solution of some classical image processing problems.

SL methods were introduced long time ago for linear advection problems. As far as we know the first example is the Courant-Isaacson and Rees scheme [19] which dates back to the 1960s. However, their application to nonlinear problems is more

E. Carlini (✉) · M. Falcone · A. Festa
Dipartimento di Matematica "G. Castelnuovo", Sapienza – Università di Roma,
P. Aldo Moro, 2 – 00185 Roma, Italy
e-mail: carlini@mat.uniroma1.it; falcone@mat.uniroma1.it; festa@mat.uniroma1.it

M. Breuß et al. (eds.), *Innovations for Shape Analysis*, Mathematics and Visualization,
DOI 10.1007/978-3-642-34141-0_9, © Springer-Verlag Berlin Heidelberg 2013

recent as well as the analysis of their mathematical properties in terms of consistency, stability, convergence and convergence rates. A comprehensive introduction to this class of schemes can be found in the book by Falcone and Ferretti [32].

Let us just recall here some of the motivations which make semi-Lagrangian schemes an interesting alternative to more classical finite differences and finite elements techniques. The first and perhaps more interesting point is that, due to the fact that they mimic the method of characteristics, they allow for large time steps in evolutive problems. This property is particularly useful when one looks for the limiting behavior for t going to infinity in order to recover the solution of the corresponding stationary problem. The second important property is that they introduce only a small numerical viscosity so that their smoothing effect on the solution is very limited. The third property to be mentioned is that they have a natural interpretation in terms of the representation formula for some nonlinear problems. One example is the solution of the eikonal equation where the Hopf-Lax formula applies (see the next section for details). Roughly speaking they can be interpreted as a discrete Hopf-Lax formula and this means that their accuracy can be increased to high-order. Finally, they can work on structured and unstructured grids. Although the latter are not very popular among the image processing community they are a natural choice in other fields of applications like fluid dynamics, meteorology, geophysics. The details regarding the analytical properties of the schemes will not be illustrated here since this survey is mainly intended to present the schemes and their potential via the solution of some test problems.

In the next section we will briefly describe how the schemes are derived for some first and second order problems which come from image processing. We will consider some simplified models in order to introduce the main ideas of SL approximation and to explain how the methods are constructed. The following sections are devoted to some classical problems in image processing: Shape from Shading, nonlinear filtering, segmentation and optical flow. For each of them we will introduce the model which we are going to approximate, present the corresponding SL scheme and illustrate some numerical experiments.

9.2 An Introduction to Semi-Lagrangian Schemes for Nonlinear PDEs

We will consider three classical examples to show the main features and properties of the SL approximation: Shape from Shading (SfS in the sequel), segmentation via the level set (LS) method and nonlinear filtering via Mean Curvature Motion (MCM). It is worth to note that they all lead to nonlinear (and possibly degenerate) partial differential models.

Let us start with the *Shape from Shading* problem. This is an inverse problem where we want to compute the surface corresponding to a single image. A simple model for this problem can be derived (see the next section for more details) assuming that the light source is at infinity in the direction ω, the surface is Lambertian, and that we can neglect perspective deformations. We consider then the equation which appears in most of the papers and corresponds to frontal light source at infinity, i.e. $\omega = (0, 0, 1)$ and

$$|\nabla u(x)| = f(x) \qquad \text{for } x \in \Omega, \tag{9.1}$$

where $I : \Omega \subset \mathbb{R}^2 \to [0, 1]$ represents the brightness function of the image and

$$f(x) = \sqrt{\frac{1}{I(x)^2} - 1}. \tag{9.2}$$

This is an *eikonal equation*, a first order stationary nonlinear pde. In order to solve (9.1) we must provide some boundary conditions on $\partial\Omega$ and/or add some informations to select a unique solution. In fact, the equation just depends on the gradient so that if u is a solution also $u + constant$ is a solution and we need to fix the height of the surface at least at the boundary to select a solution. For an image containing an "occluding boundary", it is usual to consider this boundary as $\partial\Omega$ and the domain of reconstruction will be the region Ω enclosed by the occluding boundary.

Another classical problem is the *segmentation problem*, where we want to detect the boundaries of objects represented in a picture. A very popular method for segmentation is based on the level set method, this application is often called "active contour" since the segmentation is obtained following the evolution of a simple curve (a circle for example) in its normal direction. Starting from a circle inside an object we follow its evolution until the curve touches the borders of the object. It is important to know that the LS method has had a great success for the analysis of front propagation problems for its capability to handle many different physical phenomena within the same theoretical framework. Moreover, LS methods allow to develop the analysis of interface propagation also after the onset of singularities (as for example, when there is a topological change in the front). See the books by Sethian [45] and Osher and Fedkiw [42].

In the LS methods the unknown is a "representation" function $u : \mathbb{R}^2 \times [0, T] \to \mathbb{R}$ of the interface, the position of the interface Γ_t at time t is given by the 0-level set of $u(., t)$, i.e.

$$\Gamma_t := \{x : u(x, t) = 0\}.$$

The model equation corresponding to the LS method (see e.g.[43] for details) is

$$\begin{cases} u_t + c(x, t)|\nabla u| = 0 & (x, t) \in \mathbb{R}^2 \times [0, T] \\ u(x) = u_0(x) & x \in \mathbb{R}^2 \end{cases} \tag{9.3}$$

where u_0 must be a *representation function for the initial front* $\partial\Omega_0$, where Ω_0 is a given open bounded subset of $\mathbb{R}^2$, this means

$$\begin{cases} u_0(x) > 0 & \text{for } x \in \mathbb{R}^2 \setminus \overline{\Omega_0} \\ u_0(x) = 0 & \text{for } x \in \partial\Omega_0 := \Gamma_0 \\ u_0(x) < 0 & \text{for } x \in \Omega_0. \end{cases} \tag{9.4}$$

Note that usually the scalar velocity $c : \mathbb{R}^2 \times [0, T] \to \mathbb{R}$ is given and it represents the velocity of the front in its normal direction $\eta(x,t) := \frac{\nabla u(x,t)}{|\nabla u(x,t)|}$. However, the same approach applies to other scalar velocities. For example one can use it to describe isotropic and anisotropic front propagation, Mean Curvature Motion (MCM) and other situations when the velocity depends on some geometrical properties of the front. In summary, this corresponds to the following choices:

$$c(x,t) \qquad \text{isotropic growth with time varying velocity} \tag{9.5}$$

$$c(x,t,\eta) \quad \text{anisotropic growth, i.e. dependent on the normal direction} \tag{9.6}$$

$$c(x,t,k) \qquad \text{Mean Curvature Motion} \tag{9.7}$$

where $k = k(x,t)$ is the mean curvature to the front at time t. There are even models where the velocity is obtained by convolution (nonlocal velocity).

Since the borders of objects inside a given image I are characterized by the fact that there is a rapid change in the values of I, in the segmentation problem one can take the gradient of I as an indicator of the edges. If $|\nabla I|$ is large this means that we are at an edge. For this reason, a popular choice of the velocity when we apply LS methods to the segmentation problem is

$$c(x) := (1 + |\nabla I(x)|^p)^{-1} \text{ where } p \geq 1.$$

With this choice, the velocity is 1 inside the region where the gray level is constant and rapidly decreases to 0 when x is close to an edge (which corresponds to a jump in the I derivative).

Another problem we will consider is *nonlinear filtering*. A classical model to treat this problem is to consider (9.3) with the velocity given by Mean Curvature $c(x,t) = k(x,t)$, see [34]. In this case one gets

$$\begin{cases} u_t(x,t) = \operatorname{div}\left(\dfrac{\nabla u(x,t)}{|\nabla u(x,t)|}\right)|\nabla u(x,t)| & (x,t) \in \mathbb{R}^2 \times [0,T] \\ u(x,0) \;= I_0(x) \end{cases} \tag{9.8}$$

where I_0 is a starting noisy image. This equation, in the level set formalism, represents the Mean Curvature Motion of curves, since each level set of u moves in normal direction with a speed proportional to their mean curvature. The goal is to recover a new image reducing the noise. Note that the evolutive equation is not well defined at all points where the gradient vanishes so it is a singular parabolic equation. This implies that classical finite difference and finite element methods must be adapted in order to deal with the singularities (see [43] for a finite difference scheme and [15, 41] for two convergence results).

Let us now turn our attention to the construction of the semi-Lagrangian schemes for the above model equations. For readers' convenience, we will start from the linear advection equation:

$$u_t + b(x,t)\cdot\nabla u = f(x,t) \qquad (x,t)\in\mathbb{R}^2\times(0,T). \tag{9.9}$$

Here, $b:\mathbb{R}^2\times(0,T)\to\mathbb{R}^2$ is a vectorfield (the *drift*) and $f:\mathbb{R}^2\times(0,T)\to\mathbb{R}$ is the *source* term. We look for the unique solution $u:\mathbb{R}^2\times(0,T)\to\mathbb{R}$ of (9.9) satisfying the initial condition

$$u(x,0) = u_0(x) \qquad x\in\mathbb{R}^2. \tag{9.10}$$

A simple case corresponds to $f(x,t)\equiv 0$ and $b(x,t)\equiv c$ (constant). Then, the solution u is given by the representation formula

$$u(x,t) = u_0(x - c\,t) \qquad (x,t)\in\mathbb{R}^2\times[0,T) \tag{9.11}$$

which can be easily obtained by the *method of characteristics*. The representation formula (9.11) contains an important information: the solution at the point x at time t only depends on the initial value at the point $x - ct$ which is called *the foot of the characteristic* passing through (x,t) (more details on the method of characteristics can be found on many classical books presenting the basic theory of partial differential equations, e.g. [28]). A typical choice is to build a space grid with constant step Δx

$$\mathcal{G}_{\Delta x} := \{x_j : x_j = (j_1\Delta x, j_2\Delta x) \text{ for } j_1\in\mathbb{Z},\ j_2\in\mathbb{Z}\}. \tag{9.12}$$

This grid is called *lattice*. In order to build an approximation scheme for (9.9) and more in general for evolutive problems, we need to consider a space-time grid

$$\mathcal{G}_{\Delta x,\Delta t} := \{(x_j,t_n) : x_j = (j_1\Delta x, j_2\Delta x),\ t_n = n\Delta t, \text{ for } j_1\in\mathbb{Z},\ j_2\in\mathbb{Z},\ n\in\mathbb{N}\} \tag{9.13}$$

where in the simplest case Δt is a constant time step. Naturally for numerical purposes the number of nodes of our grids has to be finite and we will consider the intersections of our grids with the computational domain Ω.

The basic idea behind all finite difference approximations is to replace every derivative by an incremental ratio. Thus, one obtains a finite dimensional problem whose unknown are the values of the numerical solution at all the nodes of the lattice, so that the value u_j^n associated to the node (x_j,t_n) should be regarded as an approximation of $u(x_j,t_n)$. For the time derivative it is natural to choose the forward incremental ratio

$$u_t(x,t) \approx \frac{u(x,t+\Delta t) - u(x,t)}{\Delta t}. \tag{9.14}$$

For finite difference, it is well known that dealing with advection problems we need to consider the incremental ratio *in the up-wind direction*, so that we must take the first node on the left (right) if the velocity is positive (negative). This choice guarantees the stability of the approximation scheme as well as a good fit with the physics behind the model. When $b(x_j, t_n)$ is positive, this corresponds to the following *up-wind* scheme for the left hand side of the advection equation

$$u_t(x_j, t_n) + b(x_j, t_n) \cdot \nabla u(x_j, t_n) \approx \frac{u(x_j, t_n + \Delta t) - u(x_j, t_n)}{\Delta t} + \\ -b_1(x_j, t_n) \frac{u(x_{j_1-1,j_2}, t_n) - u(x_{j_1,j_2}, t_n)}{\Delta x} - b_2(x_j, t_n) \frac{u(x_{j_1,j_2-1}, t_n) - u(x_{j_1,j_2}, t_n)}{\Delta x}. \tag{9.15}$$

Adopting the standard notation $u_j^n = u(x_j, t_n)$, we write the scheme in the standard explicit time marching form

$$u_j^{n+1} = u_j^n + \frac{\Delta t}{\Delta x} b(x_j, t_n) \cdot (u_{j_1-1,j_2}^n - u_j^n, u_{j_1,j_2-1}^n - u_j^n). \tag{9.16}$$

Semi-Lagrangian schemes are based on a different way to construct the approximation of (9.9), in particular this affects the discretization of the advection term $b(x,t) \cdot \nabla u$. Since this is a directional derivative, we can write

$$b(x_j, t_n) \cdot \nabla u(x_j, t_n) \approx -\frac{u^n(x_j - \delta b(x_j, t_n)) - u_j^n}{\delta}, \tag{9.17}$$

where δ is a "small" positive parameter, and u^n denotes an extension of the numerical solution (at time t_n) to be computed outside of the grid. Coupling the forward finite difference in time with this approximation we get

$$\frac{u_j^{n+1} - u_j^n}{\Delta t} - \frac{u^n(x_j - \delta b(x_j, t_n)) - u_j^n}{\delta} = 0, \tag{9.18}$$

and finally, choosing $\delta = \Delta t$, we obtain the scheme

$$u_j^{n+1} = u^n(x_j - \Delta t b(x_j, t_n)). \tag{9.19}$$

This is the semi-Lagrangian scheme for (9.9) for $f(x,t) = 0$. Note that, for $b(x,t) \equiv c$ (9.18) can be easily interpreted as a discrete version of the representation formula (9.11). Even in this simple case, the value at the point $x_j - \Delta t c$ does not belong to the grid $\mathcal{G}_{\Delta x}$ and must be obtained by interpolation on the values at the grid nodes. We will denote by $\Pi[w]$ the interpolation operator which reconstructs locally the values of a function w only known on $\mathcal{G}_{\Delta x}$. There are many possible choices for Π, e.g. linear, bilinear or cubic interpolation (see also [32] for more general choices). A very popular choice is the piecewise linear interpolation which produces a monotone first order scheme.

Now let turn our attention to the nonlinear case and consider the eikonal equation. First note that

$$|\nabla u(x)| = \max_{a \in B(0,1)} a \cdot \nabla u(x). \tag{9.20}$$

where $B(0,1)$ denotes the unit ball centered at the origin. In fact, the maximum in (9.20) is attained at $a^* = \nabla u(x)/|\nabla u(x)|$ and this shows that the right-hand side coincides with $|\nabla u(x)|$. Using now (9.20) to rewrite Eq. (9.3), we get

$$u_t + \max_{a \in B(0,1)} [c(x,t)a \cdot \nabla u(x)] = 0. \tag{9.21}$$

Since the term inside the square brackets is linear, we can apply the semi-Lagrangian scheme for the advection equation obtaining the following scheme

$$u_j^{n+1} = \min_{a \in B(0,1)} [u^n(x_j - \Delta t c(x_j, t_n)a)]. \tag{9.22}$$

Note that again u^n is computed at a point which in general is not a node of the grid, this will require a local reconstruction via the interpolation operator Π. We can write (9.22) in a more explicit way as

$$u_j^{n+1} = \min_{a \in B(0,1)} [\Pi[u^n](x_j - \Delta t c(x_j, t_n)a)]. \tag{9.23}$$

The meaning of the right-hand side is: to recover the correct solution of the nonlinear stationary equation one needs to solve a family of advection problems corresponding to a vector field pointing in every direction and then has to take the minimum value among all possible ones. This step clearly requires a constrained optimization method to be solved. In practice, for low order accuracy it suffices to compare a finite number of values corresponding to a uniform discretization of the unit ball.

It is important to note that the SL scheme is *up-wind* by construction and this explains its strong stability properties which allow to use large time steps with respect to other methods.

The derivation of an efficient discretization scheme for the MCM equation is more complicated. One can think to simply substitute $c_j^n = k_j^n$ where $k_j^n = k(x_j, t_n)$ is the curvature at the point x_j at time t_n. This can be done using a discrete version of the curvature for the level set of the solution as in [43]. However, via a stochastic interpretation of the equation, one can derive a semi-Lagrangian approximation where the role of characteristics is taken by the solution of an appropriate stochastic differential equation associated to the equation. More information on this approach for the MCM can be found in [14] and are too technical to be presented here. However, we will present the resulting SL scheme in Sect. 9.4.1.

In conclusion, the construction of SL-schemes is based on three essential building blocks: an integration backward along the characteristics (or the generalized characteristics in the second order problem), an interpolation to get the values at the foot of the characteristics and an optimization method to recover the minimum among all the values. The first step relies on standard techniques for ordinary differential equation, the interpolation can be obtained either by standard polynomial techniques or via more sophisticated techniques, for example ENO (Essentially Non Oscillatory) or WENO (Weighted ENO) methods as in [13]. Finally, for the optimization step, one can simply discretize the unit ball or use a minimization method without derivatives such as Brent algorithm [7]. An extensive presentation of those building blocks and an analysis of SL schemes can be found in [32].

9.3 Shape from Shading

As we said, we want to reconstruct a surface $u(x)$ from a single image.

This means that given an image I we are looking for the surface $u(x)$ corresponding to the object represented in it. The irradiance function I is the datum in the model since it is measured at each pixel $x := (x_1, x_2)$ of the image, for example in terms of a grey level (from 0 to 255). To construct a continuous model, we will assume that I takes real values in the interval $[0, 1]$. The height function u which is the unknown of the problem has to be reconstructed on the "reconstruction domain" Ω. The characterization of the surface via a partial differential equation relies on several assumptions. Assume that there is a unique light source at infinity whose direction is indicated by the unit vector $\omega = (\omega_1, \omega_2, \omega_3) \in \mathbb{R}^3$. Also assume for simplicity that ω is given. For a Lambertian surface of uniform albedo equal to 1, the reflectance function giving the brightness at a point on the surface is $R(\eta(x)) := \omega \cdot \eta(x)$, where $\eta(x)$ is the unit normal to the surface at $(x_1, x_2, u(x))$. This equation is called the *irradiance equation*. Recalling that the normal is given by $\eta(x) = (-u_{x_1}, -u_{x_2}, 1)$ we obtain the following equation

$$I(x) = \frac{(\omega_1, \omega_2) \cdot \nabla u(x) - \omega_3}{\sqrt{1 + |\nabla u(x)|^2}} \qquad \text{for } x \in \Omega, \tag{9.24}$$

which is a first order nonlinear partial differential equation of Hamilton-Jacobi type. We have seen that in the vertical light source case (i.e. $\omega = (0, 0, 1)$) this reduces to the eikonal equation (9.1).

As we said, this equation must be complemented with boundary conditions. A natural choice is to consider homogeneous Dirichlet type boundary conditions corresponding to flat background i.e., setting

$$u(x) = 0 \qquad \text{for } x \in \partial\Omega. \tag{9.25}$$

However, one can also choose a more general Dirichlet boundary condition

$$u(x) = g(x) \qquad \text{for } x \in \partial\Omega. \tag{9.26}$$

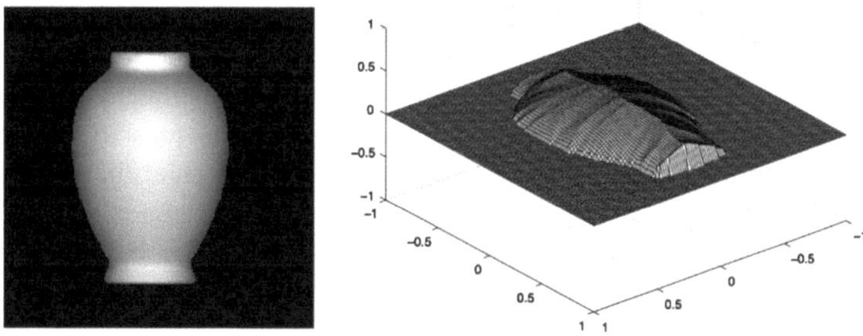

Fig. 9.1 The image and its corresponding surface (approximation)

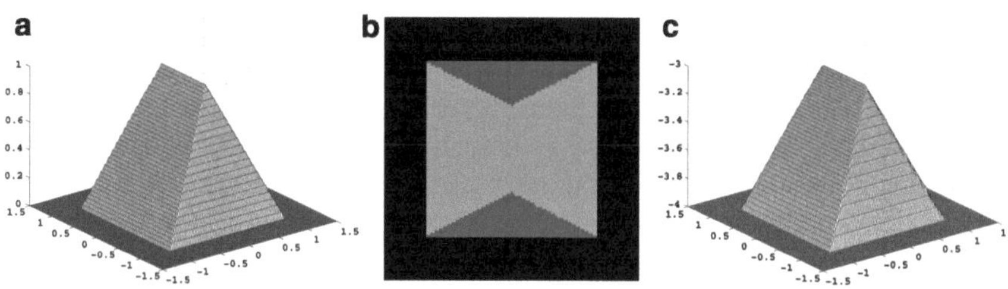

Fig. 9.2 (**a**) The original surface, (**b**) its brightness function in the plane, (**c**) the approximate surface

The solution of the above Dirichlet problems (9.24) and (9.25) or (9.24) and (9.26) will give the surface corresponding to grey level $I(x)$ measured in Ω. To set $u(x) = g(x)$ at least on a part of the boundary we must know something about the surface. This is the case for surfaces obtained by revolution around an axis (as for our vase in Fig. 9.1).

In order to illustrate one of the features of the SL scheme, let us consider a virtual image corresponding to the surface

$$u(x_1, x_2) = \begin{cases} 2(1 - |x_1|) & x_1 \in [-1, 1], \quad x_2 \in \left[-1, -\frac{1}{2}|x_1| - \frac{1}{2}\right] \\ 2(1 - |x_2|) & x_1 \in [-1, 1], \quad x_2 \in \left[\frac{1}{2}|x_1| + \frac{1}{2}, 1\right] \\ 1 - |x_1| & \text{otherwise.} \end{cases}$$

The surface looks like a "ridge tent" (see Fig. 9.2), so it is non regular but the boundary conditions are very simple: 0 on each side of the rectangle. We apply our SL scheme for the eikonal equation (9.22). Looking at the right picture of Fig. 9.2, we can see that the algorithm is accurate around the kinks and that there are no spurious oscillations where the gradient is discontinuous. Similar results have been obtained for other virtual and real images. A detailed presentation of these results as well as a comparison with other numerical methods can be found in the survey [27].

Let us also mention that the same problem can be solved by applying optimization methods based on the variational approach. The interested reader can find in Horn and Brooks' book [35] several results and references relating to the variational approach. In this class of methods two basic ingredients must be chosen: the functional which has to be optimized (in fact, minimized) and the minimization method. The first difficulty encountered in the SfS problem is the choice of unknowns. The natural unknown is of course height u as we have done before. However, this is not the only choice since u appears in the image irradiance equation only through its first derivatives $p = \partial u/\partial x_1$ and $q = \partial u/\partial x_2$, which are two non-independent functions. In fact, for $u \in C^2$ we know that

$$\partial p/\partial x_2 = \partial q/\partial x_1. \tag{9.27}$$

The only technical difficulty with these unknowns is that p or q becomes infinite at each point x belonging to an occluding boundary. This is not a real trouble if no point x in the reconstruction domain Ω is such that $I(x) = 0$. As equation (9.27) is a hard constraint on p and q, the most natural functional associated with the irradiance equation and (9.27) is

$$\mathscr{F}_1(p,q,\mu) = \int_\Omega [r(p(x),q(x)) - I(x)]^2\, dx + \int_\Omega \mu(x)\left[\frac{\partial p}{\partial x_2}(x) - \frac{\partial q}{\partial x_1}(x)\right] dx \tag{9.28}$$

where μ is a Lagrange multiplier and the function r is such that $r(p(x),q(x)) = R(n(x))$. Note that $\mathscr{F}_1$ is defined on the unknowns p, q and μ. However Horn and Brooks have shown that the three Euler equations associated with $\mathscr{F}_1$ can be reduced, for $u \in C^2$, to the Euler equation associated with the following functional:

$$\mathscr{F}_2(p,q) = \int_\Omega [r\,(p,q) - I(x)]^2\, dx. \tag{9.29}$$

The choice of p and q as unknown is due to the fact that the algorithms dealing directly with u are very slow, this choice implies that at the end we will need also another procedure to get back to u. Typically this is done integrating along paths.

As we said, in the previous model we have neglected the *perspective deformations*. New models have been studied to deal with this problem, which is usually called in the literature Perspective Shape from Shading (PSfS in short) problem. Let us define our model problem adopting the same notation used in [20] (a different model with an attenuation term has been studied in [44]). The point (X_0, Y_0) is the principal point of the image, i.e. the intersection between the optical axis of the camera and the perspective plane Π (the film). The parameters d and d' are respectively the distance of the optical lens from the perspective plane and the distance of the optical lens from the (flat) background, l and $l' = \frac{d'}{d} l$ are

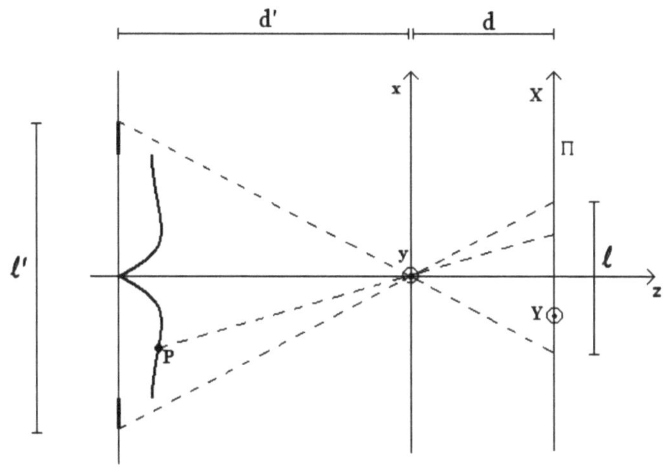

Fig. 9.3 Diagram of the optical lens and of the perspective transformation

respectively the length of a segment in the perspective plane (i.e. in the image) and the length of the real segment corresponding to it (see Fig. 9.3 for more details). The representation of the surface defined by points with coordinates (x, y, z) in terms of the (X, Y) coordinates of the points in the perspective plane is given by three parametric equations

$$x = r(X, Y), \; y = s(X, Y), \; z = t(X, Y) \tag{9.30}$$

where (see [20])

$$\begin{cases} r(X, Y) = \dfrac{X - X_0}{d} t(X, Y) \\ s(X, Y) = \dfrac{Y - Y_0}{d} t(X, Y). \end{cases} \tag{9.31}$$

The problem is then reduced to compute the third component t. Writing the classical irradiance equation, taking into account the new variables for the perspective deformation and introducing the new variable

$$\bar{t}(X, Y) := t(X, Y) + (X - X_0, Y - Y_0) \cdot \nabla t(X, Y), \tag{9.32}$$

we can get the following eikonal type equation

$$\left[\frac{d}{\bar{t}(X, Y)}\right]^2 |\nabla t(X, Y)|^2 = \frac{I_{max}^2}{I'(X, Y)^2} - 1 \quad \text{in } \Omega \tag{9.33}$$

where

$$I'(X,Y) := \frac{I(X,Y)}{cos^4\alpha(X,Y)}, \tag{9.34}$$

I_{max} is a constant depending on the parameters of the problem (typically is the maximum value of I measured on the image) and the α is the angle between the optical axis and the the ray starting at a point of the scene and passing through the optical center. Note that α varies with the point. Moreover, the coefficient

$$\cos^4(\alpha(X,Y)) = \frac{d^4}{((X-X_0)^2 + (Y-Y_0)^2 + d^2)^2}, \tag{9.35}$$

is used to decrease the brightness of the points corresponding to a large α. Note that Ω is the internal region bounded by the silhouette of the object ($\partial\Omega$ will denote its boundary) which is embedded in a rectangular domain Q, so that the set $Q \setminus \Omega$ is the background.

Defining

$$f(X,Y) := \frac{1}{d^2}\left(\frac{I_{max}^2}{I'(X,Y)^2} - 1\right) \tag{9.36}$$

we can write (9.33) in short as

$$|\nabla t(X,Y)| = \sqrt{f(X,Y)}\left|\bar{t}(X,Y)\right|. \tag{9.37}$$

We want to write (9.37) in a fixed point form and construct an approximation scheme for this equation. To this end it is important to note that $\bar{t}$ has a sign. In fact, the normal to the original surface in the point P is given by

$$\widehat{n}(P) = N(P)/|N(P)| \tag{9.38}$$

where

$$N(P) \equiv \left(\frac{\partial t}{\partial X}(X,Y), \frac{\partial t}{\partial Y}(X,Y), -\bar{t}(X,Y)\right) \tag{9.39}$$

and since $-\bar{t}$ must be positive (according to the orientation of the z axis in Fig. 9.3), $\bar{t}$ must be negative. This implies that (9.37) is in fact

$$|\nabla t(X,Y)| + \sqrt{f(X,Y)}(t(X,Y) + (X-X_0, Y-Y_0)\cdot\nabla t(X,Y)) = 0$$

which can be written in short as

$$H((X,Y), t, \nabla t) = 0, \quad \text{in } \Omega \tag{9.40}$$

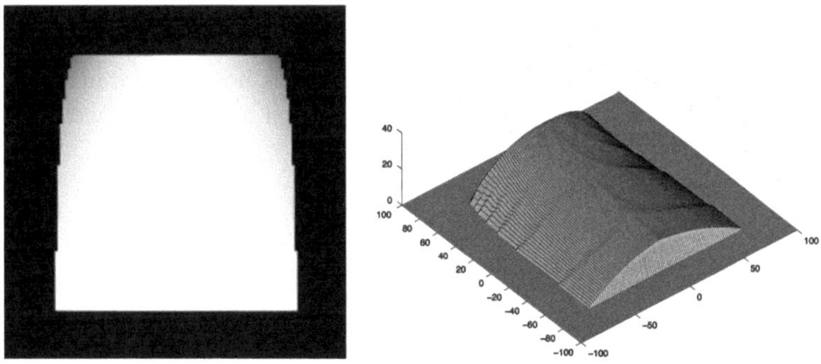

Fig. 9.4 A real photograph of one page, 128 × 128 pixels (*left*) and reconstructed surface with Dirichlet boundary conditions (*right*)

where the Hamiltonian H represents the left-hand side of (9.40). It should be noted that, with respect to the classical SfS problem without perspective, we have an additional term which is linear in ∇t.

Let us consider equation (9.40) complemented with the Dirichlet boundary condition

$$t = g(X, Y) \quad \text{on } \partial\Omega, \quad \text{where } -d' \leq g \leq 0 \tag{9.41}$$

The arguments presented in Sect. 9.2 for the approximation of the directional derivative lead to the standard semi-Lagrangian scheme for (9.40) which we write in a fixed point form as

$$t(X, Y) = F[t](X, Y), \quad \text{in } \Omega \tag{9.42}$$

where the operator F is defined as

$$F[t](X, Y) := \frac{1}{1+\delta} \inf_{a \in B(0,1)} \left\{ t\left((X, Y) + \delta\left(\frac{-a}{\sqrt{f}} - (X, Y)\right)\right)\right\}, \quad \text{on } \Omega. \tag{9.43}$$

and $B(0, 1)$ is the unit ball in $\mathbb{R}^2$. Introducing a space discretization, and defining $x_j = (X_{j_1}, Y_{j_2})$, we can write for every node $x_j \in \Omega \cap \mathcal{G}_{\Delta x}$ Eq. (9.42)

$$t(x_j) = F[t](x_j) := \frac{1}{1+\delta} \inf_{a \in B(0,1)} \left\{ \Pi[t]\left(x_j + \delta\left(\frac{-a}{\sqrt{f(x_j)}} - x_j\right)\right)\right\} \tag{9.44}$$

which gives the fully discrete semi-Lagrangian scheme for the PSfS problem. Naturally, this equation must be complemented with the boundary conditions (9.41).

In the following test we start from a real photograph where the effect of perspective is visible. The surface is a sheet of paper with the shape of a roof tile. For this image the parameter values are: $l = 6.91$ mm, $d = 5.8$ mm, $l' = 200$ mm, $d' = \frac{l'}{l} d = 167.87$ mm, $\Delta x = 0.05$ mm. We note that we performed a light correction in the preprocessing step, so we can assume $I_{max} = 1$ during computation. Figure 9.4 shows the photograph (128 × 128 pixels) and the surface reconstructed using Dirichlet boundary conditions ($t = 0$ on the left and right sides of the boundary and $t = g$ on the top and the bottom). We can see that the solution is quite good considering the fact that light source (flash camera) is not far from the object and that direction of light source is not perfectly vertical as the mathematical model would have required.

9.4 Nonlinear Filtering via MCM

The aim of this section is to present two classical methods based on Mean Curvature Motion for image restoration. These models can be written in the following general form:

$$\begin{cases} u_t(x,t) = \operatorname{div}\left(\dfrac{\nabla u(x,t)}{|\nabla u(x,t)|}\right)^{\beta} |\nabla u(x,t)| & (x,t) \in \Omega \times (0,T] \\ u(x,0) = I_0(x) & x \in \Omega \\ \dfrac{\partial u(x,t)}{\partial \eta} = 0 & (x,t) \in \partial\Omega \times (0,T] \end{cases} \tag{9.45}$$

where $\eta(x)$ is the normal direction at the point x. We will consider the case $\beta = 1$ corresponding to classical Mean Curvature Motion (MCM) and the case $\beta = 1/3$ corresponding to Affine Morphological Scale Space (AMSS). Here $\{u(x,t)\}_{t \geq 0}$ represents a family of successive restored versions of the initial noisy image I_0. The variable t is called the scale variable and, as t increases, we expect $u(x,t)$ be a smoothed image, in which the noise has been removed. The model is able to preserve significant features, like edges and corners, which are typically lost when using linear PDE-methods based on the heat equation where the smoothing effect is isotropic. In curvature-related diffusion, the smoothing effect is anisotropic, i.e. the diffusion depends on the image and is not constant in all directions. The model with $\beta = 1$ corresponds to "pure" anisotropic diffusion and it was introduced for image processing by Alvarez et al. in [1]. The term $\operatorname{div}\left(\frac{\nabla u(x,t)}{|\nabla u(x,t)|}\right) |\nabla u(x,t)|$ represents a degenerate diffusion term, which diffuses in the direction orthogonal to the gradient ∇u and does not diffuse in the direction of ∇u. In fact, denoting by $\sigma(\nabla u) = \left(\frac{\nabla u}{|\nabla u|}\right)^{\perp}$ the unit vector orthogonal to ∇u, a formulation of the previous equation with respect to $\sigma(\nabla u)$ would be:

$$u_t = \sigma(\nabla u)^{\top} D^2 u \sigma(\nabla u),$$

where $D^2 u$ is the Hessian matrix with respect to x. This equation admits a unique continuous viscosity solution, if $u(x,0)$ is bounded uniformly continuous, see [18, 29].

In the paper [2] by Alvarez et al., a connection between scale space analysis and PDEs is proved, the authors also show that the MCM operator satisfies mostly

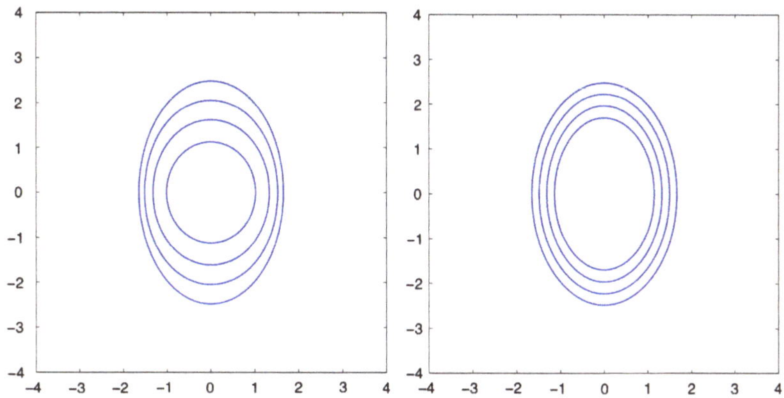

Fig. 9.5 Shrinking of an ellipse by MCM (*left*) and AMSS (*right*)

all the relevant properties in multi-scale analysis as monotonicity, invariance under grey scale change, translation invariance. Moreover, they prove that the only operator satisfying all these properties together with affine invariance is the model represented by (9.45) with $\beta = 1/3$. Affine invariance requires the solution of the equation to be invariant with respect any affine transformation: let $\{T_t\}_{t\geq 0}$ represents a family of operators which applied to the original image yields a sequence of images $u(x,t) = T_t(u)$, solution of (9.45), then the affine invariance is stated as follows:

$$T_t(u \circ \phi) = T_{t\cdot|J_\phi|} u \circ \phi$$

for any affine map ϕ, where J_ϕ represents the Jacobian of ϕ, which is an invertible matrix. Such property guarantees that shapes are better maintained during smoothing, a classical example is the evolution of an ellipse. It is well known that under MCM any convex curve will propagate to a circle and then collapse to point, instead by the AMSS model the ellipse will collapse to a point preserving the shape, we show the evolution of an ellipse by MCM and AMSS in Fig. 9.5. This property makes the AMSS model particularly attractive for image analysis. In next section, we will show a semi-Lagrangian approximation of both equations with an application to image denoising.

9.4.1 SL Approximation for the Nonlinear Filtering Problem via MCM

We present a semi-Lagrangian (SL) scheme to approximate equation (9.45) for MCM, case $\beta = 1$, and AMSS, case $\beta = 1/3$. For the first equation, we use the scheme described in [14], where a fully explicit scheme is obtained. We denote by u_j^n the numerical approximation of $u(x,t)$ at the node $(x_j, t_n) \in \mathcal{G}_{\Delta x, \Delta t}$, by

Fig. 9.6 Original, noisy ($\nu = 12$) and filtered image by MCM ($C = 0.02$, $s = 1$, $\Delta x = 1$, $\Delta t = 0.25$, $N = 5$)

D^n_j the central finite difference approximation of the space gradient, by $\mathscr{D}(j) = \{(j_1+1, j_2), (j_1-1, j_2), (j_1, j_2+1), (j_1, j_2-1)\}$ and we define $\sigma^n_j = \sigma(D^n_j)$.

Now, we can write the SL scheme for MCM:

$$u_j^{n+1} \equiv \begin{cases} \frac{1}{2}\left(\Pi[u^n](x_j + \sigma^n_j\sqrt{\Delta t}) + \Pi[u^n](x_j - \sigma^n_j\sqrt{\Delta t})\right) & \text{if } |D^n_j| > C\Delta x^s \\ \frac{1}{4}\sum_{i\in\mathscr{D}(j)} u^n_i & \text{if } |D^n_j| \le C\Delta x^s. \end{cases} \tag{9.46}$$

Here, C and s are positive constant. Let us notice that the vector $\sigma^n_j = \sigma(D^n_j)$ is defined only where the discrete gradient D^n_j is not vanishing. When the gradient is zero the MCM is consistent with the heat equation, see [14, 22]. Then, in the case $|D^n_j| \le C\Delta x^s$, the numerical solution is obtained as an average of the neighboring values, so that on these points the scheme results to be consistent with the following heat equation

$$u_t = \frac{\Delta x^2}{4\Delta t}\Delta u.$$

We show an application of the scheme to filter noisy image. The noisy image is obtained adding Gaussian white noise of mean 0 and variance $(\nu/255)^2$ to a given clean image, I_{ex} (note that the noise parameter has been denoted by ν to avoid confusions). In all the tests we have chosen a cubic interpolation, and in order to deal with non-smooth data we have regularized the gradient using $G * \nabla u$, where G is a smoothing kernel. Moreover we have used the following approximation $(G*\nabla u)(x_j, t_n) \simeq \frac{1}{4}\sum_{i\in\mathscr{D}(j)} D^n_i$. The errors are measured by a relative L_1 discrete norm $\|\cdot\|_{1,r}$, defined as follows:

$$\|I_{ex}(\cdot) - u^N_{\cdot}\|_{1,r} := \frac{\sum_{x_j\in\mathscr{G}_{\Delta x}} |I_{ex}(x_j) - u^N_j|}{\sum_{x_j\in\mathscr{G}_{\Delta x}} |I_{ex}(x_j)|}.$$

In Figs. 9.6 and 9.7 we show the clean image, the noisy image and the restored image.

Fig. 9.7 Original, noisy ($\nu = 25$) and filtered image by MCM ($C = 0.02$, $s = 1$, $\Delta x = 1$, $\Delta t = 0.25$, $N = 5$)

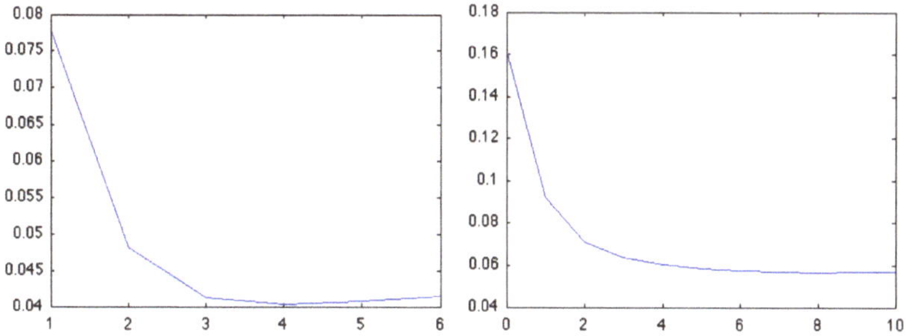

Fig. 9.8 L_1 errors ($y-$axis) and time iteration ($x-$axis) for MCM model applied to a noisy images with $\nu = 12$ (*left*) and $\nu = 25$ (*right*)

In Fig. 9.8 we show the relative L_1 discrete errors evolution with respect to t. We see that in few iterations we reached the minimization of the L_1 errors.

To describe the SL scheme for the AMSS model, we first need to remind that the equation (9.45) in this case can be rewritten as following:

$$u_t = \left(\widehat{\sigma}(\nabla u)^\top D^2 u \widehat{\sigma}(\nabla u)\right)^{1/3},$$

where $\widehat{\sigma}(\nabla u) = (\nabla u)^\perp$, see for instance [34]. Then we introduce an extra parameter ρ, we define $\widehat{\sigma}_j^n := \widehat{\sigma}(D_j^n)$ and we take a directional second finite difference:

$$\widehat{\sigma}(\nabla u)^\top D^2 u \widehat{\sigma}(\nabla u)(x_j, t_n) \simeq \frac{u(x_j + \rho\widehat{\sigma}_j^n, t_n) + u(x_j - \rho\widehat{\sigma}_j^n, t_n) - 2u_j^n}{\rho^2}. \quad (9.47)$$

Finally, the SL scheme is written as

$$u_j^{n+1} \equiv \begin{cases} w_j^{n+1} & \text{if } |D_j^n| > C\Delta x \\ \dfrac{1}{4} \displaystyle\sum_{i \in \mathscr{D}(j)} w_i^{n+1} & \text{if } |D_j^n| \leq C\Delta x \end{cases}$$

Fig. 9.9 Original, noisy ($\nu = 12$) and filtered image by AMSS ($C = 0.02$, $s = 1$, $\Delta x = 1$, $\Delta t = 0.2$, $\rho = (\Delta t)^{1/6}$, $N = 10$)

Fig. 9.10 Original, noisy ($\nu = 25$) and filtered image by AMSS ($C = 0.02$, $s = 1$, $\Delta x = 1$, $\Delta t = 0.25$, $\rho = (\Delta t)^{1/6}$, $N = 10$)

where

$$w_i^{n+1} \equiv \begin{cases} u_i^n + \Delta t \left(\dfrac{\Pi[u^n](x_i + \rho \widehat{\sigma}_i^n) + \Pi[u^n](x_i - \rho \widehat{\sigma}_i^n) - 2u_i^n}{2\rho^2} \right)^{1/3} & \text{if } |D_i^n| > C\Delta x \\ u_i^n & \text{if } |D_i^n| \leq C\Delta x. \end{cases}$$

Such scheme as been proposed in [12], where a convergence result for its monotonic version is proved.

Let us notice that in this case the vector $\widehat{\sigma}$ is always defined, anyhow we need to consider differently the case $\widehat{\sigma} = 0$. In fact, if we would use the approximation (9.47) on the points where $\widehat{\sigma} = 0$, no diffusion would happen. On these points, we extend the solution by continuity: we take an average of the numerical solution, computed at the previous step when $\widehat{\sigma}$ is small, computed using (9.47) when $\widehat{\sigma}$ is big.

As for the previous model, we have chosen a cubic interpolation and regularized the gradient.

In Fig. 9.9 and in Fig. 9.10, we show respectively the clean image, the noisy image and the restored image. In Fig. 9.11 we show the relative L_1 discrete errors evolution with respect to t. We see that in few iterations the minimum error is reached.

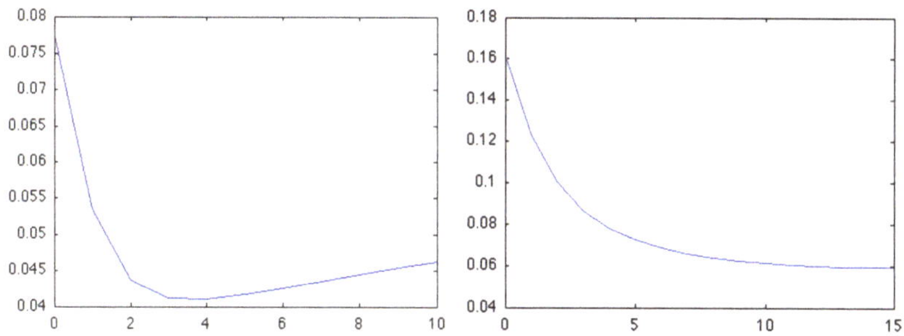

Fig. 9.11 Errors for MCM model for noisy images (*left* $\nu = 1$, *right* $\nu = 25$)

9.5 Segmentation via the LS Method

The segmentation problem in computer vision, as formulated by Mumford and Shah [40], can be defined in the following way: given an observed image $I_0 : \Omega \to [0, 1]$, find a decomposition of disjoint open sets $\Omega = \bigcup_i \overline{\Omega_i}$ and $C = \bigcup_i \partial\Omega_i$, such that I_0 varies smoothly within each Ω_i and rapidly or discontinuously across the boundaries of Ω_i. We denote by $|C|$ the length of curves which belong to C.

A classical way to solve this problem is solving the following minimization problem:

$$\inf_{I,C} F^{MS}(I, C) \tag{9.48}$$

where

$$F^{MS}(I, C) = \lambda \int_{\Omega} (I - I_0)^2 dx + \nu \int_{\Omega \setminus C} |\nabla I|^2 dx + \mu |C| \tag{9.49}$$

and μ, ν are fixed parameters of weight for the different terms of the functional. For a minimum (I^*, C^*), I^* is an optimal piecewise smooth approximation of the initial image I_0, and C^* approximates the edges of I_0. I^* will be smooth on $\Omega \setminus C$ and will have its discontinuities on C^*.

Theoretical results of existence of minimizers for this problem can be found in Mumford and Shah [40], Morel and Solimini [38] and De Giorgi et al. [26].

A reduced case can be obtained by restricting the segmentation image I to piecewise constant functions, i.e. for $c_i = mean(I_0)$ in Ω_i then $I \equiv c_i$ inside each Ω_i. In this event, for the special case $i = \{0, 1\}$ (that is the classical case of segmenting one object from the background image) the problem (9.48) becomes minimizing

$$E^{MS}(c_0, c_1, C) = \lambda \int_{\Omega_0} (I_0 - c_0)^2 dx + \lambda \int_{\Omega_1} (I_0 - c_1)^2 dx + \mu |C|. \tag{9.50}$$

We want now to use LS methods to solve this problem. LS methods are a very useful tool for computing evolving contours since they accommodate topological changes and allow to compute on a simple structured grid. For LS methods the curve C is defined as the zero-level set of a sufficiently regular function ϕ called level set function, i.e. $\phi(x) < 0$ for $x \in \Omega_0$ and $\phi(x) > 0$ for $x \in \Omega_1$.

In this formulation, denoting by $H(\phi)$ the usual Heaviside function, (9.50) becomes (see [17])

$$E^{VC}(c_0, c_1, \phi) = \lambda \int_\Omega (I_0 - c_0)^2 H(\phi)dx + \lambda \int_\Omega (I_0 - c_1)^2 (1 - H(\phi))dx$$
$$+\mu \int_\Omega |\nabla H(\phi)|dx. \tag{9.51}$$

Considering H_ϵ and δ_ϵ two C^1 regular approximations of the Heaviside function H and of the delta function so that $H'_\epsilon = \delta_\epsilon$, we can write the Euler-Lagrange equation corresponding to the energy E^{VC} obtaining

$$\frac{\partial \phi}{\partial t} = \delta_\epsilon(\phi) \left[\mu \mathrm{div} \left(\frac{\nabla \phi}{|\nabla \phi|} \right) - \lambda (I_0 - c_0)^2 + \lambda (I_0 - c_1)^2 \right]$$
$$\text{where} \quad c_0 = \frac{\int_\Omega I(x) H(\phi)dx}{\int_\Omega H(\phi)dx}, \quad c_1 = \frac{\int_\Omega I(x)(1 - H(\phi))dx}{\int_\Omega (1 - H(\phi))dx}. \tag{9.52}$$

This is a time dependent equation converging to a steady solution witch is a minimizer of the functional (9.50).

A standard rescaling can be made, as in Zhao et al. [49], by replacing $\delta(\phi)$ by $|\nabla \phi|$. This time rescaling does not affect the steady state solution, but helps removing stiffness near the zero level sets of ϕ. Finally, we get the following nonlinear evolution equation:

$$\frac{\partial \phi}{\partial t} = |\nabla \phi| \left[\mu \,\mathrm{div} \left(\frac{\nabla \phi}{|\nabla \phi|} \right) - \lambda (I_0 - c_0)^2 + \lambda (I_0 - c_1)^2 \right] \tag{9.53}$$

which is solved for t tending to $+\infty$, yielding the solution of the segmentation problem.

9.5.1 SL Scheme for Segmentation via the LS Method

In this section, we describe a SL approximation for Eq. (9.53). This equation has a first-order and a second-order term, which we approximate using SL schemes introduced in previous sections. We call $c(x)$ the speed of propagation of the first-order term,

$$c(x) = -\lambda (I_0 - c_0)^2 + \lambda (I_0 - c_1)^2, \tag{9.54}$$

and using (9.20) we obtain

$$\frac{\partial \phi}{\partial t} = \mu \,\text{div}\left(\frac{\nabla \phi}{|\nabla \phi|}\right) |\nabla \phi| + c(x) \max_{a \in B(0,1)} \{a \cdot \nabla \phi\}. \qquad (9.55)$$

As in every SL approximation, we want to build first an approximation of the directional derivative. This implies that the velocity $c(x)$ will appear directly inside the argument of the unknown function ϕ. For this particular problem, we have to take care of the change of sign of the velocity since will produce a sign change also in our scheme. We can rewrite (9.55) as

$$\frac{\partial \phi}{\partial t} = \begin{cases} \mu \,\text{div}\left(\frac{\nabla \phi}{|\nabla \phi|}\right) |\nabla \phi| + \max\limits_{a \in B(0,1)} \{c(x) a \cdot \nabla \phi\}, & \text{where } c(x) \geq 0 \\ \mu \,\text{div}\left(\frac{\nabla \phi}{|\nabla \phi|}\right) |\nabla \phi| - \max\limits_{a \in B(0,1)} \{-c(x) a \cdot \nabla \phi\}, & \text{where } c(x) < 0. \end{cases} \qquad (9.56)$$

Next, we define $\phi_j^n = \phi(x_j, t_n)$ where $(x_j, t_n) \in \mathcal{G}_{\Delta x, \Delta t}$.

Using, the schemes (9.46) for the second order term, and (9.22) for the first order term, and explicit Euler finite differences for the temporal derivative, Eq. (9.56) becomes

$$\frac{\phi_j^{n+1} - \phi_j^n}{\Delta t} = \mu \frac{\left\{\frac{1}{2} \Pi\,[\phi^n]\,(x_j + \sigma_j^n \sqrt{\Delta t}) + \frac{1}{2} \Pi\,[\phi^n]\,(x_j - \sigma_j^n \sqrt{\Delta t})\right\} - \phi_j^n}{\Delta t}$$
$$- \, sign\left(c(x_j)\right) \min_{a \in B(0,1)} \frac{\Pi[\phi^n](x_j + ac(x_j)\Delta t) - \phi_j^n}{-\Delta t}, \qquad (9.57)$$

where we have used the fact that $-\max(-p) = \min(p)$ and $\min_{a \in B(0,1)}(a \cdot f) = \min_{a \in B(0,1)}(-a \cdot f)$ to simplify the notations.

Then, we get the fully discrete SL scheme:

$$\phi_j^{n+1} = \phi_j^n + \mu \left\{\frac{1}{2} \Pi\,[\phi^n]\,(x_j + \sigma_j^n \sqrt{\Delta t}) + \frac{1}{2} \Pi\,[\phi^n]\,(x_j - \sigma_j^n \sqrt{\Delta t}) - \phi_j^n\right\}$$
$$+ sign\left(c(x_j)\right) \left\{\min_{a \in B(0,1)} \{\Pi[\phi^n](x_j + ac(x_j)\Delta t)\} - \phi_j^n\right\}, \qquad (9.58)$$

and coherently with (9.46) in the case of discrete space gradient $|D_j^n| \leq C\Delta x^s$.

We used this scheme to segment a variety of images, the results are shown in Figs. 9.12–9.14.

The most interesting feature of this approach is that the method is very rapid. While in classical finite difference schemes the information runs just one pixel for iteration, in this case we take information from a larger set. In particular to compute the value on the grid point x_j we use all the values on the grid nodes that are

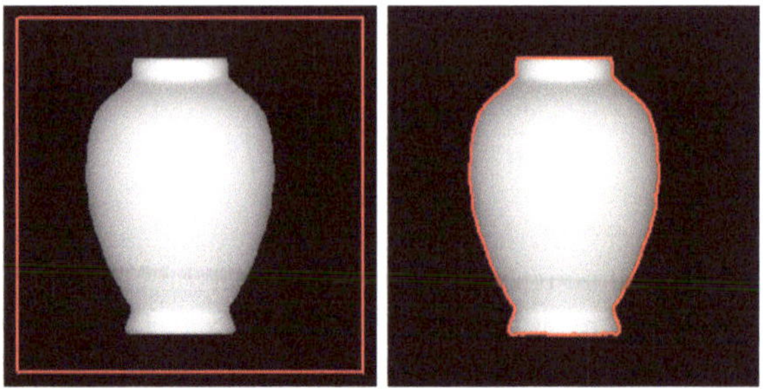

Fig. 9.12 Segmentation: $\mu = 0.1$, $\lambda = 10^4$, two iterations

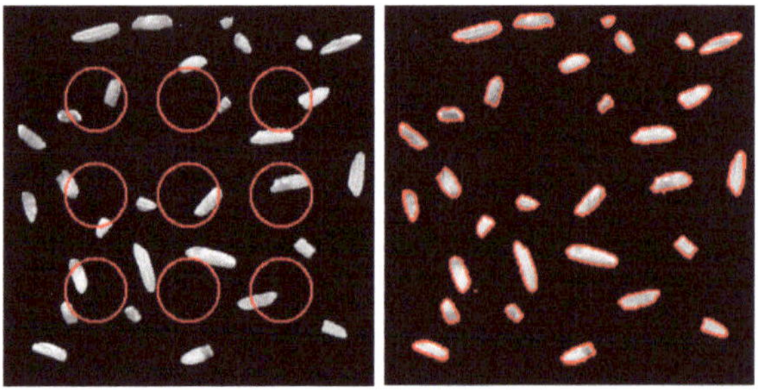

Fig. 9.13 Segmentation: $\mu = 1$, $\lambda = 0.3 * 10^4$, two iterations

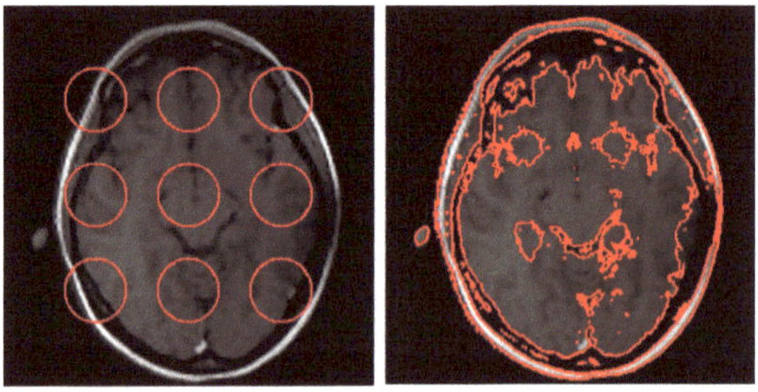

Fig. 9.14 Segmentation: $\mu = 1$, $\lambda = 0.3 * 10^4$, two iterations

contained in a ball centered in x_j with radius $c(x_j)$. This is the reason of the rapid convergence of the scheme. In our tests we found the correct segmentation of the image always after 3 or less iterations. However, this is an experimental observation, a detailed treatment of SL scheme performances on this problem is still missing.

9.6 The Motion Segmentation Problem

Optical flow is the motion field in an image sequence; this problem belongs to the more general problem of motion estimation. Motion estimation and segmentation are strongly related topics that can increase the performances from each other. In particular segmentation can help us to solve the classical ambiguity near motion boundaries.

Both topics are traditional in computer vision. Various approaches to optical flow computation have been suggested in [5], especially variational techniques based on modifications of the method of Horn and Schunck [36] have yielded very convincing results.

Also in segmentation, variational techniques perform well. As we showed in the previous section, segmentation can be obtained minimizing the functional (9.48). In recent years the functional optimization is computed using LS methods [43].

In [3] and in [10], the authors propose to compute optical flow coupling estimators for optical flow and segmentation problem in a Mumford and Shah-like functional. Due to the presence of unspecified discontinuities in the integration domain, the minimization of this kind of functionals is difficult and we are going to solve it using level set techniques.

We will use the studied semi-Lagrangian schemes for the Mean Curvature Motion in this situation, to compute the front propagation of a level set function. Later, we compare the results for this application, in terms of accuracy and speed. The variational model which we use, is based on the optical flow functional used in various works (for a short summary the dedicated section in [9]) and the segmentation model presented in [16].

Given two images $I_0(x), I_1(x) : \Omega \to \mathbb{R}$, we have to find, at each point $x \in \Omega$ the optical flow vector $w(x) : \mathbb{R}^2 \to \mathbb{R}^2$ that describes the shift of the pixel at x in the image I_0 to the new location $x + w$ in the image I_1. Let us assume the following classical constraints: $I_0(x + w) = I_1(x)$ this is *grey level constancy*. It means that a pixel does not change its brightness from an image to the next. The second important principle is *gradient constancy*, an assumption that makes the model robust with respect to global additive illumination changes in the scene. So, our term of consistency with the data is

$$f_{of}(w) = k|I_0(x + w) - I_1(x)|^2 + \gamma|\nabla I_0(x + w) - \nabla I_1(x)|^2. \tag{9.59}$$

The minimization of an optical flow functional with only the image constancy assumptions is ill-posed, to overcome this problem, a smoothness term regularizing the flow field is considered. The typical smoothness term as in [36] is

$$s_{of}(w) = |\nabla w|^2. \tag{9.60}$$

Using this term, nevertheless, we introduce regularization of the field and the known ambiguities along the discontinuities. For this reason, we place the smoothness term in a Mumford-Shah-like functional that provide to add regularization only in the regions of regularity of the solution.

Finally, calling Γ the discontinuity set of the optical flow, and $|\Gamma|$ the length of the curve, we want to minimize the following functional

$$\begin{aligned} E(w, \Gamma) = & \int_{\Omega} k|I_0(x + w) - I_1(x)|^2 + \gamma|\nabla I_0(x + w) - \nabla I_1(x)|^2 dx \\ & + \mu \int_{\Omega \setminus \Gamma} |\nabla w|^2 dx + \nu|\Gamma|. \end{aligned} \tag{9.61}$$

This energy functional follows the principle of *motion competition* proposed for example in [23] and in [3], that was inspired by the work on segmentation by Vese and Chan [48]. This energy drives the interface of discontinuity and, simultaneously, makes an estimation of the optical flow in the different regions of smoothness.

In the optimization problem related to (9.61), if at the same time the optical flow, the number or regions separated by Γ and the position of the curve Γ are all unknown, then minimization can be hard. For this reason, in the next section we will consider the following simpler case. The motion field has an interface of discontinuity that divides the domain Ω in only two regions; inside these regions the motion field is constant. This choice is made for simplicity, and because, in this section, our aim is to present an easly understandable case. However, more general situations can be considered, e.g. in [10].

9.6.1 SL Scheme for the Motion Segmentation Problem

We want to deal with the functional (9.61) in a similar way as in segmentation, so we define a level set function ϕ such that the interface of discontinuity of the motion field Γ is the 0-level set of ϕ. Then, using the Euler-Lagrange equation and considering a time dependent equation converging to a steady solution which is a minimizer of the energy functional (9.61), we get the following evolutive equation:

$$\frac{\partial \phi}{\partial t} = |\nabla \phi| \left[\nu \operatorname{div}\left(\frac{\nabla \phi}{|\nabla \phi|}\right) + [f_{of}(w^+) - f_{of}(w^-)] - \mu \left[s_{of}(w^+) - s_{of}(w^-)\right]\right] \tag{9.62}$$

where, w^+ and w^- are the constant values of w inside and outside Γ: we discretize this equation with the same technique as in (9.58). We obtain the following numerical scheme:

$$\phi_j^{n+1} = \phi_j^n + \nu \left\{ \frac{1}{2}\Pi\left[\phi^n\right](x_j + \sigma_j^n \sqrt{\Delta t}) + \frac{1}{2}\Pi\left[\phi^n\right](x_j - \sigma_j^n \sqrt{\Delta t}) - \phi_j^n \right\}$$
$$+ sign(c(x_j)) \left\{ \min_{a \in B(0,1)} \{\Pi[\phi^n](x_j + ac(x_j)\Delta t)\} - \phi_j^n \right\}, \tag{9.63}$$

where the first-order velocity term $c(x_j)$ is

$$c(x_j) = \left[f_{of}(w_j^+) - f_{of}(w_j^-)\right] - \mu \left[s_{of}(w_j^+) - s_{of}(w_j^-)\right]. \tag{9.64}$$

We used this scheme for two simple tests obtaining the results shown in Fig. 9.15 and 9.16. The first is a synthetic sequence where a ball is moving on a background, the second one is a small box taken from a more complicated real sequence (available on the site [47]). In these tests we have small displacements so we do not need a coarse-to-fine framework to capture larger ones.

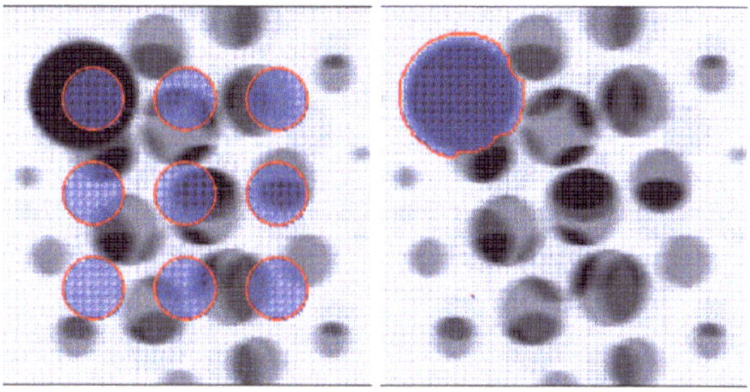

Fig. 9.15 Motion segmentation: $\mu = 0.4$, $k = 0.5 * 10^5$, $\nu = 2$, $\gamma = 0.2$, convergence in five iterations

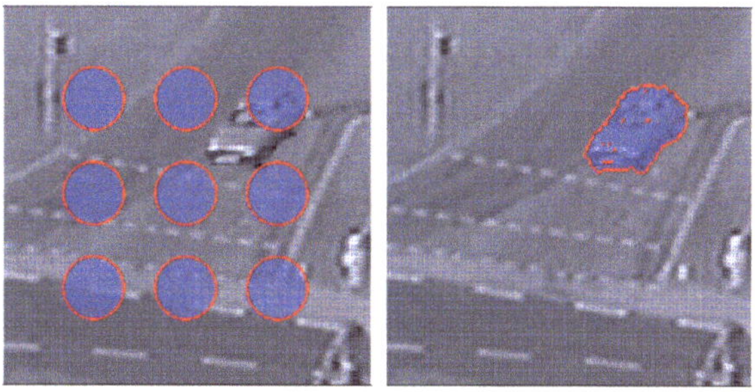

Fig. 9.16 Motion segmentation: $\mu = 0.4$, $\kappa = 0.5 * 10^5$, $\nu = 2$, $\gamma = 0.2$, convergence in four iterations

9.7 Conclusions

As we have seen in the previous sections, the approximation schemes based on the semi-Lagrangian approximation of first and second order partial differential equations can be successfully applied to the solution of classical image processing problems. The main features of SL schemes are the following:

- They mimic the method of characteristics and give even in their simplest version a more accurate approximation with respect to first order finite difference schemes;
- High-order approximation schemes can be obtained coupling high-order methods for integrating back along the characteristics (e.g. Heun, Runge-Kutta) with high-order interpolation techniques (e.g. polynomial, essentially non-oscillatory (ENO), weighted ENO);
- They can be applied to structured and unstructured grids, so they are well suited also to problems where there is a need to use triangular grids as, for example, in computer vision;
- SL approximation allows to obtain accurate schemes even for second order nonlinear degenerate equations. Such approximations are fully explicit (i.e. with moderate computing cost) and allow for large time steps. It is important to note that the last property is crucial when we look to the asymptotic behavior for t tending to $+\infty$. Since this implies a time integration over a large interval. a large time step guarantees a fast convergence to the steady state;

– SL schemes have shown to be rather robust when applied to non smooth data, as noisy images. Moreover, in the time marching evolution there is less need for a re-initialization in order to keep the level sets and acceleration techniques (such as Fast Marching methods, [24,45]) can be implemented.

A detailed analysis of these methods for first order model problems can be found in the book [32] and some of the results for second order problems can be found in [12,14]. We conclude observing that there are still many open problems in the theory of SL schemes, e.g. in the construction and analysis of high-order approximations and acceleration methods (fast sweeping, fast marching). We are confident that these methods will be also useful for new challenging applications in image processing.

References

1. Alvarez, L., Lions, P.L., Morel, J.M.: Image selective smoothing and edge detection by nonlinear diffusion. SIAM J. Numer. Anal. **29**, 845–866 (1992)
2. Alvarez, L., Guichard, F., Lions, P.L., Morel, J.M.: Axioms and fundamental equations of image processing. Arch. Ration. Mech. **123**, 199–257 (1993)
3. Amiaz, T., Kiryati, N.: Piecewise-smooth dense optical flow via level sets. Int. J. Comput. Vis. **68**(2), 111–124 (2006)
4. Barles, G.: Solutions de Viscositè des Equations d'Hamilton–Jacobi. Springer, New York (1998)
5. Barron, J.L., Fleet, D.J., Beauchemin, S.S.: Performance of optical flow techniques. Int. J. Comput. Vis. **12**, 75–104 (1996)
6. Brenner S.C., Scott, L.R.: The Mathematical Theory of Finite Element Methods, 3rd edn. Springer, New York (2008)
7. Brent, R.: Algorithms for Minimization Without Derivatives. Pentice–Hall, Englewood Cliffs (1973)
8. Breuss, M., Cristiani, E., Durou, J.D., Falcone, M., Vogel, O.: Numerical algorithms for perspective shape from shading. Kybernetika **46**, 207–225 (2010)
9. Brox, T., Bruhn, A., Papenberg, N., Weickert, J.: High Accuracy Optical Flow Estimation Based on a Theory for Warping. Lecture Notes in Computer Science, pp. 25–36. Springer, Berlin (2004)
10. Brox, T., Bruhn, A., Weickert, J.: Variational segmentation with level sets. In: Computer Vision-ECCV, Graz, pp. 471–483 (2006)
11. Camilli, F., Falcone, M.: An approximation scheme for the maximal solution of the shape-from-shading model. In: Proceedings ICIP 96, vol. I, pp. 49–52. IEEE, Piscataway (1996)
12. Carlini, E., Ferretti, R.: A semi-Lagrangian approximation for the AMSS model of image processing. submitted to Applied Numerical Mathematics (in press)
13. Carlini, E., Ferretti, R., Russo, G.: A weighted essentially non oscillatory, large time-step scheme for Hamilton Jacobi equations. SIAM J. Sci. Comput. **27**(3), 1071–1091 (2005)
14. Carlini, E., Falcone, M., Ferretti, R.: Convergence of a large time-step scheme for mean curvature motion. Interface Free Bound. **12**, 409–441 (2010)
15. Catté, F., Dibos, F., Koepfler, G.: A morphological scheme for mean curvature motion and applications to anisotropic diffusion and motion of level sets. SIAM J. Numer. Anal. **32**, 1895–1909 (1995)
16. Chan, T., Vese, L.: Active contours without edges. IEEE Trans. Image Process. **10**(2), 266–277 (2001)

17. Chan, T., Sandberg, B., Vese, L.: Active contours without edges for vector-valued images. J. Vis. Commun. Image R. **11**(2), 130–141 (2000)
18. Chen, Y.G., Giga, Y., Goto, S.: Uniqueness and existence of viscosity solutions of generalized mean curvature flow equation. J. Diff. Geom. **33**, 749–786 (1991)
19. Courant, R., Isaacson, E., Rees, M.: On the solution of nonlinear hyperbolic differential equations by finite differences. Commun. Pure Appl. Math. **5**, 243–255 (1952)
20. Courteille, F., Crouzil, A., Durou, J.D., Gurdjos, P.: Towards shape from shading under realistic photographic conditions, Int. C. Patt. Recog.- ICPR 2004, Cambridge, vol. 2, pp. 277–280 (2004)
21. Crandall, M.G., Lions, P.L.: Two approximations of solutions of Hamilton–Jacobi equations. Math. Comput. **43**, 1–19 (1984)
22. Crandall, M.G., Ishii, H., Lions, P.L.: User's guide to viscosity solutions of second order partial differential equations. Bull. Am. Math. Soc. **27**, 1–67 (1992)
23. Cremers, D., Soatto, S.: Motion competition: a variational framework for piecewise parametric motion segmentation. Int. J. Comput. Vis. **63**, 249–265 (2005)
24. Cristiani, E., Falcone, M.: Fast semi-Lagrangian schemes for the eikonal equation and applications. SIAM J. Numer. Anal. **45**(5), 1979–2011 (2007)
25. Cristiani, E., Falcone, M., Seghini, A.: Numerical solution of the shape-from-shading problem. In: Proceedings of Science POS (CSTNA2005) 008, 1–17, Electronic Journal site http://pos.sissa.it/
26. De Giorgi, E.: New functionals in calculus of variations, nonsmooth optimization and related topics. In: Proceedings of the Fourth Course of the International School of Mathematics, Erice (1988)
27. Durou, J.D., Falcone, M., Sagona, M.: Numerical methods for shape from shading: a new survey with benchmarks. Comput. Vis. Image Underst. Elsevier **109**, 22–43 (2008)
28. Evans, L.C.: Partial Differential Equations. AMS, Providence (2010)
29. Evans, L.C., Spruck, J.: Motion of level sets by mean curvature. Int. J. Diff. Geom. **33**, 635–681 (1991)
30. Falcone, M.: The minimum time problem and its applications to front propagation. In: Motion by Mean Curvature and Related Topics. De Gruyter Verlag, Berlino (1994)
31. Falcone, M., Ferretti, R.: Convergence analysis for a class of high-order semi-Lagrangian advection schemes. SIAM J. Numer. Anal. **35**(3), 909–940 (1998)
32. Falcone, M., Ferretti, R.: Semi-Lagrangian Approximation Schemes for Linear and Hamilton-Jacobi Equations. SIAM, in preparation
33. Falcone, M., Sagona, M., Seghini, A.: A global algorithm for the Shape-from-Shading problem with black shadows. Numerical Mathematics and Advanced Applications ENUMATH 2001, pp. 503–512. Springer, Milano (2003)
34. Guichard, F., Morel, J.M.: Image Analysis and P.D.E.s. IPAM GBM Tutorial, March 27–April 6 (2001)
35. Horn, B.K.P., Brooks M.J. (eds.): Shape from Shading. MIT, Cambridge (1989)
36. Horn, B.K.P., Schunck, B.: Determinig optical flow. Artif. Intell. **17**, 185–203 (1981)
37. Lions, P.L., Rouy, E., Tourin, A.: A viscosity solution approach to shape from shading. Numer. Math. **64**, 323–353 (1993)
38. Morel, J.M., Solimini, S.: Segmentation of images by variational methods: a constructive approach. Rev. Mat. Univ. Compl. de Madr. **1**, 169–182 (1988)
39. Mumford, D., Shah, J.: Boundary detection by minimizing functional. In: Proceedings of the CVPR, pp. 22–26. IEEE, Silver Spring (1985)
40. Mumford, D., Shah, J.: Optimal approximations by piecewise smooth functions and associated variational problems. Comm. Pure Appl. Math. **XLII**, 577–685 (1989)
41. Oberman, A.M.: A convergent monotone difference scheme for motion of level sets by mean curvature. Numer. Math. **99**, 365–379 (2004)
42. Osher, S.J., Fedkiw, R.P.: Level Set Methods and Dynamic Implicit Surfaces. Applied Mathematical Sciences, vol. 153. Springer, New York (2003)

43. Osher, S.J., Sethian, J.A.: Front propagating with curvature-dependent speed: algorithms based on Hamilton-Jacobi formulation. J. Comput. Phys. **79**, 12–49 (1988)
44. Prados, E., Faugeras, O.: Perspective Shape-from-Shading and viscosity solutions. In: Proceedings of ICCV'03, pp. 826–831. IEEE, Los Alamitos (2003)
45. Sethian, J.A.: Level Set Methods and Fast Marching Methods Evolving Interfaces in Computational Geometry, Fluid Mechanics, Computer Vision, and Materials Science. Cambridge Monograph on Applied and Computational Mathematics. Cambridge University Press, Cambridge (1999)
46. Strikwerda, J.C.: Finite Difference Schemes and Partial Differential Equations, 2nd edn. SIAM, Philadelphia (2004)
47. Univerität Kalsruhe. http://i21www.ira.uka.de/image_sequences/
48. Vese, L., Chan, T.: A multiphase level set framework for image segmentation using the Mumford and Shah model. Int. J. Comput. Vis. **50**, 271–293 (2002)
49. Zhao, H., Chan, T., Merriman, B., Osher, S.J.: A variational level set approach to multiphase motion. J. Comput. Phys. **127**, 179–195 (1996)

Chapter 10
Shape Reconstruction of Symmetric Surfaces Using Photometric Stereo

Roberto Mecca and Silvia Tozza

Abstract The reconstruction of a 3D surface through one gray scale digital image does not admit a unique solution in the orthographic Shape from Shading (SfS) framework. With the aim to make this type of problem well-posed it is possible to use the Photometric Stereo (PS) technique. It allows to add information about the surface introducing other images of the object taken from the same point of view but modifying, for each photo, the direction of the light source. The methods that use the PS technique with the orthographic model of SfS need of, at least, three images. However, even if three images are used, there is the possibility that the SfS-PS problem continues to be ill-posed. This is the case when the three images are taken using three coplanar light vectors. This work analyses this kind of ill-posedness in order to understand how it is possible to establish a connection among the images that do not guarantee uniqueness. A further result in this paper is given by a geometrical characterization of the surfaces for which it is possible to solve the classic SfS problem.

10.1 Introduction to the Shape from Shading: Photometric Stereo Model and Symmetric Surfaces

The Shape from Shading (SfS) problem consists on the 3D reconstruction of an object starting from only one gray scale photo. In the framework of the shape reconstruction the classical orthographic SfS problem fails its aims using only one digital

R. Mecca (✉) · S. Tozza
Dipartimento di Matematica "G. Castelnuovo", Sapienza – University of Rome, Rome, Italy
e-mail: mecca@mat.uniroma1.it; tozza@mat.uniroma1.it

M. Breuß et al. (eds.), *Innovations for Shape Analysis*, Mathematics and Visualization,
DOI 10.1007/978-3-642-34141-0_10, © Springer-Verlag Berlin Heidelberg 2013

image for the reconstruction of the corresponding surface. Our contribution is given by the SfS model where we are considering the following hypotheses:

1. The light sources are at infinity;
2. The surface is *Lambertian*;
3. The optical point is sufficiently far from the surface so that perspective deformations can be neglected.

In this case the SfS model is given by the following irradiance equation:

$$\rho(x,y)(n(x,y)\cdot\omega) = I(x,y) \tag{10.1}$$

where $\rho(x,y)$ is the albedo (supposed constant and equal to one), $n(x,y) = (n_1(x,y), n_2(x,y), n_3(x,y))$ is the unit outgoing normal to the unknown surface $z = u(x,y)$, $\omega = (\omega_1, \omega_2, \omega_3) = (\widetilde{\omega}, \omega_3)$ is the direction of the light source and $I : \overline{\Omega} \to [0,1]$ is the image function.

Under the previous assumptions this inverse problem is ill-posed due to the concave/convex ambiguity. Many papers show the impossibility to have a unique solution for this SfS model (see [2, 5]). In order to avoid this ambiguity we can add information of the surface considering other data of the same type.

The Shape from Shading problem with the Photometric-Stereo techniques (SfS-PS) uses more than one image of the same surface taken from the same point of view, but using a different light source for each image, that is:

$$n(x,y)\cdot\omega^i = I_i(x,y). \tag{10.2}$$

Let us clarify by now the meaning of coplanar light sources: since we consider each light source as a unit vector in $\mathbb{R}^3$, we will talk about coplanar light sources with the same meaning of coplanar unit vectors. That is, we will say that three or more light sources are coplanar if the representing vectors are coplanar.

There are many possible approaches to model the SfS-PS problem (e.g. [11, 12, 15]). The main two ways that allow to reconstruct the surface are the normal vector field approach [6] and the differential approach [13] which we briefly describe later.

The main difference between these approaches is that the first one is a local approach that works pixel by pixel in order to find the normal vector field of the unknown surface and then it reconstructs the surface $z = u(x,y)$ all over the domain using the gradient field [6]. The second approach instead works globally: it is based on a differential problem that, using only two images, can recognize the shape of the object but, as we will see, this differential problem is well-posed only if the Dirichlet boundary condition (i.e. the height of the surface on the boundary) is known. In order to use it we need to approximate the boundary condition and another image must be added making the problem well-posed with three images.

We emphasize one more time that both approaches need three image functions and our goal is to characterize this kind of functions with the aim to reduce the number of images needed considering the symmetry of the surface. There are several previous studies [3, 7] where this geometric constraint is used together with the photometric features. In particular there are works that focus on this field detecting the symmetries in the images [8, 14]. Other papers in which a single image of a symmetric object can be regarded as two images of another object taken from two different points of view and two different illumination directions [20] or other similar approaches as [9, 10].

In Sect. 10.2 we recall the local and the differential approach. We explain the impossibility to solve the SfS-PS problem with both approaches using three images obtained with three coplanar light source vectors. In this section we give a definition of linear dependent images used in all the subsequent sections. In Sect. 10.3 we introduce an algorithm for the "artificial" reconstruction of linear dependent images. Finally in Sect. 10.4 we study how it is possible to reduce the number of necessary images for the shape reconstruction of a symmetric surface. The main result in this section is the theorem that explains how to solve the SfS problem (with only one image) through the SfS-PS formulation when the surface under observation has at least four axes of symmetry. In the last section we present some numerical tests related to the algorithms presented.

10.2 Condition of Linear Independent Images for the SfS-PS Reconstruction

We recall briefly the mathematical formulation of the approaches introduced in the previous section.

10.2.1 Normal Vector Approach

As we have anticipated, the local approach solves the problem finding the outgoing normal vector to the surface locally, that is pixel by pixel [4, 6]. This means that, once we discretize the image domain $\overline{\Omega}$ with an uniform lattice $\overline{\Omega}_d = \Omega_d \cup \partial\Omega_d$, the following system is solved for every $(x_i, y_j) \in \overline{\Omega}_d$:

$$\begin{cases} \omega_1' n_1(x_i, y_j) + \omega_2' n_2(x_i, y_j) + \omega_3' n_3(x_i, y_j) = I_1(x_i, y_j) \\ \omega_1'' n_1(x_i, y_j) + \omega_2'' n_2(x_i, y_j) + \omega_3'' n_3(x_i, y_j) = I_2(x_i, y_j) \\ \omega_1''' n_1(x_i, y_j) + \omega_2''' n_2(x_i, y_j) + \omega_3''' n_3(x_i, y_j) = I_3(x_i, y_j). \end{cases} \tag{10.3}$$

As explained before, next step for this method is to approximate the values of the height of the surface starting from the knowledge of its gradient field (i.e. knowing $\nabla u(x_i, y_j) \forall (x_i, y_j) \in \overline{\Omega}_d$) [6]. In fact, using the definition of the outgoing normal vector:

$$n(x, y) = \frac{(-\nabla u(x, y), 1)}{\sqrt{1 + ||\nabla u(x, y)||^2}} = \frac{\left(-\frac{\partial u}{\partial x}(x, y), -\frac{\partial u}{\partial y}(x, y), 1\right)}{\sqrt{1 + ||\nabla u(x, y)||^2}}, \tag{10.4}$$

Fig. 10.1 Schematic representation of the Earth lighted up in three different times by the Sun

it follows:

$$\nabla u(x_i, y_j) = \left(-\frac{n_1(x_i, y_j)}{n_3(x_i, y_j)}, -\frac{n_2(x_i, y_j)}{n_3(x_i, y_j)} \right). \tag{10.5}$$

In the local method it is very easy to see that the uniqueness of the normal field can be reached using the non-singularity of the matrix with respect to the linear system (10.3), namely:

$$A = \begin{pmatrix} \omega_1' & \omega_2' & \omega_3' \\ \omega_1'' & \omega_2'' & \omega_3'' \\ \omega_1''' & \omega_2''' & \omega_3''' \end{pmatrix}, \tag{10.6}$$

that is supposing the non-coplanarity of the three light sources.

An important example that explains why the case of coplanar light sources is not trivial, is represented by the natural phenomenon of the sun illumination of the solar system planets (like Earth). In fact, since these planets are moving on a elliptic trajectory around the sun, we can see the sun as a light source moving on a plane (see Fig. 10.1).

10.2.2 PDE Approach

The differential approach is based on a global method to solve the problem given by the following non-linear PDEs system of the Hamilton-Jacobi type:

$$\begin{cases} \dfrac{-\nabla u(x,y)\cdot\widetilde{\omega}' + \omega_3'}{\sqrt{1+||\nabla u(x,y)||^2}} = I_1(x,y), \ \ \forall (x,y) \in \Omega \\ \dfrac{-\nabla u(x,y)\cdot\widetilde{\omega}'' + \omega_3''}{\sqrt{1+||\nabla u(x,y)||^2}} = I_2(x,y), \ \forall (x,y) \in \Omega \end{cases} \tag{10.7}$$

with the Dirichlet boundary condition $u(x, y) = g(x, y)$ known for all $(x, y) \in \partial\Omega$.

This system can also be studied eliminating the non-linearity [13]. We can then consider the following hyperbolic problem:

$$\begin{cases} b(x,y) \cdot \nabla u(x,y) = f(x,y), & \text{a.e. } (x,y) \in \Omega \\ u(x,y) = g(x,y), & \forall (x,y) \in \partial\Omega \end{cases} \tag{10.8}$$

where

$$b(x,y) = (I_2(x,y)\omega_1' - I_1(x,y)\omega_1'', I_2(x,y)\omega_2' - I_1(x,y)\omega_2'') \tag{10.9}$$

and

$$f(x,y) = I_2(x,y)\omega_3' - I_1(x,y)\omega_3''. \tag{10.10}$$

For the differential method some other passages have to be done in order to see the necessity to have non-coplanar light sources. It is clear that only taking pictures is not possible to get the height of the surface on the boundary points (i.e. the boundary condition $g(x,y)$).

This supplementary information is necessary to make (10.8) well-posed and that is why we need to compute it. In fact the approximation of the Dirichlet boundary condition $g(x,y)$ binds us to use another image (we call it I_3, as in the previous approach). This step allows us to start solving the problem with the same starting data set (i.e. ω', ω'', ω''' and the respective images I_1, I_2, I_3) for both approaches.

It is clear that we can approximate the boundary condition $g(x,y)$ using the previous local method that is using (10.3) for all pixels $(x_i,y_j) \in \partial\Omega_d$. With the aim to emphasize that also using the differential approach the condition of non-coplanarity is required, we use the combinations of the three images in the hyperbolic model in order to approximate the boundary condition.

We consider the following system of hyperbolic equations:

$$\begin{cases} b^{(1,2)}(x,y) \cdot \nabla u(x,y) = f^{(1,2)}(x,y), & \text{a.e. } (x,y) \in \Omega \\ b^{(1,3)}(x,y) \cdot \nabla u(x,y) = f^{(1,3)}(x,y), & \text{a.e. } (x,y) \in \Omega \\ b^{(2,3)}(x,y) \cdot \nabla u(x,y) = f^{(2,3)}(x,y), & \text{a.e. } (x,y) \in \Omega \end{cases} \tag{10.11}$$

where

$$b^{(h,k)}(x,y) = (I_k(x,y)\omega_1^h - I_h(x,y)\omega_1^k, I_k(x,y)\omega_2^h - I_k(x,y)\omega_2^k)$$

and

$$f(x,y)^{(h,k)} = I_k(x,y)\omega_3^h - I_h(x,y)\omega_3^k$$

where (h,k) is the combination of two of the first three natural integer without repetition. In other words (h,k) contains the indices (and the respective light sources) of the images we are using. Then, using these equations, a way to approximate the boundary condition is to choose two of the previous three equations and to consider them in a generic boundary point $(x_i, y_j) \in \partial\Omega_d$ obtaining the following linear system:

$$\begin{cases} b_1^{(1,2)}(x_i, y_j)\dfrac{\partial u}{\partial x}(x_i, y_j) + b_2^{(1,2)}(x_i, y_j)\dfrac{\partial u}{\partial y}(x_i, y_j) = f^{(1,2)}(x_i, y_j) \\ b_1^{(1,3)}(x_i, y_j)\dfrac{\partial u}{\partial x}(x_i, y_j) + b_2^{(1,3)}(x_i, y_j)\dfrac{\partial u}{\partial y}(x_i, y_j) = f^{(1,3)}(x_i, y_j). \end{cases} \quad (10.12)$$

It permits to compute the gradient field of u on the boundary points and then we can approximate the value of the height of u using some well know algorithm [4,6].

Considering the coefficient matrix of the latter linear system it is possible to find a relation between the non-singularity of the matrix A (from (10.6)) and the matrix of the linear system (10.12) proving the following proposition:

Proposition 10.1. *Let*

$$L(x_i, y_j) = \begin{pmatrix} b_1^{(1,2)}(x_i, y_j) & b_2^{(1,2)}(x_i, y_j) \\ b_1^{(1,3)}(x_i, y_j) & b_2^{(1,3)}(x_i, y_j) \end{pmatrix}. \quad (10.13)$$

Then $det(L(x_i, y_j)) = 0 \ \forall (x_i, y_j) \in \overline{\Omega}_d$ *if and only if* $\omega', \omega'', \omega'''$ *are coplanar.*

Proof. ($\Rightarrow$) Let us prove that: $det(L(x_i, y_j)) = 0$ implies $\omega', \omega'', \omega'''$ coplanar.

We generalize the proof by simplifying the notation instead of considering the dependence on a specific point (x_i, y_j). We explicit the functions that appear in the matrix L in terms of the images and of the light sources and we obtain:

$$\begin{aligned} \det(L) &= \omega_1'\omega_2' I_2 I_3 - \omega_1'\omega_2''' I_1 I_2 - \omega_1''\omega_2' I_1 I_3 + \omega_1''\omega_2'''(I_1)^2 - \\ &\quad - [\omega_1'\omega_2' I_2 I_3 - \omega_1'''\omega_2' I_1 I_2 - \omega_1'\omega_2'' I_1 I_3 + \omega_1'''\omega_2''(I_1)^2] \\ &= I_1[-\omega_1'\omega_2''' I_2 - \omega_1''\omega_2' I_3 + \omega_1''\omega_2''' I_1 + \omega_1'''\omega_2' I_2 + \omega_1'\omega_2'' I_3 - \omega_1'''\omega_2'' I_1] \\ &= I_1[(-\omega_1'\omega_2''' + \omega_1'''\omega_2')I_2 + (-\omega_1''\omega_2' + \omega_1'\omega_2'')I_3 + (-\omega_1'''\omega_2'' + \omega_1''\omega_2''')I_1]. \end{aligned} \quad (10.14)$$

Let us consider the case where the three images I_1, I_2 and I_3 are defined by the surface u and the respective light sources as follows:

$$\begin{aligned} I_1(x,y) &= \frac{-\omega_1' \frac{\partial u}{\partial x}(x,y) - \omega_2' \frac{\partial u}{\partial y}(x,y) + \omega_3'}{\sqrt{1 + ||\nabla u(x,y)||^2}}, \\ I_2(x,y) &= \frac{-\omega_1'' \frac{\partial u}{\partial x}(x,y) - \omega_2'' \frac{\partial u}{\partial y}(x,y) + \omega_3''}{\sqrt{1 + ||\nabla u(x,y)||^2}}, \\ I_3(x,y) &= \frac{-\omega_1''' \frac{\partial u}{\partial x}(x,y) - \omega_2''' \frac{\partial u}{\partial y}(x,y) + \omega_3'''}{\sqrt{1 + ||\nabla u(x,y)||^2}}. \end{aligned} \quad (10.15)$$

Replacing the previous equalities in (10.14) we have:

$$\det(L) =$$

$$I_1\left[\frac{1}{\sqrt{1+||\nabla u(x,y)||^2}}\right]\left[\frac{\partial u}{\partial x}(\omega_1'\omega_1''\omega_2''' - \omega_2'\omega_1''\omega_1''') + \frac{\partial u}{\partial y}(\omega_1'\omega_2''\omega_2''' - \omega_2'\omega_2''\omega_1''') -\right.$$

$$\omega_1'\omega_3''\omega_2''' + \omega_2'\omega_3''\omega_1''' + \frac{\partial u}{\partial x}(\omega_2'\omega_1''\omega_1''' - \omega_1'\omega_2''\omega_1''') + \frac{\partial u}{\partial y}(\omega_2'\omega_1''\omega_2''' - \omega_1'\omega_2''\omega_2''') -$$

$$\omega_2'\omega_1''\omega_3''' + \omega_1'\omega_2''\omega_3''' + \frac{\partial u}{\partial x}(\omega_1'\omega_2''\omega_1''' - \omega_1'\omega_1''\omega_2''') + \frac{\partial u}{\partial y}(\omega_2'\omega_2''\omega_1''' - \omega_2'\omega_1''\omega_2''') -$$

$$\left.\omega_3'\omega_2''\omega_1''' + \omega_3'\omega_1''\omega_2'''\right]. \tag{10.16}$$

Since the common denominator $(1+||\nabla u(x,y)||^2)$ is always different from zero, we can continue the computation considering only the numerator of (10.16). Explicit I_1 using (10.15) we can write as follows:

$$\frac{\partial u}{\partial x}\Big[-\omega_1'\big(-\omega_1'\omega_3''\omega_2''' + \omega_2'\omega_3''\omega_1''' - \omega_2'\omega_1''\omega_3'''$$

$$+\omega_1'\omega_2''\omega_3''' - \omega_3'\omega_2''\omega_1''' + \omega_3'\omega_1''\omega_2'''\big)\Big]$$

$$+\frac{\partial u}{\partial y}\Big[-\omega_2'\big(-\omega_1'\omega_3''\omega_2''' + \omega_2'\omega_3''\omega_1''' - \omega_2'\omega_1''\omega_3'''$$

$$+\omega_1'\omega_2''\omega_3''' - \omega_3'\omega_2''\omega_1''' + \omega_3'\omega_1''\omega_2'''\big)\Big]$$

$$+\omega_3'\big(-\omega_1'\omega_3''\omega_2''' + \omega_2'\omega_3''\omega_1''' - \omega_2'\omega_1''\omega_3''' + \omega_1'\omega_2''\omega_3''' - \omega_3'\omega_2''\omega_1''' + \omega_3'\omega_1''\omega_2'''\big). \tag{10.17}$$

With the aim to prove the coplanarity of ω''' with respect to ω' and ω'', we define the direction parameters of the plane generated by these last two vectors:

$$\begin{aligned} s &= \omega_2'\omega_3'' - \omega_3'\omega_2'', \\ l &= \omega_3'\omega_1'' - \omega_1'\omega_3'', \\ m &= \omega_1'\omega_2'' - \omega_2'\omega_1''. \end{aligned} \tag{10.18}$$

The direction parameters (s, l, m) of a plane represent a vector orthogonal to it, therefore, orthogonal to every vector $v = (v_1, v_2, v_3)$ belonging to the plane. This means that:

$$sv_1 + lv_2 + mv_3 = 0. \tag{10.19}$$

Now, considering $\det(L)(x, y)$ (without denominator) as a polynomial in the variables $\xi = \frac{\partial u}{\partial x}(x, y)$ and $\eta = \frac{\partial u}{\partial y}(x, y)$, we have:

$$\begin{aligned}
&\det(L)(\xi, \eta) = \\
&\xi\Big[-\omega_1'\Big(\omega_1'''(\omega_2'\omega_3'' - \omega_3'\omega_2'') + \omega_2'''(\omega_3'\omega_1'' - \omega_1'\omega_3'') + \omega_3'''(\omega_1'\omega_2'' - \omega_2'\omega_1'')\Big)\Big] + \\
&\eta\Big[-\omega_2'\Big(\omega_1'''(\omega_2'\omega_3'' - \omega_3'\omega_2'') + \omega_2'''(\omega_3'\omega_1'' - \omega_1'\omega_3'') + \omega_3'''(\omega_1'\omega_2'' - \omega_2'\omega_1'')\Big)\Big] + \\
&\quad \omega_3'\Big(\omega_1'''(\omega_2'\omega_3'' - \omega_3'\omega_2'') + \omega_2'''(\omega_3'\omega_1'' - \omega_1'\omega_3'') + \omega_3'''(\omega_1'\omega_2'' - \omega_2'\omega_1'')\Big).
\end{aligned} \tag{10.20}$$

Since $\det(L) = 0\ \forall(\xi, \eta)$ then this means that all the coefficients are zero:

$$\omega_1'''(\omega_2'\omega_3'' - \omega_3'\omega_2'') + \omega_2'''(\omega_3'\omega_1'' - \omega_1'\omega_3'') + \omega_3'''(\omega_1'\omega_2'' - \omega_2'\omega_1'') = 0 \tag{10.21}$$

which implies the coplanarity.

($\Leftarrow$) Let us prove now that: ω', ω'', ω''' coplanar implies $\det(L) = 0$.

We start seeing how, considering two images I_1 and I_2 obtained by the light vectors ω' and ω'', it is possible to get all the images obtainable with a light vector that is coplanar to the first two (i.e. considering the vectorial equality $\omega''' = \alpha\omega' + \beta\omega''$), that is $I_3 = \alpha I_1 + \beta I_2$ as α and β change. Considering the image definition of the orthographic SfS model we can write:

$$\begin{aligned}
I_3 &= \frac{-\frac{\partial u}{\partial x}\omega_1''' - \frac{\partial u}{\partial y}\omega_2''' + \omega_3'''}{\sqrt{1 + ||\nabla u||^2}} \\
&= \frac{-\frac{\partial u}{\partial x}(\alpha\omega_1' + \beta\omega_1'') - \frac{\partial u}{\partial y}(\alpha\omega_2' + \beta\omega_2'') + (\alpha\omega_3' + \beta\omega_3'')}{\sqrt{1 + ||\nabla u||^2}} \\
&= \alpha\frac{-\frac{\partial u}{\partial x}\omega_1' - \frac{\partial u}{\partial y}\omega_2' + \omega_3'}{\sqrt{1 + ||\nabla u||^2}} + \beta\frac{-\frac{\partial u}{\partial x}\omega_1'' - \frac{\partial u}{\partial y}\omega_2'' + \omega_3''}{\sqrt{1 + ||\nabla u||^2}} \\
&= \alpha I_1 + \beta I_2.
\end{aligned} \tag{10.22}$$

We continue replacing in (10.14) the coplanarity equality $\omega''' = \alpha\omega' + \beta\omega''$ and, in consequence, I_3 with $\alpha I_1 + \beta I_2$. We obtain:

$$\begin{aligned}
&\det(L) = \\
&-\omega_1'(\alpha\omega_2' + \beta\omega_2'')I_1 I_2 - \omega_1''\omega_2' I_1(\alpha I_1 + \beta I_2) + \omega_1''(\alpha\omega_2' + \beta\omega_2'')(I_1)^2 + \\
&(\alpha\omega_1' + \beta\omega_1'')\omega_2' I_1 I_2 + \omega_1'\omega_2'' I_1(\alpha I_1 + \beta I_2) - (\alpha\omega_1' + \beta\omega_1'')\omega_2''(I_1)^2 =
\end{aligned}$$

$$-\alpha\omega_1'\omega_2' I_1 I_2 - \beta\omega_1'\omega_2'' I_1 I_2 - \alpha\omega_1''\omega_2'(I_1)^2 - \beta\omega_1''\omega_2' I_1 I_2 +$$
$$\alpha\omega_1''\omega_2'(I_1)^2 + \beta\omega_1''\omega_2''(I_1)^2 + \alpha\omega_1'\omega_2' I_1 I_2 + \beta\omega_1''\omega_2' I_1 I_2 +$$
$$\alpha\omega_1'\omega_2''(I_1)^2 + \beta\omega_1'\omega_2'' I_1 I_2 - \alpha\omega_1'\omega_2''(I_1)^2 - \beta\omega_1''\omega_2''(I_1)^2 = 0.$$

□

Now we want to remark why, from the point of view of the image data, this is a required condition in order to solve this PS formulation of the SfS problem. In particular we proved that, if we consider a third image I_3 which is obtained by a light source ω''' coplanar with ω' and ω'' (that we use to obtain the first two images) we do not add any information. Rather, we will see that it is possible to compute such I_3 "artificially".

Remark 10.1. The modelization of the orthographic SfS problem allows us to define the linear dependence of the images through their associated light vectors. In fact the way we consider the vector ω''' coplanar with the other two allows us to define, in the same way, the concept of linearly dependent image functions.

In order to give a better explanation of the linearly independent images and their importance for the resolution of the problem, we start giving the following definition:

Definition 10.1. Three image functions I_1, I_2 and I_3, obtained from the orthographic SfS model, are *linearly independent* if they are generated by three non coplanar light vectors.

The idea behind this kind of definition uses the orthographic SfS model in order to compare the linear dependence/independence of vectors in $\mathbb{R}^3$ with respect to the image functions.

10.3 Linear Dependent Image Reconstruction

An interesting application regarding the orthographic SfS model is related to the possibility of generating the images of an object by varying the light source between two points ω' and ω'' for which we already have the respective images. We suppose to have two images I_1 and I_2 obtained with two light sources ω' and ω'' respectively, it is possible to deduce all the linearly dependent images. In the geometrical sense, if we consider these two light vectors like points of the upper hemisphere $B((0,0,0),1)$, then it is possible to calculate all the obtainable images with a light source in the geodesic identified by ω' and ω'' (Fig. 10.2).

The main idea is related to the approximation of the points of a geodesic defined between the two light source vectors on the unit sphere. The method that we use to calculate the vector defined from the geodesic is based on the Remark 10.1. In fact, once we fix the extremal points, we discretize the geodesic through an affine transformation that allows us to see this curve as an arc of a circumference on the

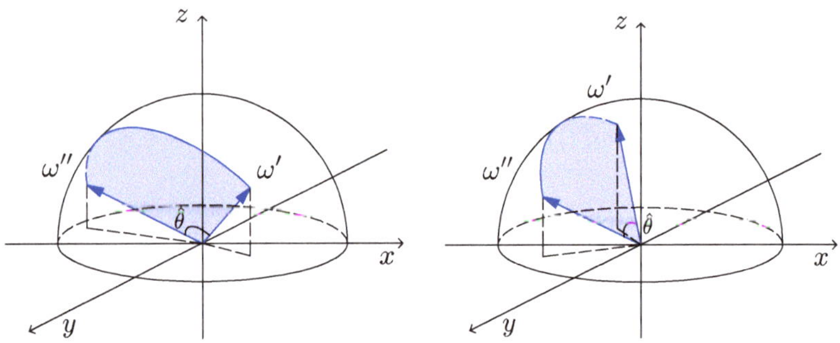

Fig. 10.2 Examples of geodesics (in *blue*) on the unit sphere between ω' and ω''

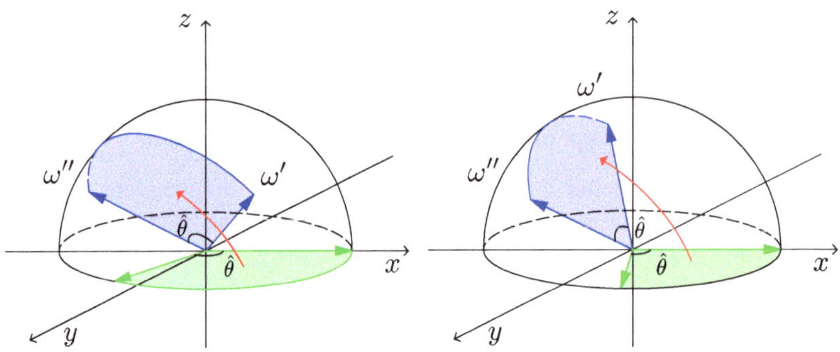

Fig. 10.3 Examples of affine transformations T between the area defined by the arc of the circumference (in *green*) and the one relates to the geodesic (in *blue*)

xy plane. Let us start by observing that the angle between the two unit light vectors $\hat{\theta}$ is computable using the dot product between them. In fact:

$$\hat{\theta} = \arccos(\omega' \cdot \omega'') \tag{10.23}$$

Now, we determine the affine transformation which allows the use of one angular parameter for the discretization of all the curve. For this purpose, we specify the linear mapping T like a matrix in $\mathbb{R}^{3x3}$. In order to compute the coefficients of this matrix (i.e. the degree of freedom) we fix the following three conditions (Fig. 10.3):

$$\begin{aligned} T(1,0,0)^T &= \omega' \\ T(\cos\hat{\theta}, \sin\hat{\theta}, 0)^T &= \omega'' \\ T(0,0,1)^T &= q. \end{aligned} \tag{10.24}$$

We denote q as the unit vector (opportunely oriented) orthogonal to the plane where the vectors ω' and ω'' lie (such that they are outgoing from the plane, i.e.

$q_3 > 0$). We consider the plane, generated by the vectors ω' and ω'', to have the following direction parameters:

$$\begin{aligned} s &= \omega_2'\omega_3'' - \omega_3'\omega_2'' \\ l &= \omega_3'\omega_1'' - \omega_1'\omega_3'' \\ m &= \omega_1'\omega_2'' - \omega_2'\omega_1''. \end{aligned} \tag{10.25}$$

We consider the normalization of the vector (s, l, m) like a possible vector q depending on the sign of m. If $m < 0$ we take the opposite direction to the normalization.

Let us make explicit the condition imposed by (10.24) writing the coefficients of the matrix that represents the transformation:

$$T = \begin{pmatrix} \omega_1' & \frac{\omega_1''-\omega_1'\cos\hat{\theta}}{\sin\hat{\theta}} & q_1 \\ \omega_2' & \frac{\omega_2''-\omega_2'\cos\hat{\theta}}{\sin\hat{\theta}} & q_2 \\ \omega_3' & \frac{\omega_3''-\omega_3'\cos\hat{\theta}}{\sin\hat{\theta}} & q_3 \end{pmatrix}. \tag{10.26}$$

If we consider an angle θ such that $0 < \theta < \hat{\theta}$, we obtain a third light source $\bar{\omega}$ using just the previous application:

$$T(\cos\theta, \sin\theta, 0)^T = \bar{\omega}. \tag{10.27}$$

Now, if we have built $\bar{\omega}$ coplanar to the previous two light vectors, then we have $\bar{\omega} = \alpha\omega' + \beta\omega''$. The next step is to determine the coefficients $\alpha, \beta \in \mathbb{R}$. We write the previous equality and we obtain the following overdetermined linear system:

$$\begin{cases} \bar{\omega}_1 = \alpha\omega_1' + \beta\omega_1'' \\ \bar{\omega}_2 = \alpha\omega_2' + \beta\omega_2'' \\ \bar{\omega}_3 = \alpha\omega_3' + \beta\omega_3'' \end{cases} \tag{10.28}$$

With the aim to calculate the coefficients that determine the third light source, we solve the following linear system of two equations in two unknowns:

$$C\begin{pmatrix} \alpha \\ \beta \end{pmatrix} = b \tag{10.29}$$

where $C \in \mathbb{R}^{2\times 2}$ and clearly $b \in \mathbb{R}^2$.

Selecting two of the three equations of (10.28) in order to avoid singularity, the matrix C depends on the first two light vectors that we can express in spherical coordinates as follows:

$$\omega' = (\cos\theta_1 \sin\varphi_1, \sin\theta_1 \sin\varphi_1, \cos\varphi_1)$$

and

$$\omega'' = (\cos\theta_2 \sin\varphi_2, \sin\theta_2 \sin\varphi_2, \cos\varphi_2)$$

(with $\theta_1, \theta_2 \in [0, 2\pi]$ and $\varphi_1, \varphi_2 \in [0, \frac{\pi}{2})$). Our aim is to solve this overdetermined linear system considering only two of the three equations in order to avoid that one of the first two equations has all coefficients close or equal to zero (for example it can happen if we have $\theta_1 = \theta_2 = 0$ with respect to the first equation or if we have $\theta_1 = \theta_2 = \frac{\pi}{2}$ with respect to the second equation). Then, in order to avoid a bad conditioned linear system, if we consider the following set:

$$\bar{A} = \left[0, \frac{\pi}{4}\right] \cup \left[\frac{3\pi}{4}, \frac{5\pi}{4}\right] \cup \left[\frac{7\pi}{4}, 2\pi\right] \tag{10.30}$$

we take

$$\begin{pmatrix} \omega_2' & \omega_2'' \\ \omega_3' & \omega_3'' \end{pmatrix} \begin{pmatrix} \alpha \\ \beta \end{pmatrix} = \begin{pmatrix} \bar{\omega}_2 \\ \bar{\omega}_3 \end{pmatrix} \quad \text{if} \quad \theta_1, \theta_2 \in \bar{A} \tag{10.31}$$

while

$$\begin{pmatrix} \omega_1' & \omega_1'' \\ \omega_3' & \omega_3'' \end{pmatrix} \begin{pmatrix} \alpha \\ \beta \end{pmatrix} = \begin{pmatrix} \bar{\omega}_1 \\ \bar{\omega}_3 \end{pmatrix} \quad \text{otherwise.} \tag{10.32}$$

Once we determine the coefficients α and β, we can compute the linear dependent image as follows:

$$\bar{I} = \alpha I_1 + \beta I_2. \tag{10.33}$$

It is important to note that there is no approximation when we compute the linear dependent image $\bar{I}$. In fact, the only computation that can introduce an error in the computation of the third image is related to the approximation of the linear system (10.29) for which we consider two different formulations ((10.31) and (10.32)) in order to make unimportant the error committed.

10.4 Reduction of the Number of the Images Using Symmetries

Supposing that we are able to determine univocally the surface $u(x, y)$ of the SfS-PS problem using three images obtained with three non coplanar light sources, now we want to consider some classes of particular surfaces for which it is possible to resolve the SfS-PS problem using less than three images.

10.4.1 Symmetric Surfaces

We will prove a theorem that gives us the possibility to produce an image (of a particular surface) by using another image of the same surface, obtained from a

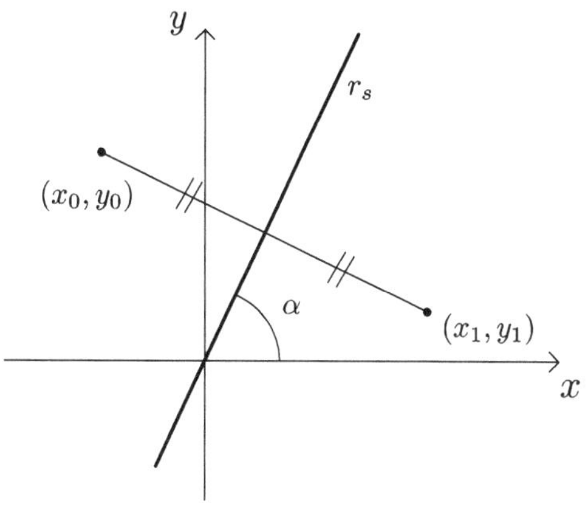

Fig. 10.4 Orthogonal view of the plane (x, y): example of a positioning of the points (x_0, y_0) and (x_1, y_1) with respect to the *symmetry straight line* r_s

light source appropriately chosen. Let us start to define the type of surface we can use in order to apply that.

Definition 10.2. Let $z = u(x, y)$ be a function defined in $\Omega \subset \mathbb{R}^2$. Let π_s (symmetry plane) be a plane passing through the z axis and let r_s (symmetry straight line) be its intersection with the xy plane. We say that u is *symmetric* with respect to the plane π_s if, for every point $(x_0, y_0) \in \Omega$, it results:

$$u(x_0, y_0) = u(x_1, y_1) \tag{10.34}$$

where $(x_1, y_1) \in \Omega$ is the symmetric point of (x_0, y_0) with respect to the straight line r_s (Fig. 10.4).

With regard to the orthographic SfS problem, since the model has not a direct dependence on the value of the function u in a point, but only on its gradient, it is essential to determine a symmetry relation with respect to the partial derivatives of the surface u. We identify r_s like a particular straight line passing through the origin of the reference system xyz, that is

$$x \sin\alpha = y \cos\alpha, \text{ with } \alpha \in [0, \pi]. \tag{10.35}$$

We consider this straight line like a subspace generated by the vector $(\cos\alpha, \sin\alpha)$. It is possible to identify the orthogonal straight line to r_s from its base vector. Then, let

$$v = \left(\cos\left(\alpha + \frac{\pi}{2}\right), \sin\left(\alpha + \frac{\pi}{2}\right)\right) = (-\sin\alpha, \cos\alpha) \tag{10.36}$$

be this orthogonal straight line. The main relation about the symmetry that we use in the SfS model is the following:

$$v \cdot \nabla u(x_0, y_0) = -v \cdot \nabla u(x_1, y_1). \tag{10.37}$$

10.4.2 Uniqueness Theorem for the Symmetric Surfaces

We aim at obtaining further information relative to the symmetric surfaces by deducing the grey scale values of an image corresponding to a particular light source whose position depends on the symmetry straight line r_s. These information, together with a third image, allow us to resolve the SfS-PS problem using three images linearly independent.

Theorem 10.1. *Let $u(x, y)$ be a symmetric surface with respect to any straight line r_s such that $(0,0) \in r_s$. Let ω' and ω''_{ded} be two light vectors such that:*

1. *$\widetilde{\omega}'$ is orthogonal with respect to the symmetry straight line r_s;*
2. *$\widetilde{\omega}''_{ded} = -\widetilde{\omega}'$ (that is $\omega''_{ded} = (-\widetilde{\omega}', \omega'_3)$ since the light vectors are constrained to be taken in the upper hemisphere).*

Then it is possible to deduce the image associated to the light source ω''_{ded} only by I_1 as follows:

$$I_2^{ded}(x_0, y_0) = I_1(x_1, y_1). \tag{10.38}$$

Proof. Considering the image definition of the SfS problem, we have:

$$I_2^{ded}(x_0, y_0) = \frac{-\widetilde{\omega}''_{ded} \cdot \nabla u(x_0, y_0) + \omega''_{ded\,3}}{\sqrt{1 + ||\nabla u(x_0, y_0)||^2}} \stackrel{**}{=} \frac{\widetilde{\omega}' \cdot \nabla u(x_0, y_0) + \omega'_3}{\sqrt{1 + ||\nabla u(x_0, y_0)||^2}} \stackrel{*}{=}$$
$$= \frac{-\widetilde{\omega}' \cdot \nabla u(x_1, y_1) + \omega'_3}{\sqrt{1 + ||\nabla u(x_0, y_0)||^2}} = \frac{-\widetilde{\omega}' \cdot \nabla u(x_1, y_1) + \omega'_3}{\sqrt{1 + ||\nabla u(x_1, y_1)||^2}} = I_1(x_1, y_1) \tag{10.39}$$

where in ** we use the assumption 2 of the theorem, whereas in *, as the surface is symmetric, we exploit (10.37). In the last equality we consider furthermore that the normal vector in the point (x_0, y_0) has the same norm as in (x_1, y_1). □

Let us explain with an example the previous theorem considering the symmetric surface of Fig. 10.5 that we will use again in the numerical test section.

It is clear that its symmetry straight line r_s drown in Fig. 10.5b defined by the angle $\alpha = \frac{3\pi}{4}$, permits us to apply the previous theorem starting with a first light source ω' having its projection on the xy plane $\widetilde{\omega}'$ orthogonal to r_s. For example we can choose $\theta_1 = \frac{\pi}{4}$ and, with the only aim to avoid shadows, we choose $\varphi_1 = 0.1$ obtaining the image I_1 shown in Fig. 10.6 (on the left). Theorem 10.1 says that we can deduce (and not approximate) a second image I_2^{ded} shown in Fig. 10.6 (on the right) associated to the light source $\omega''_{ded} = (-\omega', \omega'_3)$. In this example the angles that define the polar coordinates of ω''_{ded} are $\theta_2^{ded} = \theta_1 + \pi = \frac{5\pi}{4}$ and $\varphi_2^{ded} = \varphi_1$.

As we will remark the deduced image will help to have the sufficient information (about the three linear independent images) in order to solve the SfS-PS problem using less than three images.

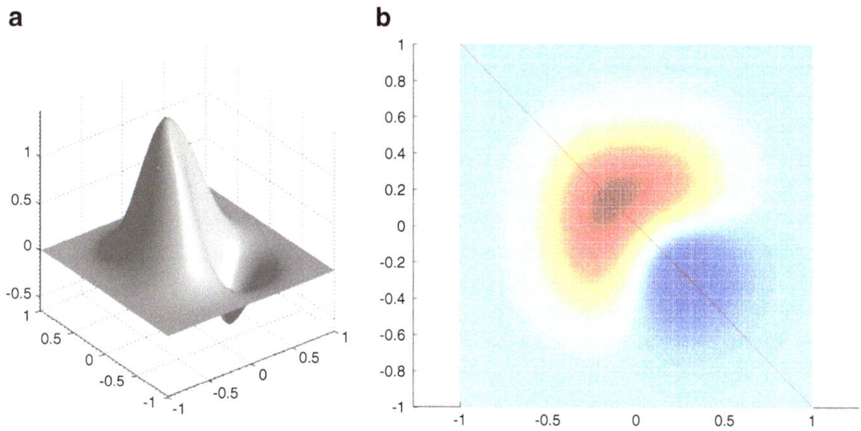

Fig. 10.5 On the *left* the 3D shape of the symmetric surface taken into account. On the *right* the orthographic view of the surface is shown in order to make clear the symmetry (the *red line*). The color is proportional to surface height, starting from the *cold color* (for the lowest values) to the *hotter color*

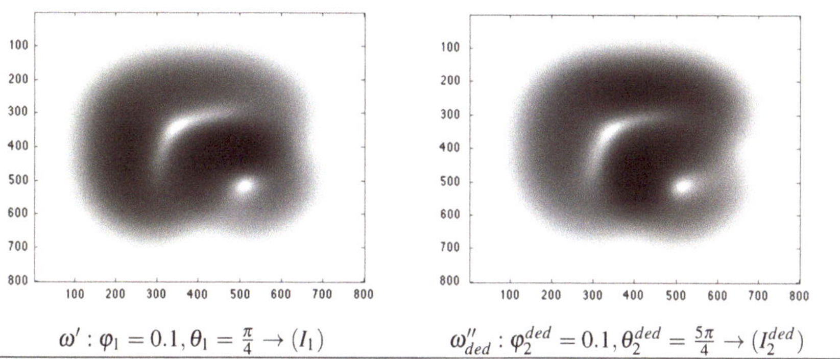

Fig. 10.6 On the *left* the starting image obtained by the light source ω' and on the *right* the deduced image computed using the light source ω''_{ded}

Once we choose $(x_0, y_0) \in \Omega$, the problem consists in determining the coordinates of the symmetric point (x_1, y_1) with respect to r_s. The point (x_1, y_1) belongs to the straight line $r_s^{\perp}$, orthogonal to r_s and passing through the point (x_0, y_0), whose equation is

$$(x - x_0)\cos\alpha = -(y - y_0)\sin\alpha. \tag{10.40}$$

Finally, observing that the symmetric straight line r_s passes through the origin, it results necessarily that the distances with respect to $(0, 0)$ of both points are the same.

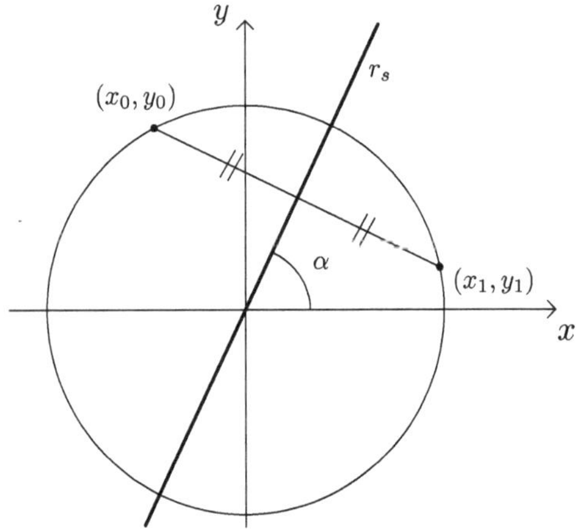

Fig. 10.7 In this schematic example is shown the circumference containing the two symmetric points (x_0, y_0) and (x_1, y_1) with respect to the *straight line* r_s

That is, both points belong to the same circumference centered in the origin of the axis as shown in Fig. 10.7. This formulation of the problem allows that the following non-linear system

$$\begin{cases} (x - x_0)\cos\alpha = -(y - y_0)\sin\alpha \\ x^2 + y^2 = x_0^2 + y_0^2 \end{cases} \tag{10.41}$$

admits only two solutions, that is (x_0, y_0) and (x_1, y_1).

The procedure we use to solve the system (10.41) is related to an algorithm that takes into account the numerical instability. To this purpose, some particularities appear.

For first in the computation of the solutions of (10.41) we have to take into account the angle α. In order to avoid numerical instability, we consider two possible sets to which this angle could belong:

$$A = \left[0, \frac{\pi}{4}\right] \cup \left[\frac{3\pi}{4}, \pi\right] \text{ and } B = \left(\frac{\pi}{4}, \frac{3\pi}{4}\right). \tag{10.42}$$

Let us suppose that $\alpha \in A$; we use the first equation of (10.41), and we write the first coordinate of the point (since the division by $\cos\alpha$ does not involve numerical instability). We have then:

$$x = (y_0 - y)\frac{\sin\alpha}{\cos\alpha} + x_0$$

which, replaced in the second equation of (10.41), gives a second order equation (in the general case written as $ay^2 + by + c = 0$) where the coefficients are:

$$\begin{aligned} a &= 1 + (\tan\alpha)^2 \\ b &= -2(y_0(\tan\alpha)^2 + x_0\tan\alpha) \\ c &= y_0^2(\tan\alpha)^2 + 2y_0x_0\tan\alpha - y_0^2. \end{aligned} \tag{10.43}$$

The two solutions of this equation are y_0 and y_1. It is possible to determine the exact solution since y_0 is known.

In the same way, if $\alpha \in B$ it is possible to divide by $\sin\alpha$. We write then the second coordinate of the point:

$$y = (x_0 - x)\frac{\cos\alpha}{\sin\alpha} + y_0$$

and, by substitution, we solve the second order equation in the variable x with the following coefficients:

$$\begin{aligned} a &= 1 + (\cot\alpha)^2 \\ b &= -2(x_0(\cot\alpha)^2 + y_0\cot\alpha) \\ c &= x_0^2(\cot\alpha)^2 + 2x_0y_0\cot\alpha - x_0^2. \end{aligned} \tag{10.44}$$

The choice of the right solution can be done exactly as explained before, that is excluding the value x_0 already known.

In the end, if we consider an image I_1 obtained with a light source that respects the hypotheses of the Theorem 10.1, it is possible to deduce another image I_2 in the way we explained before.

Corollary 10.1. *Let $u(x, y)$ be a symmetric surface with respect to a straight line r_s passing through the point* $(0, 0)$.

1. *Let I_1 be an image obtained with a light vector $\omega' = (\widetilde{\omega}', \omega_3')$ such that $\widetilde{\omega}'$ is orthogonal to the symmetry straight line of $u(x, y)$,*
2. *Let I_3 be an image obtained with a light source $\omega''' = (\widetilde{\omega}''', \omega_3''')$ such that $\widetilde{\omega}' \cdot \widetilde{\omega}''' \neq \pm||\widetilde{\omega}'||||\widetilde{\omega}'''||$ (that is $\widetilde{\omega}'$ and $\widetilde{\omega}'''$ have not the same direction).*

Then it is possible to univocally determine the surface $u(x, y)$ using the SfS-PS model.

The use of the image I_2 defined in the Theorem 10.1 is necessary to the application of the reconstruction technique using three images obtained with non coplanar light vectors.

Other simplifications in the calculus of the point (x_1, y_1) are possible when the surface $u(x, y)$ has a particular kind of symmetry. As a first particular case we consider now a symmetry with respect to the orthogonal straight line: let r_s' and r_s'' be two symmetry straight lines of u orthogonal between them. Thus we have the following relation:

$$\nabla u(x_0, y_0) = -\nabla u(-x_0, -y_0) \tag{10.45}$$

for which we note the facility of calculus in the point previously called $(x_1, y_1) = (-x_0, -y_0)$.

10.4.3 Surfaces with Four Symmetry Straight Lines

A very interesting particular case is related to the surfaces with four symmetry straight lines. In relation to these surfaces, in fact, it is possible to solve the classical SfS problem considering only one image generated from any light source. For this case the constraint concerning the light sources direction vanishes. As before, we follow the procedure of reconstructing the missing information (that is the three linear independent images). Now, because of the particular surface geometry, we are able to obtain the necessary three images only from one. That is we can reconstruct the two missing images (generated with non coplanar light vectors) starting from only one image of the surface lighted up by any light source.

Let us consider the image I_1 and let ω' be its relative light source. The other two images can be calculated in the following way:

$$\begin{aligned} I_2(x,y) &= I_1(-x,-y) \text{ with } \omega'' = (-\widetilde{\omega}', \omega'_3) \\ I_3(x,y) &= I_1(-y,x) \text{ with } \omega''' = (\omega'_2, -\omega'_1, \omega'_3). \end{aligned} \tag{10.46}$$

Relating to the third image, we use $(v_1, v_2) \cdot \nabla u(x,y) = -(v_2, -v_1) \cdot \nabla u(-y,x)$, where (v_1, v_2) is any vector of the plane (x,y). The choice of the light placing, together with the way of determining the images, is not unique. The aim of this construction is to obtain linearly independent reconstructed images for which the uniqueness of solution for the SfS problem stands.

We can adapt to this particular situation based on symmetries the uniqueness theorem in [13] that proves the uniqueness of a Lipschitz solution (i.e. a solution in the Sobolev space $W^{1,\infty}(\overline{\Omega})$) relatively to the differential approach:

Theorem 10.2. *Let $I(x,y)$ be an image of the surface $u(x,y) \in W^{1,\infty}(\overline{\Omega})$, with four straight lines of symmetry obtained by the SfS model such that $0 < I(x,y) \leq 1$ $\forall (x,y) \in \overline{\Omega}$ with the light source ω. Then it is possible to determine this surface univocally solving the SfS classic problem.*

The procedure followed until now is a constructive proof of the previous theorem and allows to solve, for a particular class of surfaces, the classical SfS problem passing through the SfS-PS one. Furthermore, our analysis of the problem allows us to give also a numerical method for the approximation of this unique weak solution.

Remark 10.2. Following this procedure it is possible to prove the uniqueness of a weak solution of the classic SfS problem even the surface is radial by symmetries.

Remark 10.3. For both cases (one or four axes) the construction of the other images can be done only if the light source of the first image is not vertical. This means that in this case, even using this image deduction technique, the concave/convex ambiguity cannot be solved.

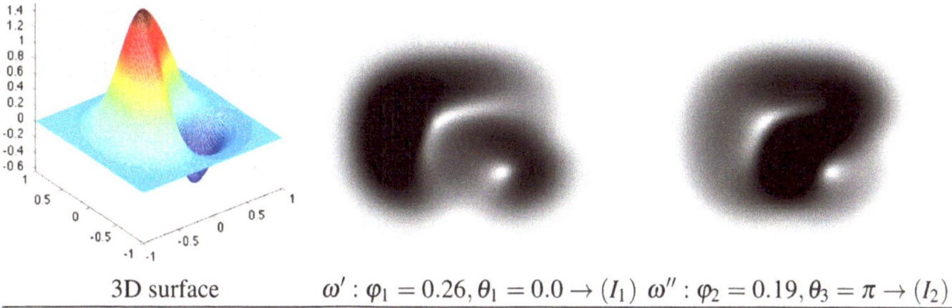

Fig. 10.8 From *left* to *right*: the 3D shape of the surface and the starting images with the respective light sources

Fig. 10.9 Between I_1 and I_2 there are three reconstructed images with the intermediate light sources taken equidistant in the geodesic

10.5 Numerical Tests

This section is divided in two parts. The first one is related to the computation of the linear dependent images explained in Sect. 10.3. In the second part we present some shape recovery tests relative to the local and differential method for the SfS-PS using symmetric surfaces in order to compute the error of both methods using less than three images.

10.5.1 Numerical Computation of Linear Dependent Images

The tests on the linear independent images concept is carried out on synthetic and real surfaces.

For the synthetic case we start using the surface (with the respective initial images) shown in Fig. 10.8. In Fig. 10.9 it is possible to see the images of the same surface for which the light sources are equidistant in the geodesic defined by ω' and ω''.

For the real case we start from the images shown in Fig. 10.10, namely I_1^{real} and I_2^{real}. In Fig. 10.11 it is possible to see three linear dependent images obtained with

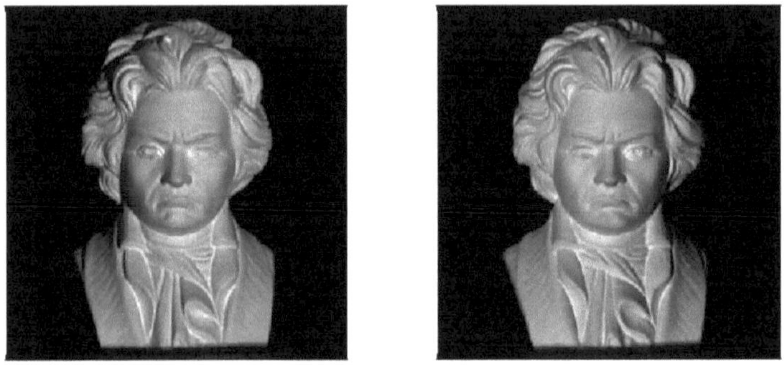

$\omega'_{real} : \varphi_1 = 0.263, \theta_1 = -0.305 \to (I_1^{real})$ $\omega''_{real} : \varphi_2 = 0.281, \theta_2 = 3.502 \to (I_2^{real})$

Fig. 10.10 Real starting data

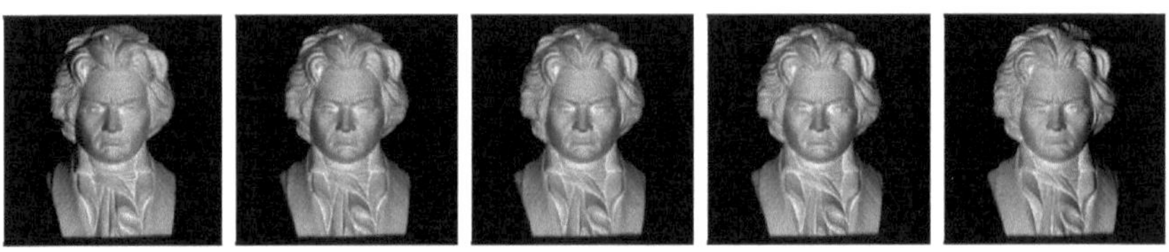

I_1^{real} I_2^{real}

Fig. 10.11 The results of the computation using our algorithm. The three images on the middle are lighted up from equidistant light sources in the geodesic defined by the starting images

the algorithm explained in the previous section. Also in this case we are considering three light sources equidistant in the geodesic defined by ω'_{real} and ω''_{real}.

10.5.2 Shape Reconstruction for Symmetric Surfaces

Let us start the numerical tests by studying several surfaces. Every single one is of a typology that allows us to apply each variant of the previous results explained. These surfaces have symmetries that allows us to apply all the previous techniques that solve the SfS-PS problem.

The different typologies of the three surfaces represented in Fig. 10.12 are such that their geometrical characteristics differentiate the number of necessary images for their tridimensional reconstruction. In fact:

- $v_{symm}(x, y)$, used also in the previous numerical test, is a surface with one symmetry axis (the straight line $y + x = 0$) for which we need two images;
- $v_{lip}(x, y)$ is a Lipschitz surface with four axis of symmetry (that is $y = \pm x$, $y = 0$, $x = 0$). For its reconstruction we need only one image;
- $v_{rad}(x, y)$ is a particular case of the previous type of surfaces. In this radial case the curves where it is not differentiable are concentric circles.

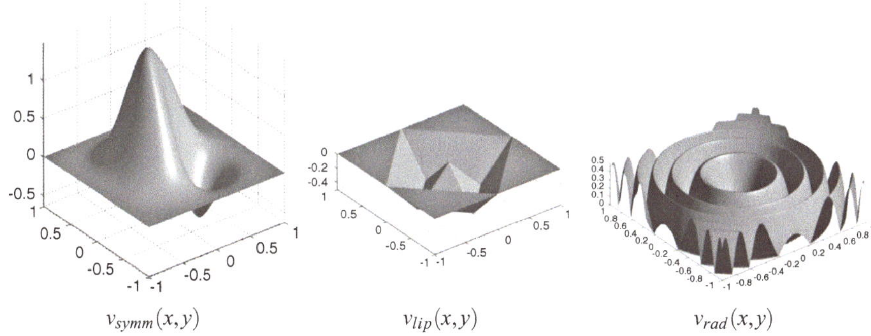

Fig. 10.12 Set of surfaces used for the numerical tests

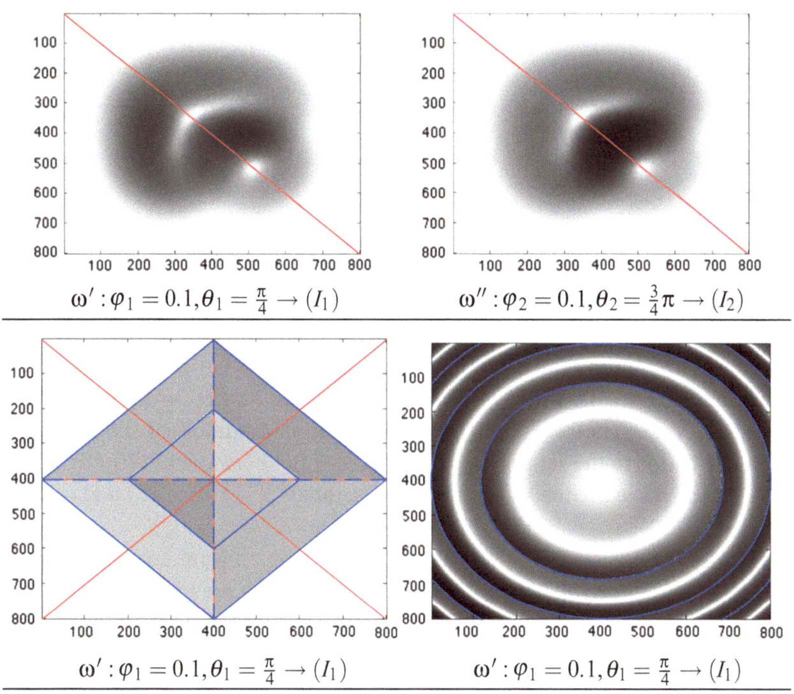

Fig. 10.13 Set of images used with the respective light sources described by their spherical coordinates. For each surface it is possible to see the *straight line* of symmetry (in *red*) and the curve where the surface is not differentiable (in *blue*). In the *first line* we have the only two images used to compute the shape of v_{sym}. On the *second line* we have the only image needed to compute the surfaces v_{lip} and v_{rad} (respectively from *left* to *right*)

Table 10.1 Error in $L^{\infty}(\Omega_d)$ norm and order of convergence for v_{sym} with angles: $\theta_1 = \frac{\pi}{4}$ and $\theta_2 = \frac{3}{4}\pi$; $\varphi_1 = \varphi_2 = 0.1$

Δ	SL forward error	SL forward order	SL backward error	SL backward order	Local normal
0.02	5.254×10^{-2}		6.856×10^{-2}		1.63×10^{-2}
0.01	2.618×10^{-2}	1.0055	3.428×10^{-2}	1.0021	1.74×10^{-2}
0.005	1.307×10^{-2}	1.0033	1.713×10^{-2}	1.0008	6.75×10^{-2}
0.0025	6.530×10^{-3}	1.0011	8.570×10^{-3}	0.9992	1.289×10^{-1}

For these starting data, summarized in Fig. 10.13, we make different tests modifying the size of the grid Δ, uniform for all the domain Ω.

In order to solve the differential problem (10.8), once boundary condition $g(x_i, y_j) = u(x_i, y_j)\ \forall (x_i, y_j) \in \partial\Omega_d$ is computed, we use the two following semi-lagrangian fix point schemes [13]:

$$u_{i,j}^{n+1} = u^n(x_i - h\gamma_1(x_i, y_j), y_j - h\gamma_2(x_i, y_j)) + \frac{f_{i,j}}{||b_{i,j}||}h, \quad \forall (x_i, y_j) \in \Omega_d \tag{10.47}$$

and

$$u_{i,j}^{n+1} = u^n(x_i + h\gamma_1(x_i, y_j), y_j + h\gamma_2(x_i, y_j)) - \frac{f_{i,j}}{||b_{i,j}||}h, \quad \forall (x_i, y_j) \in \Omega_d \tag{10.48}$$

where $\gamma(x, y) = (\gamma_1(x, y), \gamma_2(x, y)) = \frac{b(x,y)}{||b(x,y)||}$, $f_{i,j}$ is the discrete version of (10.10) and h is a small parameter that discretizes the directional derivative. Using a bilinear interpolation with respect to the points of the lattice Ω_d at point $(x_i \pm h\gamma_1(x_i, y_j), y_j \pm h\gamma_2(x_i, y_j)) \notin \Omega_d$, we can reach the first order of consistency choosing $h = \Delta$. The previous two semi-lagrangian numerical schemes will be respectively called in the sequel as *forward* (10.47) and *backward* (10.48). For these iterative schemes we use the same initial guess, that is $u_{i,j}^0 = 0$ in Ω_d and $u_{i,j}^0 = g_{i,j}$ in $\partial\Omega_d$.

The numerical approximation is carried out with both semi-lagrangian schemes and we calculate the errors in $L^{\infty}(\Omega_d)$ norm by varying the uniform discretization step Δ. In particular we consider a domain $\overline{\Omega_d} = [-1, 1]^2$ using synthetic images of size starting from 100×100 pixels ($\Delta = 0.02$) to 800×800 pixels ($\Delta = 0.0025$). In all the following tables we compute also the convergence order with the aim to verify the convergence of the methods. In the last column we show the error committed using the approximation by the local normal approach [6] always in the same $L^{\infty}(\Omega_d)$ norm (Tables 10.1–10.3).

In all the tables it is possible to see how both the numerical schemes work well for all the cases considered. In particular it is possible to see how the semi-lagrangian iterative schemes have an order of about one for all the tests. This means that the approximation of the symmetric surfaces can be done without loss of precision considering less than three images.

Table 10.2 Error in $L^{\infty}(\Omega_d)$ norm and order of convergence for v_{lip} with angles: $\theta_1 = \frac{\pi}{4}$ and $\varphi_1 = 0.1$

Δ	SL forward error	SL forward order	SL backward error error	SL backward order	Local normal
0.02	2.731×10^{-2}		2.731×10^{-2}		1.01×10^{-2}
0.01	1.466×10^{-2}	0.8975	1.466×10^{-2}	0.8975	5.1×10^{-3}
0.005	8.060×10^{-3}	0.8630	8.060×10^{-3}	0.8630	3.0×10^{-3}
0.0025	4.270×10^{-3}	0.9165	4.270×10^{-3}	0.9165	4.32×10^{-2}

Table 10.3 Error in $L^{\infty}(\Omega_d)$ norm and order of convergence for v_{rad} with angles: $\theta_1 = \frac{\pi}{4}$ and $\varphi_1 = 0.1$

Δ	SL forward error	SL forward order	SL backward error	SL backward order	Local normal
0.02	2.5491×10^{-1}		3.7057×10^{-1}		1.136×10^{-1}
0.01	1.2472×10^{-1}	1.0313	1.7993×10^{-1}	1.0423	4.53×10^{-2}
0.005	6.2360×10^{-2}	1	9.1010×10^{-2}	0.9833	2.0×10^{-2}
0.0025	3.1280×10^{-2}	0.9963	4.5890×10^{-2}	0.9878	8.3×10^{-3}

10.6 Conclusion and Perspectives

We have seen how it is possible to resolve the orthographic SfS problem with the PS technique uniquely, getting over the concave/convex ambiguity. We have shown how the linear independence of images is fundamental to obtain a unique reconstruction of the surface. Also an interesting application is explained: the possibility to generate linear dependent images of an object changing the light source between two points ω' and ω'' for which we already have the relative images. The numerical tests give good results for both the two approaches: normal vector field approach and differential one.

The main perspective is to apply these two results to a weaker model that take into account a less number of assumptions. For example, the well-know perspective SfS model [17–19], represented by the following equation

$$\frac{f(\widetilde{\omega} \cdot \nabla u(x)) - \omega_3(x \cdot \nabla u(x) + u(x))}{\sqrt{f^2|\nabla u(x)|^2 + (x \cdot \nabla u(x) + u(x))^2}} = I(x)$$

(where f is the focal distance), contains some degree of ill-posedness. A further step forward can be done solving the perspective problem with the same technique explained before instead of the orthographic projection assumption.

With the aim to eliminate the Lambertian surface assumption the specular model can be considered (see [1, 16, 21–23]).

References

1. Blinn, J.F.: Models of light reflection for computer synthesized pictures. Comput. Graph. **11**(2), 192–198 (1977)
2. Brooks, M.J., Chojnacki, W., Kozera, R.: Impossible and ambiguous shading patterns. Int. J. Comput. Vis. **7**(2), 119–126 (1992)
3. Cham, T.-J., Cipolla, R.: Symmetry detection through local skewed symmetries. Image Vis. Comput. **13**(5), 439–450 (1995)
4. Durou, J.-D., Courteille, F.: Integration of a normal field without boundary condition. In: Proceedings of the First International Workshop on Photometric Analysis For Computer Vision, Rio de Janeiro (2007)
5. Durou, J.-D., Falcone, M., Sagona, M.: Numerical methods for shape from shading: a new survey with benchmarks. Comput. Vis. Image Underst. **109**(1), 22–43 (2008)
6. Durou, J.-D., Aujol, J.-F., Courteille, F.: Integrating the normal field of a surface in the presence of discontinuities. In: Proceedings of the Energy Minimization Method in Computer Vision and Pattern Recognition 2009. Lecture Notes in Computer Science, vol. 5681, pp. 261–273. Springer, Berlin (2009)
7. Friedberg, S.A.: Finding axes of skewed symmetry. Comput. Vis. Graph. Image Process. **34**(2), 138–155 (1986)
8. Gofman, Y., Kiryati, N.: Detecting symmetries in grey level images: the global optimization approach. In: Proceedings of the International Conference on Pattern Recognition, San Francisco (1996)
9. Gordon, G.G.: Shape from symmetry. In: Proceedings of SPIE Intellent Robots and Computer Vision VIII, Philadelphia (1989)
10. Gordon, G.G.: 3d reconstruction from a single view of an object ant its image in a plane mirror. In: Proceedings of International Conference on Pattern Recognition, Brisbane (1998)
11. Kozera, R.: Existence and uniqueness in photometric stereo. Appl. Math. Comput. **44**(1), 103 (1991)
12. Kozera, R.: Uniqueness in shape from shading revisited. J. Math. Imaging Vis. **7**(2), 123–138 (1997)
13. Mecca, R.: Uniqueness for shape from shading via photometric stereo technique. In: Proceedings of the 18th IEEE International Conference on Image Processing, Brussels, pp. 2994–2997 (2011)
14. Mukherjee, D.P., Zisserman, A., Brady, J.M.: Shape from symmetry – detecting and exploiting symmetry in affine images. Technical report, report no. OUEL 1988/93. Department of Engineering Science, University of Oxford
15. Onn, R., Bruckstein, A.M.: Integrability disambiguates surface recovery in two-image photometric stereo. Int. J. Comput. Vis. **5**(1), 105–113 (1990)
16. Phong, B.T.: Illumination for computer generated pictures. Commun. ACM **18**(6), 311–317 (1975)
17. Prados, E., Faugeras, O.: Perspective shape from shading and viscosity solutions. In: Proceedings of the 9th IEEE International Conference on Computer Vision, Nice, vol. 2, pp. 826–831 (2003)
18. Prados, E., Faugeras, O.: Shape from shading: a well-posed problem? In: Proceedings of the IEEE International Conference on Computer Vision and Pattern Recognition, San Diego, vol. 2, pp. 870–877 (2005)
19. Prados, E., Camilli, F., Faugeras, O.: A viscosity solution method for shape-from-shading without image boundary data. Math. Model. Numer. Anal. **40**, 393–412 (2006)
20. Shimshoni, I., Moses, Y., Lindenbaum, M.: Shape reconstruction of 3d bilaterally symmetric surfaces. Int. J. Comput. Vis. **39**(2), 97–110 (2000)
21. Torrance, K.E., Sparrow, E.M.: Theory for off-specular reflecion from roughened surfaces. J. Opt. Soc. Am. **57**(9), 1105–1114 (1967)

22. Tsai, P.S., Shah, M.: Shape from shading using linear approximation. Image Vis. Comput. J. (Elsevier, Oxford) **12**(8), 487–498 (1994)
23. Vogel, O., Breuss, M., Weickert, J.: Perspective shape from shading with non-lambertian reflectance. In: Rigoll, G. (ed.) Pattern Recognition. Lecture Notes in Computer Science, vol. 5096, pp. 517–526. Springer, Berlin (2008)

Chapter 11
Remeshing by Curvature Driven Diffusion

Serena Morigi and Marco Rucci

Abstract We present a method to regularize an arbitrary topology mesh M, which defines a piecewise linear approximation of a surface $\mathscr{M}$, with the purpose of having an accurate representation of $\mathscr{M}$: the density of the nodes should correlate with the regularity of $\mathscr{M}$. We use the mean curvature as an intrinsic measure of regularity. Unlike sophisticated parameterization-dependent techniques, our parameterization-free method directly redistributes the vertices on the surface mesh to obtain a good quality sampling with edges on element stars approximately of the same size, and areas proportional to the curvature surface features. First, an appropriate area distribution function is computed by solving a partial differential equation (PDE) model on the surface mesh, using discrete differential geometry operators suitably weighted to preserve surface curvatures. Then, an iterative relaxation scheme incrementally redistributes the vertices according to the computed area distribution, to adapt the size of the elements to the underlying surface features, while obtaining a good mesh quality. Several examples demonstrate that the proposed approach is simple, efficient and gives very desirable results especially for curved surface models with sharp creases and corners.

11.1 Introduction

The 3D geometry commonly used for shape representation in geometric modeling, physical simulation and scientific visualization is mainly based on meshes. The 3D scanning devices, medical equipments and computer vision systems often perform a uniform acquisition without any a priori knowledge of the surface properties. This may lead to raw meshes with a sampling quality usually far away from the desired sampling distribution needed for subsequent processing. Algorithms

---

S. Morigi (✉) · M. Rucci
Department of Mathematics-CIRAM, University of Bologna, Bologna, Italy
e-mail: serena.morigi@unibo.it; marco.rucci@unibo.it

M. Breuß et al. (eds.), *Innovations for Shape Analysis*, Mathematics and Visualization,
DOI 10.1007/978-3-642-34141-0_11,

for mesh simplification, denoising (fairing), decimation and remeshing represent fundamental preliminary steps in mesh processing.

In particular, remeshing refers to the improvement process of the mesh quality in terms of redistribution of the sampling, connectivity of the geometry, and triangle quality, in order to satisfy mesh property requirements while maintaining surface features. The reader is referred to [1] and the references therein for a survey on remeshing techniques.

Some remeshing techniques are parameterization-dependent, i.e. they associate the mesh with a planar parameterization, and apply the algorithms on this plane. For arbitrary genus objects, this involves also the creation of an atlas of parametrization, a well known complex process that inevitably introduces some metric distortion and may lead to the loss of important feature information [2, 4, 9].

In contrast, parameterization-free methods avoid these problems by working directly on the surface mesh and performing local modifications on the mesh. A parameterization-free method has been proposed in [5] for direct remeshing using area-equalizing weights in multiresolution modeling, and in [11], several tangential velocity strategies are introduced to regularize geometric surface flows.

Isotropic remeshing methods based on Centroidal Voronoi Tessellation (CVT) require to repeatedly compute a geodesic Voronoi diagram which is a complex and time-consuming step in this approach [4]. Several interesting proposals have been presented for this type of methods, both parameterization-based strategies [3], which compute the CVT on the 2D parametric domain, and parameterization-free methods, based on the intersection between a 3D Voronoi diagram and an input mesh surface, see [18].

In [1] the remeshing methods are classified by their end goal rather than by the algorithmic strategy they employ. The techniques are classified into five categories: structured, compatible, high quality, feature and error-driven remeshing. The structured and compatible remeshing methods aim to obtain a given connectivity structure, the main goals for the high quality remeshing methods are the shape of the elements as well as the vertex distribution, while the end goal of feature remeshing is to preserve sharp features when producing the resulting meshes.

The proposed remeshing strategy is an adaptive, parameterization-free technique designed to produce a good compromise of high quality and feature remeshing techniques. High quality remeshing amounts to generating a mesh with well-shaped elements, uniform or isotropic sampling and smooth gradation sampling. Good quality elements mainly lead to minimizing numerical instabilities in subsequent computations. However, we relaxed the uniform sampling property in order to adapt the size of the elements to the underlying surface features.

The proposed remeshing algorithm alternates equalization of edge lengths and vertex valence, which generate a new connectivity, with mesh regularization, which modifies the distribution of the vertices on the surface to satisfy given mesh quality requirements. While the techniques that explicitly modify the connectivity, such as e.g. edge split, collapse, and flip, are widely used, the potential of the regularization step is still not much investigated.

We present a new method to regularize an unstructured triangle mesh M, which defines a piecewise linear approximation of a curved surface $\mathcal{M}$, with the purpose of having an accurate representation of $\mathcal{M}$: the density of the vertices should correlate with the regularity of $\mathcal{M}$. We cannot rely on parameterization to quantify regularity of $\mathcal{M}$ because this concept would not be invariant under reparameterization and furthermore we do not assume any parameterization is given. Therefore, we used the mean curvature as a measure of regularity. To improve the regularization of the mesh M, the points $X(t)$ on the surface $\mathcal{M}$ are geometrically evolved using a tangential flow

$$\frac{\partial X}{\partial t} = \gamma \overrightarrow{T}, \qquad (11.1)$$

where γ is the velocity in the tangent direction $\overrightarrow{T}$. The new resulting sampling adapts itself to the sharper features of the surface. This motivates us to name the proposed method Adaptive Remeshing (AR).

In order to satisfy high quality remeshing we investigate the design of tangential velocities that aim to keep all edges on element stars approximatively of the same size and all areas proportional to the surface features. To achieve this aim, we use a two-step approach. First, we compute an area distribution function driven by a mean curvature map of the surface mesh. Then the mesh vertices are moved on the tangential plane to satisfy edge equalization and area distribution quality requirements. The process is iterated until a significant improvement in triangle shape is obtained.

The organization of this chapter is as follows. The adaptive mesh regularization is described in Sect. 11.2, and the proposed adaptive remeshing algorithm together with its numerical aspects is discussed in Sect. 11.3. Numerical examples and comments are provided in Sect. 11.4, and the paper is concluded in Sect. 11.5.

11.2 Adaptive Mesh Regularization

The AR method alternates equalization of edge lengths and vertex valence, which generate a new connectivity, with adaptive mesh regularization, which modifies the distribution of the vertices on the surface.

The mesh connectivity regularization is based on classical tools for meshes and it will be briefly discussed in Sect. 11.3, while, in the following, we will focus on the proposed adaptive mesh regularization method, which consists of a two-step PDE model.

In the first step, the vertex area distribution function $A(X)$ defined on the mesh M with vertex set $X = \{X_i\}_{i=1}^{n_v}$, is diffused over the mesh, constrained by the mean curvature map. In the second step, the vertex position is tangentially relocated to obtain edges on element stars approximatively of the same size, and all the vertex areas proportional to the surface features.

Let A_0 be the initial vertex area distribution function computed as the Voronoi area at each vertex on the mesh M, with vertex set X_0. Then in STEP 1, the vertex area distribution function $A(X)$ is diffused on $\mathscr{M}$ by solving

$$\frac{\partial A}{\partial t} = \Delta^{w_H}_{\mathscr{M}} A(X), \quad A(0) = A_0. \tag{11.2}$$

In (11.2) the operator $\Delta^{w_H}_{\mathscr{M}}$ is the *weighted Laplace-Beltrami operator* discretized on the mesh M. The space discretization on M of the Laplace-Beltrami $\Delta_{\mathscr{M}}$ is the connectivity matrix $L \in R^{(n_v \times n_v)}$ with elements

$$L^{ij} = \frac{1}{\sum_{j \in N(i)} w_{ij}} \begin{cases} -\sum_{j \in N(i)} w_{ij} & i = j \\ +w_{ij} & i \neq j, j \in N(i) \\ 0 & otherwise \end{cases} \tag{11.3}$$

where $N(i)$ is the set of one-ring neighbor vertices of vertex X_i. The weights w_{ij} are positive numbers defined as in [14]

$$w_{ij} = (\cot \alpha_{ij} + \cot \beta_{ij}), \tag{11.4}$$

where α_{ij} and β_{ij} are the two angles opposite to the edge in the two triangles sharing the edge (X_j, X_i), see Fig. 11.1.

In order to solve (11.2), the weighted Laplace-Beltrami operator $\Delta^{w_H}_{\mathscr{M}}$ is discretized by the matrix L_{w_H} with elements

$$L^{ij}_{w_H} = \frac{1}{\sum_{j \in N(i)} w_{ij}} \begin{cases} -\sum_{j \in N(i)} w_{ij} W_{ij} & i = j \\ +w_{ij} W_{ij} & i \neq j, j \in N(i) \\ 0 & otherwise. \end{cases} \tag{11.5}$$

The weight W_{ij} depends on a similarity measure between the ith and the jth vertex, and it is defined in terms of mean curvature values H on the mesh M as follows

$$\begin{aligned} W_{ij} = & \frac{1}{\sum_{j \in N(i)} W_{ij}} e^{-D(X_i, X_j)/\sigma}, \\ D(X_i, X_j) = & \|H(X_i) - H(X_j)\|_2^2, \quad j \in N(i), \end{aligned} \tag{11.6}$$

where σ is a filtering parameter which controls the decay of the exponential function and therefore the decay of the weights as function of the Euclidean distance between mean curvature values. Since $H(X)$ is normalized to one, we can fix the value for σ in order to identify a significant change in the curvature between vertices X_i and X_j. For example, for $\sigma = 0.1$ we identify as a curvature change when the mean curvature values in the two vertices differ more than 10 %. The use of smaller σ leads to detect more sharper features.

More details on the weights defined in (11.6) are provided in [12], where the authors apply a weighted Laplace-Beltrami operator in the context of surface fairing.

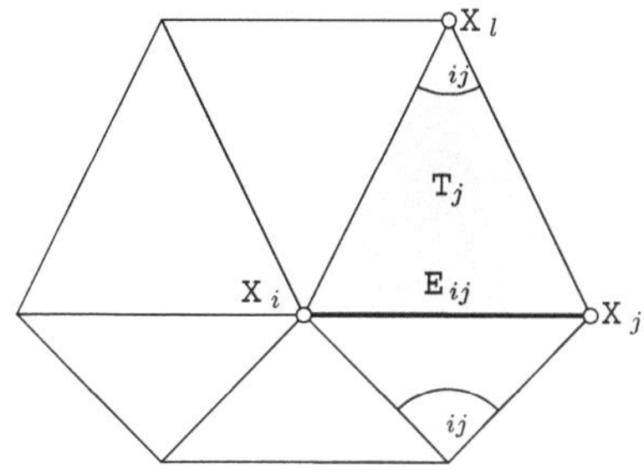

Fig. 11.1 Stencil of the first ring neighborhood of the vertex X_i. The *triangle* T_j is defined by the vertices X_i,X_j, and X_ℓ

The mean curvature $H(X)$ in (11.6) is defined as the sum of the two principal curvatures, see [10], and we compute it as described in Sect. 11.3.1. The mean curvature attribute tends to be dominated by the maximum curvature and consequently it is visually similar to it. We chose the mean curvature attribute to determine the characteristics of the underlying surface, rather than the Gaussian curvature attribute, since many shapes cannot be differentiated by Gaussian curvature alone.

The weights (11.6) used in (11.2) prevents the area diffusion in high curvature regions. The method tends to adapt the areas to the object features: high curvature regions will be covered by small area elements, while flat regions will be covered by faces with larger areas. Figure 11.2 shows the benefit of the weights in STEP 1. The result of applying a few time steps of (11.2) without the help of the weights (11.6) on the irregular initial mesh shown in Fig. 11.2a, is illustrated in Fig. 11.2b, while the contribution in (11.2) of the weights (11.6) is shown in Fig. 11.2c. The area diffusion function is represented by false colors, red colors for big areas, blue color for small areas. Increasing the number of time steps, the diffusion of (11.2) without weights converges to a constant area all over the entire mesh.

In STEP 2 the vertex position X is updated, taking into account the resulting $A(X)$ area distribution obtained in STEP 1, by solving the following constrained curvature diffusion equation

$$\frac{\partial X}{\partial t} = \nabla^{w_A}_{\mathcal{M}} \cdot (g(|H(X)|)\nabla^{w_A}_{\mathcal{M}} X), \quad X(0) = X_0, \tag{11.7}$$

where the function $g(\cdot)$, referred to as the diffusivity, is defined as

$$g(s) := \frac{1}{(1 + s^\alpha)}, \tag{11.8}$$

where $\alpha > 0$ is a small positive constant value. The geometric evolution driven by (11.7) constrains the movement of vertices with high mean curvature values, that is, belonging to sharp creases and corners.

At each vertex X_i, linearizing (11.7) by evaluating $g(|H(X_i)|)$ with X_i from the previous time-step, the right-hand side of (11.7) is rewritten as

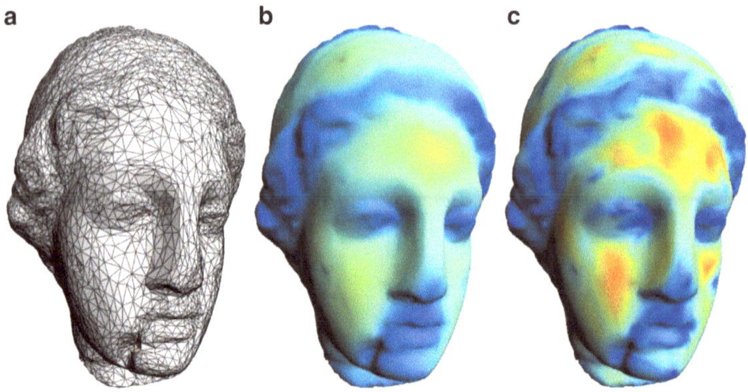

Fig. 11.2 Area diffusion in AR STEP 1: (**a**) the flat shaded mesh; (**b**) vertex area distribution function without weights (11.6); (**c**) vertex area distribution function with weights (11.6)

$$g(|H(X_i^{old})|)\Delta_{\mathcal{M}}^{w_A} X_i^{new}. \tag{11.9}$$

We denote by L_{w_A} the discretization of the weighted Laplace Beltrami operator $\Delta_{\mathcal{M}}^{w_A}$ at vertex X_i defined as in (11.5) with w_{ij} given in (11.4) and the equalization weights W_{ij}^A defined by the sigmoid function

$$W_{ij}^A = \frac{1}{\sum_{j\in N(i)} W_{ij}^A} \frac{1}{1+e^{-f(X_i,X_j)/\sigma}} \tag{11.10}$$

$$f(X_i, X_j) = \lambda_1 \frac{\Delta A(X_j) - \Delta A(X_i)}{\overline{A}} + \lambda_2 (\frac{E_{ij} - \overline{E_i}}{\overline{E}}), \tag{11.11}$$

where $\Delta A(X) = A(X) - A^d(X)$ is the offset between the vertex area $A(X)$ and the ideal vertex area $A^d(X)$ resulting from STEP 1, $\overline{A}$ is the mean of the mesh vertex areas, $\overline{E_i}$ is the local average edge length, and $\overline{E}$ is the mean of the mesh edge lengths. The coefficients $\lambda_i > 0$, $i = 1, 2$, sum up to 1 and determine how much respectively the area gap and the local edge length difference influence the movement of vertex X_i toward the neighborhood vertex X_j. In other words, considering only the area gap influence (i.e. $\lambda_1 = 1$), the vertex X_i is attracted by X_j when the area of X_i needs to grow more than the area of X_j. On the other hand, the vertex X_i does not move towards X_j when the area of both vertices do not need either to shrink or to grow or when they both need to shrink or grow by the same amount.

Finally, the displacement of the vertex X_i is in the tangent plane if we replace (11.7) with

$$\frac{\partial X_i}{\partial t} = (I - \overrightarrow{N}_i \overrightarrow{N}_i^T) g(|H(X_i)|)\Delta_{\mathcal{M}}^{w_A} X_i, \quad X(0) = X_0, \tag{11.12}$$

where $\overrightarrow{N}$ is the unit normal to the surface.

11.3 Adaptive Remeshing (AR) Algorithm

The AR algorithm iterates on the two stages of mesh connectivity regularization and adaptive mesh regularization approaching to a mesh with a smooth gradation of vertex density depending on mean curvature values and represented by well-shaped triangles.

We terminate the process and accept $X^{(i)}$ as the resulting mesh as soon as the difference of area variance between consecutive iterations is sufficiently small; specifically, we accept $X^{(i)}$ when for the first time

$$\Delta Var(A) := |Var(A(X^{(i)})) - Var(A(X^{(i-1)}))| < 1 \cdot 10^{-6}.$$

The regularization of mesh connectivity (named STEP 0) aims to perform an adjustment of edge lengths and vertex valences, and is implemented by applying the following basic tools:

1. Specify target edge length $L \in [L_{min}, L_{max}]$
2. Split all edges longer than L_{max}
3. Collapse all edges shorter than L_{min}
4. Flip edges to promote valence 6 (or 4 on boundaries).

These are commonly used tools to obtain mesh connectivity regularization, and we refer the reader to [5] for more details.

The adaptive mesh regularization stage is a two-step process which implements the PDEs (11.2) and (11.12), named in the sequel STEP 1 and STEP 2, to relax the vertex position according to a computed area distribution.

The following algorithm summarizes the computations required by our method.

Adaptive Remeshing Algorithm

Given an initial position vector X_0,
Compute $L_{w_H}(X_0), A(X_0)$, set $X^{(0)} = X_0$, i=1
While $\Delta Var(A) < 1 \cdot 10^{-6}$
  STEP 0: MESH CONNECTIVITY REGULARIZATION
  STEP 1: AREA REGULARIZATION:
  Set $A^{(0)} = A(X^{(i-1)})$
  $(I - dtL_{w_H})A^{(n+1)} = A^{(n)}$
  Compute $L_{w_A}(X^{(i-1)})$, $g(|H^{(i-1)}|)$
  STEP 2: VERTEX TANGENTIAL UPDATE:
  Set $X^{(0)} = X^{(i-1)}$
  For $n = 1, \cdots, n_{MAX}$
    $X^{(n+1)} = (I + dt(I - \vec{N}\vec{N}^T)g(|H^{(i-1)}|)L_{w_A})X^{(n)}$
  end for
  i=i+1
end while

Considering a uniform discretization of the time interval $[0, T], T > 0$, with a temporal time step dt, then (11.2) and (11.12) can be fully discretized using the forward Euler scheme which yields a first order scheme in time. We applied an implicit time scheme to (11.2), and an explicit time scheme to (11.12) with initial condition $A^{(0)}$ determined from X_0. From our experimental work we tuned up the maximum number of time iterations to be $n_{MAX} \leq 10$.

The tangential smoothing approach used in several remeshing works is a simple Laplacian smoothing discretized as in (11.3) applied to the three vertex coordinates components $X = (x, y, z)$ and then projected back into the tangent plane, see [13, 17]. Thus the tangential movement $\gamma \overrightarrow{T}$ in (11.1) at the surface vertex X_i is given by $(I - \overrightarrow{N}_i \overrightarrow{N}_i^T) L(X_i)$. In [5] the authors proposed an improvement, considering the tangential movement $\gamma \overrightarrow{T}$ in (11.1) at the surface vertex X_i as

$$(I - \overrightarrow{N}_i \overrightarrow{N}_i^T) L_w(X_i) \tag{11.13}$$

with

$$L_w(X_i) = \frac{1}{\sum_{j \in N(i)} A(X_j)} \sum_{j \in N(i)} (A(X_j)X_j - A(X_i)X_i), \tag{11.14}$$

and $A(X_j)$ represents the Voronoi area of vertex X_j. Vertices with large Voronoi area have a higher weight ('gravity') and attract other vertices, thereby reducing their own area. We call this method the Laplacian Flow (LF) scheme and we compare LF with our proposal in Sect. 11.4.

The LF scheme presented in [5] is integrated into an iterative remeshing procedure similar to the AR algorithm which alternates mesh connectivity regularization (like the STEP 0 in AR method) with the LF mesh regularization given in (11.14) (which is replaced by STEP 1 and STEP 2 in AR method).

Both the LF scheme and the STEP 2 of the AR algorithm are discretized in time using explicit integration schemes. If we let dt satisfy the stability criterion for the diffusion PDE in (11.12), then $dt \leq \frac{min(|E|)^2}{2}$, that is, it depends on the square of the smallest edge length, which is a very restrictive criterion involving an enormous number of integration steps [7].

We instead propose a "geometric" criterion on dt, which is defined by the formula

$$dt = 0.1 \frac{\overline{E}}{\max_i (\| L_{w_A} X_i \|_2)}.$$

In this way each vertex will never be moved by a distance greater than 10 % of the average edge length $\overline{E}$. Even if the time step computed by the geometric criterion is slightly larger than the dt obtained by stability requirements, in our computational experiments we always converged to an acceptable solution.

The implicit integration scheme used in the discretization of STEP 1 of the AR algorithm does not suffer from numerical instability problems, anyway to avoid triangle flips dt is chosen such that:

$$dt = 10\frac{\overline{A}}{\max_i(\|L_{w_H}A(X_i)\|_2)}.$$

This choice makes the remeshing procedure independent on the mesh area distribution. The solution of STEP 1 is obtained by performing one implicit integration step, that is by solving a linear system of equations, with a very sparse matrix of coefficients. Therefore, an efficient iterative method can be applied, such as *gmres* method for sparse matrices, see [15].

11.3.1 Calculating Gradient and Divergence Operators

Let $\mathcal{M} \subset R^3$ be a surface, we can naturally consider the three coordinate functions $(x, y, z) : \mathcal{M} \to R^3$ as three scalar functions on the surface $\mathcal{M}$.

We use a discretization of the differential operators $\nabla_{\mathcal{M}}$ and $div_{\mathcal{M}}$ for the space discretization of $\mathcal{M}$ given by the mesh M in a matrix form similar to the Laplace-Beltrami discretization in [10].

The intrinsic gradient and divergence operator of a function $f : \mathcal{M} \to R$, $f \in C^1(\mathcal{M})$ and a vector field $V \in T\mathcal{M}$ in local coordinates $V = v_1\frac{\partial}{\partial\xi_1} + v_2\frac{\partial}{\partial\xi_2}$ are defined as

$$\nabla_{\mathcal{M}} f = \sum_{i,j=1}^{2} g^{ij}\frac{\partial f}{\partial\xi_j}\frac{\partial}{\partial\xi_i}, \quad div_{\mathcal{M}} V = \frac{1}{\sqrt{det(G)}}\sum_{i=1}^{2}\frac{\partial}{\partial\xi_i}(\sqrt{det(G)}v_i), \tag{11.15}$$

where $g_{ij} = \frac{\partial}{\partial\xi_i} \cdot \frac{\partial}{\partial\xi_j}$ ($\cdot$ indicates the inner product) are the coefficients of the metric matrix G and $(g^{ij})_{i,j=1,2}$ are the elements of the inverse matrix G^{-1}. We consider a weighted average in the first ring neighbors of a vertex X_i in terms of the triangle area, which naturally leads to the following discretization

$$\nabla_{\mathcal{M}} f(X_i) = \frac{1}{\sum_{j\in N(i)} A_j}\sum_{j\in N(i)} A_j \nabla_{T_j} f(X_i) \tag{11.16}$$

$$div_{\mathcal{M}} V(X_i) = \frac{1}{\sum_{j\in N(i)} A_j}\sum_{j\in N(i)} A_j div_{T_j} V(X_i), \tag{11.17}$$

with A_j area of the triangle jth in the first ring neighbor of vertex X_i. For a triangle T of vertices X_i, X_j, X_ℓ, with $j \in N(i), \ell \in N(i)$ we define

$$\nabla_{T_j} f(X_i) = \sum_{k \in i,j,\ell \prec T_j} \overrightarrow{\omega}_{T_j}^{X_k} f(X_k), \tag{11.18}$$

where k varies in the set of the vertices of T_j, and

$$\begin{aligned} \overrightarrow{\omega}_T^{X_j} &= g^{11}(X_j - X_i) + g^{12}(X_\ell - X_i), \\ \overrightarrow{\omega}_T^{X_\ell} &= g^{21}(X_j - X_i) + g^{22}(X_\ell - X_i), \\ \overrightarrow{\omega}_T^{X_i} &= -(\overrightarrow{\omega}_T^{X_j} + \overrightarrow{\omega}_T^{X_\ell}). \end{aligned} \tag{11.19}$$

The mean curvature vector field, used in the similarity weights (11.6), is approximated on a vertex X_i of the mesh, by $H(X_i) = div_{\mathcal{M}}(\nabla_{\mathcal{M}}(X_i))$. The discretization of the Laplace, gradient and divergence operators at the vertex X_i depends on the elements of its first ring neighborhood. The measure of mean curvature at the vertex X_i depends on its two-ring neighbors. Therefore, this measure of mean curvature is more reliable than the mean curvature computed by discretizing the relation $\Delta_{\mathcal{M}}(X) = -2H(X)\overrightarrow{N}(X)$, mostly because it is computed on a wider stencil which allows to better identify the features of the mesh, thus improving the efficacy of the weights in (11.6) and the quality of the diffusivity function in (11.7).

Moreover, instead of the linearization defined in (11.9), we can compute a more accurate approximation of the non-linear diffusion equation (11.7) by applying the discretization of the gradient and divergence operators on the mesh proposed in (11.16) and (11.17), respectively.

11.4 Remeshing Results

To our knowledge, there are no standard measures for evaluating and comparing the efficacy of remeshing techniques. Instead, comparison is often based on a qualitative evaluation of the final results and it is strictly related to the specific end goal the remeshing is used for. In [16] the authors measure the quality of a remeshed model by measuring the geometric properties of the resulting triangles, but this approach slightly limits the global overview on the benefits of the remeshing algorithm.

In our work, to assess the quality of the mesh generated by the proposed AR algorithm, we introduce the following measures:

Area Variance,

$$Var(A(X)) := \sum_{i=1}^{n_v} (A(X_i) - \overline{A})^2, \tag{11.20}$$

where $\overline{A}$ is the average vertex area for a mesh with n_v vertices.

Mean curvature variation,

$$\Delta H = \|H - H_0\|_2 / \|H_0\|_2, \quad (11.21)$$

where H and H_0 are the mean curvature maps of the remeshed and the original meshes, respectively. For each vertex X_i on the original mesh, the difference $H(X_j) - H_0(X_i)$ is computed with respect to the nearest vertex X_j on the remeshed mesh.

Variation of local edge length variance,

$$\Delta Var(E) = \|Var(E) - Var(E_0)\|_2 / \|Var(E_0)\|_2, \quad (11.22)$$

where the local edge length variance of the vertex X_i is computed as $Var(E)_i = \frac{1}{d_i} \sum_{j \in N(i)} (E_{ij} - \overline{E}_i)^2$.

Hausdorff distance,

$$d_H(X_0, X) = \max\{ \sup_{x \in X_0} \inf_{y \in X} d(x, y), \sup_{y \in X} \inf_{x \in X_0} d(x, y) \}, \quad (11.23)$$

is a measure of distance between the remeshed and the original meshes, see [6].

The goal of our remeshing strategy is to minimize the area variance ($Var(A(X))$), while preserving the mean curvature values and the shape of the original mesh, that is minimizing (ΔH) and the average one-sided Hausdorff distance $\overline{d}_h$. Moreover, the variation of mean curvature variance ($\Delta Var(H)$) is a value that should be preserved, and the variation of area variance ($\Delta Var(A)$) and the variation of edge variance ($\Delta Var(E)$) provide a measure of the quality of the resulting mesh.

The parameters α in the diffusion function $g(\cdot)$ in (11.8), and σ in (11.6) are chosen to be 0.05, while the parameter σ in (11.10) is 0.1. All experiments were performed on Intel Core i7-720QM Quad-Core 1.6 GHz machine, with 4 GB/RAM and Nvidia G335M graphics card in a linux environment. The visualization is based on OpenGL library and the application is written using C language. Numerical computations are performed using the "meschach" matrix library.

We compare the AR algorithm with the LF method defined in [5], when applied to the meshes shown in Figs. 11.3a–11.7a. In the reported experiments we omit the mesh connectivity regularization stage integrated in both AR (STEP 0) and LF methods in order to effectively point out the performances of the mesh regularization stage in the AR algorithm. Moreover, we noticed that the features of the mesh get easily compromised if the operations involved in the mesh connectivity stage are not performed adaptively, as we do in STEP 0 of the AR algorithm. In particular, in STEP 0 of the AR method the flip tool is applied only when the deviation in curvature normals is acceptable. This improves further on the results obtained by the AR method, but it makes more difficult to distinguish the benefit of the mesh regularization stage inside the entire iterative process.

A qualitative comparison is illustrated in Figs. 11.3–11.7. In Figs. 11.3–11.6 the area distribution function A is visualized using false colors superimposed onto

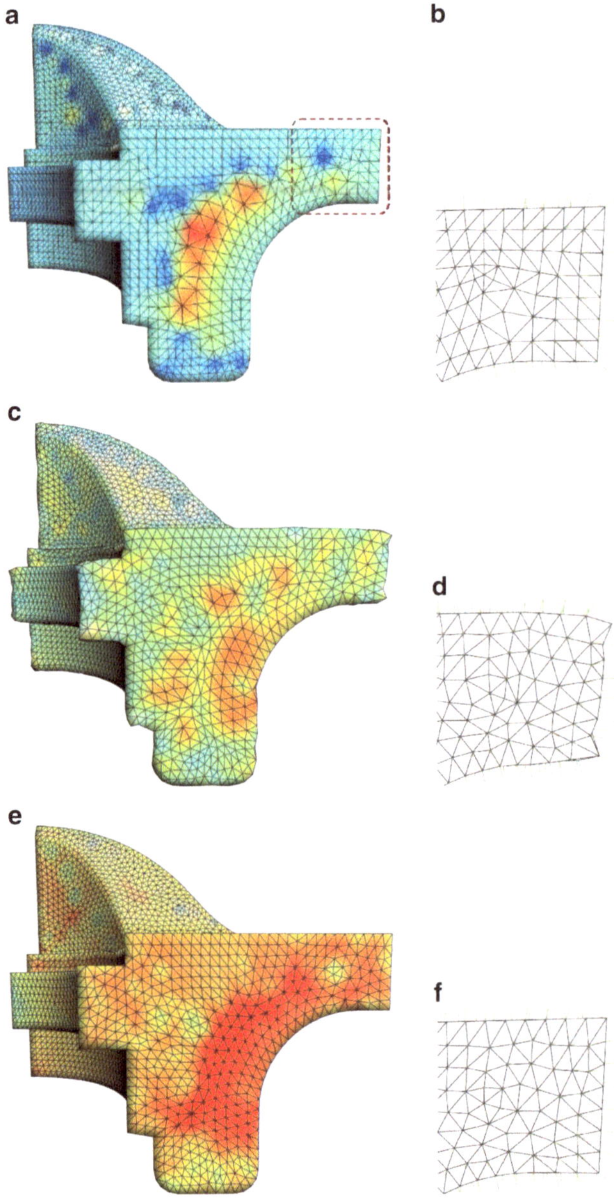

Fig. 11.3 (**a**) and (**b**) The original mesh and a zoomed detail; (**c**) and (**d**) the result of LF and a zoomed detail; (**e**) and (**f**) the result of AR and a zoomed detail from the rightmost part of `fandisk` mesh

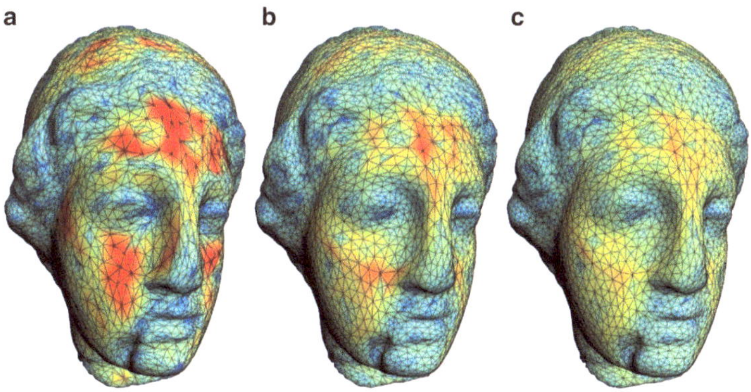

Fig. 11.4 (**a**) The original mesh with area distribution function superimposed using false color; (**b**) LF remeshing (**c**) AR remeshing

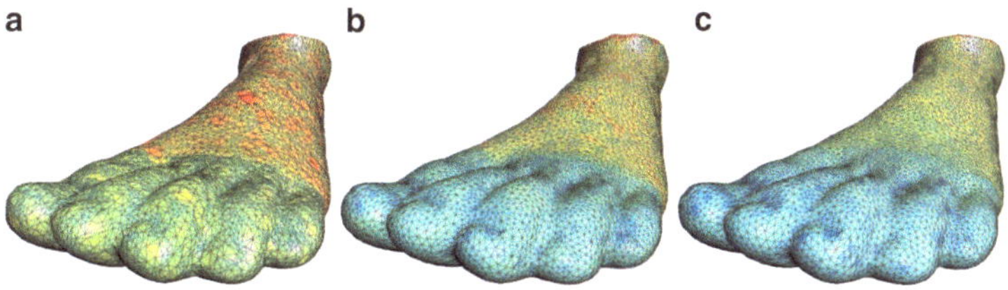

Fig. 11.5 (**a**) The original mesh with area distribution function superimposed in false color; (**b**) LF remeshing and (**c**) AR remeshing

the meshes. In Fig. 11.7 the curvature map is superimposed on the `gargoyle` mesh using false colors where red colors represent high curvatures and blue colors low curvature values. Figures 11.3 and 11.7 show two examples of applying our AR method to meshes with sharp features to highlight the weaknesses of the LF remeshing when applied to regions with high curvature values and high triangle density.

The `fandisk` mesh, illustrated in Fig. 11.3a, presents a medium quality vertex area distribution, sharp edges and corners which allow us to demonstrate the capacity of our AR algorithm to adaptively distribute the vertex areas in zones of high curvature, preserving the sharp features of the mesh. The superiority of the AR approach w.r.t. the LF method can be visually appreciated in Fig. 11.3 comparing the resulting mesh by 20 remeshing iterations of LF algorithm (Fig. 11.3c) and the mesh obtained by 20 remeshing iterations of AR (Fig. 11.3e). A detail of the rightmost part (see the dashed rectangular box in Fig. 11.3a) is shown to enhance the area equalization and preservation of features obtained by the AR method.

In the `gargoyle` model of Fig. 11.7 the wings of the gargoyle are completely ruined by the LF method (see Fig. 11.7b) while they are well preserved by the AR method (Fig. 11.7c).

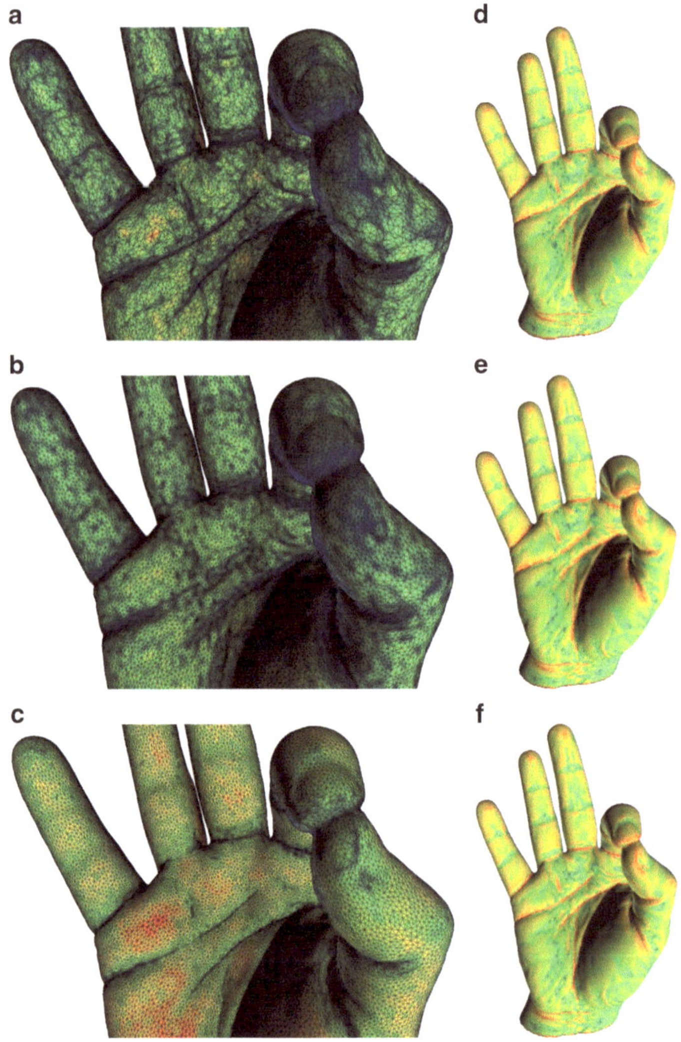

Fig. 11.6 *First column*: area distribution function superimposed on the original mesh (**a**), the LF (**b**) and AR (**c**) remeshing respectively. *Second column*: the mean curvature map on the same meshes

The irregular models illustrated in Figs. 11.4–11.6 are characterized by a vertex area distribution particularly corrupted and many badly shaped triangles. In all the examples AR performs better than LF in both the task of area distribution and mean curvature preservation.

Table 11.1 summarizes the quality measurements obtained by the illustrated experiments. In Table 11.1 for each mesh (first column) the applied algorithm is shown in the second column and the corresponding number of remeshing iterations

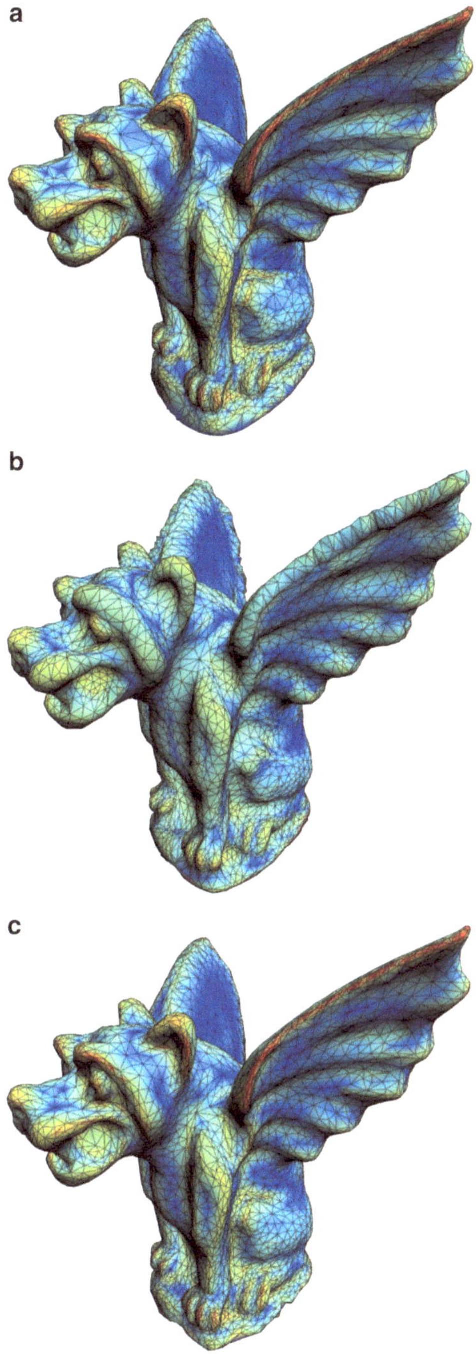

Fig. 11.7 (**a**) The original mesh, (**b**) LF remeshing and (**c**) AR remeshing. *Red* represents high curvature, *blue* represents low curvature

Table 11.1 Quality measurements for the remeshing tests

Mesh	Method	# its	$Var(A)$	ΔH %	$\Delta Var(E)$ %	$\Delta Var(A)$ %	$\Delta Var(H)$ %	$\overline{d}_h$
Hand	AR	5	0.0000077	17.8	77.2	61.2	5.9	0.002
Hand	LF	5	0.0000096	40.3	78.4	51.6	47.1	0.003
Gargoyle	AR	10	0.0015316	26.4	63.7	47.5	14.3	0.007
Gargoyle	LF	10	0.0015180	69.0	81.6	48.0	59.7	0.016
Fandisk	AR	20	0.0002583	4.7	51.7	55.2	0.8	0.005
fandisk	LF	20	0.0003982	30.3	74.4	31.0	5.9	0.013
Igea	AR	20	0.0001252	12.9	57.3	41.4	5.2	0.004
Igea	LF	20	0.0001248	27.5	64.9	41.5	34.3	0.005
Foot	AR	10	0.0007397	13.1	92.8	65.4	3.7	0.007
Foot	LF	10	0.0007995	28.2	93.0	62.7	22.5	0.008

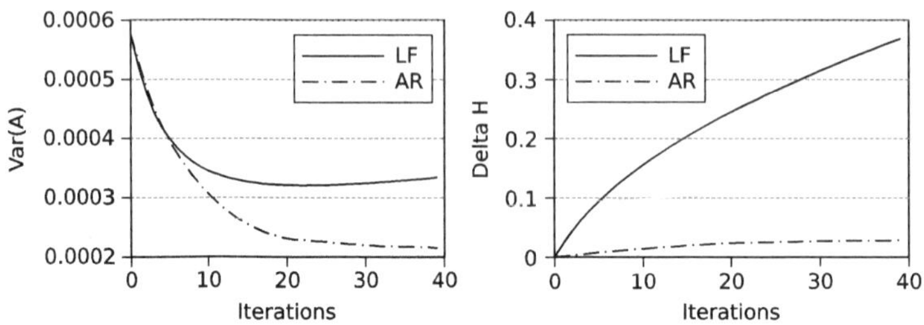

Fig. 11.8 The area variance $Var(A)$ (*left*) and the mean curvature changing ΔH (*right*) as function of the remeshing iterations

(*its*) is in the third column. The results reported in Table 11.1 show that the AR method successfully produces well-shaped triangles while preserving the mean curvature map of the original mesh better than the LF method.

The plots in Fig. 11.8 (left and right) show the area variance $Var(A)$ and the mean curvature variation ΔH as functions of the number of remeshing iterations when the LF method and the AR method are applied to the `fandisk` mesh. When the AR algorithm is applied, the area variance $Var(A)$ rapidly decreases, while the mean curvature of the mesh is preserved. The comparison with the LF method highlights a strong effect on the mean curvature of the resulting mesh and a non-convergent behavior in the minimization of the area variance. This aspect requires further theoretical investigation.

11.5 Conclusions

We presented a new adaptive remeshing scheme based on the idea of improving mesh quality by a series of local modifications of the mesh geometry and connectivity. The fundamental quality criteria we pursued are: faces of good quality, and adaptive mesh resolution to preserve the surface structural characteristics.

To satisfy the last requirement the size of the faces is adapted to the local mean curvature (big enough for smoothed parts and smaller for sharp features). The proposed AR algorithm iterates a connectivity regularization, which is merely based on elementary mesh operations such as edge collapse, flip, and split, with a mesh regularization method. Our contribution to the family of parameterization-free remeshing techniques is a curvature-based mesh regularization technique which allows the control of both triangle quality and vertex sampling over the mesh, as a function of the surface mesh curvature. The first step approaches to an appropriate area distribution function computed by solving an evolutive PDE diffusion model using discrete differential geometry operators suitably weighted to preserve surface curvatures. The area distribution is then applied in the second step of the AR algorithm to redistribute the vertices.

Future work will include operations on the mesh in order to produce a mesh that has a semi-regular structure, namely, with subdivision connectivity. This operations will be integrated in STEP 0 of the AR algorithm.

Acknowledgements This work is supported by MIUR-Prin Grant 20083KLJEZ and by ex 60 % project by the University of Bologna: Funds for selected research topics.

References

1. Alliez, P., Ucelli, G., Gotsman, C., Attene, M.: Recent advances in remeshing of surfaces. In: De Floriani, L., Spagnuolo, M. (eds.) Shape Analysis and Structuring. Mathematics and Visualization, pp. 53–82. Springer (2008)
2. Alliez, P., Meyer, M., Desbrun, M.: Interactive geometry remeshing. ACM Trans. Graph. **21**(3), 347–354 (2002)
3. Alliez, P., de Verdiere, E.C., Devillers, O., Isenburg, M.: Centroidal Voronoi Diagrams for isotropic surface remeshing. Graph. Models **67**(3), 204–231 (2005)
4. Alliez, P., de Verdiere, E.C., Devillers, O., Isenburg, M.: Isotropic surface remeshing. In: Proceedings of the Shape Modeling International 2003 (SMI '03), pp. 49–58. IEEE Computer Society, Washington, DC (2003)
5. Botsch, M., Kobbelt, L.: A remeshing approach to multiresolution modeling. In: Scopigno, R., Zorin, D. (eds.) Proceedings of 2nd Eurographics Symposium on Geometry Processing, Eurographics, pp. 189–196. ACM, New York (2004)
6. Cignoni, P., Callieri, M., Corsini, M., Dellepiane, M., Ganovelli, F., Ranzuglia, G., MeshLab, G.: An open-source mesh processing tool. In: Sixth Eurographics, pp. 47–48, (2008)
7. M. Desbrun, M. Meyer, P. Schroder, A.H. Barr, Implicit Fairing of Irregular Meshes using Diffusion and Curvature Flow, Computer Graphics (SIGGRAPH 99 Proceedings), 317-324, (1999)
8. do Carmo, M.P.: Riemannian Geometry. Birkhauser, Boston/Basel/Berlin (1993)
9. Hormann, K., Labsik, U., Greiner, G.: Remeshing triangulated surfaces with optimal parameterizations. Comput. Aided Des. **33**(11), 779–788 (2001)
10. Meyer, M., Desbrun, M., Schroeder, P., Barr, A.: Discrete differential geometry operators for triangulated 2-manifolds. In: Proceedings of the VisMath '02, Berlin-Dahlem, Germany, pp. 237–242 (2002)
11. Morigi, S.: Geometric surface evolution with tangential contribution. J. Comput. Appl. Math. **233**, 1277–1287 (2010)

12. Morigi, S., Rucci, M., Sgallari, F.: Nonlocal surface fairing. In: Bruckstein, A.M., et al. (eds.): SSVM 2011, LNCS 6667, pp. 38–49. Springer, Berlin/Heidelberg (2011)
13. Ohtake, Y., Belyaev, A.G., Bogaevski, I.A.: Mesh regularization and adaptive smoothing. Comput. Aided Des. **33**, 789–800 (2001)
14. Pinkall, U., Polthier, K.: Computing discrete minimal surfaces and their conjugates. Exp. Math. **2**, 15–36 (1993)
15. Saad, Y.: Iterative Methods for Sparse Linear Systems. SIAM, Philadelphia (2003)
16. Surazhsky, V., Gotsman, C.: Explicit surface remeshing. In: Proceedings of the 2003 Eurographics/ACM SIGGRAPH Symposium on Geometry Processing, Aachen, Germany, pp. 20–30 (2003)
17. Xu, G., Pan, Q., Bajaj, C.L.: Discrete surface modelling using partial differential equations. CAGD **23**(2), 125–145 (2006)
18. Yan, D.-M., Levy, B., Liu, Y., Sun, F., Wang, W.: Isotropic Remeshing with fast and exact computation of restricted Voronoi Diagram. Proc. Eurograph. Symp. Geom. Process. **28**(5), 1445–1454 (2009)

Chapter 12
Group-Valued Regularization for Motion Segmentation of Articulated Shapes

Guy Rosman, Michael M. Bronstein, Alexander M. Bronstein, Alon Wolf, and Ron Kimmel

Abstract Motion-based segmentation is an important tool for the analysis of articulated shapes. As such, it plays an important role in mechanical engineering, computer graphics, and computer vision. In this chapter, we study motion-based segmentation of 3D articulated shapes. We formulate motion-based surface segmentation as a piecewise-smooth regularization problem for the transformations between several poses. Using Lie-group representation for the transformation at each surface point, we obtain a simple regularized fitting problem. An Ambrosio-Tortorelli scheme of a generalized Mumford-Shah model gives us the segmentation functional without assuming prior knowledge on the number of parts or even the articulated nature of the object. Experiments on several standard datasets compare the results of the proposed method to state-of-the-art algorithms.

G. Rosman (✉) · R. Kimmel
Department of Computer Science, Israel Institute of Technology, 32000, Haifa, Israel
e-mail: rosman@cs.technion.ac.il; ron@cs.technion.ac.il

M.M. Bronstein
Institute of Computational Science, Faculty of Informatics, Universita della Svizzera italiana, Lugano, Switzerland
e-mail: michael.bronstein@usi.ch

A.M. Bronstein
School of Electrical Engineering, Faculty of Engineering, Tel Aviv University, Ramat Aviv 69978, Israel
e-mail: bron@eng.tau.ac.il

A. Wolf
Department of Mechanical Engineering, Technion – Israel Institute of Technology, Haifa, 32000, Israel
e-mail: alonw@tx.technion.ac.il

M. Breuß et al. (eds.), *Innovations for Shape Analysis*, Mathematics and Visualization, DOI 10.1007/978-3-642-34141-0_12, © Springer-Verlag Berlin Heidelberg 2013

12.1 Introduction

Articulated objects segmentation is a key problem in biomechanics [1], mechanical engineering, computer vision [8, 30, 38, 41, 54], and computer graphics [6, 35, 37, 39, 59, 64, 65]. Related problems of deformation analysis [4, 63] and motion segmentation [5, 22] have also been studied extensively in these disciplines. Algorithms solving these problems try to infer the articulated motion of an object, given several instances of the object in different poses. Simultaneously, the segmentation of the object into rigid parts takes place along with motion estimation between the corresponding parts in the various poses.

Most motion analysis techniques make some assumptions on the object to be segmented. These usually concern the number or location of rigid parts in the articulated object. This can be in the form of a skeleton describing the topology of the shape, or some other prior on the object structure. Such priors are usually formulated in an ad hoc manner, but not based on the kinematic model commonly assumed for near-rigid objects [1, 4]. In cases where such a prior is not available for the objects in question, or where assumptions about the data are only approximate, this can lead to errors in the segmentation and motion estimation.

Another common assumption, especially in graphics applications, is that of known correspondences. In computer graphics, the problem is usually referred to as *dynamic mesh segmentation*. While a matching technique between poses can be combined with existing motion segmentation tools, a more complete formulation for motion segmentation should handle the correspondence problem implicitly.

Clearly, the above assumptions are often too limiting in real-world applications, and should be avoided as part of the basic problem formulation. We would like instead to apply the intuition often used when studying real-life near-rigid objects, about the existence of a representative rigid motion existing for each body part. We wish, however, to avoid detecting the articulated parts in advance. Furthermore, in some object, a clear partition into rigid parts may not exist for all of the surface. We wish to obtain reasonable results in such a case. In other words, we would like to obtain a "soft" segmentation of the surface, without knowing the number or location of regions in advance, an explicit analysis of the surface features, or having additional priors on the various object parts. Also, we strive towards a formulation of motion segmentation that incorporates an implicit handling of the correspondence problem, given a reasonable initialization.

12.1.1 Main Contribution

In this chapter we try to remedy the shortcoming of existing approaches to articulated motion estimation by combining the two tasks of motion estimation and segmentation into a single functional. This scheme has been described in a recent conference paper [50] and we now slightly expand upon it. Unlike existing methods, we propose a principled variational approach, attempting to find a rigid transformation at each surface point, between the instance surfaces, such that the overall transformation is described by a relatively sparse set of such transformations, each matching a rigid part of the object. The functional we propose regularizes the motion between the surfaces, and is guided by the fact that the parameters of the motion transformations

(i) Should describe the motion at each point with sufficient accuracy.
(ii) Should vary smoothly within the (unknown) rigid parts.
(iii) Can vary abruptly between rigid parts.

We see our main contribution in these:

A new framework: First, we propose an axiomatic variational framework for articulated motion segmentation. While focusing on the segmentation problem in this chapter, our framework is more general and the proposed functionals can be easily incorporated into other applications such as motion estimation, tracking, and surface denoising.

Variational segmentation: We claim that using the right parameterization, taken from the specific domain of rigid motion analysis, we can formulate the articulated motion segmentation problem as a generalization of classical tools in variational computer vision. This allows for an elegant and simple solution within the proposed framework, obtaining results competitive with domain-specific state-of-the-art tools.

A novel visualization algorithm: Third, we suggest a spatially-coherent algorithm for spatial visualization of group valued data on manifolds, which draws from the same variational principles.

12.1.2 Relation to Prior Work

Several previous works have attempted motion based segmentation of surfaces. We mention but a few of these. Kompatsiaris et al. [38] use an estimation of the rigid motion at each segment in order to segment the visible surface in a coarse-to-fine manner. Arcila et al. [6] iteratively refine the segmentation for segments whose transformation error is too large. Wuhrer and Brunton [64] use a dual tree representation of the surface with weights between triangles set according to the dihedral angles. Lee et al. [39] use a similar graph-based formulation, looking at deformation matrices around each triangle.

The scheme we propose involves diffusing the transformations between poses along the surface, in the spirit of the Ambrosio-Tortorelli scheme [2] for Mumford-Shah segmentation [43]. The diffusion component of our scheme is a diffusion process of Lie-group elements, which has recently attracted significant attention in other applications [25, 28, 56]. In diffusing transformations on the surface, our work is similar to that of Litke et al. [40], although the parameterization of the motion and of the surface is different. In addition, we do not make an assumption on the surface topology; to that end, the proposed method diffuses transformations along the surface, rather than representing the surface in an evenly sampled 2D parametrization plane. When dealing with real-life deformable objects that seldom admit regular global parametrization, such an assumption could be too restrictive.

The idea of combining soft segmentation and motion estimation has been attempted before in the case of optical flow computation (see, e.g., [3,18]). In optical flow fields, however, the motion field is merely expected to be piecewise smooth. For truly articulated objects one would expect piecewise-constant flow fields, when expressed in the correct parametrization.

Finally, the functional can be extended with priors from general mesh segmentation techniques. These are usually based on the geometry of the surface itself, and obtain remarkable results for a variety of objects. We point the reader to [9,21,36,55], and references therein, for additional examples of mesh segmentation algorithms. We do not, however, use an additional prior as such an addition will prevent the isolated examination of the principles shown in this chapter.

12.2 Problem Formulation

We now proceed to define the problem we try to solve and the proposed model.

12.2.1 Articulation Model

We denote by X a two-manifold representing a three-dimensional shape. We assume X to have several embeddings into $\mathbb{R}^3$. Each of these embedding constitutes a pose of the surface. In the following, we will denote by $\mathbf{x} : X \to \mathbb{R}^3$ the embedding of X into $\mathbb{R}^3$, and use synonymously the notation x and $\mathbf{x}$ referring to a point on the manifold and its Euclidean embedding coordinates, for a specific pose.

In the setting of rigid motion segmentation, we assume that X represents an *articulated* shape, i.e., it can be decomposed into *rigid parts* $S_1, \dots, S_p$. These are transformed between different poses of the objects by a rigid transformation. This transformation, a rotation and a translation, is an isometry of $\mathbb{R}^3$. The rigid parts are connected by nonrigid *joints* $J_1, \dots, J_q$, such that $X = \bigcup_{i=1}^p S_i \cup \bigcup_{k=1}^q J_k$. An *articulation* $Y = \mathbf{A}X$ is obtained by applying rigid motions $\mathbf{T}_i \in \text{Iso}(\mathbb{R}^3)$ to the rigid parts, and non-rigid deformations $\mathbf{Q}_k$ to the joints, such that $\mathbf{A}X = \bigcup_{i=1}^p \mathbf{T}_i S_i \cup \bigcup_{k=1}^q \mathbf{Q}_k J_k$.

12.2.2 Motion Segmentation

The problem of *motion-based segmentation* can be described as follows: given two articulations of the shape, X and Y, extract its rigid parts. An extension to the case of multiple shape poses is straightforward. We therefore consider in the following only a pair of shapes for the sake of simplicity and without loss of generality. A strongly related question attempts to determine, given these articulations, the motion parameters linking the poses of the object.

Assuming that the correspondence between the two poses X and Y is known, given a point $x \in X$ and its correspondent point $y(x) \in Y$, we can find a motion $g \in \mathbb{G}$ such that $g\mathbf{x} = \mathbf{y}$, where $\mathbb{G}$ is some representation of coordinate transformations in $\mathbb{R}^3$. This motion g may change, in the setting described above, for each surface point. We therefore consider g to be a function $g : X \to \mathbb{G}$. We will simultaneously use $g\mathbf{x} \in \mathbb{R}^3$ to denote the action of $g(x)$ on the coordinates of the point x, as well as consider the mapping given by $g : X \to \mathbb{G}$ and its properties.

We note that typical representations of motion in $\mathbb{R}^3$ contain more than three degrees of freedom. In this sense, they are over-parameterized [45], and thus some measure of regularity is required in order to avoid ambiguity as well as favor a meaningful solution. On the other hand, we note that since the articulated parts of the shape move rigidly, if we choose an appropriate motion representation (as detailed below), two points $x, x' \in S_i$ will undergo the same transformation, from which it follows that $g(x)|_{x \in S_i} = \text{const}$. One possibility is to adopt a constrained minimization approach, forcing $g(X) = Y$, where $g(X)$ is a notation for the set $g(x)\mathbf{x}(x)$ for all $x \in X$. This approach, however, needs to somehow handle the set of joints, for which such a constraint may be meaningless. In general, restricting the feasible set of solutions by such constraints or even constraints involving an error in the data may be harmful for the overall result. In order to avoid this, another possible approach is to take an unconstrained, yet regularized, variational formulation,

$$\min_{g:X \to \mathbb{G}} \lambda E_{\text{D}}(g) + \rho(g), \tag{12.1}$$

where ρ denotes a smoothness term operating on the motion parameters field. This term is expected to be small for fields g which are piecewise constant on the manifold X. While an appropriate parameterization of motion g, and regularization term $\rho(g)$ are crucial, we also require a data term that will encourage consistency of the transformation field g with the known surface poses. Specifically, we wish to favor a transformation field where the point $\mathbf{x}$ is taken by its transformation $g(x)$ to a point on the other surface. $E_{\text{D}}(g)$ is our fitting term which measures this consistency with the data.

$$E_{\text{D}}(g) = \int_X \|g(x)\mathbf{x} - \mathbf{y}(\mathbf{x})\|^2 da, \tag{12.2}$$

where $\mathbf{y}(x) \in \mathbb{R}^3$ denotes the coordinate of the point $y(x) \in Y$ corresponding to x, $g(x)$ is the transformation at x, and da is a measure on X. We have assumed in the discussion so far that the correspondence between X and Y is known, which is usually not true. We can solve for the correspondence as part of the optimization in an efficient manner. We will mention this issue in Sect. 12.4.1. We use the term corresponding point $y(x)$ since, as in the case of *iterative closest point* (ICP) algorithms [13,67], several approaches for pruning erroneous or ineffective matches exist [53].

Minimizing the functional with respect to $g, y(x)$ from a reasonable initial solution allows recovery of the articulated parts by clustering g into regions of equal value. Yet another choice of a data term is a semi-local fitting term, is a semi-local one,

$$E_{\mathrm{D,SL}}(g) = \int_X \int_{y \in \mathscr{N}(x)} \|g(x)\mathbf{x}' - \mathbf{y}(\mathbf{x}')\|^2 da' da, \qquad (12.3)$$

where $\mathscr{N}(x)$ denotes a small neighborhood around the point x (we took $\mathscr{N}(\mathbf{x})$ to be the 12 nearest neighbors). This fitting term, by itself, formulates a local ICP process. The functional (12.1) equipped with the semi-local data term can be considered as the geometrical fitting equivalent of the combined global-local approach for optic flow estimation [17].

The simplest representation of motion is a *linear motion* model, affectively setting $\mathbb{G}$ to be the group of translation, or $\mathbb{G} = \mathbb{R}^3$. This results in the motion model $g\mathbf{x} = \mathbf{x} + \mathbf{t} = \mathbf{y}$ for some $\mathbf{t} \in \mathbb{R}^3$. However, such a simplistic model fails to capture the piecewise constancy of the motion field in most cases. Instead of turning to a higher order approximation model such as the affine over-parameterized model [46], or to more elaborate smoothness priors [60], we look for a relatively simple model that will capture natural motions with a simple smoothness prior. Thus we turn to a slightly different motion model, naturally occurring in motion research.

12.2.3 Lie-Groups

One parametrization often used in computer vision and robotics [28, 37, 42, 61] is the representation of rigid motions by the Lie-group $SE(3)$ and the corresponding Lie-algebra $se(3)$, respectively. In general, given two coordinate frame, an element in $SE(3)$ describes the transformation between them. Works by Brockett [14], Park et al. [47] and Zefran et al. [69, 70] strongly relate Lie-groups, both in their global and differential description, to robotics and articulated motions. We give a very brief introduction to the subject and refer the reader to standard literature on the subject (e.g., [29, 44]) for more information.

Lie-groups are topological groups with a smooth manifold structure such that the group action $\mathbb{G} \times \mathbb{G} \mapsto \mathbb{G}$ and the group inverse are differentiable maps.

For every Lie-group, we can canonically associate a *Lie-algebra* $\mathfrak{g}$. A Lie-algebra is as a vector space endowed with a *Lie brackets operator* $[\cdot,\cdot] : \mathbb{G} \times \mathbb{G} \to \mathbb{G}$, describing the local structure of the group. The Lie-algebra associated with a Lie-group can be mapped diffeomorphically via the *exponential map* onto a neighborhood of the identity operator and its tangent space.

This property will allow us express neighboring elements in the Lie-group in a vector space, and thereby define derivatives, regularity, and diffusion operators on the group valued data.

In this chapter, we are specifically interested in the special orthogonal (rotation) matrix group $SO(3)$ and the Euclidean group $SE(3)$ to represent rigid motions. These can be represented in matrix forms, where $SO(3)$ is given as

$$SO(3) = \left\{\mathbf{R} \in \mathbb{R}_{3\times 3}, \mathbf{R}^T\mathbf{R} = \mathbf{I}\right\}, \tag{12.4}$$

and $SE(3)$ is given by

$$SE(3) = \left\{\begin{pmatrix}\mathbf{R} & \\ \mathbf{0} & 1\end{pmatrix}, \mathbf{R} \in SO(3), \mathbf{t} \in \mathbb{R}^3\right\}. \tag{12.5}$$

The Lie-algebra of $SO(3)$, $so(3)$ consists of skew-symmetric matrices,

$$so(3) = \left\{\mathbf{A} \in \mathbb{R}_{3\times 3}, \mathbf{A}^T = -\mathbf{A}\right\}, \tag{12.6}$$

whereas the Lie-algebra of $SE(3)$ can be identified with the group of 4×4 matrices of the form

$$se(3) = \left\{\begin{pmatrix}\mathbf{A} & \\ \mathbf{0} & 0\end{pmatrix}, \mathbf{A} \in so(3), \mathbf{t} \in \mathbb{R}^3\right\}, \tag{12.7}$$

where $so(3)$ is the set of 3×3 skew-symmetric matrices.

In order to obtain piecewise constant description over the surface for the relatively simple case of articulated object, we would like the points at each object part to have the same representative. Under the assumption of $\mathbb{G} = SE(3)$, this desired property holds. We note, however, that the standard parameterization of small rigid motions has six degrees of freedom, while the number of degrees of freedom required to describe the motion of point is mere three. Thus, this parameterization clearly constitutes an over-parameterized motion field [46] for articulated surfaces.

We now turn to the regularization term, $\rho(g)$, and note that the formulation given in Eq. (12.1) bears much resemblance to *total variation* (TV) regularization common in signal and image processing [51]. Total variation regularization does not, however, favor distinct discontinuity sets. This property of TV regularization is related to the *staircasing effect*. Furthermore, in the scalar case, discontinuity sets form closed curves, which may not be the case in some surfaces with large joint areas. Instead, a model that better suits our segmentation problem is the Mumford-Shah segmentation model [43]. This model can be implemented using an Ambrosio-Tortorelli scheme [2], which can be easily generalized for the case of maps between general manifolds such as maps from surfaces into motion manifolds. We further describe the regularization chosen in Sect. 12.3.

We also note that due to the non-Euclidean structure of the group, special care should be taken when parameterizing such a representation [28, 37, 42, 56], as discussed in Sect. 12.4.2.

12.3 Regularization of Group-Valued Functions on Surfaces

Ideally, we would like the transformation field defined on the articulated surface to be piecewise smooth, if not piecewise constant. Therefore, a suitable regularization of the transformation parameters is required. Since the Lie-group $\mathbb{G}$ as a Riemannian manifold, it is only natural to turn to regularization functionals defined on maps between manifolds of the form $g : X \rightarrow \mathbb{G}$.

A classical functional defined over such maps is the well-known *Dirichlet energy* [26],

$$\rho_{\mathrm{DIR}}(g) = \frac{1}{2}\int_X \langle \nabla g, \nabla g\rangle_{g(x)} da = \frac{1}{2}\int_X tr\left(g^{-1}\nabla g\right)^2 da, \qquad (12.8)$$

where ∇g denotes the *intrinsic gradient* of g on X, $\langle \cdot, \cdot \rangle_{g(x)}$ is the Riemannian metric on $\mathbb{G}$ at a point $g(x)$, and *da* is the area element of X. This functional is the more general form of the Tikhonov regularization (for Euclidean spaces X and $\mathbb{G}$), and its properties are well defined for general manifolds, as studied by Eells [26].

Minimizers of the Dirichlet energy are called *harmonic maps*. These result from a diffusion process, and are often used for surface matching [62, 68].

12.3.1 Ambrosio-Tortorelli Scheme

Unfortunately, the Dirichlet energy favors smooth maps defined on X, whereas our desired solution has discontinuities at the boundaries of rigid parts. We would, intuitively, want to prevent diffusion across these discontinuity curves. This can be obtained by adding a diffusivity function $v : X \rightarrow [0, 1]$ to the Dirichlet functional, leading to the generalized Ambrosio-Tortorelli scheme [2] for Mumford-Shah regularization [43].

$$\rho_{\mathrm{AT}}(g) = \int_X \left(\frac{1}{2}v^2\langle \nabla g, \nabla g\rangle_g + \epsilon\langle \nabla v, \nabla v\rangle + \frac{(1-v)^2}{4\epsilon}\right) da, \qquad (12.9)$$

where ϵ is a small positive constant. This allows us to extend our outlook in several ways. The Mumford-Shah functional replaces the notion of a set of regions with closed simple boundary curves with general discontinuity sets. It furthermore generalizes our notion of constant value regions with that of favored smoothness inside the areas defined by these discontinuity curves. This is in order to handle objects which deviate from articulated motion, for example in flexible regions or joints.

Furthermore, the generalized Ambrosio-Tortorelli scheme allows us to explicitly reason about places in the flow where the nonlinear nature of the data manifold manifests itself. Suppose we have a solution (g^*, v^*) satisfying our piecewise-constancy assumptions of g, and a diffusivity function with 0 at region boundaries and 1 elsewhere. At such a solution, we expect two neighboring points which belong to different regions to have a very small diffusivity value v connecting them, effectively nullifying the interaction between far-away group elements which is dependent on the mapping used for the logarithm map at each point, and hence can be inaccurate [32, 42]. While such a solution (g^*, v^*) may not be a minimizer of the functional, it serves well to explain the intuition motivating the choice of the functional.

12.3.2 *Diffusion of Lie-Group Elements*

In order to efficiently compute the Euler-Lagrange equation corresponding to the generalized Ambrosio-Tortorelli functional (12.9), we transform the neighborhood of each point into the corresponding Lie-algebra elements before applying the diffusion operator. Using Lie-algebra representation of differential operators for rigid motion has been used before in computer vision [56], numerical PDE computations [32], path planning and optimal control theory [37, 42].

The Euler-Lagrange equation for the generalized Dirichlet energy measuring the map between two manifolds is given as [26]

$$\Delta_X g^\alpha + \Gamma^\alpha_{\beta\gamma} \left\langle \nabla g^\beta, \nabla g^\gamma \right\rangle_{g(x)} = 0, \tag{12.10}$$

where $\alpha, \beta\ \gamma$ enumerate the local coordinates of our group manifold, $se(3)$, and we use Einstein's notation according to which corresponding indices are summed over. $\Gamma^\alpha_{\beta\gamma}$ are the *Christoffel symbols* of $SE(3)$, which express the Riemannian metric's local derivatives. We refer the reader to [24] for an introduction to Riemannian geometry. Finally, Δ_X denotes the Laplace-Beltrami operator on the surface X.

In order to avoid computation of the Christoffel symbols, we transform the point and its neighbors using the logarithm map at that point in $SE(3)$. The diffusion operation is now affected only by the structure of the surface X. After applying the diffusion operator, we use the exponential map in order to return to the usual representation of the transformation. While this approach may suffer at discontinuities, where the logarithm and exponential maps are less accurate, it is at these continuities that we expect the diffusivity function v to be very small, preventing numerical instability. In practice, as we will demonstrate, this did not pose a significant problem.

Algorithm 1 Articulated surface segmentation and matching

Given an initial correspondence.
for $k = 1, 2, \ldots,$ until convergence **do**
  Update $g^{k+1/2}, v^{k+1}$ w.r.t. the diffusion term, according to Eq. (12.12).
  Obtain g^{k+1} according to the data term, using Eq. (12.15).
  Update $y^{k+1}(x)$, the current estimated correspondence of the deformed surface.
end for

12.4 Numerical Considerations

We now describe the algorithm for articulated motion estimation based on the minimization of the functional

$$E(g, v) = \lambda E_{\text{DATA}}(g) + \rho_{AT}(g, v), \tag{12.11}$$

where $E_{\text{DATA}}(g)$ is the matching term defined by Eq. (12.2), and $\rho_{AT}(g, v)$ is defined in Eq. (12.9). The main steps of the algorithm are outlined as Algorithm 1. Throughout the algorithm we parameterize $g(x)$ based on the first surface, given as a triangulated mesh, with vertices $\{x_i\}_{i=1}^{N}$, and an element from $SE(3)$ defined at each vertex. The triangulation is used merely to obtain a more consistent numerical diffusion operator, and can be avoided, for example by point-cloud based Laplacian approximations [10]. Special care is made in the choice of coordinates during the optimization as explained in Sect. 12.4.2.

12.4.1 Initial Correspondence Estimation

As in other motion segmentation and registration algorithms, some initialization of the matching between the surfaces must be used. One approach [6] is to use nonrigid surface matching for initialization. Another possibility, in the case of high framerate range scanners [65], is to exploit temporal consistency by 3D tracking. Yet another possible source for initial matches incorporates motion capture marker systems. Such sparse initial correspondence lends itself to interpolation of the motion field, in order to initialize a local ICP algorithm, and match the patch around each source point to the target mesh. In Fig. 12.4, we use 30 matched points for initialization. This number of points is within the scope of current motion capture marker systems, or of algorithms for global nonrigid surface matching such as spectral methods [34,41,49,52], or the *generalized multidimensional scaling* (GMDS) algorithm [15].

We expect that a better initial registration, as can be obtained e.g. using a smoothness assumption, or by pruning unsuitable candidates [53], will reduce the number of markers needed.

12.4.2 Diffusion of Lie-Group Elements

Rewriting the optimization over the functional in Eq. (12.11) in a fractional step approach [66], we update the parameters w.r.t. each term of the functional in a suitable representation. The treatment of regularized data fitting in a fractional step approach with respect to different forces has been used before for rigid body motion [20], and is also similar to the approach taken by Thirion's demons algorithm [48, 58] for registration.

Using the transformation described in Sect. 12.3, the update step with respect to the regularization now becomes

$$g^{k+1/2} = g^k \exp\left(-dt\frac{\delta\rho_{\mathrm{AT}}}{\delta\tilde{g}}\right), v^{k+1} = v^k - dt\frac{\delta\rho_{\mathrm{AT}}}{\delta v} \tag{12.12}$$

where $\exp(A) = I + A + A^2/2! + A^3/3! + \ldots$ denotes the matrix exponential, $\tilde{g}$ denotes the logarithm transform of g, and dt denotes the time step. $\frac{\delta\rho_{\mathrm{AT}}}{\delta\tilde{g}}$ denotes the variation of the regularization term $\rho_{AT}(g)$ w.r.t. the Lie-algebra local representation of the solution, describing the Euler-Lagrange descent direction. $g(x)$ and the neighboring transformations are parameterized by a basis for matrices in $se(3)$, after applying the logarithm map at $g(x)$. The descent directions are given by

$$\begin{aligned} \frac{\delta\rho_{\mathrm{AT}}}{\delta\tilde{g}_i} &= v^2\Delta_X(\tilde{g}_i) + v\left\langle\nabla v, \nabla\tilde{g}_i\right\rangle \\ \frac{\delta\rho_{\mathrm{AT}}}{\delta v} &= \left\langle\nabla g, \nabla g\right\rangle_{g(x)} v + 2\epsilon\Delta_X(v) + \frac{(v-1)}{2\epsilon}, \end{aligned} \tag{12.13}$$

where $\tilde{g}_i$ denote the components of the logarithmic representation of g. The discretization we use for Δ_X is a cotangent one suggested by [23], which has been shown to be convergent for relatively smooth and well-parameterized surfaces. It is expressed as

$$\Delta_X(u) \approx \frac{3}{\mathscr{A}_i}\sum_{j\in\mathscr{N}_1(i)} \frac{\cot\alpha_{ij} + \cot\beta_{ij}}{2}\left[u_j - u_i\right], \tag{12.14}$$

for a given function u on the surface X, where $\mathscr{N}_1(i)$ denotes the mesh neighbors of point i, and α_{ij}, β_{ij} are the angles opposing the edge ij in its neighboring faces. $\mathscr{A}_i$ denotes the area of the one-ring around i in the mesh. After a gradient descent step w.r.t. the diffusion term, we take a step w.r.t. the data term.

$$g^{k+1} = P_{SE(3)}\left(g^{k+1/2} - dt\frac{\delta E_{DATA}}{\delta g}\right), \tag{12.15}$$

where $P_{SE(3)}(\cdot)$ denotes a projection onto the group $SE(3)$ obtained by correcting the singular values of the rotation matrix [12]. We compute the gradient w.r.t. a basis for small rotation and translation matrices comprised of the regular basis for translation and the skew-matrix approximation of small rotations. We then reproject the update onto the manifold. This keeps the inaccuracies associated with the projecting manifold-constrained data [19, 28, 32, 42] at a reasonable level, and leads to a first-order accuracy method. As noted by Belta and Kumar [11] in the context of trajectory planning and ODEs over Lie-groups, this method is reasonably accurate. In practice the time-step is limited in our case by the data-fitting ICP term and the explicit diffusion scheme. We expect improved handling of these terms to allow faster implementation of the proposed method.

Finally, we note that we may not know in advance the points $y(x)$ which match X in Y. The correspondence can be updated based on the current transformations in an efficient manner similarly to the ICP algorithm. In our implementation we used the ANN library [7] for approximate nearest-neighbor search queries. We did not incorporate, however, any selective pruning of the matching candidates. These are often used in order to robustify such the ICP algorithm against ill-suited matches but are beyond the scope of this chapter.

12.4.3 Visualizing Lie-Group Clustering on Surfaces

Finally, we need to mention the approach taken to visualize the transformations as the latter belong to a six-dimensional non-Euclidean manifold. Motivated by the widespread use of vector quantization in such visualizations, we use a clustering algorithm with spatial regularization. Instead of minimizing the Lloyd-Max quantization [33] cost function, we minimize the function

$$E_{VIS}(g_i, R_i) = \sum_i \int_{R_i} \|g - g_i\|^2 da + \int_{\partial R_i} v^2(s) ds, \qquad (12.16)$$

where ∂R_i denotes the set of boundaries between partition regions $\{R_i\}_{i=1}^N$, g_i are the group representatives for each region, and $v^2(s)$ denotes the diffusivity term along the region boundary. The representation of members in $SE(3)$ is done via its embedding into $\mathbb{R}^{12}$, with some weight given to spatial location, by looking at the product space $\mathbb{R}^3 \times SE(3) \subset \mathbb{R}^{15}$. Several (about 50) initializations are performed, as is often customary in clustering, with the lowest cost hypothesis kept. The visualization is detailed as Algorithm 2

While this visualization algorithm coupled with a good initialization at each point can be considered as a segmentation algorithm in its own right, it is less general as it assumes a strict separation between the parts. One possible question that can be raise concerned the meaning behind vector quantization of points belonging to a manifold through its embedding into Euclidean space. In our case, since we are dealing with

Algorithm 2 Spatially-consistent clustering algorithm

1: **for** $j = 1, 2, \ldots,$ for a certain number of attempts **do**
2: Use k-means on the spatial-feature space embedding, $\mathbb{R}^3 \times SE(3) \subset \mathbb{R}^{15}$, to get an initial clustering.
3: Use the clusters in order to optimize a spatially-regularized vector quantization measure,

$$C = \min_{g_i, \partial R_i} \int_X \|g - g_i\|^2 da + \int_{\partial R_i} v^2(s) ds,$$

where ∂R_i denotes the set of boundaries between clustered regions, g_i are the transformation representatives for each region, and $v^2(s)$ denotes the diffusivity term along the region boundary.
4: If C is lower than the lowest C found so far, keep the hypothesis.
5: **end for**
6: **return** current best hypothesis.

relatively well-clustered points (most of the points in a part move according to a single transformation in $SE(3)$), the distances on the manifold are not large and are therefore well-approximated by Euclidean ones. We further note, however, that the diffusion process lowered the score obtained in Eq. (12.16) in the experiments we conducted, indicating a consistency between the two algorithms in objects with well-defined rigid parts.

12.5 Results

We now show the results of our method, in terms of the obtained transformations clusters and the Ambrosio-Tortorelli diffusivity function. In Fig. 12.1 we show the segmentation obtained by matching two human body poses taken from the TOSCA dataset [16]. We visualize the transformations obtained using the clustering algorithm described in Sect. 12.4.3. We initialized the transformations on the surface by matching the neighborhood of each surface point to the other poses using the true initial correspondence. The results of our method seem plausible, except for the missing identification of the right leg, which is due to the fact that its motion is limited between the two poses.

Figure 12.1 also demonstrates the results of comparing four poses of the same surface, this time with the patch-based data term described by (12.3). In our experiments the patch-based term gave a cleaner estimation of the motion, as is observed in the diffusivity function. We therefore demonstrate the results of minimizing the functional incorporating this data term. We also show the diffusivity function, which hints at the location of boundaries between parts, and thus justifies the assumption underlying Algorithm 2.

In Figs. 12.3 and 12.2 we show the results of our algorithm on a set of six poses of a horse and camel surfaces taken from [57]. In this figure we compare our results

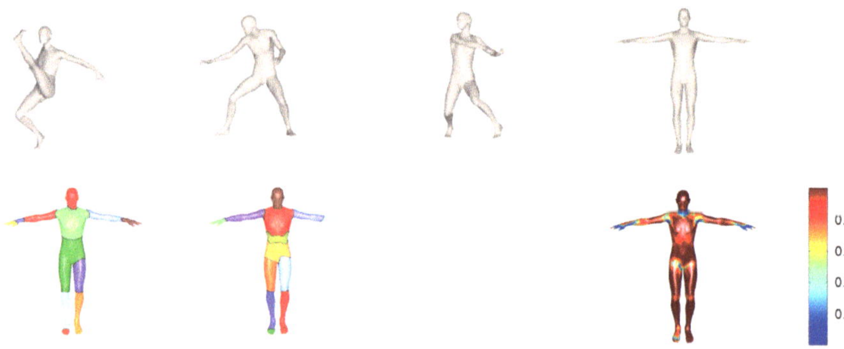

Fig. 12.1 Segmenting a human figure. *Top row*: the set of poses used. *Bottom row*, left to right: the transformations obtained from the two left most poses, the transformations obtained from all four poses using Eq. (12.3) as a data term, and the Ambrosio-Tortorelli diffusivity function based on four poses

Fig. 12.2 Segmenting a camel dynamic surface motion based on six different poses. *Top row*: the poses used. *Bottom row*, left to right: a visualization of the transformations of the surface obtained by our method and the diffusivity function v

to those of Wuhrer and Brunton [64], obtained on a similar set of poses with ten frames. The results of our method seem to be quite comparable to those obtained by Wuhrer and Brunton, despite the fact that we use only six poses. We also note that both the diffusion scheme and the visualization algorithm gave a meaningful result for the tail part, which is not rigid and does not have a piecewise-rigid motion model.

In Fig. 12.4 we demonstrate our algorithm, with an initialization of 30 simulated motion capture marker points, where the displacement is known. The relatively monotonous motion range available in the dynamic mesh sequence leads to a less complete, but still quite meaningful, segmentation of the horse, except for its head.

We also note the relatively low number of poses required for segmentation – in both Figs. 12.3 and 12.4 we obtain good results despite the fact that we use only a few poses, six and eight respectively.

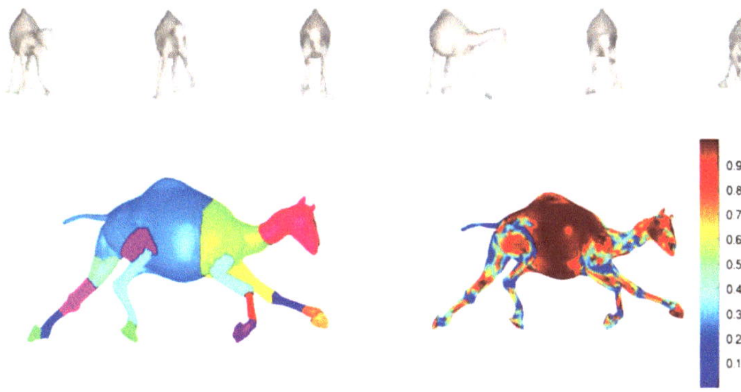

Fig. 12.3 Segmenting a horse dynamic surface motion based on six different poses. *Top row*: the poses used. *Bottom row*, left to right: a visualization of the transformations of the surface obtained by our method, and the segmentation results obtained by [64], and the diffusivity function v

Fig. 12.4 Segmenting a horse dynamic surface motion with a given sparse initial correspondences. *Top row*: the eight random poses used. *Bottom row*, left to right: the set of points used for initializing the transformations, and a visualization of the transformations obtained, and the diffusivity function v

Finally, in Fig. 12.4 we demonstrate initialization of our method based on a sparse point set, with 30 known correspondence points. The points are arbitrarily placed using farthest point sampling [27, 31]. This demonstrates a possibility of initializing the algorithm using motion capture markers, coupled with a 3D reconstruction pipeline, for object part analysis. While the examples shown in this chapter are synthetic, this example shows that the algorithm can be initialized with data obtained in a realistic setup.

12.6 Conclusion

In this chapter we present a new method for motion-based segmentation of articulated objects, in a variational framework. The method is based on minimizing a generalized Ambrosio-Tortorelli functional regularizing a map from the surface onto the Lie-group $SE(3)$. The results shown demonstrate the method's effectiveness, and compare it with state-of-the-art articulated motion segmentation algorithms. The functional we suggest can be easily tailored to specific problems where it can be contrasted and combined with domain-specific algorithms for articulated object analysis. In future work we intend to adapt the proposed algorithm to real data from range scanners, and explore initialization methods as well as use the proposed framework in other applications such as articulated surfaces tracking and denoising.

References

1. Alexander, E.J., Bregler, C., Andriacchi, T.P.: Non-rigid modeling of body segments for improved skeletal motion estimation. Comput. Model. Eng. Sci. **4**, 351–364 (2003)
2. Ambrosio, L., Tortorelli, V.M.: Approximation of functional depending on jumps by elliptic functional via Γ-convergence. Commun. Pure Appl. Math. **43**(8), 999–1036 (1990)
3. Amiaz, T., Kiryati, N.: Piecewise-smooth dense optical flow via level sets. Int. J. Comput. Vis. **68**(2), 111–124 (2006)
4. Andersen, M.S., Benoit, D.L., Damsgaarda, M., Ramsey, D.K., Rasmussen, J.: Do kinematic models reduce the effects of soft tissue artefacts in skin marker-based motion analysis? An in vivo study of knee kinematics. J. Biomech. **43**, 268–273 (2010)
5. Anguelov, D., Koller, D., Pang, H.-C., Srinivasan, P., Thrun, S.: Recovering articulated object models from 3D range data. In: Proceedings of the Conference on Uncertainty in Artificial Intelligence, pp. 18–26. AUAI Press, Arlington (2004)
6. Arcila, R., Kartik Buddha, S., Hétroy, F., Denis, F., Dupont, F.: A framework for motion-based mesh sequence segmentation. In: International Conference on Computer Graphics, Visualisation and Compter Vision, Plzeň, Czech Republic, 2010
7. Arya, S., Mount, D.M., Netanyahu, N.S., Silverman, R., Wu, A.Y.: An optimal algorithm for approximate nearest neighbor searching fixed dimensions. J. ACM **45**(6), 891–923 (1998)
8. Ashbrook, A.P., Fisher, R.B., Robertson, C., Werghi, N.: Segmentation of range data into rigid subsets using surface patches. In International Conference on Computer Vision, pp. 201–206. IEEE Computer Society, Washington, DC (1998)
9. Attene, M., Katz, S., Mortara, M., Patane, G., Spagnuolo, M., Tal, A.: Mesh segmentation – a comparative study. In: Proceedings of the IEEE International Conference on Shape Modeling and Applications, pp. 7–18. IEEE Computer Society, Washington, DC (2006)
10. Belkin, M., Sun, J., Wang, Y.: Constructing Laplace operator from point clouds in $\mathbb{R}^d$. In: Proceedings of the Twentieth Annual ACM-SIAM Symposium on Discrete Algorithms, SODA '09, pp. 1031–1040. Society for Industrial and Applied Mathematics, Philadelphia (2009)
11. Belta, C., Kumar, V.: On the computation of rigid body motion. Electron. J. Comput. Kinemat. **1** (2001)
12. Belta, C., Kumar, V.: An SVD-based project method for interpolation on se(3). IEEE Trans. Robot. Autom. **18**(3), 334–345 (2002)
13. Besl, P.J., McKay, N.D.: A method for registration of 3D shapes. IEEE Trans. Pattern Anal. Mach. Intell. **14**(2), 239–256 (1992)
14. Brockett, R.: Robotic manipulators and the product of exponential formula. In: Fuhrman, P. (ed.) Proceedings of the Mathematical Theory of Networks and Systems, pp. 120–129. Springer (1984)
15. Bronstein, A.M., Bronstein, M.M., Kimmel, R.: Generalized multidimensional scaling: a framework for isometry-invariant partial surface matching. Proc. Natl. Acad. Sci. USA **103**(5), 1168–1172 (2006)
16. Bronstein, A.M., Bronstein, M.M., Kimmel, R.: Numerical geometry of non-rigid shapes. Springer, New York (2008)

17. Bruhn, A., Weickert, J., Schnörr, C.: Lucas/Kanade meets Horn/Schunck: combining local and global optic flow methods. Int. J. Comput. Vis. **61**(3), 211–231 (2005)
18. Brune, C., Maurer, H., Wagner, M.: Detection of intensity and motion edges within optical flow via multidimensional control. SIAM J. Imag. Sci. **2**(4), 1190–1210 (2009)
19. Celledoni, E., Owren, B.: Lie group methods for rigid body dynamics and time integration on manifolds. Comput. Methods Appl. Mech. Eng. **19**, 421–438 (1999)
20. Celledoni, E., Fassò, F., Säfström, N., Zanna, A.: The exact computation of the free rigid body motion and its use in splitting methods. SIAM J. Sci. Comput. **30**(4), 2084–2112 (2008)
21. Chen, X., Golovinskiy, A., Funkhouser, T.: A benchmark for 3D mesh segmentation. ACM Trans. Graph. **28**(3), Article 73 (2009)
22. Daniel Cremers and Steffano Soatto: Motion competition: a variational framework for piecewise parametric motion segmentation. Int. J. Comput. Vis. **62**(3), 249–265 (2005)
23. Desbrun, M., Meyer, M., Schroder, P., Barr, A.H.: Implicit fairing of irregular meshes using diffusion and curvature flow. SIGGRAPH Proceedings of the 26th Annual Conference on Computer Graphics and Interactive Techniques, pp. 317–324. ACM/Addison-Wesley, New York (1999)
24. do Carmo, M.P.: Riemannian Geometry. Birkhäuser Verlag, Boston (1992)
25. Duits, R., Burgeth, B.: Scale spaces on Lie groups. In: Scale Space and Variational Methods in Computer Vision. LNCS, vol. 4485, pp. 300–312. Springer, Berlin/Heidelberg (2007)
26. Eells, J. Jr., Sampson, J.H.: Harmonic mappings of Riemannian manifolds. Am. J. Math. **86**(1), 106–160 (1964)
27. Gonzalez T.F.: Clustering to minimize the maximum intercluster distance. Theor. Comput. Sci. **38**, 293–306 (1985)
28. Gur, Y., Sochen, N.A.: Regularizing flows over lie groups. J. Math. Imag. Vis. **33**(2), 195–208 (2009)
29. Hall, B.C.: Lie Groups, Lie Algebras,and Representations, An Elementary Introduction. Springer-Verlag, 2004.
30. Hauberg, S., Sommer, S., Pedersen, K.S.: Gaussian-like spatial priors for articulated tracking. In: European Conference on Computer Vision. LNCS, vol. 6311, pp. 425–437. Springer, Berlin/New York (2010)
31. Hochbaum, D., Shmoys, D.: A best possible approximation for the k-center problem. Math. Oper. Res. **10**(2), 180–184 (1985)
32. Iserles, A., Munthe-Kaas, H.Z., Nørsett, S.P., Zanna, A.: Lie group methods. Acta Numer. **9**, 215–365 (2000)
33. Jain, A.K.: Fundamentals of Digital Image Processing. Prentice-Hall, Upper Saddle River (1989)
34. Jain, V., Zhang, H.: Robust 3D shape correspondence in the spectral domain. In: Proceedings of Shape Modeling International, pp. 118–129. IEEE Computer Society, Washington DC (2006)
35. James, D.L., Twigg, C.D.: Skinning mesh animations. SIGGRAPH **24**(3), 399–407 (2005)
36. Kalogerakis, E., Hertzmann, A., Singh, K.: Learning 3D mesh segmentation and labeling. ACM Trans. Graph. **29**(4), Article 102 (2010)
37. Kobilarov, M., Crane, K., Desbrun, M.: Lie group integrators for animation and control of vehicles. ACM Trans. Graph. **28**(2), 1–14 (2009)
38. Kompatsiaris, I., Tzovaras, D., Strintzis, M.G.: Object articulation based on local 3D motion estimation. In: Proceedings of ECMAST, pp. 378–391. Springer, London (1999)
39. Lee, T.-Y., Wang, Y.-S., Chen, T.-G.: Segmenting a deforming mesh into near-rigid components. Vis. Comput. **22**(9), 729–739 (2006)
40. Litke, N., Droske, M., Rumpf, M., Schröder, P.: An image processing approach to surface matching. In: Proceedings of the Eurographics Symposium on Geometry processing, pp. 207–216. Eurographics Association, Aire-la-Ville (2005)
41. Mateus, D., Horaud, R., Knossow, D., Cuzzolin, F., Boyer, E.: Articulated shape matching using Laplacian eigenfunctions and unsupervised point registration. In: CVPR. IEEE Computer Society (2008)
42. Müller, A., Terze, Z.: Differential-geometric modelling and dynamic simulation of multibody systems. Strojarstvo **51**(6), 597–612 (2009)

43. Mumford, D., Shah, J.: Optimal approximations by piecewise smooth functions and associated variational problems. Commun. Pure Appl. Math. **42**(5), 577–685 (1989)
44. Murray, R.M., Li, Z., Sastry, S.S.: A Mathematical Introduction to Robotic Manipulation, 1 edn. CRC, Boca Raton (1994)
45. Nir, T., Kimmel, R., Bruckstein, A.M.: Variational approach for joint optic-flow computation and video restoration. Technical Report CIS-05, Technion, 2005
46. Nir, T., Bruckstein, A.M., Kimmel, R.: Over-parameterized variational optical flow. Int. J. Comput. Vis. **76**(2), 205–216 (2008)
47. Park, F.C., Bobrow, J.E., Ploen, S.R.: A Lie group formulation of robot dynamics. Int. J. Robot. Res. **14**, 609–618 (1995)
48. Pennec, X., Cachier, P., Ayache, N.: Understanding the "demon's algorithm": 3D non-rigid registration by gradient descent. In: MICCAI Proceedings of the Second International Conference on Medical Image Computing and Computer-Assisted Intervention, pp. 597–605. Springer, London (1999)
49. Raviv, D., Dubrovina, A., Kimmel, R.: Hierarchical shape matching. In: Bruckstein, A.M., ter Haar Romeny, B., Bronstein, A.M., Bronstein, M.M. (eds.) Scale Space and Variational Methods in Computer Vision. LNCS, vol. 6667, pp. 604–615 (2011)
50. Rosman, G., Bronstein, M.M., Bronstein, A.M., Wolf, A., Kimmel, R.: Group-valued regularization framework for motion segmentation of dynamic non-rigid shapes. In: Scale Space and Variational Methods in Computer Vision, pp. 725–736 (2011)
51. Rudin, L.I., Osher, S., Fatemi, E.: Nonlinear total variation based noise removal algorithms. Phys. D Lett. **60**, 259–268 (1992)
52. Ruggeri, M.R., Patanè, G., Spagnuolo, M., Saupe, D.: Spectral-driven isometry-invariant matching of 3D shapes. Int. J. Comput. Vis. **89**(2–3), 248–265 (2010)
53. Szymon Rusinkiewicz and Marc Levoy: Efficient variants of the ICP algorithm. In: Third International Conference on 3D Digital Imaging and Modeling (3DIM), pp. 145–152. IEEE Computer Society (2001)
54. Sapp, B., Toshev, A., Taskar, B.: Cascaded models for articulated pose estimation. In: European Conference Computer Vision, pp. 406–420. Springer, Berlin/New York (2010)
55. Shamir, A.: A survey on mesh segmentation techniques. Comput. Graph. Forum **27**(6), 1539–1556 (2008)
56. Subbarao, R., Meer, P.: Nonlinear mean shift over Riemannian manifolds. Int. J Comput. Vis. **84**(1), 1–20 (2009)
57. Sumner, R.W., Popović, J.: Deformation transfer for triangle meshes. In: SIGGRAPH, pp. 399–405. ACM, New York (2004)
58. Thirion, J.P.: Image matching as a diffusion process: an analogy with Maxwell's demons. Med. Image Anal. **2**(3), 243–260 (1998)
59. Tierny, J., Vandeborre, J.-P., Daoudi, M.: Fast and precise kinematic skeleton extraction of 3D dynamic meshes. In: ICPR. IEEE Computer Society (2008)
60. Trobin, W., Pock, T., Cremers, D., Bischof, H.: An Unbiased Second-Order Prior for High-Accuracy Motion Estimation, pp. 396–405. Springer, Berlin/New York (2008)
61. Tuzel, O., Porikli, F., Meer, P.: Learning on lie groups for invariant detection and tracking. In: CVPR. IEEE Computer Society (2008)
62. Wang, S., Wang, Y., Jin, M., Gu, X.D., Samaras, D.: Conformal geometry and its applications on 3D shape matching, recognition, and stitching. IEEE Trans. Pattern Anal. Mach. Intell. **29**(7), 1209–1220 (2007)
63. Wolf, A., Sharf, I., Rubin, M.B.: Using Cosserat point theory for estimating kinematics and soft-tissue deformation during gait analysis. In: Lenarcic, J., Stanisic, M.M. (eds.) Advances in Robot Kinematics: Motion in Man and Machine, pp. 63–70. Springer, Netherlands (2010)
64. Wuhrer, S., Brunton, A.: Segmenting animated objects into near-rigid components. Vis. Comput. **26**, 147–155 (2010)
65. Yamasaki, T., Aizawa, K.: Motion segmentation for time-varying mesh sequences based on spherical registration. EURASIP J. Appl. Signal Process. 1–9, (2009). Article 346425

66. Yanenko, N.N.: The Method of Fractional Steps: Solution of Problems of Mathematical Physics in Several Variables. Springer, Berlin/New York (1971). Translated from Russian
67. Yang, C., Medioni, G.: Object modelling by registration of multiple range images. Image Vis. Comput. **10**, 145–155 (1992)
68. Zhang, D., Hebert, M.: Harmonic maps and their applications in surface matching. In: CVPR, vol. 2, pp. 524–530. IEEE Computer Society (1999)
69. Žefran, M., Kumar, V., Croke, C.: On the generation of smooth three-dimensional rigid body motions. IEEE Trans. Robot. Autom. **14**(4), 576–589 (1998)
70. Žefran, M., Kumar, V., Croke, C.: Metrics and connections for rigid-body kinematics. Int. J. Robot. Res. **18**(2), 242 (1999)

Chapter 13
Point Cloud Segmentation and Denoising via Constrained Nonlinear Least Squares Normal Estimates

Edward Castillo, Jian Liang, and Hongkai Zhao

Abstract We first introduce a surface normal estimation procedure for point clouds capable of handling geometric singularities in the data, such as edges and corners. Our formulation is based on recasting the popular Principal Component Analysis (PCA) method as a constrained nonlinear least squares (NLSQ) problem. In contrast to traditional PCA, the new formulation assigns appropriate weights to neighboring points automatically during the optimization process in order to minimize the contributions of points located across singularities. We extend this strategy to point cloud denoising by combining normal estimation, point projection, and declustering into one NLSQ formulation. Finally, we propose a point cloud segmentation technique based on surface normal estimates and local point connectivity. In addition to producing consistently oriented surface normals, the process segments the point cloud into disconnected components that can each be segmented further into piecewise smooth components as needed.

13.1 Introduction

Two key processing tools for unstructured point cloud data are surface normal estimation and segmentation. Surface normal estimation is utilized by several applications such as reconstruction [5, 6, 20], local feature size estimation [10, 26], computer aided design [24], and inside-outside queries, while segmentation is

E. Castillo (✉)
Department of Radiation Oncology, University of Texas, MD Anderson Cancer Center, Houston, TX, USA

Department of Computational and Applied Mathematics, Rice University, Houston, TX, USA
e-mail: ecastillo3@mdanderson.org

J. Liang · H. Zhao
Department of Mathematics, University of California, Irvine, CA, USA
e-mail: jianl@uci.edu; zhao@math.uci.edu

M. Breuß et al. (eds.), *Innovations for Shape Analysis*, Mathematics and Visualization, DOI 10.1007/978-3-642-34141-0_13, © Springer-Verlag Berlin Heidelberg 2013

useful for efficient point cloud representation and classification [15, 30]. However, depending on the method, segmentation often requires surface normal estimates as a prerequisite.

As described in [11], surface normal estimation methods can be classified algorithmically as either numerical or combinatorial. Combinatorial methods are based on the information obtained from a Delaunay triangulation procedure applied to the point cloud. Consequently, combinatorial methods such as the algorithm proposed in [2], are known to be sensitive to noise. Though noise robustness has been incorporated into the combinatorial methodology [9], the approach in general becomes infeasible for large data sets due to the computational cost required by the Delaunay triangulation [11]. On the other hand, numerical methods are based on the optimization of a fitting model defined on a subset of point contained in a local neighborhood.

The most common approach employed by numerical methods is to first determine a subset of data points representing the local neighbors of a given point p. As described in [13], the estimated surface normal for the point p is then the normal to the plane that best fits the neighboring data points in the least squares sense. The least squares fitting plane is determined using Principal Component Analysis (PCA). As such, the entire procedure is often referred to as PCA. Since the method is based on a least squares formulation, PCA is known to be stable in the presence of noise. However, near geometric singularities such as corners or edges, where the normal is discontinuous, the PCA fails to produce sharp normal estimates since neighboring points are used nondistinctively to compute the planar fit. The effect is rounded or smoothed normal estimates along edges. Adjustments to the basic PCA procedure, such as distance weighting [1, 27], varying neighborhood sizes [23, 26], and higher-order fitting [18, 24, 32], have been suggested to improve the accuracy near geometric discontinuities. However, such approaches fail to address the fundamental problem of determining which points contained in a given neighborhood should contribute to the normal computation.

In this work, we first introduce a surface normal approximation method based on recasting the popular PCA formulation as a constrained nonlinear least squares (NLSQ) problem that is designed to produce accurate surface normal estimates near geometric singularities. Our constrained NLSQ method incorporates the weight of each point contribution into the normal estimate problem formulation as an unknown quantity. In another word, our formulation can be viewed as a robust statistic strategy, i.e., points belonging to a different smooth patch across a geometric singularity can be seen as outliers in the sense of robust statistics [17]. By simultaneously optimizing the weights and the normal, essentially only neighboring points from the same smooth piece contribute to the normal estimate. This constrained NLSQ formulation has the benefit of being simpler than neighborhood reorganization methods such as [6] and can be implemented with almost the same efficiency as the standard PCA. In addition, we extend the constrained NLSQ methodology to address point cloud denoising by combining normal estimation, point projection, and declustering together into one formulation.

Next we introduce a point cloud segmentation scheme that utilizes the estimated surface normals computed by our constrained NLSQ approach. Point cloud segmentation methods are typically categorized as either a region growing method or border-based method. Region growing methods, such as those described in [4, 28, 30, 31], are based on growing segmented regions from preselected seed point locations according to a given consistency or continuity rule. Though known to be robust to noise, the main difficulties associated with region growing are the seed point selection criteria and the monitoring of the growing process. Border-based methods on the other hand [7, 8, 15, 19], rely on identifying the points located along edges. The full segmentation is then computed from the detected edges by applying a filling process or meshing. However, due to noise or spatially uneven point distributions, such methods often detect disconnected edges that make it difficult for a filling procedure to identify closed segments.

Our segmentation procedure is based on first constructing an adjacency matrix representing the point connectivity of the full data set from unoriented surface normal estimates according to an angular continuity measure. The connected components present in the adjacency matrix constitute the initial segmentation. As a result, the method does not require a filling process or region growing procedure. Additionally, the point connectivity information represented by the adjacency matrix provides a means for consistently orienting the surface normal estimates, even for data sets representing multiple closed surfaces. A finer segmentation can then be obtained from the initial connected components by classifying the oriented surface normals according to both their direction and the local geometry information contained in the adjacency matrix.

Here is the outline. First, we reformulate the standard PCA as a constrained least squares problem in Sect. 13.2. Then, we derive our constrained NLSQ formulation for normal estimation in Sect. 13.3. This formulation is extended to point cloud denoising in Sects. 13.4–13.6. Our point cloud segmentation algorithm based on surface normal estimates and point connectivity is described in Sect. 13.7.

13.2 PCA as Constrained Linear Least Squares

PCA is a popular method for computing surface normal approximations from point cloud data [13]. Given a point cloud data set $D = \{x_i\}_{i=1}^{n}$, the PCA surface normal approximation for the point $p \in D$ is typically computed by first determining the *K-nearest neighbors*, $x_k \in D$, of p. Given the *K-neighbors*, the approximate surface normal is then the eigenvector associated with the smallest eigenvalue of the symmetric positive semi-definite matrix:

$$P = \sum_{k=1}^{K} (x_k - \bar{p})^T (x_k - \bar{p}), \tag{13.1}$$

where $\bar{p}$ is the local data centroid, $\bar{p} = (\frac{1}{K}) \sum_{k=1}^{K} x_k$.

The PCA normal approximation, also referred to as total least squares [23], is accurate when the underlying surface is smooth, but tends to smear across singularities, such as corners or edges as illustrated in Fig. 13.1. The smearing is

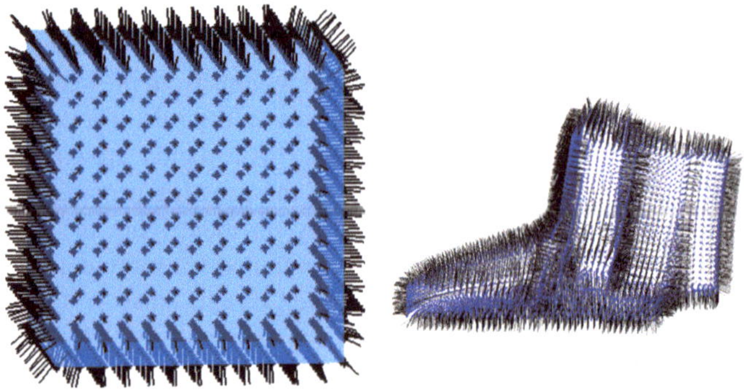

Fig. 13.1 PCA normal estimation for a cube and a fan disk. Notice the smeared normals along the edges of the cube

caused by the contribution of neighboring data points from across the singularity to the PCA covariance matrix P. Hence, the standard PCA normal estimate can be improved if one can automatically detect singularities and choose an appropriate set of neighboring points from the same smooth piece. However, Eq. (13.1) alone does not provide much insight into determining which data point contributions should be included or excluded. However, the PCA normal approximation can be described equivalently as the solution to the following equality constrained least squares problem:

$$\arg\min_{\eta} \frac{1}{2}\|V\eta\|_2^2 \qquad (13.2)$$

$$\text{s.t.}\|\eta\|_2^2 = 1,$$

where the rows of the matrix $V \in \mathbb{R}^{K\times 3}$ are the difference vectors $v_k = x_k - \bar{p}$ and η represents the unknown normal vector. The single equality constraint is necessary to ensure a nontrivial solution.

Examination of the by the first order Karush-Kuhn-Tucker (KKT) conditions for problem 13.2:

$$V^T V\eta - \mu\eta = 0, \qquad (13.3)$$

reveals that the stationary points for problem (13.2) are the eigenpairs of $P = V^T V$, where the matrix P is exactly the PCA covariance matrix in (13.1) and μ is the Lagrange multiplier corresponding to the equality constraint. The global minimum is achieved at the smallest eigenvalue and therefore the solution to problem (13.2) is equivalent to the PCA approximation. However, in contrast to the standard PCA formulation, formulation (13.2) provides insight into how to improve the normal estimation near singularities.

Problem (13.2), and consequently the PCA method, describe the normal approximation as the vector that is most orthogonal to the collection of difference vectors v_k. In the simplest case where all v_k live in the same plane, the value of the objective function $V^T\eta$ at the solution η^* is zero, and η^* is orthogonal to the plane. From this perspective, it is not necessary to compute the local centroid, since utilizing the difference vectors $a_k = x_k - p$, as opposed to the vectors $v_k = x_k - \bar{p}$, yields the same result. Moreover, the magnitude of the inner product between η^* and each difference vector quantifies the "planar deviation" of each local neighbor. Thus, monitoring this quantity provides a means for determining which data points, if any, are contributing from across a discontinuity.

13.3 Normal Estimation via Constrained Nonlinear Least Squares

The PCA method produces poor normal approximations near surface discontinuities because the method equally weights the erroneous contributions of points positioned across the singularity into the normal calculation. However, utilizing the PCA equivalent, constrained least squares formulation (13.2), such contributions can be detected by examining the magnitude of the orthogonality mismatch, $|a_k^T\eta|$, where $a_k = x_k - p$. In the language of robust statistics, points with large orthogonality mismatch are considered outliers.

Assuming noise free data, large orthogonality mismatches correspond to neighboring points that highly deviate from the local least squares fitted plane. Consequently, large orthogonality mismatches indicate that the corresponding neighbors are potentially positioned across a discontinuity. Thus, we propose a dynamic weighting scheme that adaptively determines the weight of each point contribution as a function of the orthogonality mismatch. Incorporating this strategy into formulation (13.2) results in the following constrained nonlinear least squares formulation:

$$\begin{aligned} \arg\min_{\eta} \ & \frac{1}{2}\sum_{k=1}^{K} e^{-\lambda(a_k^T\eta)^2}\left(a_k^T\eta\right)^2 \\ \text{s.t.} \ & \|\eta\|_2 = 1. \end{aligned} \tag{13.4}$$

While traditional weighting schemes place emphasis on proximity [1, 27], the weighting term $e^{-\lambda(a_k^T\eta)^2}$ adaptively deflates the contribution of terms with high orthogonality mismatch at a rate defined by the parameter λ. In this way, formulation (13.4) is designed to compute the surface normal estimate using only the neighboring points positioned on the same smooth surface component. Naturally, setting λ to zero results in the original PCA linear least squares problem (13.2), with the exception that the difference vectors are taken from the data point p and not the centroid $\bar{p}$.

At first glance, formulation (13.4) appears to be substantially more difficult to solve than the small eigenvalue problem associated with the PCA approach.

However, the equality constraint is easily absorbed into the objective function by representing the unknown surface normal η in spherical coordinates with magnitude set to unity:

$$\eta(\phi,\theta) = \begin{bmatrix} \cos(\phi)\cos(\theta) \\ \cos(\phi)\sin(\theta) \\ \sin(\phi) \end{bmatrix}.$$

Substituting the spherical coordinate representation of the unknown surface normal into formulation (13.4) results in the following unconstrained nonlinear least squares problem for only two variables:

$$\min_{\phi,\theta} \frac{1}{2} \sum_{k=1}^{K} e^{-\lambda(a_k^T \eta(\phi,\theta))^2} \left[a_k^T \eta(\phi,\theta)\right]^2. \tag{13.5}$$

Applying formulation (13.4) to the examples shown in Fig. 13.1 results in significant improvement in the surface normal estimates at singularities, as illustrated by Fig. 13.2. Moreover, the computational cost is comparable to that of standard PCA.

In general, computing the k-nearest neighbors represents the main computational cost associated with normal estimation. For the city example containing 1.99 million points and illustrated in Fig. 13.8, the nearest neighbor calculation (tree based sorting method with $O(n \log n)$ complexity) required 66 s on a Dell Precision M6400 laptop. The normal calculations utilizing our nonlinear least squares method required 39 s whereas the PCA normal calculation required 24 s.

13.4 Incorporating Point Cloud Denoising into the NLSQ Normal Estimate

Point cloud denoising is naturally an important task in many applications. The noise associated with a data point can be viewed as a perturbation of the point's true location. Hence, for a given noise polluted point $\tilde{p}$, the task of denoising can be viewed as finding the projection of the point $\tilde{p}$ onto the underlying surface. However, the underlying surface is unknown and has to be inferred from the noisy point cloud itself. Developing methods for projecting points onto curves and surfaces is an active area of research with applications in computer graphics and vision (see [14]). Work has also been done for the problem of projecting points onto point cloud represented surfaces [3, 22], based on first determining a "projection direction". The goal here is to couple point cloud denoising with the surface normal estimation together. The motivation for this approach is the idea that a denoised point cloud improves surface normal estimation, which in turn provides a better projection direction for point cloud denoising.

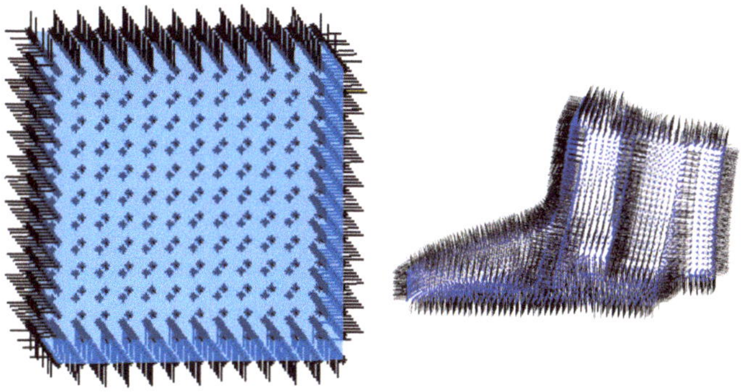

Fig. 13.2 Nonlinear least squares normal estimation for a cube and a fan disk. Notice the improvement in the normal estimation along the edges of the cube compared to standard PCA (Fig. 13.1)

Given a surface normal estimate η at $\tilde{p}$, we wish to determine a projection along η that minimizes the orthogonality mismatch energy (formulation (13.5) with $\lambda = 0$):

$$\arg\min_t \frac{1}{2} \sum_{k=1}^{K} \left[(x_k - (\tilde{p} + t\eta))^T \eta \right]^2. \tag{13.6}$$

Rearranging terms and taking advantage of the fact that $\|\eta\| = 1$, the problem reduces to a simple least squares problem for the scalar value t:

$$\arg\min_t \frac{1}{2} \|A\eta - t\|_2^2, \tag{13.7}$$

where the rows of A are the difference vectors $a_k = x_k - \tilde{p}$, and the denoised location is given by the solution t^* as $p^* = \tilde{p} + (t^*)\eta$.

As is the case for most denoising procedures, naively applying (13.7) to clean data will result in over smoothing and the loss of sharp surface features. However, given a bound T for the distance between $\tilde{p}$ and the underlying surface, data fidelity can easily be incorporated into the formulation by way of linear inequality constraints. Specifically,

$$\arg\min_t \frac{1}{2} \|A\eta - t\|_2^2, \quad \text{s.t. } |t| < T. \tag{13.8}$$

Single variable calculus dictates that the solution to problem (13.8) is achieved at either the unconstrained minimum $t_{unc} = (\frac{1}{K}) \sum_{k=1}^{K} a_k^T \eta$, or at one of the bounds $-T$ or T.

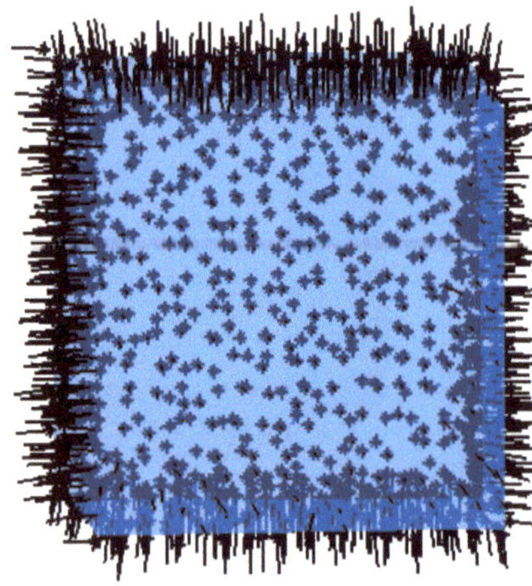

Fig. 13.3 Constrained nonlinear least squares normal estimation for a noisy cube. Three percent (in term of cube size) random perturbation was added to the original data. The estimated normals are still quite accurate even at edges

For the typical case where η is not known, the noise in the data can lead to inaccurate PCA or NLSQ surface normal approximations. However, both noise robustness and robustness to surface discontinuities can be incorporated into one NLSQ framework by coupling the denoising formulation (13.8) with the NLSQ formulation (13.5) (Fig. 13.3):

$$\arg\min_{\phi,\theta,t} \frac{1}{2} \sum_{k=1}^{K} e^{-\lambda(a_k^T \eta(\phi,\theta)-t)^2} \left[a_k^T \eta(\phi,\theta) - t\right]^2 \quad (13.9)$$

$$\text{s.t. } |t| \leq T.$$

Our numerical approach for solving Problem (13.9) is based on reformulating the two linear constraints as a pair of log-barrier functions:

$$L_1(t) = -\ln\left(\frac{T-t}{T}\right),$$

$$L_2(t) = -\ln\left(\frac{T+t}{T}\right),$$

designed so that $L_1(0) = L_2(0) = 0$ and $L_1(T) = L_2(-T) = \infty$. Problem (13.9) with the inequality constraints replaced by the log-barrier functions, comprise a standard, unconstrained, and nonlinear least squares problem. Our current implementation utilizes the *MINPACK* software package, which employs the Levenberg-Marquardt Quasi-Newton trust region algorithm, using exact values for the gradient of the objective function. The initial guess used for the optimization routine is computed by solving problems (13.5) and (13.8) for λ, $h = 0$, respectively. It is however possible to extend formulation (13.9) into a more general point cloud denoising scheme.

13.5 Generalized Point Cloud Denoising and NLSQ Normal Estimation

In formulation (13.6), we make the rather restrictive assumption that the point $\tilde{p}$ can only move along the normal direction η. We relax this constraint to allow the point to move to a new position p^* for the following two reasons: (1) perturbation in the point location could be in both tangential and normal direction, and (2) it is also desirable that points are evenly distributed after denoising. Substituting $a_k^T \eta(\phi, \theta) - t$ for $(x_k - p^*)^T \eta(\phi, \theta)$ within the basic NLSQ adaptive weighting formulation (13.5) results in the following:

$$\arg\min_{\phi,\theta,p^*} \frac{1}{2} \sum_{k=1}^{K} e^{-\lambda((x_k - p^*)^T \eta)^2} \left[(x_k - p^*)^T \eta\right]^2,$$

where $\lambda \geq 0, \eta = \eta(\phi, \theta)$. The simplicity of the denoising NLSQ formulation (13.9) allowed for the utilization of linear inequality constraints to maintain data fidelity. However, in this case, applying the same strategy results in a single nonlinear constraint: $||\tilde{p} - p^*||_2 < T$. Rather than employ the nonlinear constraint, we instead simplify the overall formulation by introducing an exponential distance penalty term to maintain data fidelity. In addition, a standard proximity weighting term, such as those applied in [1, 27], is incorporated to ensure that points x_k contained in the *K-neighbors* that lie further away from $\tilde{p}$ do not contribute as heavily to the minimization energy. The resulting formulation is then:

$$\arg\min_{\phi,\theta,p^*} \frac{1}{2} e^{\alpha||\tilde{p} - p^*||_2^2} \sum_{k=1}^{K} e^{-\beta||x_k - \tilde{p}||_2^2} e^{-\lambda((x_k - p^*)^T \eta)^2} \left[(x_k - p^*)^T \eta\right]^2 \qquad (13.10)$$

where $\alpha > 0, \beta > 0, \lambda \geq 0, \eta = \eta(\phi, \theta)$. Although formulation (13.10) bears some resemblance to the bilateral filtering method introduced in [12], the key difference is that the bilateral filtering method computes the normal direction at point $\tilde{p}$ a priori using PCA.

A drawback of the combined surface normal estimation and point cloud denoising formulation (13.10) is the potential for producing poor spatial point distributions. For example, a clustering effect can occur near high curvature locations when the points move predominately along the normal directions. To avoid this, we introduce the idea of electrostatic repulsion into the denoising model.

13.6 Combined Point Cloud Declustering, Denoising, and NLSQ Normal Estimation

The main concept behind electrostatic repulsion is that each point x_k in the *K-neighbors* emits a repulsion force that is defined at any location x as the inverse ratio of the squared distance between x_k and x. Repulsion has previously been shown to have utility in image processing [29], and is commonly employed in different forms by point cloud consolidation methods such as those in [16, 21, 25]. Incorporating electrostatic repulsion into formulation (13.10) yields:

$$\arg\min_{\phi,\theta,p^*} \frac{1}{2} e^{\alpha||\tilde{p}-p^*||_2^2} \sum_{k=1}^{K} E_k + \frac{\mu}{2} \sum_{k=1}^{K} \frac{1}{||x_k - p^*||_2^2} \tag{13.11}$$

where $E_k = e^{-\beta||x_k-\tilde{p}||_2^2} e^{-\lambda((x_k-p^*)^T\eta)^2} \left[(x_k - p^*)^T\eta\right]^2, \alpha > 0, \beta > 0, \lambda \geq 0, \eta = \eta(\phi, \theta), \mu \geq 0$. Equation (13.11) is designed to produce both the surface normal estimate and the denoised position by balancing the following three factors: fidelity, smoothness, and even distribution. In comparison to the motion restricted NLSQ formulation (13.9), formulation (13.11) introduces two more degrees of freedom but is unconstrained. As such, formulation (13.11) is easy to implement and can be solved via any standard nonlinear least squares solver package. Our current implementation again utilizes the *MINPACK* library.

Two examples are illustrated in Figs. 13.4 and 13.5. In both examples, Gaussian noise with mean 0 and a variance of 1 % of the length of the diagonal of the box bounding the data is added. A comparison of the results produced by applying formulation (13.10) and the bilateral filtering method (introduced in [12]) to the test data reveals that both methods denoise the point clouds well. However, in contrast to the bilateral filtering method, formulation (13.11) produces a much more evenly distributed set of denoised points.

13.7 Segmentation Based on Point Connectivity

The overall segmentation strategy is two-fold. First, an initial segmentation is obtained from an adjacency matrix, constructed from unoriented surface normal estimates, describing the point connectivity for the entire data set. The final and finer segmentation is determined from the initial segmentation, the point connectivity information contained in the adjacency matrix, and the oriented surface normal estimates.

Our strategy for the initial segmentation is built upon the assumption that one can obtain a reasonable approximation to the underlying unoriented surface normals (presumably by applying the methodology described in Sect. 13.3, but this is not necessary) at each point in the cloud. Based on the approximated normals and the

Fig. 13.4 Nonlinear least squares denoising for a noisy sphere. (*Upper left*) is the noisy data, (*upper middle*) is the denoised result using bilateral filter, (*upper right*) is the zoom in front part of (*upper middle*), (*lower left*) is the denoised result using formulation (13.10), (*lower middle*) is the denoised result using formulation (13.11) and (*lower right*) is the zoom in front part of (*lower middle*)

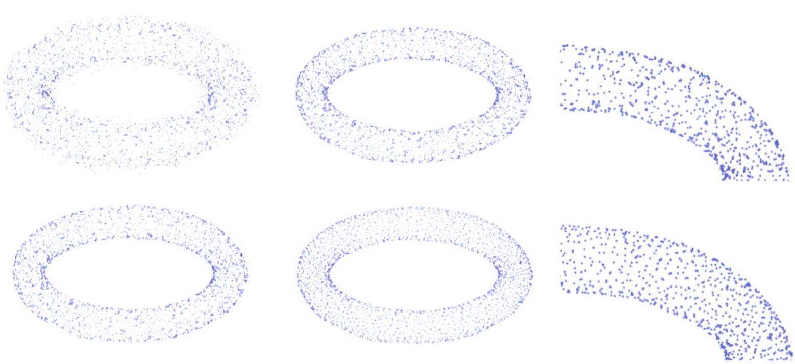

Fig. 13.5 Nonlinear least squares denoising for a noisy torus. (*Upper left*) is the noisy data, (*upper middle*) is the denoised result using bilateral filter, (*upper right*) is the zoom in *upper right* quarter of (*upper middle*), (*lower left*) is the denoised result using formulation (13.10), (*lower middle*) is the denoised result using formulation (13.11) and (*lower right*) is the zoom in *upper right* quarter of (*lower middle*)

K-nearest neighbor information, a symmetric adjacency matrix A is built such that $A_{i,j} = 1$ if the data points i, j are "connected", or in terms of graph theory, there exist an edge between the two points. $A_{i,j} = 0$ otherwise. The initial segmentation then is given by the connected components of the undirected graph represented by A.

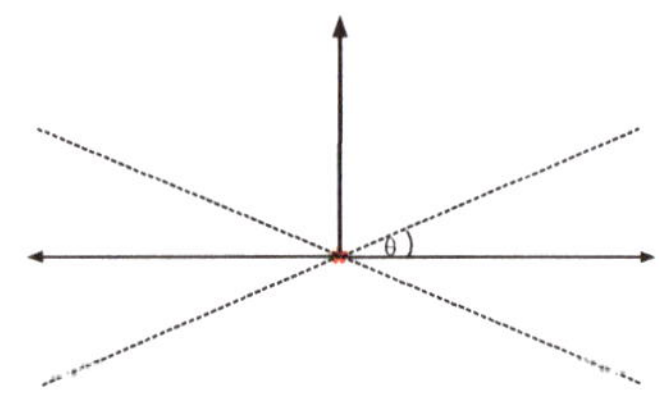

Fig. 13.6 Two dimensional representation of a data point and its tangent plane

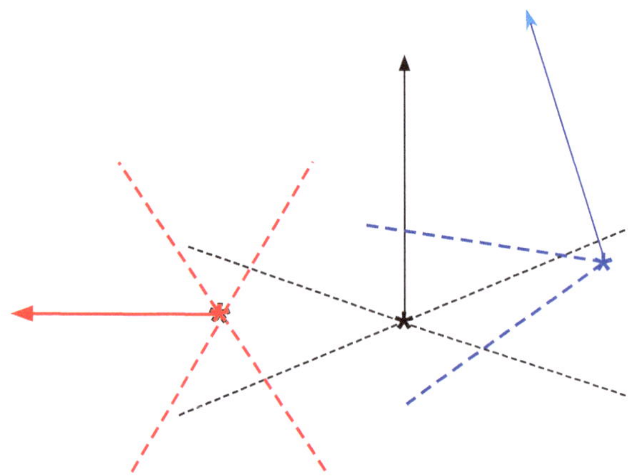

Fig. 13.7 The *black point* and *blue point* are connected to one another. The *red point* is not connected to either point

Of course, the nature of "connected component segmentation" is dictated by the criteria governing the placement of edges between points. Specifically, given points x_i and x_j, the corresponding adjacency matrix entry $A_{i,j} = 1$ if both are a k-nearest neighbor of each other, and if the line $x_j - x_i$ lies within a user defined angle θ (see Fig. 13.6) of the tangent plane associated with the unoriented surface normal approximation for x_i, and vice versa (see Fig. 13.7). In this way, point proximity, as well as local geometry are taken into account when determining point connectivity. Or put another way, neighboring points are connected if the local geometry estimate, represented by the associated surface normal approximations, are compatible with the location of x_i and x_j in $\mathbb{R}^3$.

Based on this criteria, the process of segmenting the point cloud according to the connected components of A can be considered an edge detection scheme. However, a serendipitous characteristic of this methodology is its innate ability to account for the possibility of multiple closed surfaces, or disconnected components, within a given data set. For a smoothly varying surface, such as a sphere, the corresponding adjacency matrix is comprised of one connected component. However, for the cylinder or city data depicted in Figs. 13.8 and 13.9, this is not the case. Consequently, the associated adjacency matrices contain several connected components (point cloud segments) and the resulting segmentations break the data sets into piecewise smooth components.

A finer segmentation can be obtained from the connected component segmentation by first traversing the connectivity information contained in A to orient the approximated surface normals in the same manner as the minimum spanning tree

Fig. 13.8 Point cloud representation of a city (*right*) and its connected component segmentation (*left*) represented by different colors

Fig. 13.9 Point cloud representation of a cylinder with computed normals (*right*) and its piecewise smooth segmentation (*left*) represented by different colors

method described in [13]. However, the benefit of employing the graph represented by A is the ability to orient data sets containing more than one body. The orientation procedure is carried out on one individual connected component segment at a time. The orientation within a body is kept consistent by identifying points with nearest neighbors belonging to two connected component segments, and then propagating the information from one connected component to another.

Given the oriented surface normal approximations, the data set is further segmented by classifying points with similar surface normals. Specifically, given a set of N classification vectors, each data point is categorized into one of N groups according to the similarity between the surface normal and the classification vectors. Consider the example of a sphere with the classification vectors chosen to be the six outward normals corresponding to the faces of a cube. Since the sphere is a smoothly varying surface, there is only one connected component present in the

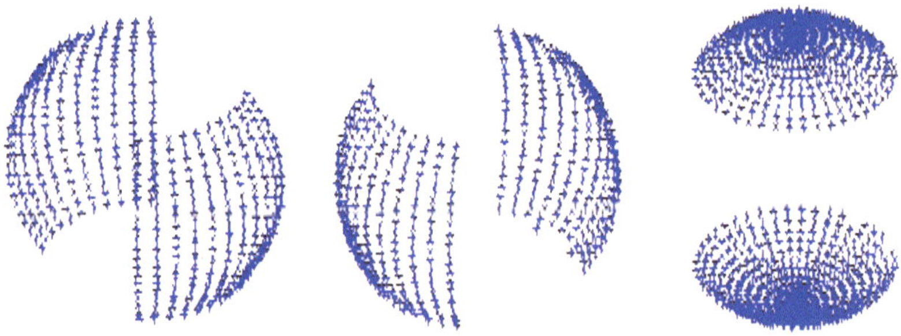

Fig. 13.10 Segmentation results for a sphere using the six outward normals for the faces of a cube as classification vector

Fig. 13.11 Piecewise smooth segmentation for a mechanical part. (*Left*) The original point cloud. (*Right*) Segmentation result: different colors correspond to different segments

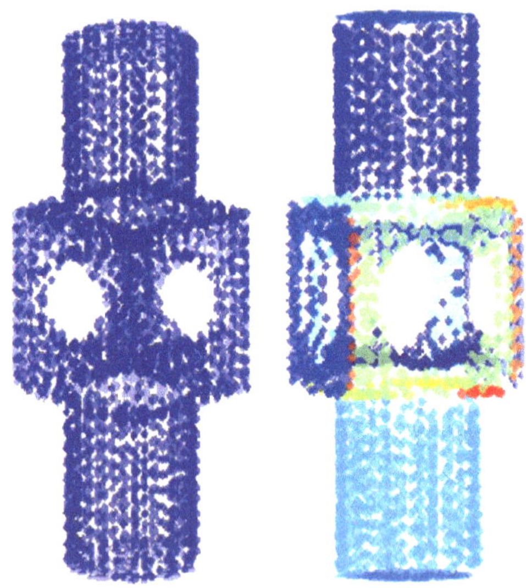

data set (see Fig. 13.10). Assuming properly oriented surface normals, each point is categorized according to their closeness (in terms of angular difference) to one of the six classification vectors and then grouped together based on the point connectivity. The full algorithm is as follows (Fig. 13.11):

Algorithm

1. Find the K-nearest neighbors of each data point.
2. Calculate the surface normal approximations.
3. Construct the adjacency matrix A by determining the edges between points.

(continued)

(continued)

4. Determine the initial segmentation by finding the connected components of A.
5. Orient the surface normal approximations.
6. (Optional) Refine the segmentation according to N classification vectors.

13.8 Conclusions

A robust surface normal approximation strategy for point clouds using a constrained nonlinear least squares (NLSQ) formulation is developed. The formulation is designed to improve normal estimation near geometric singularities, such as edges and corners, by employing a nonlinear adaptive weighting scheme. The weighting scheme is designed to incorporate only neighboring points positioned on the same smooth surface component into the surface normal computation. This approach generalizes to the broader problem of outlier detection. More specifically, the NLSQ formulation finds neighboring points that are the most consistent with one another in terms of the orthogonality mismatch. A byproduct of this procedure is then the determination of the inconsistent points, which can be considered as outlier data.

Using the NLSQ surface normal formulation as a foundation, we also derive a new point cloud processing strategy that combines denoising, declustering, and surface normal estimation into one formulation.

The point cloud segmentation procedure introduced in this paper is based on a simple point connectivity criteria and the connected components present in the representative adjacency matrix. This approach allows for surface normal orientation even when the data set contains more than one closed surface. Furthermore, a finer segmentation that splits the point cloud into piecewise smooth regions can be achieved.

Acknowledgements The authors acknowledge ARO/MURI award W911NF-07-1-0185 and NGA NURI Award HM1582-10-1-0012 for support with this work.

References

1. Alexa, M., Behr, J., Cohen-Or, D., Fleishman, S., Levin, D., Silva, C.T.: Computing and rendering point set surfaces. IEEE Trans. Vis. Comput. Graph. **9**, 3–15 (2003)
2. Amenta, N., Bern, M.: Surface reconstruction by voronoi filtering. In: Proceedings of the Fourteenth Annual Symposium on Computational Geometry. ACM, New York (1998)

3. Azariadis, P.: Parameterization of clouds of unorganized points using dynamic base surfaces. Comput. Aided Des. **36**, 607–623 (2004)
4. Besl, P.J., Jain, R.C.: Segmentation through variable-order surface fitting. IEEE Trans. Pattern Anal. Mach. Intell. **10**, 167–192 (1988)
5. Boissonnat, J.D., Cazals, F.: Smooth surface reconstruction via natural neighbor interpolation of distance functions. In: Proceedings of 16th Annual Symposium on Computational Geometry, pp. 223–232. ACM, New York (2000)
6. Calderon, F., Ruiz, U., Rivera, M.: Surface-normal estimation with neighborhood reorganization for 3d reconstruction. In: Proceedings of the Congress on Pattern Recognition 12th Iberoamerican Conference on Progress in Pattern Recognition, Image Analysis and Applications, pp. 321–330. Springer, Berlin/Heidelberg (2007)
7. Canny, J.: A computational approach to edge detection. IEEE Trans. Pattern Anal. Mach. Intell. **8**, 679–698 (1986)
8. Demarsin, K., Vanderstraeten, D., Volodine, T., Roose, D.: Detection of closed sharp edges in point clouds using normal estimation and graph theory. Comput. Aided Des. **39**, 276–283 (2007)
9. Dey, T.K., Goswami, S.: Provable surface reconstruction from noisy samples. Comput. Geom. Theory Appl. **35**, 124–141 (2006)
10. Dey, T., Sun, J.: Normal and feature approximations from noisy point clouds. In: FSTTCS 2006: Foundations of Software Technology and Theoretical Computer Science, vol. 4337, pp. 21–32. Springer, Berlin/Heidelberg (2006)
11. Dey, T.K., Li, G., Sun, J.: Normal estimation for point clouds: a comparison study for a voronoi based method. In: Proceedings Eurographics/IEEE VGTC Symposium Point-Based Graphics, vol. 0, pp. 39–46 (2005)
12. Fleishman, S., Drori, I., Cohen-or, D.: Bilateral mesh denoising. ACM Trans. Graph. **22**, 950–953 (2003)
13. Hoppe, H., DeRose, T., Duchamp, T., McDonald, J., Stuetzle, W.: Surface reconstruction from unorganized points. In: Proceedings of ACM Siggraph, pp. 71–78. ACM, New York (1992)
14. Hu, S., Wallner, J.: A second order algorithm for orthogonal projection onto curves and surfaces. Comput. Aided Graph. Des. **22**(3), 251–260 (2005)
15. Huang, J., Menq, C.-H.: Automatic data segmentation for geometric feature extraction from unorganized 3-D coordinate points. IEEE Trans. Robot. Autom. **17**, 268–278 (2001)
16. Huang, H., Li, D., Zhang, H., Ascher, U., Cohen-Or, D.: Consolidation of unorganized point clouds for surface reconstruction. ACM Trans. Graph. **28**, 176:1–176:7 (2009)
17. Huber, P.J.: Robust Statistics. Wiley, New York (2004)
18. Klasing, K., Althoff, D., Wollherr, D., Buss, M.: Comparison of surface normal estimation methods for range sensing applications. In: IEEE International Conference on Robotics and Automation, 2009. ICRA '09. IEEE, Piscataway (2009)
19. Lee, Y., Park, S., Jun, Y., Choi, W.C.: A robust approach to edge detection of scanned point data. Int. J. Adv. Manuf. Technol. **23**, 263–271 (2004)
20. Leung, S., Zhao, H.: A grid based particle method for evolution of open curves and surfaces. J. Comput. Phys. **228**, 2993–3024 (2009)
21. Lipman, Y., Cohen-Or, D., Levin, D., Tal-Ezer, H.: Parameterization-free projection for geometry reconstruction. ACM Trans. Graph. **26**, Article no 22 (2007)
22. Liu, Y.S., Paul, J.C., Yong, J.H., Yu, P.Q., Zhang, H., Sun, J.G., Ramani, K.: Automatic least-squares projection of points onto point clouds with applications in reverse engineering. Comput. Aided Des. **38**, 1251–1263 (2006)
23. Mitra, N., Nguyen, A., Guibas, L.: Estimating surface normals in noisy point cloud data. In: Spec. Issue Int. J. Comput. Geom. Appl. **14**, 261–276 (2004)
24. OuYang, D., Feng, H.-Y.: On the normal vector estimation for point cloud data from smooth surfaces. Comput. Aided Des. **37**, 1071–1079 (2005)
25. Pauly, M., Gross, M., Kobbelt, L.P.: Efficient simplification of point-sampled surfaces. In: Proceedings of the conference on Visualization 2002. VIS 2002. IEEE Computer Society, Washington, DC (2002)

26. Pauly, M., Keiser, R., Gross, M.: Multi-scale feature extraction on point-sampled surfaces. Comput. Graph. Forum **22**, 281–289 (2003)
27. Pauly, M., Keiser, R., Kobbelt, L.P., Gross, M.: Shape modeling with point-sampled geometry. ACM Trans. Graph. **22**, 641–650 (2003)
28. Rabbani, T., Van Den Heuvel, F., Vosselmann, G.: Segmentation of point clouds using smoothness constrain. ISPRS Comm. V Symp. Image Eng. Vis. Metrol. **36**, 248–253 (2006)
29. Schmaltz, C., Gwosdek, P., Bruhn, A., Weickert, J.: Electrostatic halftoning. Comput. Graph. Forum **29**, 2313–2327 (2010)
30. Vosselman, G.: Advanced point cloud processing. In: Fritsch, D. (ed.) Photogrammetric Week '09, pp. 137–146. Wichmann, Heidelberg (2009)
31. Vosselman, G., Gorte, B., Rabbani, T., Sithole, G.: Recongnising structure in laser scanner point clouds. Int. Arch. Photogramm. Remote Sens. Spat. Inf. Sci. **46**, 33–38 (2004)
32. Yang, M., Lee, E.: Segmentation of measured point data using parametric quadric surface approximation. Computer A **31**, 449–457 (1999)

Chapter 14
Distance Images and the Enclosure Field: Applications in Intermediate-Level Computer and Biological Vision

Steven W. Zucker

Abstract Early computer vision is dominated by image patches or features derived from them; high-level vision is dominated by shape representation and recognition. However there is almost no work between these two levels, which creates a problem when trying to recognize complex categories such as "airports" for which natural feature clusters are ineffective. In contrast, the neurobiology of vision indicates a very rich interaction between low- and high-level constructs, because there is a rich system of feedback connections. Based on this, we argue that an intermediate-level representation is necessary for computer vision and that it should incorporate certain high-level notions of distance and geometric arrangement into a form derivable from images. We propose an algorithm based on a reaction-diffusion equation that meets these criteria; we prove that it reveals (global) aspects of the distance map locally; and illustrate its performance on airport and other imagery, including visual illusions. Finally, we conjecture that these ideas also can inform the neurobiology of vision, by providing a novel basis for neural computation.

14.1 Introduction

Consider the problem of finding complex man-made structures, such as airports or medical or industrial complexes, within urban, suburban, and even rural environments from satellite imagery. Such problems are different from the object recognition tasks normally addressed in computer vision. Even though there is significant variation among people or chairs, this variation seems small in comparison with the variation among the complex structures listed above. People have

---

S.W. Zucker (✉)
Computer Science, Biomedical Engineering and Applied Mathematics, Yale University, New Haven, CT, USA
e-mail: steven.zucker@yale.edu

M. Breuß et al. (eds.), *Innovations for Shape Analysis*, Mathematics and Visualization, DOI 10.1007/978-3-642-34141-0_14,

arms and legs and heads; airports have runways and buildings and access roads. Arms and legs have bilateral symmetry; airports do not. In fact, the wide scope for arrangement variation is what makes airports so much more complex than faces, say, as a category. Humans, however, can readily detect airports as well as faces, which suggests that there is a level (or several levels) of structure to be found at which objects such as airports can be described. We discuss, in this paper, one such structure: distance measures derived from arrangement information about edge elements. It captures the notion that airports consist of elongated structures that are separated from other, possibly more dense structure. For computer vision applications we realize this structure as a *distance image*.

Another motivation for considering these distance images derives from visual psychophysics and physiology. Gestalt psychologists [21] identified the concepts of *figure* and *ground*, and it is now common in computer vision to seek image "segmentations" that delimit such figures. Typically in computer vision a boundary completely encloses the figure, so these boundaries are defined by closed curves.

Gestalt psychologists have refined this notion in two subtle but important ways. First, they showed that a key property of the figure is that it "owns" the boundary. Secondly, they showed that the boundary need not be closed, continuous, or even connected; this suggests more of a field of points, an *enclosure field*, than the rigid characteristic function defining inside/outside sought in computer vision.

To unify these two ideas – distance images in computer vision and enclosure fields in visual psychophysics – we first develop the concept of *border-ownership*, a Gestalt notion of how the border belongs to the figure and not to the background. To compute it there is an "information at a distance" problem: how do distant edges "inform" whether a given edge is part of the figure or part of the background? Computer vision seeks to solve this with a global segmentation; we take a different tack, by deriving a partial differential equation that summarizes aspects of the distance property inherent in border ownership. Several of its relevant properties are sketched, and its usefulness is demonstrated on the airport problem.

14.1.1 Figure, Ground, and Border Ownership

Rubin, in a seminal 1915 publication [32], introduced the notions of *figure* and *ground*; see Fig. 14.1a. Note: as the figure shifts from the vase to the faces, the border appears to shift as well; it "belongs" to the figure.

It is now known that such border-ownership effects exist in neurobiology. While many in computer vision are aware of the existence of neurons early in the cortical visual system that are selective for "edges" and their orientation, it is not widely known that, for many such cells, the response can be modulated depending on the border-ownership sense. To explain this, recall that so-called "edge detecting neurons" were first identified using stimuli of light and dark regions as they would arise from ideal, step edges [19]. Now, consider a neuron that responds vigorously

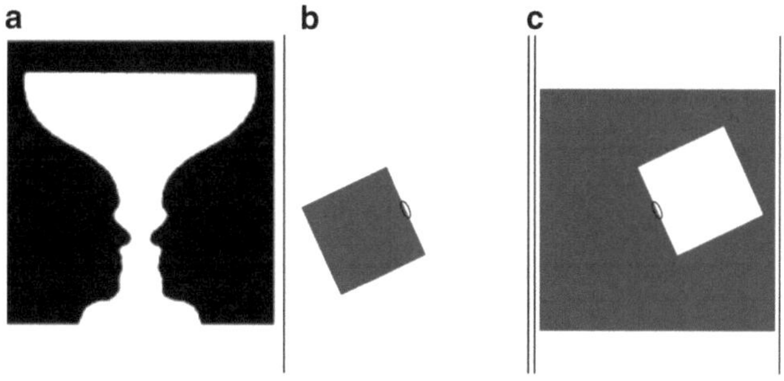

Fig. 14.1 Illustration of border ownership. (**a**) The classical Rubin vase, in which a closed figure alternates with a pair of "faces." The border separating the *dark* region from the *light* region belongs to the figure. (**b**,**c**) Border ownership is signalled by the response of certain neurons. The stimuli consist of a pair of images, the first of which shows a *dark square* on a *white* background (**b**), and the second (**c**) a *white square* on a dark background. Recordings were made from a neuron whose receptive field is shown as the small ellipse, and the stimuli were aligned so that the receptive field was optimally situated. Notice that, in both cases the neuron "sees" a dark (*left*)/bright (*right*) edge pattern. However, for some neurons, the response is more vigorous to pattern (**b**) than to (**c**); for others it might be the reverse; and for still others it might respond equally to both configurations. The interpretation is that this neuron prefers e.g. "light" figures against a dark background, as in (**c**), and is signaling not only the boundary but also the fact that it is part of a light figure. (Individual responses not shown.) Note that the light-dark pattern within the receptive field does not change, only the global arrangement of which it is a part (Figure after [42])

to a particular bright/dark configuration about, say, 1° of visual angle in diameter. This small, local configuration could indicate a border which belongs to two classes of figures: a dark one on a light background, or a bright one on a dark background. Border ownership would distinguish these two possibilities.

In a remarkable series of experiments, R. von der Heydt and colleagues [42] discovered that there are neural responses that distinguish these two cases: some cells respond to the pattern in Fig. 14.1a but not to Fig. 14.1b; while others respond in the reverse fashion. This is interpreted as a *border ownership response*, in the sense that the cell is signalling both that it is detecting an edge and that this edge belongs to the figure (and not the background).

Not all cells show a border-ownership response. Many in the first cortical visual area, V1, do not; they respond mainly to the edge brightness configuration. However the fraction of border-ownership responding cells increases significantly in the next higher visual areas (V2 and V4); it is for this reason that intermediate-level effects are implicated.

The challenge is to explain those circuits responsible for computing the border ownership signal, which involves integrating information about boundaries from a good distance away. It is tempting to assert that the border ownership signal propagates along closed borders, as in computer vision segmentation algorithms, but visual psychophysics suggests that this is not the case, as we review next.

14.1.2 Soft Closure in Visual Psychophysics

Visual search tasks, in which the subject seeks to find that figure which differs from an array of background distractors, provides another, "softer" view of closure; see Fig. 14.2.

While closure makes the search easy – closed figures often *pop out* from the others [37] – the effect persists even if the boundary is broken (Fig. 14.2b). Even a small amount of contour "pointing" toward it's continuation helps to support the perceptual organization of closure.

In effect there is a local/global stress implicit in computing closure and border-ownership. Global figural effects influence local border ownership, but figure implies borders are known. Models have recently been developed that reflect this stress, in which an approximate skeleton of the (global) shape influences the local calculation of border ownership. To relax the need for exact boundaries to reveal the skeleton, a Bayesian approach has been attempted [17], although how this might be realized biologically remains unclear.

The concept of closure needs to be relaxed to reflect that perceptual closure only hints at mathematical closure. And this holds in both directions: while one might predict that mathematical closure implies perceptual closure, this is not necessarily the case. Closed contours exist that are too complex [12] to assess without tracing along [38]. Moreover, border ownership can shift along a contour [17]. More than topology is required.

14.1.3 Intermediate-Level Computer Vision

Border ownership computations have not been a focus in computer vision. Rather the emphasis has been on object recognition systems and, e.g., the tradeoff between within-class or category variation relative to between-class/category variation. While scale-invariant features (e.g. [24]) and interest detectors can limit some of the within-class variation, an important trend is revealing that edge and shape features working together can improve performance; see e.g. [14, 15]. These may involve not only organizing edge fragments into object boundary parts, but also their relative arrangement as captured by the centroid [27, 28]. Such techniques follow a "bottom up" strategy, by which increasingly more context is involved in the recognition [39].

Centroids suggest involving higher-level shape features directly, and it is this observation that reveals a connection to border ownership. Skeleton points are extrema of the distance map and the centroid is related to shock-based formulations [35]. Computing such skeletons requires a (relatively) closed boundary, which is barely plausible for shape categories such as cups and airplanes given photometric variation, and relates back to Bayesian perceptual models [17]. But man-made structures, such as airports and sports complexes are much less structured: although runways are straight, there are huge variations in the buildings, parking facilities

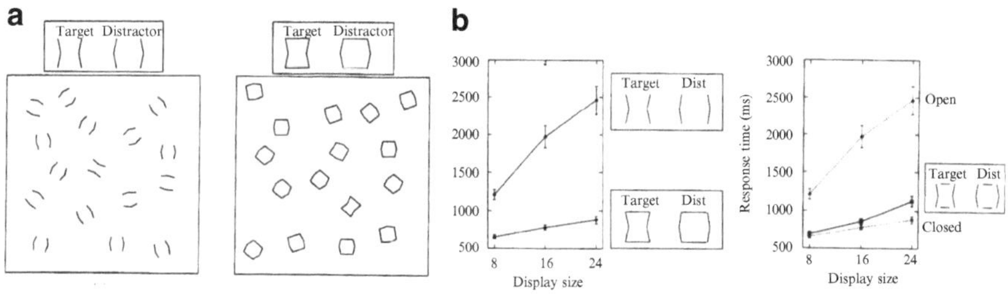

Fig. 14.2 The visual search task. Subjects are shown an example figure to find among a field of distractors. The time taken to find the distractor is a measure of difficulty. (**a**) Examples of two displays with a figure embedded among distractors. Notice how much easier the task is for the closed rather than the open figures. This suggests the power of closure. (**b**) Data showing that nearly closed figures are effectively the same as closed figures, and that the arrangement of contour fragments is key to the effect (Figures after [13])

and supply roads that flank them. How can a boundary be computed, bottom-up, around an airport? Within-class variation among airports exceeds the between-class variation with highways. Attempts to build templates for them failed, and researchers resorted early to rule-based systems [25]. But the variation among such complex features precludes such systems: the rules for defining airports in urban areas are quite similar to the rules for defining freeway exchanges; and the rules for defining airports in developing countries are significantly different. Similar problem plague recognition of other socially-developed structures, such as medical complexes and sports arenas, and organically developing biological compounds.

We explore the position that the airport recognition problem and the border ownership problem are not unrelated, and are examples that expose the need for new intermediate-level visual structures.

Mathematically the isoperimetric inequality, (perimeter)^2/area, has something of the flavor we seek, because it integrates a boundary property with a shape property. Although this can be a useful feature, operationally defining the perimeter and the area can be difficult. The problem is illustrated in Fig. 14.3; to summarize: edge maps are too local, too broken, and too rigid. High-level features, such as the skeleton, are too global, too susceptible to boundary detail, and too sensitive to closure and interior features. We seek something in between, that extends naturally the unification of top-down shape with bottom-up features [4], and that is reflective of the better parts of both.

14.2 Global Distance Information Signaled Locally

The key idea behind this paper is to represent locally certain aspects of the distance map; that is,certain global aspects of shape, so that they can be used in an intermediate-level manner. This provides a middle-ground between abstract

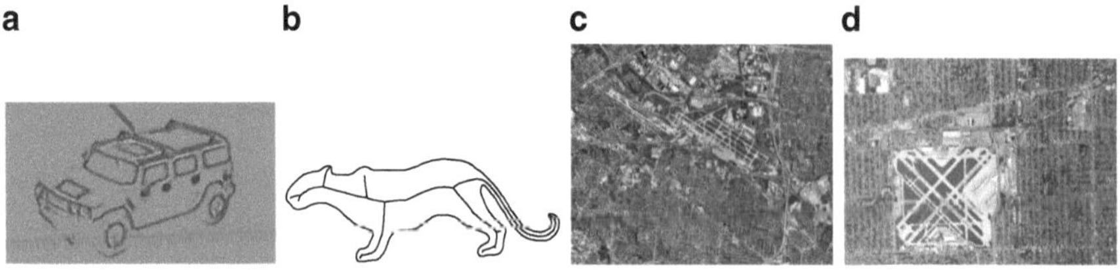

Fig. 14.3 The quest for intermediate-level vision is to find useful representational structures between edges (**a**), reflecting local, bottom-up processing, and global shape features such as the medial axis (**b**). While informative, edges are incomplete; ideally, skeletons need complete boundaries. We recall that each medial axis, or skeleton point, is the center of a maximal inscribed disc and also a singularity of the distance map. Both of these properties will be exploited by different approaches to border ownership. We seek a representation that captures aspects of both boundaries and their arrangement. For organically-and industrially-developed structures, such as airports (**c**, **d**), the relevant structure is captured by an abstract combination of edge and distance effects, rather than only local image properties

high-level representations such as skeletons and templates and the lower-levels of layered images. The representation is derived from a partial differential equation, and leads to a non-linear scale space for distances, estimated over increasingly larger domains. We note that there are many applications of pde's in scale space analysis (e.g., [16]) but none, to our knowledge, that relate the solutions to properties of the distance map.

As a warm-up, we note that there are many problems in developmental biology that involve signalling of distant events, and we borrow heavily from a plant example: A young leaf in a developing plant consists of a network of veins that form cycles surrounding domains of cells. A problem arises when the domain of cells enlarges to exceed the nutrient delivery capability of the existing vasculature: how do the "starving" cells in the center of the domain signal the need to form new veins? What is the nature of the signal, how is it generated, and what is the value that can be "read out" as a new vein instruction. The local task (a particular cell must transform from a ground cell to a vascular one) requires global information (distance to the nearest vein). To make this even more complicated, the signal must be read by those cells adjacent to existing veins, so that the network remains connected.

A theoretical solution to this problem has been developed in [8–10], and we take their model as a starting point for this paper; the intuition is shown in Fig. 14.4. The key idea is that cells in the developing leaf all produce a hormone called *auxin* at the same rate. This hormone then diffuses from cell to cell and is cleared away at the existing veins. The result is a differential equation (stated in the next section), the equilibria of which carry information about the distance from the veins to the furthest cell. Two properties are salient: the concentration of the hormone peaks at the furthest cells; and the magnitude of the gradient peaks at the existing vasculature. It is this gradient peak that provides the signal for plant development.

We interpret the hormone concentration function in [8–10] as a kind of *distance image*; that is, an image whose value at certain points corresponds to properties of

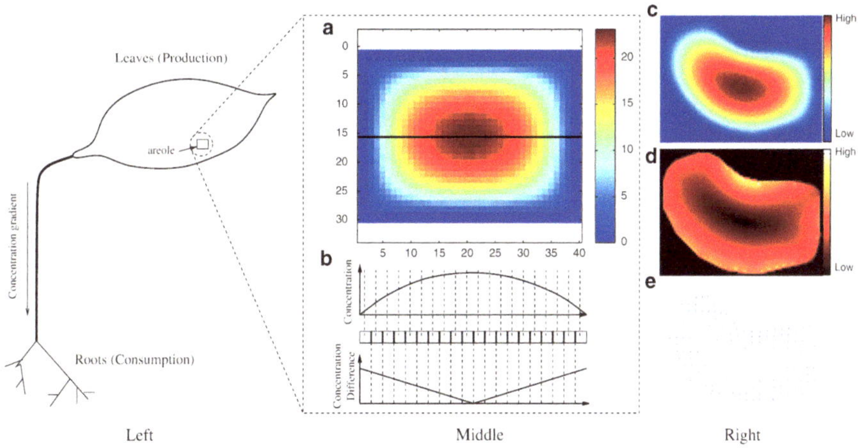

Fig. 14.4 How do young plants determine where the next vein shoot should go? Consider a rectangular portion of a leaf, surrounded by existing veins. If each cell (or pixel in the *rectangle*) produces a hormone at a constant rate, the hormone diffuses to neighboring cells and is cleared away by the existing vasculature (boundary condition = 0), the equilibrium distribution shown in (**a**) results. Taking a cross section through it, the peak in hormone concentration is at the center (**b**) and the peak in the gradient of concentration at equilibrium is at the existing veins (boundary); this last peak informs the developing leaf about where to start a new vein fragment and in which direction. (**c**) Concentration and (**d** and **e**) gradient of concentration in a real portion of a developing leaf (Figure after [8])

the distance map (distance to nearest vein). But in this form it is not at all clear how to apply it to vision problems.

A clue comes from another famous psychological display. Although in computer vision we take edge and line locations to be calibrated projections of certain positions in space, the human visual system is not so veridical. Arrangements of edge elements can effect apparent global shape properties, as in the Muller-Lyer illusion (Fig. 14.5). It is known that optical blur, as first hypothesized by Helmholtz, do not explain all of the illusion [40], nor do cognitive effects [18].

We interpret the Muller-Lyer illusion by observing that the "wings" at the ends of the horizontal lines effectively define an area context, and this area context is larger when the wings point out than when they point in; it is within this context that the lines appear to be of different lengths. So we conclude that line and edge arrangements can effect certain aspects of global shape, such as distance, at least perceptually. Returning to the airport example, we notice an analogy: the arrangement of boundaries, and the spaces between them, are the common thread through the different attempts to define them. Runways are straight and not too close to the buildings around them.

Our goal is to combine edge representations with information about their arrangement, resulting in an enhanced edge map called the distance image. Dimensionality-reduction techniques distill out some major components from these distance images

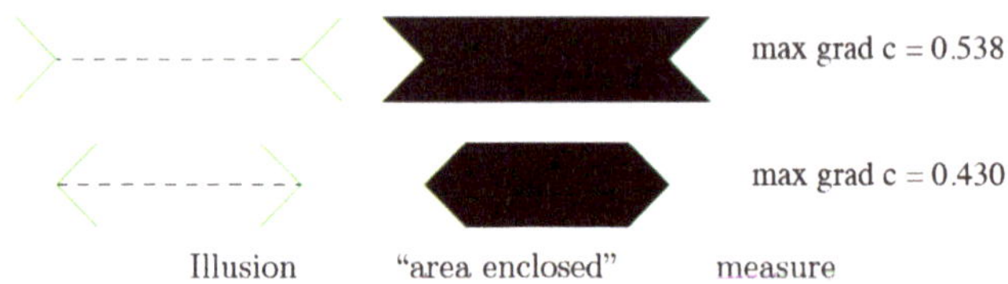

Fig. 14.5 (*Left*) The Muller-Lyer Illusion: are the horizontal lines of equal length? Notice how the outward "wings" provide a context in which the line appears longer than for the inward "wings," even though they are equal in length. (*middle*) The area enclosed by the wings, here shown in *black*, is enlarged by the outward "wings." (*right*) The Muller-Lyer illusion is predicted by Theorem 14.2; the gradient of concentration values are as shown

and a curious property of airport definitions emerges: that the distribution in orientation, arrangement, and density of edge elements can be key to defining classifiers. Finally, we relate distance images to modern notions in neurobiology.

14.3 Mathematical Formulation

We begin with the formalization of the model for plants, even though it is unrealistic for images, to introduce the type of result we seek. For concreteness, consider the example of the Muller-Lyer illusion in Fig. 14.5. Imagine that there exists a substance to report distance information, and that it is produced by all of the black pixels at the constant rate K. The set of black pixels, Ω, is a shape and $c : \Omega \to \Re$ denotes the concentration of the distance substance. Since it diffuses from pixel to pixel, it obeys:

$$c_t = D\nabla^2 c + K \tag{14.1}$$

where c_t is the derivative of concentration, D is the diffusion constant, and K is the constant production. The *Euclidean distance function* on Ω, denoted $\mathscr{E}_\Omega$, is $\mathscr{E}_\Omega(P) = \inf_{Q\in\partial\Omega} ||P - Q||_2$. The *boundary support* of P, denoted $\text{bsupp}(P; \Omega)$, is $\text{bsupp}(P; \Omega) = \{Q \in \partial\Omega : ||P - Q|| = \mathscr{E}_\Omega(P)\}$.

At equilibrium we have:

Theorem 14.1. *Let Ω be a shape and $c : \Omega \to \Re$ the unique function satisfying $c(x, y) = 0$ on $(x, y) \in \partial\Omega$ and $\nabla^2 c = -\frac{K}{D}$.*

Suppose $P \in \Omega$ is such that $\mathscr{E}_\Omega(P) = L = \sup_\Omega \mathscr{E}_\Omega$ and $Q \in \text{bsupp}(P; \partial\Omega)$. Suppose the smallest concave curvature radius is pL with $p > 0$. Then,

(a) $c(P) \in \Theta(L^2)$,
(b) $\frac{K}{2D} L \leq |\nabla c| \leq \frac{K}{D} L \frac{2p+1}{p}$,
(c) $\sup_{\partial\Omega} |\nabla c| = \sup_{\Omega-\partial\Omega} |\nabla c|$

That is, (a) the peak in concentration at P is proportional to the distance squared between the closest boundary point Q and P; (b) the gradient of the concentration reports the (approximate) length between P and Q; and (c) the largest gradient value is on the boundary. In other words, properties of the global distance map are represented locally by the concentration of the hormone.

Uniqueness follows from classical results, and the proof is based on two functions that sandwich c, one from above and one from below, and that take on the same values at the boundary. Shapes can be thought of as bounded by two contours, an "inner edge" and an "outer edge." The results hold even when the initializing function is non-smooth, a point that is important below, although the gradient of the concentration and sandwiching functions are shown to remain perpendicular to the boundary. Not all first derivatives exist (consider medial axis points) although divergence conditions can be defined. The full proof of Theorem 14.1 is in [9]. Discretization and numerical issues are also important: c is defined on a continuous domain although cells are discrete; see [9, 10].

One more conceptual point from the plant model is important. From a mathematical perspective we typically think of the boundary (vein cells) as fixing the concentration $c = 0$. However for plants we can think of the vein cells as having a larger diffusion coefficient between them than between the ground cells, so that in effect the concentration remains at 0 because the hormone diffuses away so fast that its concentrations is effectively 0. Such non-isotropic diffusions are also developed in [9, 10].

14.4 Edge Producing Model

We are now ready to develop the model for computer vision applications, and we do this in two steps, the first with plants and the second with images

In the first step we imagine a dual model to the one above for plants: instead of having all of the ground cells produce the hormone and the veins clear it away, now imagine that only the veins produce the hormone and that it diffuses into the interior tissue. (This is relevant in the development of the root.) To ensure a finite equilibrium concentration, suppose the substance is destroyed (metabolized) everywhere proportional to concentration.

This results in a reaction diffusion equation (writing the continuous version), $c_t = D\nabla^2 c + \rho - \alpha c$, with three terms: the change in concentration at a pixel depends on the amount that enters by diffusion, with the amount produced there ($\rho : \Omega \rightarrow \Re$ is the production rate) and with the amount destroyed there ($\alpha > 0$ is the destruction constant).

It is shown in [9, 10] that this dual model enjoys all of the properties from Theorem 14.1, and this is the model that we develop for computational vision.

We proceed as follows (step 2). Imagine that cells are pixels, and the vein cells are pixels at which edges have been detected. Thus the network of veins is replaced by a network of edges, and it is these edge pixels that each produce a unit of the substance per unit of time; non-edge pixels produce no substance. It is important to realize that this "substance" is not a simple measure of distance; it is an abstraction of a function that integrates distance and arrangement information from the edge pixels. Nevertheless, we will picture the equilibrium of this substance as a "distance image."

Proposition 14.1. *Consider the dynamical system*

$$\frac{\partial c}{\partial t} = D\nabla^2 c + \rho_\Omega - \alpha c. \tag{14.2}$$

Suppose that it acts over a domain Ω which is a shape as in Theorem 14.1 and on which we impose a zero-flux boundary condition (Neumann). Let $\rho_\Omega : \Omega \to \Re$. In this version the diffusion is constant between regular and edge pixels, although the production term is not. Then the following holds.

(a) If $\alpha > 0$, then $\lim_{t\to\infty} c = c_\alpha$ for a unique steady-state c_α.
(b) Let $\alpha = 0$ and $R = \int \rho_\Omega \mathrm{d}\Omega / \int \mathrm{d}\Omega$ be the average production. Then $\lim_{t\to\infty} c_t = R$ and c converges to $c_\alpha + \mathrm{cst.}$ whenever $R = 0$. Further, ∇c_α is unique even when $R \neq 0$.
(c) If $A, B \in \Re$, then the transformation $\rho_\Omega \mapsto A\rho_\Omega + \alpha B$ induces a unique transformation of the steady state $c_\alpha \mapsto Ac_\alpha + B$ and vice versa. It follows that the gradient of c_α is only affected if $A \neq 1$: $\nabla c_\alpha \mapsto A\nabla c_\alpha$.

Remarks. (i) There is a variational form which may be related to the Ambrosio-Tortorelli functional [1, 36]. (ii) In part (c), if the destruction term is not linear, e.g. $\alpha c + \beta c^2$, then the gradient might be affected by B as well.

Proof. Parts (a) and (b). To show existence we prove that the dynamical system achieves $c_t = 0$ as $t \to \infty$. Consider the dynamical system $c_{tt} = D\nabla^2 c_t - \alpha c_t$. The boundary conditions are inherited: since no flux goes through the boundary, there must be no change of concentration in time, i.e. $\nabla c_t \cdot \mathbf{n} = 0$ on $\partial\Omega$. The unique solution of this system is $c_t = 0$.

To prove uniqueness, suppose u_1 and u_2 both satisfy the equation given the boundary conditions and $c_t = 0$. Thus $D\nabla^2 u_1 + \rho_\Omega - \alpha u_1 = D\nabla^2 u_2 + \rho_\Omega - \alpha u_2$ which gives rise to $D\nabla^2 v - \alpha v = 0$ where $v = u_1 - u_2$ and $\nabla v \cdot \mathbf{n} = 0$ where $\mathbf{n}$ is the normal to the boundary. Since v is elliptic and $\alpha > 0$, v vanishes everywhere and uniqueness follows (see [6, pp. 329 and 321]). The same reference shows that if $\alpha = 0$, then this uniqueness is up to an additive constant $u = u_1 + \mathrm{cst}$; that is, only ∇u is unique.

Now to show the convergence in (b) whenever $R = 0$, note that $c_{tt} = D\nabla^2 c_t$ assuming $\alpha = 0$. This has a steady-state s.t. $c_t = \mathrm{cst.}$ everywhere. Also, $\int c_t = \int \rho_\Omega \mathrm{d}\Omega$ which shows that $c_t = R$.

Part (c). Let c_α satisfy Eq. (14.2) for $c_t = 0$ and a production function $\rho_\Omega^{(\alpha)}$. Then, $D\nabla^2 c_\alpha - \alpha c_\alpha = -\rho_\Omega^{(\alpha)}$. Suppose $c = Ac_\alpha + B$ satisfies the equation for some ρ_Ω. Since this c is unique, the following verification proves the claim.

$$\begin{aligned} & D\nabla^2 c - \alpha c = -\rho_\Omega = D\nabla^2(Ac_\alpha + B) - \alpha(Ac_\alpha + B) \\ \Rightarrow\ & AD\nabla^2 c_\alpha - A\alpha c_\alpha - \alpha B = -\rho_\Omega \\ \Rightarrow\ & A(-\rho_\Omega^{(\alpha)}) = -\rho_\Omega + \alpha B \\ \Rightarrow\ & \rho_\Omega = A(\rho_\Omega^{(\alpha)}) + \alpha B \end{aligned}$$

The other direction is derived similarly and the result follows. □

We now generalize from a single, constant diffusion constant D to one defined on edge pixels and the another on non-edge pixels. Note that the Laplacian operator now is sub-Riemannian; i.e., it is non-homogeneous in the diffusion constants.

Proposition 14.2. *Let Ω be a shape with two components $\Omega = \Omega_0 \cup \Omega_1$ such that $\Omega_0 \cap \Omega_1 = \partial\Omega_0$. Let D_0 and D_1 be the diffusion coefficients inside Ω_0 and Ω_1 respectively. If $\int_{\Omega_0} \rho_\Omega \mathrm{dv} + \int_{\Omega_1} \rho_\Omega \mathrm{dv} = 0$ and $\rho_\Omega(\Omega_0) = K \int_{\Omega_0} \mathrm{dv} > 0$, then*

$$\lim_{D_0/D_1 \to 0} c_\alpha = c_K$$

where c_K satisfies Theorem 14.1 for the shape Ω_0 by setting $c_K(\partial\Omega_0) = 0$.

Proof. The convergence of the system derives from Proposition 14.1(b). As $D_0/D_1 \to 0$ the relative speed of diffusion in Ω_1 increases to infinity. Thus, the concentration over Ω_1 will tend to a constant and, consequently, so will $c(\partial\Omega_0) = c(\Omega_0 \cap \Omega_1)$. The conditions of Theorem 14.1 are therefore satisfied and the claim follows. □

Theorem 14.2. *Suppose that Ω is a region in an image and that ρ_Ω takes a value of 1 at edge pixels and 0 everywhere else. Let the perimeter P be the number of edge pixels and the area A be the total number of pixels in Ω, i.e. $\int_\Omega \mathbf{d}\Omega = A$. Denote by $c_\infty = \lim_{\alpha\to 0} c_\alpha$ and assume that the diffusion coefficient between non-edge pixels $D = 1$ and that the diffusion coefficient between edge pixels is much larger than D. Then, for each pixel Q that is not an edge pixel*

$$|\nabla c_\infty(Q)| = \frac{P}{A} L \quad and \quad |\nabla^2 c_\infty(Q)| = \frac{P}{A}$$

Proof. The derivatives of c_∞ are well defined and unique as Proposition 14.1 shows. They are approximated by c_α to any precision provided that a sufficiently small α is chosen. Thus, given an arbitrary but fixed precision, suppose that α satisfies that requirement. According to Proposition 14.1(c), we may transform the production function by writing: $\rho_{new} = -\rho_\Omega + \alpha B$ where $\alpha B = \frac{P}{A}$. Thus, $\int_\Omega \rho_{new} \mathbf{d}\Omega = -\int_\Omega \rho_\Omega \,\mathbf{d}\Omega + \int_\Omega \frac{P}{A} \,\mathbf{d}\Omega = -P + P = 0$. Hence, according to Proposition 14.2, this transformed setup is equivalent to c_K where $K = \frac{P}{A}$ and the claims are true for c_K due to Theorem 14.1. The result for c_∞ follows from Proposition 14.1(c) by observing that $\nabla c_\infty = -\nabla c_K$. □

The gradient of concentration that emerges in this last result, and that scales with L, is precisely what was shown in Fig. 14.5.

14.4.1 Density Scale Space

Displaying the concentration function $c(x, y)$ reveals our "distance image". By varying the destruction constant, α, a scale space is created; notice how these capture several aspects of edge density (Fig. 14.6). Although this bears some resemblance to image scale spaces [20], there are fundamental differences. In particular, the interpretation of the gradient and the Laplacian of concentration in isoperimetric terms is completely novel.

To demonstrate that the Laplacian of concentration across scale can be useful for classification, we build a vector of four values of α as input to a standard linear classifier. The result for this toy experiment is in Fig. 14.7. Although experiments remain to be done on recognition of standard object databases, we believe these will demonstrate a role for distance images. The advantage of using the centroid already points in this direction. For computer vision applications it is more interesting to demonstrate that distance images are relevant for airport and other complex feature recognition.

14.5 Distance Images Support Airport Recognition

We now demonstrate our first application of distance images by applying them to airport recognition. Following a standard approach in machine learning, we use Laplacian eigenmaps [2] to reveal structure in a dataset of distance images from airports. (To remain focused on distance images, we ignore other sources of information which could also be useful.)

Laplacian eigenmaps are applied as follows. Let a data point be a vectorized patch of regular or distance images at three scales. The edge map was obtained with the Matlab Canny operator, and the scales were ($\alpha = 1.0, 0.01, 0.0033$; the large $\alpha = 1.0$ is essentially the original edge map); patch size $= (17 \times 17)$. Formally, let $X = \{x_1, x_2, \dots, x_N\}$ be the set of data points (typically N = 10,000), with each $x_i \in \Re^{n=867}$. We seek to find a projection of these data into much lower dimension, under the assumption that they are not randomly distributed throughout $\Re^n$ but rather that they lie on (or near) a lower-dimensional manifold embedded in $\Re^n$.

The structure of the data are revealed via a symmetric, positivity-preserving, and positive semi-definite *kernel* $k(x, y)$, which provides a measure of similarity between data points. (We use a Gaussian kernel, the value for which can be truncated to 0 for all but very similar points.) The intuition is that the natural structure among these "distance images" can be revealed by examining their low-dimensional embedding in significant eigenfunction coordinates. Then nearby points can be clustered to reveal airport structure.

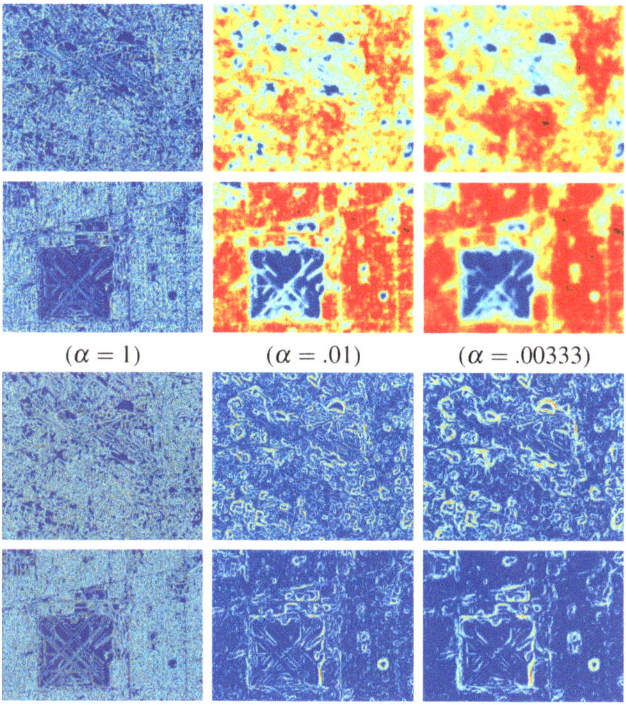

Fig. 14.6 (*Top*) A concentration scale space for edge density, computed according to Eq. (14.3). Note how decreasing α permits the "substance" to live longer and hence allow integration of information over a larger area. When α is large, in the limit the result is formally the edge map convolved against a small Gaussian. (*bottom*) The gradient of concentration. Notice how this concentrates "signals" about edge density very close to the edge locations

The diffusion map is obtained by the following Algorithm 1:

Algorithm 1 Given a set of n input image vectors $x_i \in \Re^d$

- `Step 1:` $K_0(i,j) \leftarrow e^{-\frac{\|x_i - x_j\|^2}{\sigma^2}}$;
- `Step 2:` $p(i) \leftarrow \sum_{j=1}^{n} K_0(i,j)$ approximates the density at x_i;
- `Step 3:` $\widetilde{K}(i,j) \leftarrow \frac{K_0(i,j)}{p(i)p(j)}$;
- `Step 4:` $d(i) \leftarrow \sum_{j=1}^{n} \widetilde{K}(i,j)$;
- `Step 5:` $K(i,j) \leftarrow \frac{\widetilde{K}(i,j)}{\sqrt{d(i)}\sqrt{d(j)}}$;
- `Step 6:` $USV^T = K$ (by SVD of K);

Steps 2 and 3 normalize for the density of sampling from the manifold, whereas Steps 4 and 5 perform the graph Laplacian normalization; see [5].

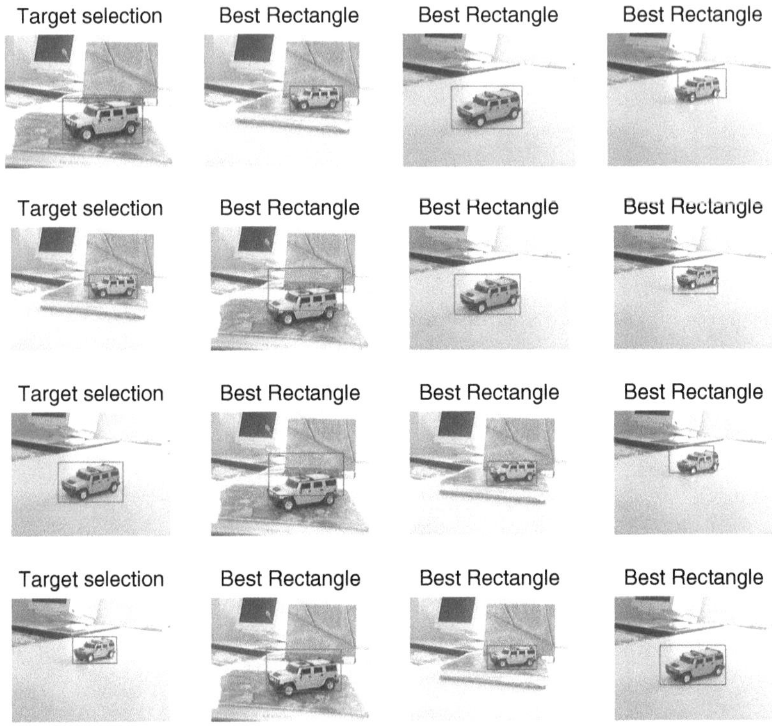

Fig. 14.7 Classification of the truck image with pose, orientation, and scale differences using the P/A measure, which is related to the Laplacian of concentration

The result of applying Algorithm 1 to the distance map images is illustrated in Fig. 14.8. Notice how a boomerang-shaped "manifold" is revealed, the basic coordinates of which are edge density (going along it) and edge orientation (going around it). These two dimensions codify our intuition that airports are defined by a certain collection of oriented edges (runways, etc.) arranged in a particular fashion relative to surrounding context.

To test the usefulness of the distance-images, we collected a set of 20 airport images from Google images by randomly spinning the globe, half of which were for training and half for testing. Our goal is to compare distance images against standard image patches for airport detection to demonstrate "proof of concept." Since normal image blur also collects information from a neighborhood around a point, our three-scale distance image patches were compared against the original image patch plus two blurred versions (comparable scales).

We interactively placed a rough outline around airports in the ten training images; patches from within this outline are the red points in Fig. 14.8 for the distance images. (The traditional blurred images are not shown.) To use this training information operationally, we built a characteristic function in embedded patch coordinates that defined "airport."

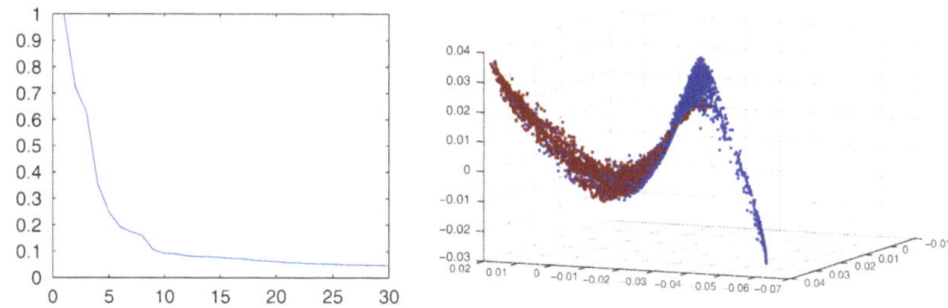

	1 patch	5 patches	10 patches	50 patches
DISTANCE mean	0.60 ± .48	0.62 ± .39	0.60 ± .34	0.55 ± .31
DISTANCE median	1.00	0.80	0.70	0.59
IMAGE mean	0.30 ± .48	0.32 ± .47	0.34 ± .46	0.48 ± .42
IMAGE median	0.00	0.00	0.01	0.16

Fig. 14.8 Structure of data patches after projection into diffusion coordinates (*top*) and classification results (*bottom*). (*left*) Eigenvalue spectrum shows that the first few eigenfunctions capture most of the structure. (*right*) Data patches plotted in top three eigenfunction coordinates; note "manifold" from which they are drawn. The primary coordinate captures edge density; the next dimensions information about orientation. *Red* points are patches from airport training images; they cluster around the sparse end of the "manifold" which defines a function on the embedded patch data that defines "airport." Nystrom extensions of new patches onto these coordinates determine if they fall in the (*red*) airport cluster. (*bottom*) Table showing performance (mean and median of classification for DISTANCE image patches vs IMAGE blur patches). Columns are the number of patches tested and entries show the airport fraction correctly classified

To test the quality of these patches for finding airports in test images, we used ten new images. The airport characteristic function was then Nystrom extended onto the embedding of the new patch and scored according to whether or not it was in the airport lobe. The results over all patches (5,000 patches/image; training set = 10,000 patches) are shown in the table (Fig. 14.8 (bottom)). Counting the mean and the median number of patches that were correctly scored as airport shows that the distance images significantly outperformed the intensity information on this task; some sample results for both cases are shown in Fig. 14.9. Although the results need to be repeated with larger datasets, in this experiment the blurred image scale space was totally ineffective.

14.6 The Enclosure Field Conjecture

In the early sections of this paper we reviewed the neurobiological notion of border ownership and the (associated) psychophysical notion of closure. Both involved a kind of "feedback" of global shape information onto the computation of local border

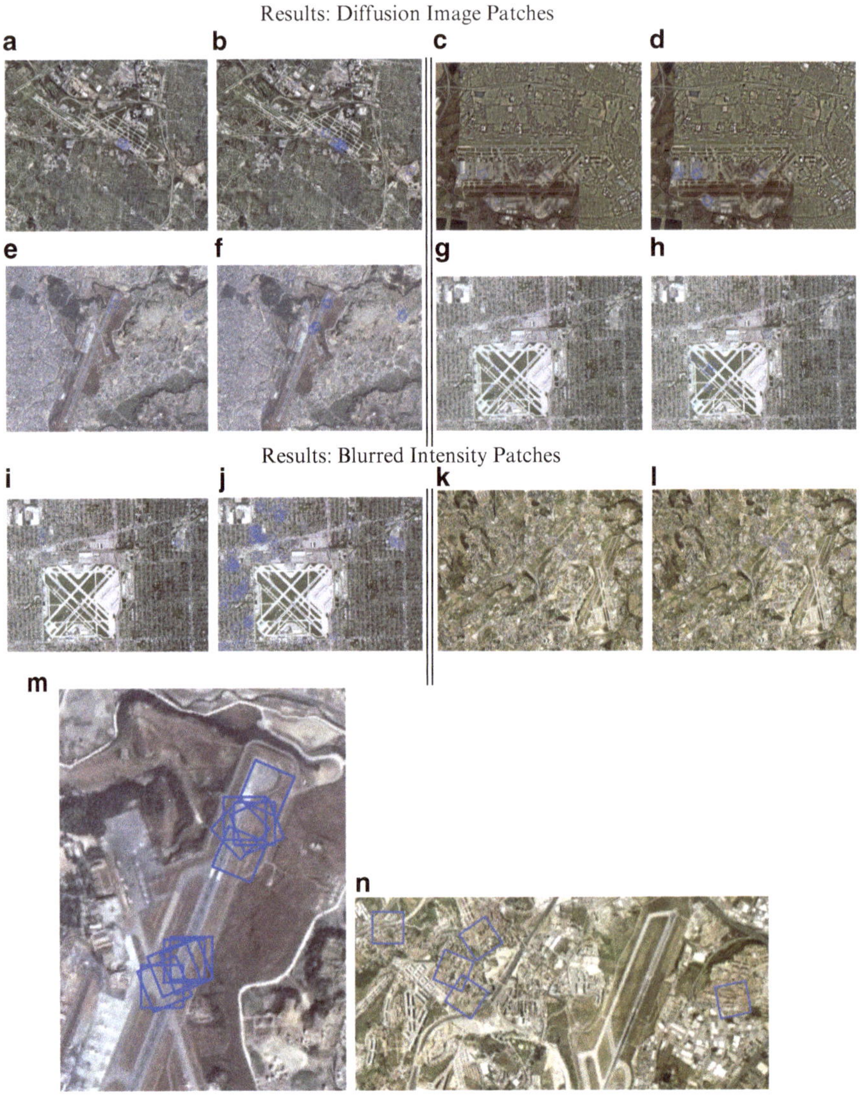

Fig. 14.9 Results of airport classification experiment. Patches are shown as boxes superimposed on the original images. Results are illustrated with two images: the first 5 patches on the *left* and the next 20 on the *right*, for each example. (*top*) DISTANCE image results for four example airport images. (*middle*) Blurred INTENSITY image results for two airport images, also shown as pairs. (*bottom*) Enlargements for viewing: (**m**) is a zoom on (**f**) and (**n**) is a zoom on (**l**). Note how the DISTANCE image boxes fall on/near the airports, but the INTENSITY image boxes do not

inferences. With the background we have now developed, it becomes possible to assert that there may be a relationship between the equilibrium solution of the distance pde and neural computation. This is our second application of the distance image idea. To develop it, we first briefly describe a traditional view, which serves as a basis for the more elaborate conjecture.

14.6.1 *Inferring Coherent Borders*

As shown in Fig. 14.1, visual cortex contains neurons selective for local edge orientation. These cells are arranged in networks that could implement a geometric notion of border coherence. Since the large number of neurons that are sensitive to boundary segments cover all (discretely sampled) positions and orientations, these can be arranged as *columns* of cells selective for all orientations at a point; then interactions between the cells in these columns could excite those cells that form smooth boundaries and inhibit others. Since borders of objects are mostly smooth, locally defined orientations can be interpreted as contour tangents. Fitting tangents together requires a notion of curvature: how much does the tangent *rotate* as it is transported from one position along a curve to a nearby one. One model [3] for such curvature-mediated connections is outlined in Fig. 14.10, where the circuits running between columns reinforce one another to provide the "flow" of tangents along the curve.

Such models of neural computation provide consistent local edge elements, but do not yet incorporate the more global inferences involved in glueing them together to form shapes. Nevertheless, there is substantial evidence that border ownership and other shape related signals are available as early as V2 and, in some cases, V1 [22,23,29,43].

To establish the shape context feedback, there are two possibilities. First, one can assert that the same class of long-range horizontal connections incorporated for border consistency above provide the signals, as in [34,41], but the timing of border ownership signals argues strongly against this class of model; see [7]. Instead they (and others, e.g. [33]) argue that higher-level vision areas are involved. Cells in the superficial layers of V1, for example, project to V2, and V2 projects to V4 (among other areas). Finally V4 projects to inferotemporal (IT) cortex, where much of visual shape analysis is believed to reside. Most importantly, there is a back projection for each of these forward projections; see Fig. 14.11a. We now turn to a description of such feedback.

14.6.2 *Feedback Projections via Specialized Interneurons*

Feedback projections could provide the more global information, and these are shown in Fig. 14.11 as arriving in layer I and forming synapses with apical dendrites

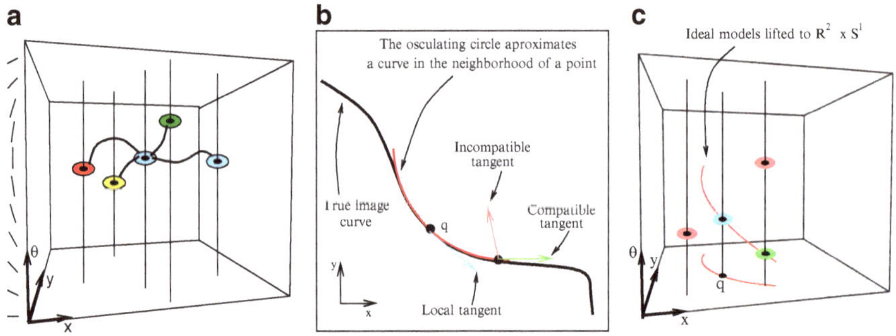

Fig. 14.10 The relationship between individual orientationally-selective neurons and coherent responses along a boundary can be developed geometrically. (**a**) The superficial (*upper*) layers of visual cortex, V1, can be abstracted via orientation columns. These consist of groups of neurons (a column) selective for every possible orientation at each position in the visual array. These columns are denoted by the vertical lines, indicating that at each (x,y)-position there are all (θ)-orientations. Long range horizontal connections define circuits among these neurons (there are also many more local circuits) so that consistent firing among those neurons in such excitatory circuits specifies the activity along a putative contour. (**b**) Differential geometry specifies how orientations align. Interpreting the orientationally-selective cell's response as signaling the tangent to a curve, this tangent can be transported along an approximation to the curve (indicated as the osculating *circle*) to a nearby position. Compatible tangents are those that agree with sufficient accuracy in position and orientation following transport. (**c**) The transport operation can be embedded in long range connections. Here this is shown as the "lift" of an arc of (osculating) *circle* in the (x,y)-plane into a length of helix in (x, y, θ) coordinates. Many models of border ownership are based on similar ideas, although it is the topological orientation (toward inside or outside of the figure) that is communicated via the long-range horizontal projections (Figure after [3])

of other neurons. To incorporate such signals, however, specialized interneurons must be postulated to accomplish the various "shape" functions. Von der Heydt's group, for example, postulates a kind of skeleton neuron [7] that integrates boundary information from far away, but there are no data supporting the existence of such cells. In effect these so-called "grouping cells" implement a receptive field that integrates boundary information within an annular distance; different cells are tuned to different distances. This specialized circuitry implements a portion of the skeleton computation by measuring a value for the distance map in a quantized fashion. However, the specificity of these grouping cells makes this approach difficult to organize for arbitrary, general figures.

A second way to organize the feedback is via surface computations, and this approach is developed in [33]. While there is little doubt that various surface properties are inferred in the higher-level areas, and while these must be integrated with stereo and border ownership [31], this amounts to a much bigger solution than is required for establishing border ownership. As we show next, by extending the class of neural codes it becomes possible to integrate our theory of the distance image for this purpose.

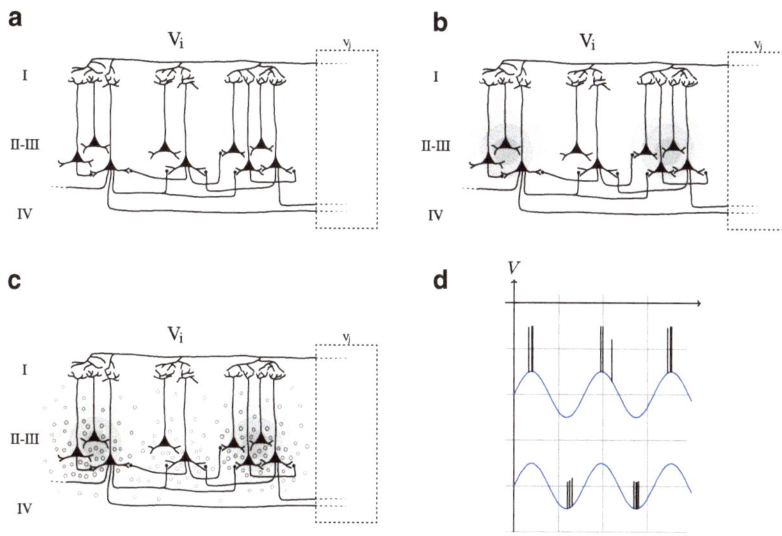

Fig. 14.11 Traditional and elaborated views of neural coding that could underlie neural computation. Neurons are shown as small triangles, with both lateral and apical dendrites taking inputs from other neurons. Axons are shown forming both local circuits, and as projecting up to higher visual areas and projecting back from these areas. We concentrate on neural circuits in the superficial layers *(II-III)*, and note these project to layer *IV* in higher areas and receive backprojections in layers *I* and *II-III* from those areas. (**a**) Networks of neurons in visual area V_i could infer consistent boundary signals, and these could project to higher visual areas (V_j), shown as a rectangle, in which shape (or other higher-level) properties are inferred. The result then projects back to the lower area (via layer *I*) to influence the earlier computation. This is also the way that neural circuits are being developed for border ownership, except that now different classes of neurons, in addition to the orientationally-selective ones, are postulated. (**b**) An elaborated view in which the local field potential surrounding each neuron is taken into account. This field potential could correspond to the "distance" signal in the paper. (**c**) An even more elaborated view of the anatomy, in which glial cells (shown as *small circles*) participate in establishing the lfp. (**d**) Illustration that the firing rate of neurons is modulated by local field potentials. The lfp is cartooned as a sinusoid (*blue curves*), indicating that it fluctuates in time. Action potentials are shown as *black lines*. The *upper trace* shows a neuron that tends to fire when the lfp is high, and the *lower trace* shows a neuron that tends to fire when it is low

14.6.3 Local Field Potentials Carry the Enclosure Field

The model of distance images suggests a different tack. It builds on the idea of signals propagating, but through the physical tissue as well as through the neural connections. And it incorporates a new type of neural code as well. Basically the backward projection is regularized by the diffusion term, which takes care of broken boundaries and blurs the requirement that different shapes must be considered individually.

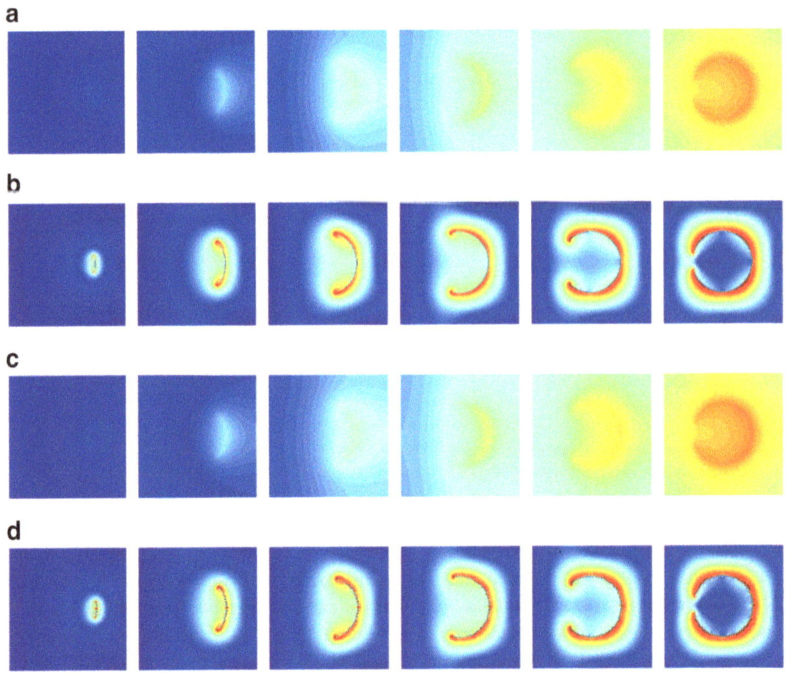

Fig. 14.12 Examples of the enclosure field computation (distance images) for the development of a circle. The panels along each row contain an increasing fraction of the boundary of a *circle*, showing the build up of concentration on the "inside" of the circle, even for short segments. Concentration here is taken as a proxy for local field potential. (**a**) Plot of concentration for a complete set of "producing" pixels defining the *circle* fragment. (**b**) The gradient of the concentration; this provides explicit information about the density and (rough) arrangement of boundary activity some distance away. (**c**) Concentration field for 20 % of the pixels in (**a**); (**d**) gradient of concentration for 20 % of the pixels. The similarity between (**a**) and (**c**), and (**b**) and (**d**) illustrates how the lfp regularizes differences in edge arrangement as required for border ownership. Standard "heat" color scale

Neurons are not situated in isolation; rather, they are located in tissue that provides a conductive substrate. Simplifying enormously (see e.g. [30]), current passes into both synapses and the physical substrate giving rise to an extra-cellular potential; these potentials can be recorded with extracellular electrodes. When low-pass filtered to remove the (high frequency) action potential component, the result is called a *local field potential*, and it reflects both processing in the local network and input fluctuations; Fig. 14.11b. Most importantly for us, there are neurons whose activity is more robust at certain points on the lfp; such phase of firing codes are reviewed in [26]; see Fig. 14.11d.

Our conjecture can now be stated: *the arrangement of (boundary-based) neural activity gives rise to a component of the local field potential that modulates border-ownership neurons. This activity could arise from feedback projections as well as local neurons.*

We stress that this is a novel view of neural computation, in which the lfp signal carries different information (coding distant activity about shape) from the spiking network signal (about local edges); the resulting system requires both.

The connection between this idea and distance maps is that the concentration equation derives from neurons embedded in a linear resistive medium:

$$c_t = \underbrace{\rho}_{\text{charge/ion injection}} + \underbrace{D\nabla^2 c}_{\text{diffusion}} - \underbrace{\alpha c}_{\text{resistive loss}} \tag{14.3}$$

Clearly neural tissue is non-linear so that more biophysically plausible models will be necessary, but this is sufficient to at least illustrate the principles; see Fig. 14.12. Finally, we note that the glial cells, which richly surround the pyramidal neurons and outnumber them by perhaps an order of magnitude, are now known to have channels that allow them to participate in regulating the lfp. This further suggests an important computational role for it. See [44].

14.7 Summary and Conclusions

We developed a structural coupling between local edge information and more global arrangement information by postulating a pde whose equilibria provided local signals about global properties of the distance map. A computer vision application of these ideas to locating airports demonstrated that these properties are useful, and a biological interpretation suggested a richer view of neural computation involving the local field potential for border ownership.

Much more remains to be done with both the recognition of complex structures and understanding border ownership. Nevertheless, by considering these two problems together we emphasize how ideas from biology can influence computer vision, and vice versa.

Acknowledgements The computer vision content in this paper was in [11]; I thank P. Dimitrov and M. Lawlor for permission to reuse that material here. S. Tari pointed out the possible connection to Ambrosio-Tortorelli. Research supported by AFOSR, ARO, NIH/NIAAA and NSF.

References

1. Ambrosio, L., Tortorelli, V.: On the approximation of functionals depending on jumps by elliptic functionals via? γ-convergence. Commun. Pure Appl. Math. **43**(8), 999–1036 (1990)
2. Belkin, M., Niyogi, P.: Laplacian eigenmaps for dimensionality reduction and data representation. Neural Comput. **6**(15), 1373–1396 (2003)
3. Ben-Shahar, O., Zucker, S.W.: Geometrical computations explain projection patterns of long-range horizontal connections in visual cortex. Neural Comput. **16**(3), 445–4476 (2003)
4. Borenstein, E., Ullman, S.: Combined top-down/bottom-up segmentation. IEEE Trans. Pattern Anal. Mach. Intell. **30**(12), 2109–2125 (2008)

5. Coifman, R., Lafon, S., Lee, A., Maggioni, M., Nadler, B., Warner, F., Zucker, S.W.: Geometric diffusions as a tool for harmonic analysis and structure definition of data: Diffusion maps. Proc. Nat. Acad. Sci. USA **102**, 7426–7431 (2005)
6. Courant, R., Hilbert, D.: Methods of Mathematical Physics, vol. 2. Interscience, New York (1962)
7. Craft, E., Schutze, H., Niebur, E., von der Heydt, R.: A neural model of figure-ground organization. J Neurophysiol. **97**(6), 4310–4326 (2007)
8. Dimitrov, P., Zucker, S.W.: A constant production hypothesis that predicts the dynamics of leaf venation patterning. Proc. Nat. Acad. Sci. USA **13**(24), 9363–9368 (2006)
9. Dimitrov, P., Zucker, S.W.: Distance maps and plant development #1: Uniform production and proportional destruction. arXiv.org, arXiv:0905.4446v1 (q-bio.QM), 1–39 (2009)
10. Dimitrov, P., Zucker, S.W.: Distance maps and plant development #2: Facilitated transport and uniform gradient. arXiv.org, arXiv:0905.4662v1 (q-bio.QM)(24), 1–46 (2009)
11. Dimitrov, P., Lawlor, M., Zucker, S.W.: Distance images and intermediate-level vision. In: Bruckstein, A.M., ter Haar Romeny, B.M., Bronstein, A.M., Bronstein, M.M. (eds.). Third International Conference on Scale Space and Variational Methods in Computer Vision. Lecture Notes in Computer Science, vol. 1, pp. 653664. Springer (2011)
12. Dubuc, B., Zucker, S.W.: Complexity, confusion, and perceptual grouping. part i: the curve like representation. Int. J. Comput. Vis. **42**(1), 55–82 (2001)
13. Elder, J., Zucker, S.W.: Contour closure and the perception of shape. Vis. Res. **33**(7), 981–991 (1993)
14. Ferrari, V., Jurie, F., Schmid, C.: From images to shape models for object detection. Int. J. Comput. Vis. **87**(3), 284–303 (2010)
15. Fidler, S., Leonardis, A.: Towards scalable representations of object categories: learning a hierarchy of parts. In: Proceedings of the 2007 IEEE Computer Society Conference on Computer Vision and Pattern Recognition (CVPR 2007), 1823 June 2007, Minneapolis, MN IEEE Computer Society (2007)
16. Florack, L.M.J., ter Haar Romeny, B.M., Viergever, M.A., Koenderink, J.J.: The Gaussian scale-space paradigm and the multiscale local jet. Int. J. Comput. Vis. **18**(1), 61–75 (1996)
17. Froyen, V., Feldman, J., Singh, M.: A bayesian framework for figure-ground interpretation. Adv. Neural Inf. Process. Syst. **23**, 631–639 (2010).
18. Gregory, R.L.: Eye and Brain. McGraw Hill, New York (1966)
19. Hubel, D.H., Wiesel, T.N.: Functional architecture of macaque monkey visual cortex. Proc. R. Soc. Lond. B **198**, 1–59 (1977)
20. Koenderink, J.J.: The structure of images. Biol. Cybern. **50**, 363–370 (1984)
21. Koffka, K.: Principles of Gestalt Psychology. Harcourt, Brace & World, New York (1935)
22. Lamme, V.A.F.: The neurophysiology of figure ground segregation in primary visual cortex. J. Neurosci. **15**, 1605–1615 (1995)
23. Lee, T.S., Mumford, D., Romeo, R., Lamme, V.A.F.: The role of the primary visual cortex in higher level vision. Vis. Res. **38**, 2429–2454 (1998)
24. Lowe, D.: Distinctive image features from scale-invariant keypoints. Int. J. Comput. Vis. **60**, 91–110 (2004)
25. McKeown, D.M., Harvey, W.A., McDermott, J.: Rule-based interpretation of aerial imagery. IEEE Trans. Pattern Anal. Mach. Intell. **7**(5), 570–585 (1985)
26. Montemurro, M.A., Rasch, M.J., Murayama, Y., Logothetis, N.K., Panzeri, S.: Phase-of-firing coding of natural visual stimuli in primary visual cortex. Curr. Biol. **18**(5), 375–380 (2008)
27. Opelt, A., Pinz, A., Zisserman, A.: A boundary-fragment-model for object detection. In: Leonardis, A., Bischof, H., Pinz, A. (eds.) ECCV (2). Lecture Notes in Computer Science, vol. 3952, pp. 575–588. Springer, Berlin/Heidelberg (2006)
28. Opelt, A., Pinz, A., Zisserman, A.: Learning an alphabet of shape and appearance for multi-class object detection. Int. J. Comput. Vis. **80**(1), 16–44 (2008)
29. Orban, G.A.: Higher order visual processing in macaque extrastriate cortex. Physiol. Rev. **88**(1), 59–89 (2008)
30. Plonsey, R.: Bioelectric Phenomena. McGraw-Hill, New York (1969)

31. Qiu, F., von der Heydt, R.: Figure and ground in the visual cortex: V2 combines stereoscopic cues with gestalt rules. Neuron **47**, 155–166 (2005)
32. Rubin, E.: Synsoplevede Figurer. Gyldendalske Boghandel, Nordisk Forlag, Denmark (1915)
33. Sajda, P., Finkel, L.: Intermediate-level visual representations and the construction of surface perception. J. Cognit. Neurosci. **7**, 267–291 (1995)
34. Sakai, K., Nishimura, H.: Determination of border ownership based on the surround context of contrast. Neurocomputing **58**, 843–848 (2004)
35. Siddiqi, K., Shokoufandeh, A., Dickinson, S.J., Zucker, S.W.: Shock graphs and shape matching. Int. J. Comput. Vis. **35**(1), 13–32 (1999)
36. Tari, S., Genctav, M.: From a modified ambrosio-tortorelli to a randomized part hierarchy tree. In: Third International Conference on Scale Space and Variational Methods in Computer Vision, vol. 1. Springer, Berlin/Heidelberg (2011)
37. Treisman, A., Gelade, G.: A feature-integration theory of attention. Cognit. Psychol. **12**(1), 97–136 (1980)
38. Ullman, S.: Visual routines. In: Fischler, M.A., Firschein, O. (eds.) Readings in Computer Vision: Issues, Problems, Principles, and Paradigms, pp. 298–328. Morgan Kaufmann, San Francisco (1987)
39. Ullman, S., Epshtein, B.: Visual classification by a hierarchy of extended fragments. In: Ponce, J., Hebert, M., Schmid, C., Zisserman, A. (eds.) Toward Category-Level Object Recognition. Lecture Notes in Computer Science, vol. 4170, pp. 321–344. Springer, Berlin/New York (2006)
40. Ward, L.M., Coren, S.: The effect of optically induced blur on the magnitude of the mueller-lyer illusion. Bull. Psychon. Soc. **7**(5), 483–484 (1976)
41. Zhaoping, L.: Border ownership from intracortical interactions in visual area v2. Neuron **47**, 143–153 (2005)
42. Zhou, H., Friedman, H.S., von der Heydt, R.: Coding of border ownership in monkey visual cortex. J. Neurosci. **20**, 6594–6611 (2000)
43. Zipser, K., Lamme, V.A.F., Schiller, P.H.: Contextual modulation in primary visual cortex. J. Neurosci. **16**(22), 7376–7389 (1996)
44. Zucker, S.W.: Local field potentials and border ownership: a conjecture about computation in visual cortex. J. Physiol. (Paris) **106**, 297–315 (2012)

Part III
Optimization Methods for Shape Analysis

Chapter 15
Non-rigid Shape Correspondence Using Pointwise Surface Descriptors and Metric Structures

Anastasia Dubrovina, Dan Raviv, and Ron Kimmel

Abstract Finding a correspondence between two non-rigid shapes is one of the cornerstone problems in the field of three-dimensional shape processing. We describe a framework for marker-less non-rigid shape correspondence, based on matching intrinsic invariant surface descriptors, and the metric structures of the shapes. The matching task is formulated as a quadratic optimization problem that can be used with any type of descriptors and metric. We minimize it using a hierarchical matching algorithm, to obtain a set of accurate correspondences. Further, we present the correspondence ambiguity problem arising when matching intrinsically symmetric shapes using only intrinsic surface properties. We show that when using isometry invariant surface descriptors based on eigendecomposition of the Laplace-Beltrami operator, it is possible to construct distinctive sets of surface descriptors for different possible correspondences. When used in a proper minimization problem, those descriptors allow us to explore a number of possible correspondences between two given shapes.

15.1 Introduction

Three-dimensional shape processing became increasingly popular in the last decade. One of its corner-stone tasks is detecting a correspondence between two given shapes. It is essential for shape comparison, retrieval, shape morphing and deformation, or shape calculus [5], etc. The most interesting yet complex task is automatic non-rigid shape matching. In this work we address the problem of

A. Dubrovina (✉) · D. Raviv
Technion, Israel Institute of Technology, Haifa, Israel
e-mail: nastyad@cs.technion.ac.il,darav@cs.technion.ac.il

R. Kimmel
Technion, Department of Computer Science, Israel Institute of Technology, Haifa, Israel
e-mail: ron@cs.technion.ac.il

M. Breuß et al. (eds.), *Innovations for Shape Analysis*, Mathematics and Visualization, DOI 10.1007/978-3-642-34141-0_15, © Springer-Verlag Berlin Heidelberg 2013

matching non-rigid approximately isometric shapes. We perform the matching using certain surface properties that remain invariant under isometric transformations. In particular, we use two types of such properties – pointwise surface descriptors, and distances measured between pairs of points on the surface. We show how these two properties can be incorporated into a measure of dissimilarity between the shapes, which can be written as a quadratic function of the correspondence. We then minimize this dissimilarity measure in order to find the minimal dissimilarity correspondence.

Another important issue we address here is the correspondence ambiguity present when matching intrinsically symmetric shapes. In this case, there may exist several correspondences minimizing the proposed dissimilarity measure. We show that this ambiguity can be resolved by constructing distinct sets of symmetry-aware surface descriptors. By employing them within the proposed framework it is possible to find several matchings between the shapes.

The rest of the paper is organized as follows: in the next section we review the related work on matching non-rigid shapes. In Sect. 15.3 we describe the proposed problem formulation. In Sect. 15.4 we describe the possible choices of metric and descriptors. In Sect. 15.5 we describe the correspondence ambiguity problem and the construction of the symmetry-aware surface descriptors. In Sect. 15.6 we present the matching results obtained with the proposed framework combined with different descriptors and distances measures. We summarize the paper and discuss future research directions in Sect. 15.7.

15.2 Related Work

Zigelman et al. [43] and Elad and Kimmel [9] suggested a method for matching isometric shapes by embedding them into a Euclidian space using multidimensional scaling (MDS), thus obtaining isometry invariant representations, followed by rigid shape matching in that space. Since it is generally impossible to embed a non-flat 2D manifold into a flat Euclidean domain without introducing some errors, the inherited embedding error affects the matching accuracy of all methods of this type. For that end, Jain et al. [13], Mateus et al. [19] and Sharma and Horaud [34] suggested alternative isometry-invariant shape representations, obtained by using eigendecomposition of discrete Laplace operators. The Global Point Signature (GPS) suggested by Rustamov [33] for shape comparison employs the discrete Laplace-Beltrami operator, which, at least theoretically, captures the shape's geometry more faithfully. The Laplace-Beltrami operator was later employed by Sun et al. [35], and Ovsjanikov et al. [25], to construct their Heat Kernel Signature (HKS) and Heat Kernel Maps, respectively. Zaharescu et al. [41] suggested an extension of 2D descriptors for surfaces, and used them to perform the matching. While linear methods, such as [25, 41] produce good results, once distortions start to appear, ambiguity increases, and alternative formulations should be thought of. Adding the proposed approach as a first step in one of the above linear dense matching algorithms can improve the final results. Hu and Hua [12] used the Laplace-Beltrami operator for matching using prominent features, and Dubrovina and Kimmel [8] suggested employing surface descriptors based on its eigendecomposition, combined with geodesic distances, in a quadratic optimization formulation of the matching problem. The above methods, incorporating pairwise constraints, tend to be slow due to high computational complexity. Wang et al. [40] used a similar problem formulation, casted as a graph labeling problem, and experimented with different surface descriptors and metrics.

Memoli and Sapiro [22], Bronstein et al. [4], and Memoli [20, 21] compared shapes using different approximations of the Gromov-Hausdorff distance [10]. Bronstein et al. [6] used the approach suggested in [4] with diffusion geometry, in order to match shapes with topological noise, and Thorstensen and Keriven [37] extended it to handle surfaces with textures. The methods in [20–22] were intended for surface comparison rather than matching, and as such they do not produce correspondence between shapes. At the other end, the GMDS algorithm [6] results in a non-convex optimization problem, therefore it requires good initializations in order to obtain meaningful solutions, and can be used as a refinement step for most other shape matching algorithms. Other algorithms employing geodesic distances to perform the matching were suggested by Anguelov et al. [1], who optimized a joint probabilistic model over the set of all possible correspondences to obtain a sparse set of corresponding points, and by Tevs et al. [36] who proposed a randomized algorithm for matching feature points based on geodesic distances between them. Zhang et al. [42] performed the matching using extremal curvature feature points and a combinatorial tree traversal algorithm, but its high complexity allowed them to match only a small number of points. Lipman and Funkhouser [18] used the fact that isometric transformation between two shapes is equivalent to a Möbius transformation between their conformal mappings, and obtained this transformation by comparing the respective conformal factors. However, there is no guarantee that this result minimizes the difference between pairwise geodesic distances of matched points.

Self-similarity and symmetry detection are particular cases of the correspondence detection problem. Instead of detecting the non-rigid mapping between two shapes, [14, 17, 24, 28] search for a mapping from the shape to itself, and thus are able to detect intrinsic symmetries.

15.3 Matching Problem Formulation

The suggested problem formulation is based on comparison of shape properties that remain approximately invariant under non-rigid ϵ-isometric transformations, specifically – distances between the points on the shape, and pointwise surface descriptors defined at every point of the shape. We assume to be given shapes represented by sampled surfaces, which is one of the common 3D shape representations.

In this work shapes were represented by triangular meshes, but the following discussion is not limited to some specific sampled surface representation.

Given a shape X, we assume that is endowed with a distance measure $d_X : X \times X \to \mathbb{R}^+ \cup \{0\}$, and a set of pointwise d-dimensional surface descriptors $f_X : X \to \mathbb{R}^d$. Given two shapes X and Y, we define a correspondence between them by a mapping $P : X \times Y \to \{0, 1\}$, such that

$$P(x, y) = \begin{cases} 1, & x \in X \text{ corresponds to } y \in Y, \\ 0, & \text{otherwise} \end{cases} \tag{15.1}$$

We can measure the dissimilarity introduced by the mapping P into the surface descriptors and the metric structures by

$$\text{dis}(P) = \sum_{\substack{x \in X \\ y \in Y}} \| f_X(x) - f_Y(y) \|_F \, P(x, y) + \alpha \cdot \sum_{\substack{x, \tilde{x} \in X \\ y, \tilde{y} \in Y}} |d_X(x, \tilde{x}) - d_Y(y, \tilde{y}))| \, P(x, y) P(\tilde{x}, \tilde{y}), \tag{15.2}$$

where $\|\cdot\|_F$ is a norm in the descriptor space. The first term of the dissimilarity measure is a linear function of the mapping P, and it expresses the pointwise surface descriptor dissimilarity. This term provides a The second term of dis(P) is a quadratic function of the mapping P, and it expresses the metric structure dissimilarity. The parameter $\alpha \geq 0$ determines the relative weight of the second term in the total dissimilarity measure.

Note that by setting $\alpha = 0$ we obtain the linear matching method used by [25, 41]. When the descriptors of different points on the shape are not unique or sufficiently distinct, say due to numerical inaccuracies, the correspondences obtained by minimizing only the linear part of the dissimilarity measure may not be consistent in terms of pairwise relationships between f the matched points. By adding the quadratic term in Eq. (15.2) we ensure that the optimal correspondence preserves also these pairwise relationships. On the other hand, by choosing $\alpha \gg 1$ we obtain a problem similar to the one addressed in [4, 20, 22], and, since the problem is non-convex, a good initialization is required in order to obtain a close-to-global minimizer. This is achieved by adding the linear term as in Eq. (15.2).

The optimal matching, which we denote by P^*, is obtained by minimizing the dissimilarity

$$P^* = \text{argmin}_{P: X \times Y \to \{0,1\}} \{\text{dis}(P)\}. \tag{15.3}$$

In order to avoid a trivial solution $P^* \equiv 0$, we constrain P to the space of valid correspondences. Note that the above problem formulation allows us to consider different types of possible correspondences between the two shapes. For example, when a bijective mapping from X to Y is required, the constraints on P are

$$\sum_{x \in X} P(x, y) = 1, \forall y \in Y, \quad \sum_{y \in Y} P(x, y) = 1, \forall x \in X. \tag{15.4}$$

For a surjective mapping we relax the constraints to be

$$\sum_{x \in X} P(x, y) = 1, \forall y \in Y. \tag{15.5}$$

Thus, the resulting optimization problem is

$$P^* = \text{argmin}_{P:X \times Y \to \{0,1\}} \{\text{dis}(P)\} \quad s.t. \quad \text{suitable constrains on P.} \tag{15.6}$$

15.3.1 Quadratic Programming Formulation

When the two shapes X and Y are represented by a finite number of points N, the mapping P is a binary matrix of size N^2. In order to convert the problem into a more convenient quadratic programming form we reshape the matrix P by taking its columns and concatenating them, thus obtaining a vector p of dimension N^2, referred to as a *correspondence vector*. Thus, k-th entry of the vector p corresponds to some entry (i, j) in the matrix P – we will denote these corresponding indices by (i_k, j_k). The vector entry p_k encodes the correspondence between the points x_{i_k} and y_{j_k}.

Similarly, we introduce the following notations for the metric and the descriptor dissimilarity

$$b_k = \left\| f_X(x_{i_k}) - f_Y(y_{j_k}) \right\|_F \quad , \quad Q_{kl} = \left\| d_X(x_{i_k}, x_{m_k}) - d_Y(y_{j_l}, y_{n_l})) \right\| . \tag{15.7}$$

The vector $b \in \mathbb{R}^{N^2}$ represents the pointwise descriptor dissimilarity, and the matrix $Q \in \mathbb{R}^{N^2 \times N^2}$ represents the metric structure dissimilarity.

Lastly, we re-write optimization problem (15.6) in the quadratic programming form

$$p^* = \text{argmin}_{p \in \{0,1\}^{N^2}} \left\{ b^T p + \lambda \cdot p^T Q p \right\} \quad \text{s.t.} \quad Sp = \mathbb{1}, \tag{15.8}$$

where $Sp = \mathbb{1}$ is the matrix form of the constraints in Eqs. (15.4) or (15.5).

15.3.2 Hierarchical Matching

The optimization problem in Eq. (15.8) belongs to the class of NP-hard Integer Quadratic Programming (IQP) problems. There exist different techniques for approximating its solution, [2, 38] among them, which are able to solve only moderate size IQPs. The implication on the matching problem is that the algorithm will be able to find only small number of correspondences – up to several tens

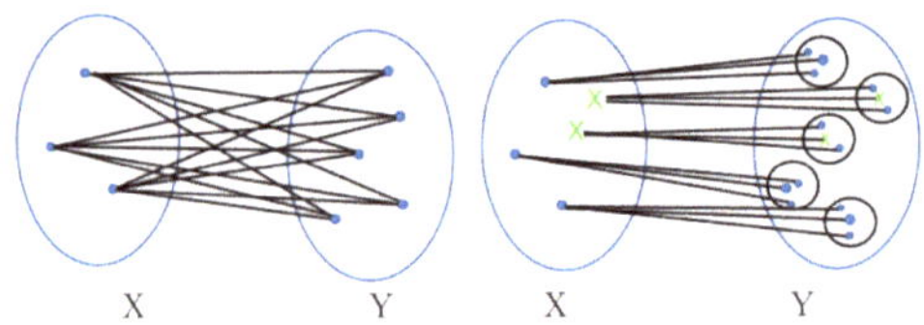

Fig. 15.1 In the first step (*left*) we construct a quadratic correspondence matrix from all points in X into all points in Y. In each iteration (*right*) we search for possible matches between points in X from the previous iteration (*blue circle*) and new sampled points in X (*green* Xs) and their corresponding neighborhoods (*black circles*) in Y

of points. In addition, prior to the matching the algorithm has to choose the initial set of N candidate points on each one of the shapes X and Y. The simplest way to choose these points is by using the Farthest Point Sampling technique [11], and the sampling density will determine the accuracy of the matching.

In order to overcome these limitations we use the hierarchical matching technique introduced in [29]. It exploits the shapes' geometric structures to reduce the number of potential correspondences, and thus is able to find a denser matching, with improved accuracy. Since the problem is not strictly combinatorial by nature, but rather derived from a smooth geometric measure. At the first step we follow [8] and solve (15.6) using a branch-and-bound procedure [2]. Each point $x \in X$ is now matched to a point $p(x) \in Y$ by the mapping P. We denote $y = p(x)$ if $P(x, y) = 1$. In each iteration we search for the best correspondence between x and $p(x)$ neighborhood, instead of all points $y \in Y$, in a manner similar to [39]. Between iterations we add points $x \in X$ and $y \in Y$ using the two-optimal *Farthest Point Sampling* (FPS) strategy [11], evaluate the neighborhood in Y of the new points, reevaluate the neighborhood of the old points, and continue until convergence. In Fig. 15.1 we show a diagram of the process.

We solve the relaxed version of (15.6), using quazi-Newton optimization, and project the solution to integers between iterations. Convergence is guaranteed, but only to a local minimum, as for all QAP problems.

A different approach for approximating the solution of the IQP in Eq. (15.8) can be, for instance, using the relaxation technique of Bronstein et al. [4] and solving the problem on a continuous domain. The optimization problem can also be solved using the approach for graph matching by Torresani et al. [38]. Both can reduce the complexity of the solution. We will explore these directions in the future research.

15.4 On the Choice of Metric and Descriptors

The above formulation of the matching problem can be used with any type of surface descriptors or distance measure. Below we describe different descriptors and metrics that can be employed in the proposed framework.We start with a brief review of the Laplace-Beltrami operator, and later use concepts related to it for both metric and descriptor definition. Note that both metric definitions and some of the descriptor definitions are given in terms of continuous surface representation (or 2D Riemannian manifolds). For each one of them we state the discrete approximation we used for numerical evaluation.

15.4.1 Laplace-Beltrami Operator

The Laplace-Beltrami operator is a generalization of the Laplacian operator from flat domain to compact Riemannian manifolds. Given a manifold M, its Laplace-Beltrami operator Δ_M is given by

$$\Delta_X f = -\mathrm{div}_X\left(\nabla_X f\right), \quad f : X \to \mathbb{R}. \tag{15.9}$$

The divergence and the gradient operators, div_X and ∇_X respectively, are defined by the intrinsic geometry of the manifold X. Explicitly, the Laplace-Beltrami operator of a function $f : X \to \mathbb{R}$ defined on the manifold X equipped with a Riemannian metric g is given by

$$\Delta_X f = -\frac{1}{\sqrt{\det g}} \sum_{j,k} \frac{\partial}{\partial x_j}\left(g^{jk}\sqrt{\det g}\frac{\partial f}{\partial x_k}\right). \tag{15.10}$$

In the above equation, $\det g = \det(g_{ij})$ and the g^{jk} are the elements of g^{-1}. For more details see [30].

Consider the Laplace-Beltrami operator eigenvalue problem given by

$$\Delta_X \phi_i = \lambda_i \phi_i. \tag{15.11}$$

$\{\phi_i\}$ are the eigenfunctions of Δ_X, corresponding to the eigenvalues $\{\lambda_i\}$. The spectrum of the Laplace-Beltrami operator consists of positive eigenvalues (see, for example, [30]). When X is a connected manifold without boundary, then Δ_X has additional eigenvalue equal to zero, with corresponding constant eigenfunction. We can order the eigenvalues as follows

$$0 = \lambda_0 < \lambda_1 \leq \lambda_2 \leq \lambda_3 \leq \ldots \tag{15.12}$$

The set of corresponding eigenfunctions given by

$$\{\phi_1, \phi_2, \phi_3, \ldots\} \tag{15.13}$$

forms an orthonormal basis defined on X with inner product induced by the metric g.

There exist various approximations for the Laplace-Beltrami operator. In this work we used the cotangent weight scheme [23, 26].

15.4.2 Choice of Metric

Geodesic distance: The simplest *intrinsic* metric defined on a surface X is the *geodesic metric*. It measures the lengths of the shortest paths on the surface X

$$d_X(x, x') = \inf_{\gamma \in \Gamma(x,x')} \ell(\gamma). \tag{15.14}$$

$\Gamma(x, x')$ is the set of all admissible paths between the points x and x' on the surface X, with a length of a path γ given by $\ell(\gamma)$. In order to calculate the geodesic distances we used the *fast marching method* [15], which simulates a wavefront propagation on a triangular mesh, and associates the front arrival time with the distance traveled by it.

Diffusion geometry: The diffusion of heat on surface X is governed by the heat equation,

$$\left(\Delta_X + \frac{\partial}{\partial t}\right) u(x, t) = 0, \tag{15.15}$$

where a scalar field $u : X \times [0, \infty) \to \mathbb{R}$ is the heat profile at location x and time t, and Δ_X is the Laplace-Beltrami operator.

The *heat kernel* $h_t(x, z)$ describes the amount of heat transferred from a point heat source located at x to another point z at time t, and can be written as

$$h_t(x, z) = \sum_{i=0}^{\infty} e^{-\lambda_i t} \phi_i(x)\phi_i(z). \tag{15.16}$$

The *diffusion distance* can then be defined as a cross-talk between two heat kernels [3, 7]

$$\begin{aligned} d_{X,t}^2(x, y) &= \|h_t(x, \cdot) - h_t(y, \cdot)\|_{L_2(X)}^2 \\ &= \int_X |h_t(x, z) - h_t(y, z)|^2 dz \\ &= \sum_{i=0}^{\infty} e^{-2\lambda_i t} (\phi_i(x) - \phi_i(y))^2 . \end{aligned} \tag{15.17}$$

Since the heat flow on the surface is governed entirely by its intrinsic geometry, the diffusion distance defined above is an intrinsic property of the surface, and, according to [3, 7], also fulfills the metric axioms.

We approximate the diffusion distances using a finite number of the eigenvalues and the eigenvectors of the discretized Laplace-Beltrami operator. Specifically, we used several hundred eigenvalues with the smallest magnitude and their corresponding eigenvectors.

15.4.3 Choice of Descriptors

Distance histograms: Given a shape X and the corresponding distance d_X, the distance histogram descriptor [27, 28, 32] is constructed as follows

$$f_X(x) = \text{hist}\{d_X(x,\tilde{x}) \mid d_X(x,\tilde{x}) \le d_{max},\ \tilde{x} \in X\}, \tag{15.18}$$

where d_{max} controls the local support of the descriptor. If two shapes are represented by differently sampled surfaces, the descriptors can be normalized to have L_1-norm equal to one. The descriptor comparison can be performed using either an L_p-norm, or some measure of distances between histograms, such as the *earth moving distances* (EMD) [31].

Heat kernel signatures and heat kernel maps: Local descriptors based on the *heat equation* were presented by Sun et al. in [35] and Ovsjanikov et al. in [25]. The *heat kernel signature* (HKS) is constructed using the diagonal of the heat kernel $h_t(x,x)$ (15.16) at multiple times t

$$f_X(x) = \left[h_{t_1}(x,x), h_{t_2}(x,x), \ldots, h_{t_d}(x,x)\right]. \tag{15.19}$$

The *heat kernel map* (HKM) is constructed using the heat kernel values with a pre-specified heat source x_0

$$f_X(x) = \left[h_{t_1}(x_0,x), h_{t_2}(x_0,x), \ldots, h_{t_d}(x_0,x)\right]. \tag{15.20}$$

For the latter descriptors, the heat sources chosen for the two shapes we want to match must be in correspondence in order to produce consistent descriptors. One can choose the heat source x_0 either as proposed by the authors of [25], or by some different method. Both HKS and HKM remains invariant under isometric deformations of X, and are insensitive to topological noise at small scales.

To compute HKS and HKM we used eigenvalues and eigenfunction of the discretized Laplace-Beltrami operator, similar to the diffusion distance calculation.

15.5 Matching Ambiguity Problem

The matching ambiguity problem arises when matching intrinsically symmetric shapes [24, 27, 28]. Given a shape X, we say that it is intrinsically symmetric if there exists a mapping $S : X \to X$ that preserves all the geodesic distances between the corresponding points

$$d_X(x,\tilde{x}) = d_X(S(x), S(\tilde{x})), \quad \forall x,\ \tilde{x} \in X. \tag{15.21}$$

If the shape X is intrinsically symmetric, and $S : X \to X$ is its intrinsic symmetry, then the surface descriptors mentioned in the previous section are also symmetric functions with respect to S. That is, for each of their components $f_X^{(i)}$ the following holds

$$f_X^{(i)}(x) = f_X^{(i)}(S(x)). \tag{15.22}$$

From the Eq. (15.21) and the above property of the descriptors it follows that if $P^*(x, y) = \text{argmin}\{\text{dis}(P)\}$, then $P^*(S(x), y)$ also minimizes the dissimilarity $\text{dis}(P)$, with the same minimal value. Thus, when matching intrinsically symmetric shapes, the optimization problem (15.8) has multiple solutions, and by minimizing $\text{dis}(P)$ we can obtain only one of them.

In order to overcome the above problem, a technique for construction of symmetry-aware surface descriptors was suggested in [8]. These descriptors are based on the eigendecomposition of the Laplace-Beltrami operator, and exploit the important property of the eigenfunctions of Δ_X, described by Ovsjanikov et al. in [24]. As stated in Theorem 3.1. of [24], eigenfunctions corresponding to non-repeating eigenvalues of the Laplace-Beltrami operator of an intrinsically symmetric shape exhibit reflection symmetry, with respect to the shape's intrinsic symmetry. That is, such an eigenfunction ϕ can be either symmetric of anti-symmetric with respect to S

$$\phi(x) = \phi(S(x)) \text{ or } \phi(x) = -\phi(S(x)). \tag{15.23}$$

As described in [8], the symmetry-aware surface descriptors are constructed as follows

$$f_X(x) = [\phi_1^X(x), \phi_2^X(x), \ldots, \phi_d^X(x)], \tag{15.24}$$

and

$$f_Y(y) = [s_1\phi_1^Y(y), s_2\phi_2^Y(y), \ldots, s_3\phi_d^Y(y)]. \tag{15.25}$$

In the above, $\{\phi_i^X\}_{i=1}^d$ and $\left\{\phi_j^Y\right\}_{j=1}^d$ are the eigenfunctions corresponding to the first d non-repeating eigenvalues of the Laplace-Beltrami operators of the two shapes, respectively. The values of $\left\{\phi_j^Y\right\}_{j=1}^d$ are then multiplied by the sign sequence $\{s_j\}_{j=1}^d$, to obtain consistent descriptors for X and Y.

Figure 15.2 shows an example of two human shapes colored according to the values of the first three eigenfunctions of their corresponding Laplace-Beltrami operators. It is easy to see that the eigenfunctions of the lower shape have to be multiplied by a sequence $[+, -, +]$, in order to be equal to the eigenfunctions of the upper shape in the corresponding points. But it is also possible to multiply them by a sequence $[+, +, -]$, and thus obtain eigenfunctions reflectionally symmetric to the eigenfunctions of the upper shape. In general, the number of different sign sequences, and thus different sets of descriptors for the shape Y, is determined by the number of intrinsic symmetries of the shape. Using these sets of descriptors in the optimization problem (15.6) allows us to find several different correspondences

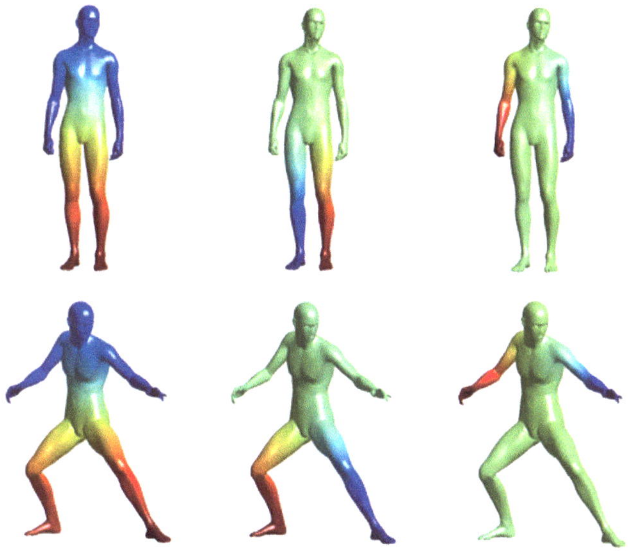

Fig. 15.2 Two articulations of a human shape, colored according to the values of the first three eigenfunctions of their Laplace-Beltrami operators, from left to right. The two possible sign sequence relating the two groups of the eigenfunctions are $[+, -, +]$ and $[+, +, -]$

between the two shapes. The exact algorithm for the sign sequence detection and its limitations are presented in details in [8].

15.6 Results

In this section we provide several matching results obtained with the proposed framework. All the shapes we used in our tests were represented by triangulated meshes with several thousand vertices. We further sub-sampled the shapes using the Farthest Points Sampling algorithm [11], to obtain sets of matching candidate points. In each one of our tests, we performed the matching using ten points at the coarse scale, and 30–64 points at the finest scale. Note that the later sub-sampling affects the accuracy of the matching, and the denser the sub-sampling is the more accurate the obtained correspondences are.

Figures 15.3 and 15.4 present the results of matching ϵ-isometric shapes using the proposed framework combined with different distance measured and descriptors, at several hierarchies, where the correspondences are shown by Voronoi cells of the matched points, corresponding patches having the same color. The matches in Fig. 15.3a–c are the symmetrical ones, which is one of the possible matchings in this case, as explained in Sect. 15.5. Some matching inaccuracies, e.g. inaccurate

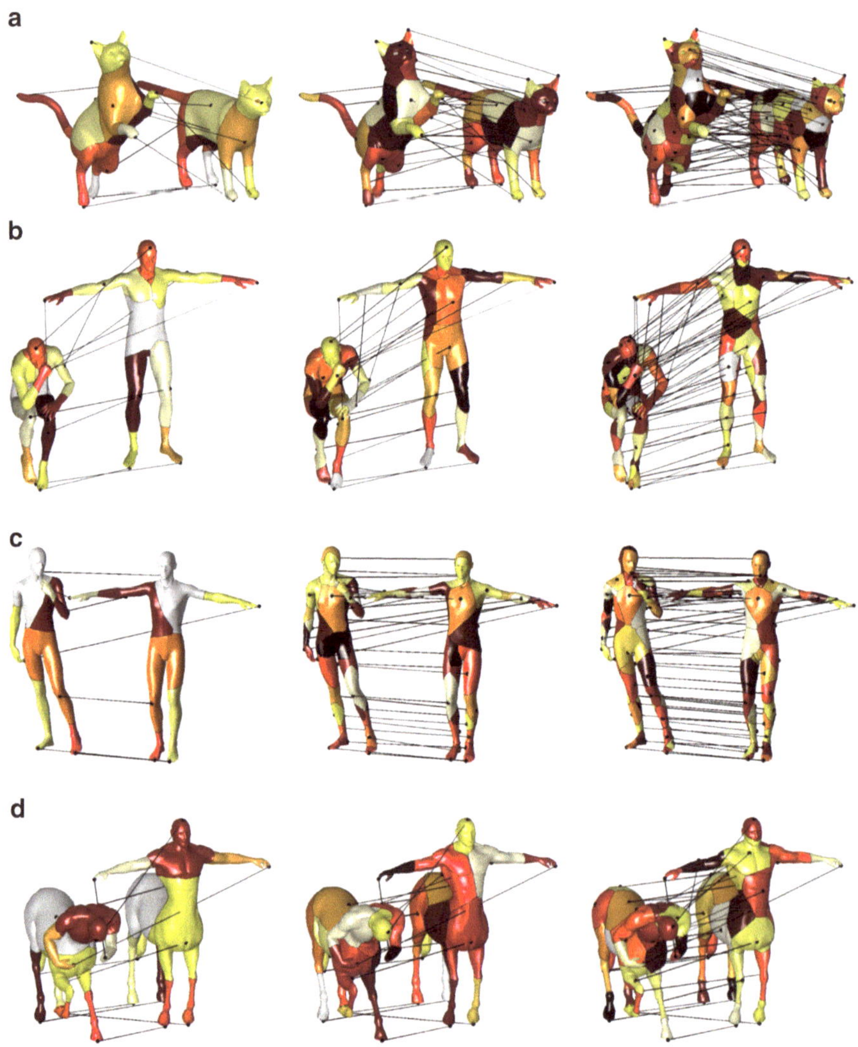

Fig. 15.3 Matching results obtained with the proposed framework combined with different descriptors and metrics, at several hierarchies. (**a**) and (**b**) Geodesic distance and geodesic distance-based histogram descriptor; (**c**) diffusion distance and diffusion distance-based histogram descriptor; (**d**) Diffusion distance and Heat Kernel Signatures

correspondences between the cats' ears in Fig. 15.4a, appear when the algorithm converges to local minima.

In order to find dense correspondence between all the points on the shapes, the above matching results can be used as a input for algorithms such as described in [25] or [16].

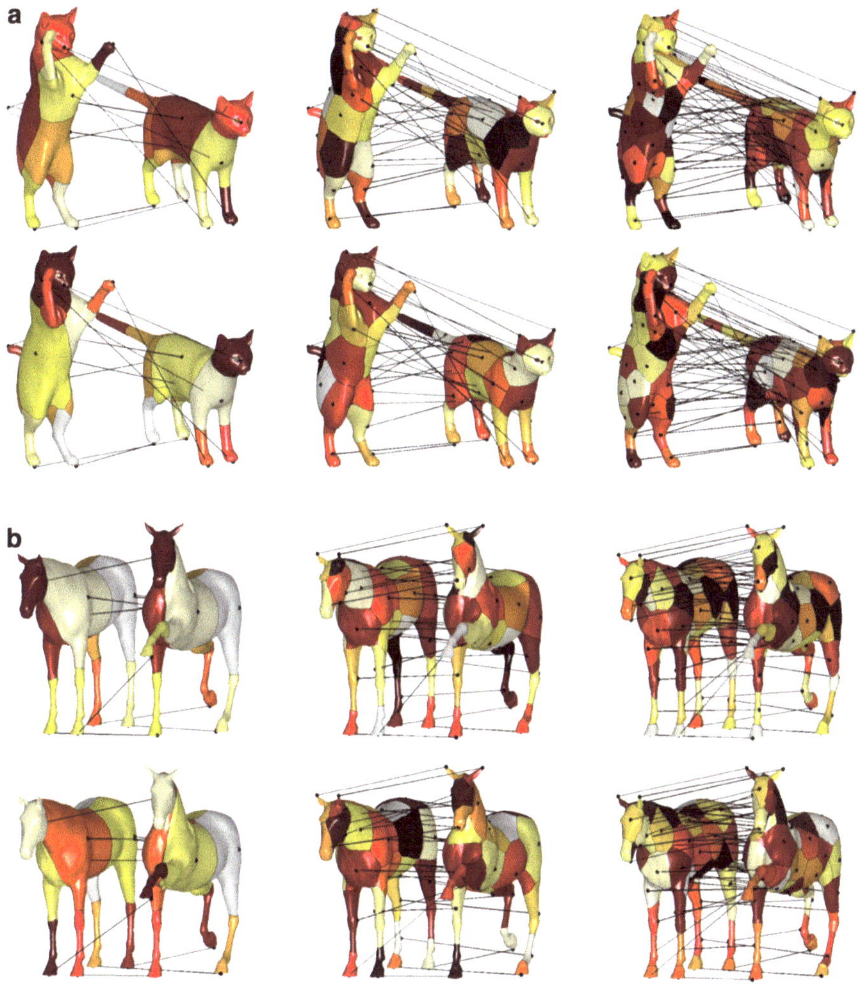

Fig. 15.4 Matching results obtained with the proposed framework combined with geodesic distance metric and the Laplace-Beltrami operator-based descriptors; *upper row* – same orientation correspondence, *lower row* – the reflected one

15.7 Conclusions

In this paper we have presented a method for automatic correspondence of non-rigid ϵ-isometric shapes, based on comparison of surface descriptors and distance structures, and tested it with different choices of the latter. In addition, we have formulated the matching ambiguity problem arising when matching intrinsically symmetric shapes, and showed that the proposed framework combined with certain descriptors allows us to detect multiple possible correspondences. In future work we intend to adapt the proposed framework for partial shape matching and extend it to shapes that are not necessarily ϵ-isometric.

Acknowledgements This research was supported by European Community's FP7- ERC program, grant agreement no. 267414.

References

1. Anguelov, D., Srinivasan, P., Pang, H.-C., Koller, D., Thrun, S.: The correlated correspondence algorithm for unsupervised registration of nonrigid surfaces. In: Proceedings of the Neural Information Processing Systems (NIPS) Conference, vol. 17, pp. 33–40. MIT Press, Cambridge (2004)
2. Bemporad, A.: Hybrid Toolbox - User's Guide (2004). http://www.dii.unisi.it/hybrid/toolbox.
3. Bérard, P., Besson, G., Gallot, S.: Embedding Riemannian manifolds by their heat kernel. Geom. Funct. Anal. **4**(4), 373–398 (1994)
4. Bronstein, A.M., Bronstein, M.M., Kimmel, R.: Generalized multidimensional scaling: a framework for isometry-invariant partial surface matching. Proc. Natl. Acad. Sci. (PNAS) **103**(5), 1168–1172 (2006)
5. Bronstein, A.M., Bronstein, M.M., Kimmel, R.: Numerical Geometry of Non-rigid Shapes. Springer, New York (2008)
6. Bronstein, A.M., Bronstein, M.M., Kimmel, R., Mahmoudi, M., Sapiro, G.: A Gromov-Hausdorff framework with diffusion geometry for topologically-robust non-rigid shape matching. Int. J. Comput. Vis. (IJCV) **89**(2–3), 266–286 (2009)
7. Coifman, R.R., Lafon, S.: Diffusion maps. Appl. Comput. Harmon. Anal. **21**, 5–30 (2006)
8. Dubrovina, A., Kimmel, R.: Matching shapes by eigendecomposition of the Laplace-Beltrami operator. In: International Symposium on 3D Data Processing Visualization and Transmission (3DPVT) (2010)
9. Elad, A., Kimmel, R.: On bending invariant signatures for surfaces. IEEE Trans. Pattern Anal. Mach. Intell. (PAMI) **25**(10), 1285–1295 (2003)
10. Gromov, M.: Structures Metriques Pour Les Varietes Riemanniennes. Textes Math. **1** (1981). Cedic
11. Hochbaum, D.S., Shmoys, D.B.: A best possible heuristic for the k-center problem. Math. Oper. Res. **10**(2), 180–184 (1985)
12. Hu, J., Hua, J.: Salient spectral geometric features for shape matching and retrieval. Vis. Comput. **25**(5–7), 667–675 (2009)
13. Jain, V., Zhang, H., Van Kaick, O.: Non-rigid spectral correspondence of triangle meshes. Int. J. Shape Model. **13**(1), 101–124 (2007)
14. Kim, V., Lipman, Y., Chen, X., Funkhouser, T.: Mobius transformations for global intrinsic symmetry analysis. In: Proceedings of the Eurographics Symposium on Geometry Processing (SGP). Wiley Online Library (2010)
15. Kimmel, R., Sethian, J.A.: Computing geodesic paths on manifolds. Proc. Natl. Acad. Sci. (PNAS) **95**, 8431–8435 (1998)
16. Kraevoy, V., Sheffer, A.: Cross-parameterization and compatible remeshing of 3D models. ACM Trans. Graph. (Proc. SIGGRAPH) **23**(3), 861–869 (2004)
17. Lipman, Y., Chen, X., Daubechies, I., Funkhouser, T.: Symmetry factored embedding and distance. In ACM Transactions on Graphics (Proc. SIGGRAPH) **29**(4), 103 (2010)
18. Lipman, Y., Funkhouser, T.: Mobius voting for surface correspondence. ACM Trans. Graph. (Proc. SIGGRAPH) **28**(3), 72:1–72:12 (2009). Article number 72. ACM, New York

19. Mateus, D., Horaud, R.P., Knossow, D., Cuzzolin, F., Boyer, E.: Articulated shape matching using Laplacian eigenfunctions and unsupervised point registration. In: Proceedings of the IEEE Conference on Computer Vision and Pattern Recognition (CVPR) (2008)
20. Mémoli, F.: On the use of Gromov-Hausdorff distances for shape comparison. In: Point Based Graphics 2007, pp. 81–90. The Eurographics Association (2007)
21. Mémoli, F.: Spectral Gromov-Wasserstein distances for shape matching. In: Workshop on Non-Rigid Shape Analysis and Deformable Image Alignment (ICCV workshop, NORDIA'09), Kyoto, Japan 2009
22. Mémoli, F., Sapiro, G.G.: A theoretical and computational framework for isometry invariant recognition of point cloud data. Found. Comput. Math. **5**(3), 313–347 (2005)
23. Meyer, M., Desbrun, M., Schröder, P., Barr, A.: Discrete differential geometry operators for triangulated 2-manifolds. In: Hege, H.-C., Polthier, K. (eds.) Visualization and Mathematics III, pp. 35–57 (2003)
24. Ovsjanikov, M., Sun, J., Guibas, L.: Global intrinsic symmetries of shapes. Comput. Graph. Forum **27**(5), 1341–1348 (2008)
25. Ovsjanikov, M., Mérigot, Q., Mémoli, F., Guibas, L.: One point isometric matching with the heat kernel. In: Eurographics Symposium on Geometry Processing (SGP). Wiley Online Library (2010)
26. Pinkall, U., Polthier, K.: Computing discrete minimal surfaces and their conjugates. Exp. Math. **2**, 15–36 (1993)
27. Raviv, D., Bronstein, A.M., Bronstein, M.M., Kimmel, R.: Symmetries of non-rigid shapes. In: IEEE 11th International Conference on Computer Vision (ICCV 2007), pp. 1–7 (2007)
28. Raviv, D., Bronstein, A.M., Bronstein, M.M., Kimmel, R.: Full and partial symmetries of non-rigid shapes. Int. J. Comput. Vis. (IJCV) **89**(1), 18–39 (2010)
29. Raviv, D., Dubrovina, A., Kimmel, R.: Hierarchical matching of non-rigid shapes. In: Internation Conference on Scale Space and Variational Methods (SSVM). Springer (2011)
30. Rosenberg, S.: The Laplacian on a Riemannian manifold: an Introduction to Analysis on Manifolds. Cambridge University Press, Cambridge (1997)
31. Rubner, Y., Tomasi, C., Guibas, L.J.: The earth mover's distance as a metric for image retrieval. Int. J. Comput. Vision **40**, 99–121 (2000)
32. Ruggeri, M.R., Saupe, D.: Isometry-invariant matching of point set surfaces. In: Proceedings of the Eurographics 2008 Workshop on 3D Object Retrieval. The Eurographics Association (2008)
33. Rustamov, R.M.: Laplace-Beltrami eigenfunctions for deformation invariant shape representation. In Proceedings of SGP, pp. 225–233 Eurographics Association, Aire-la-Ville (2007)
34. Sharma, A., Horaud, R.P.: Shape matching based on diffusion embedding and on mutual isometric consistency. In: Proceedings of the Workshop on Nonrigid Shape Analysis and Deformable Image Alignment (NORDIA) (2010)
35. Sun, J., Ovsjanikov, M., Guibas, L.: A concise and provably informative multi-scale signature based on heat diffusion. In: Proceedings of the Eurographics Symposium on Geometry Processing (SGP). Wiley Online Library (2009)
36. Tevs, A., Bokeloh, M., Wand, M., Schilling, A., Seidel, H.-P.: Isometric registration of ambiguous and partial data. In: Proceedings of the IEEE Conference on Computer Vision and Pattern Recognition (CVPR) pp. 1185–1192 (2009)
37. Thorstensen, N., Keriven, R.: Non-rigid shape matching using geometry and photometry. In: Asian Conference on Computer Vision, pp. 1–12. Springer (2009)
38. Torresani, L., Kolmogorov, V., Rother, C.: Feature correspondence via graph matching: Models and global optimization. In: Proceedings of the 10th European Conference on Computer Vision (ECCV '08), pp. 596–609. Springer, Berlin/Heidelberg (2008)
39. Wang, C., Bronstein, M.M., Paragios, N.: Discrete minimum distortion correspondence problems for non-rigid shape matching. Technical report, INRIA Research Report 7333, Mathématiques Appliquées aux Systèmes, École Centrale Paris, 2010

40. Wang, C., Bronstein, M.M., Paragios, N.: Discrete minimum distortion correspondence problems for non-rigid shape matching. Technical report, Int. Conf. Scale Space and Variational Methods (SSVM) (2011)
41. Zaharescu, A., Boyer, E., Varanasi, K., Horaud, R.P.: Surface feature detection and description with applications to mesh matching. In: Proceedings of the IEEE Conference on Computer Vision and Pattern Recognition (CVPR), (2009)
42. Zhang, H., Sheffer, A., Cohen-Or, D., Zhou, Q., van Kaick, O., Tagliasacchi, A.: Deformation-driven shape correspondence. Comput. Graph. Forum (Proc. SGP) **27**(5), 1431–1439 (2008)
43. Zigelman, G., Kimmel, R., Kiryati, N.: Texture mapping using surface flattening via multidimensional scaling. IEEE Trans. Vis. Comput. Graph. **8**(2), 198–207 (2002)

Chapter 16
A Review of Geometry Recovery from a Single Image Focusing on Curved Object Reconstruction

Martin R. Oswald, Eno Töppe, Claudia Nieuwenhuis, and Daniel Cremers

Abstract Single view reconstruction approaches infer the structure of 3D objects or scenes from 2D images. This is an inherently ill-posed problem. An abundance of reconstruction approaches has been proposed in the literature, which can be characterized by the additional assumptions they impose to make the reconstruction feasible. These assumptions are either formulated by restrictions on the reconstructable object domain, by geometric or learned shape priors or by requiring user input. In this chapter, we examine a representative set of state-of-the-art reconstruction approaches, which are applicable to real-world images. We classify the approaches according to their reconstruction objective and compare them based on a variety of important criteria. Finally, we show experimental comparisons for five curved object reconstruction approaches.

16.1 Introduction

Estimating 3D geometry from images has been a core research topic in Computer Vision for several decades. For the case of multiple input images a large variety of methods has been developed which are able to deliver high quality reconstruction results. For the special case that only a single still image is available the problem gets considerably more difficult. For specific assumptions imposed on the image a variety of methods to estimate 3D geometry exist in literature. However, a thorough comparison has not been carried out so far.

The reason for this lies partly in the significant diversity of existing approaches which in turn is due to the inherent ill-posedness of the underlying problem: during image formation, depth is irrecoverably lost. In their effort to make the problem

M.R. Oswald (✉) · E. Töppe · C. Nieuwenhuis · D. Cremers
Department of Computer Science, Institut für Informatik, TU München, Boltzmannstr. 3, Garching bei München, Germany
e-mail: oswaldm@in.tum.de; toeppe@in.tum.de; nieuwenhuis@in.tum.de; cremers@in.tum.de

M. Breuß et al. (eds.), *Innovations for Shape Analysis*, Mathematics and Visualization, DOI 10.1007/978-3-642-34141-0_16, © Springer-Verlag Berlin Heidelberg 2013

tractable, single view methods have come up with an abundance of very different assumptions, methods and priors to infer the geometry of a depicted scene or object. The reconstruction precision of such approaches exceeds that of plausible estimates only in very few cases. Consequently, the reconstruction objectives are of very different nature, which makes a comparison difficult.

In this chapter we give a brief survey on the subject of single view reconstruction. We provide an introduction to the field and examine basic image information and assumptions that are used in order to compensate for ill-posedness. We then review, categorize and compare existing state-of-the-art approaches. Finally, we give a detailed theoretical and experimental comparison of five single view reconstruction methods for curved surfaces.

16.2 Single View Reconstruction

Single view reconstruction has the objective of recovering geometric information from a single photograph or synthetic image. The geometric information that is to be retrieved can be of very different manifestation reaching from purely relational information, sparse metrics or dense depth information to a complete 3D model of a single object or even a scene. This circumstance in combination with the inherent ill-posedness of the problem is the main reason for the strong diversity that is witnessed among the works in single view reconstruction and it is by no means a straightforward task to develop a taxonomy let alone a comparison.

A first approach will therefore be on the one hand to give an overview on the different types of image information (‘*image cues*’) used in the different reconstruction processes. On the other hand we will review the *priors* that are assumed in order to overcome the ill-posedness. This will also serve as a brief survey on existing works. Later in Sect. 16.3, we will classify a number of high-level approaches and compare them in Sect. 16.4. We will then narrow down further consideration to a subclass of methods that concentrates on reconstructing curved objects which will form the basis for our comparison in Sect. 16.5.

16.2.1 Image Cues

Approaches to single view reconstruction extract specific higher or lower level information contained in the input image either automatically or with the help of user input. This information then is interpreted to infer geometric relations of the depicted object or scene. In the following we list the most important categories and give prominent references.

Shading The problem of *Shape from Shading* (SfS) is to infer a surface (height field) from a single gray level image by using the gradual variation of shading that is induced by the surface interaction of light. Some approaches also co-estimate lighting conditions and reflection properties. In general, the goal is to find a solution to the following image formation model

$$R(n(x)) = I(x) \ , \qquad (16.1)$$

where I is the image, n is the normal field of the surface and R is the reflectance function which is dependent on the object. In most SfS approaches a Lambertian reflectance model is assumed. There are, however, other models which also include the specular case (e.g. Wang et al. [56]). SfS is an ill-posed problem, although there has been progress on deriving conditions for unique solutions by Prados and Faugeras [42].

As shown by Durou et al. [12] and Zhang et al. [62] reconstruction from real world images is limited in the sense that each approach exhibits special and sometimes unrealistic requirements on lighting conditions or reflection properties. Especially the presence of texture is an issue. Work has been done, however, to incorporate interreflection [39], shadowing and perspective projection [7] just to name a few. One of the first minimization approaches to SfS is by Ikeuchi and Horn [24]. For a current survey see Durou et al. [12].

Shadow The shadow that is thrown by objects conveys geometric information relative to the viewpoint of the light source. This information can be used to narrow down possible reconstruction results. Often point light sources have to be assumed as soft shadows do not provide enough information. Furthermore, shadow is not always thrown on known geometry, which makes the problem even more complex. Apart from reconstruction ambiguities, it is not straightforward to extract shadow borders from images. References include works by Daum and Dudek [9], Kender and Smith [29], Yu and Chang [60], and Hatzitheodorou [16].

Contour Edges Contour edges are salient structures in the image that are induced by surface discontinuities, occlusion, object boundaries or reflectance changes. They give evidence for geometry and relative pose/position of objects. *Junction points* or *corners*, where multiple contour edges meet or end, are also of importance for single view reconstruction.

Subclasses of contour edge-based methods are contour-based and silhouette-based reconstruction methods. *Shape from Contour* approaches try to infer geometry given the object contours alone. However, reconstructions are in most cases ambiguous. Especially, smooth surfaces often do not exhibit sufficient contour lines in the image. Shape from Contour approaches based on closed contour drawings include Horaud et al. [19], Ulupinar et al. [53], and Li et al. [33]. Karpenko et al. [27, 28] interpret user line drawings. Other single view reconstruction approaches that use contour edges for reconstruction include [10, 14, 17, 30, 47, 48].

Silhouette Closely related to Shape from Contour are approaches that infer geometry given the object silhouette. The silhouette is the image of the contour generator and the contour generator is the set of visible points on a surface, whose image rays are tangent to the surface.

The goal of silhouette based approaches is to find a geometric reconstruction, whose projection into the image plane agrees with the silhouette. As there are always infinitely many objects that are silhouette consistent this cue suffers from inherent ambiguity if used alone.

There are several silhouette based single view reconstruction algorithms that we will consider in more detail later. These include works by Prasad et al. [44, 45], Oswald et al. [41], and Töppe et al. [52]. Related to these approaches are a class of sketch based modeling tools e.g. by Igarashi et al. [23], Karpenko et al. [27], and Nealen et al. [40].

Texture Besides geometry, the appearance of real world objects is also determined by texture. Although a complete distinction from shading is not possible, texture is considered as an inherent property of an object rather than a result of an interaction of light and geometry.

If one assumes objects to have a regular and known texture it is possible to infer their geometry from the way the texture is deformed after image projection. These *Shape from Texture* approaches, obviously, impose strong constraints on the reconstructable objects. An example constitutes the work of Malik and Rosenholtz [37].

Further single view reconstruction algorithms that use texture cues include Super et al. [50], Hassner and Basri [15], and Vetter et al. [55]. Approaches that combine texture and contour edges for reconstruction by considering so-called 'superpixels' are Hoiem et al. [17] and Saxena et al. [48].

Defocus Due to physical aspects of image formation, the sharpness of a depicted object correlates with its distance to the camera. This fact can be used to infer a dense depth map from an image. However, the accuracy of such methods is limited and camera calibration is necessary. References include works from Levin [32] and Bae and Durand [1].

Location The location of objects in the image can infer semantic knowledge of the objects. For example, ground, floor or sky can be identified more easily from their location in the image. This information can be helpful for 3D reconstructions. Hoiem et al. [17] reconstruct vertical objects by distinguishing them from the ground and the sky. Delage et al. [10] use a Bayesian network to identify floor pixels.

16.2.2 Priors

Priors are of utter importance in single view reconstruction in order to compensate for the problem of ill-posedness. Depending on the ultimate reconstruction goal and the target group of objects, different priors or a combination of them can be applied. Priors can either be chosen in fixed form, or they can be gained by learning. Furthermore, there are low-level and high-level priors. In the following we will list priors that are most frequently assumed in single view reconstruction.

Smoothness Smoothness can be defined as the small spatial change of some property. In single view reconstruction we are often not able to infer a dense reconstruction. It is therefore good practice to choose among the possible reconstruction surfaces those which tend to be smooth. Smoothness naturally plays a significant role in the reconstruction of curved surfaces as in [41, 45, 52, 63].

Smoothness in a wider sense can also be learned as the consistency of object surfaces. Hoiem et al. [17] use a machine learning approach to find image features indicating the assignment of neighboring superpixels to the same object. Saxena et al. [48] use image cues and geometric relations to learn the relative depth of neighboring superpixels.

Geometric Relations Basic geometric relations are often encountered specifically in man-made environments. As a prior they can help to dissolve ambiguities in the reconstruction process. As part of those basic geometric relations we consider e.g. *coplanarity*, *collinearity*, *perpendicularity* and *symmetry*. An early work which makes use of such simple rules is the one of Lowe [36]. By assuming planes to be parallel or perpendicular one can also derive camera parameters (see Criminisi et al. [8]). This is even more important, as perspective projection is not angle-preserving and the image of parallel lines will not necessarily be parallel. We can often assume objects to stand vertically on the ground plane [10, 14, 17].

Symmetric objects exhibit identical sections, which are projected to different locations in the input image. Knowing that these parts possess similar geometric and reflectance properties one can interpret their respective projections as views of the same object part from different observation points. This can be seen as a weak multiview scenario providing more information for reconstruction [18]. Also, occluded geometry can be inferred from this prior [14].

Volume/Area With smoothness as a prior on its own, solutions tend to be trivial or flat depending on the reconstruction approach. Adding a volume prior to the reconstruction process will ensure an inflation of the object surface and will still result in a certain compactness of the solution due to the smoothness assumption. Volume priors can be found in Li et al. [33] and Töppe et al. [52].

Semantic Relations Semantic relations infer high-level knowledge on the relative position and inner structure of different objects and their depth values. Han and Zhu [14], for example, infer occluded points based on semantic human knowledge, e.g. that leaves are connected to the plant. Koutsourakis et al. [30] introduce semantic knowledge to ensure the consistency of different floors. Finally, knowledge on the location of the ground and the sky represents an important cue for 3D reconstruction. The ground is often used as starting point for the reconstruction as objects, especially walls, are usually perpendicular to this plane [10, 14, 17].

Shape Priors Shape priors impose high-level knowledge on the objects to be reconstructed. Among commonly used priors, full shape priors usually impose the strongest restrictions. On the one hand, this leads to a rather limited applicability of the approach. On the other hand, the reconstructions are usually of high quality and work automatically without user input.

Shape priors can be defined or learned. In [30], Koutsourakis et al. define a full shape grammar for the reconstruction of facades. This limits the approach to the reconstruction of buildings in urban environments. In contrast, Rother and Sapiro [46] and Chen and Cipolla [4] shape priors are learned from a database of sample objects. Hence, they are not a-priori limited to a specific object class. However, their choice of samples intrinsically limits their approach to the object classes represented in the database.

The representation of shape priors ranges from specified sets of grammar rules over parametric models to probabilistic priors. In [4], Chen and Cipolla learn depth maps of human bodies by means of principal component analysis. This model imposes strong assumptions on the 3D object, but the dimension of the state space is reduced and only valid 3D reconstructions are obtained. In contrast, Rother and Sapiro [46] impose less strong assumptions on the learned model. For each object class a shape prior is learned as the relative occupancy frequency of each voxel in the object.

16.3 Classification of High-Level Approaches

In this section we will examine selected works in the field of single view reconstruction. Due to the abundance and diversity of approaches we selected algorithms with respect to the following criteria: the chosen approaches are applicable to real world images that are not taken under special or unrealistic material or lighting conditions. We rather concentrate on approaches inferring objects from ordinary photographs where reconstruction plausibility is more important than precision. The selection, furthermore, focuses on works that are representative and state-of-the-art. We provide a classification and examine differences and similarities.

For classification we found several categories ranging from application domain over image cues and shape priors to user input and 3D representation (see below). However, these categories are not suitable to partition the set of approaches due to strong overlap. Instead, we think that the most important information for each single view reconstruction approach is its application domain, i.e. the set of objects and scenes, which can be reconstructed. We introduce the literature sorted by the following reconstruction categories:

- Curved Objects
- Piecewise Planar Objects
- Learning Specific Objects
- 3D Impression from Scenes

More categories will follow in the next subsection. We distinguish between objects and scenes. Reconstructions of scenes aim at producing 3D impressions or depth maps from the whole scene contained in the image. In contrast, object reconstruction approaches concentrate on single objects within the scene. Approaches that reconstruct *curved objects* principally aim at producing arbitrary, mostly smooth objects.

Often, minimal surface approaches are used, which try to minimize the surface of the object given a fixed area or volume. The second class consists of methods that concentrate on *piecewise planar objects* such as buildings and man-made environments. Furthermore, we distinguish arbitrary curved and planar objects from *learning specific objects*. Approaches in this class cannot reconstruct arbitrary objects, but are inherently limited to specific object classes by shape information learned from sample databases. Finally, we discuss methods that do not aim to reconstruct exact or plausible 3D geometry but rather provide a pleasing *3D Impression from Scenes*. In the following, we will present and classify related work on single view reconstruction.

16.3.1 Curved Objects

16.3.1.1 Zhang et al.

Zhang et al. [63] proposed a method for interactive depth map editing based on an input image. The depth map reconstruction is the result of minimizing a thin plate energy [11], which favors smooth surfaces and penalizes bending. User input comes as a variety of constraints on the thin plate energy that can be placed interactively into the depth map. These comprise of position constraints, surface normals, surface or normal discontinuities, planar region constraints or curves on which curvature or torsion is minimized.

From a mathematical point of view the thin plate energy for a continuous function f on a two dimensional rectangular domain $[0,1]^2$ is defined as:

$$E(f) = \int_0^1 \int_0^1 \left[\alpha(u,v) \left\| \frac{\partial^2 f}{\partial u^2} \right\|^2 + 2\beta(u,v) \left\| \frac{\partial^2 f}{\partial uv} \right\|^2 + \gamma(u,v) \left\| \frac{\partial^2 f}{\partial v^2} \right\|^2 \right] du\, dv \ , \tag{16.2}$$

where functions $\alpha, \beta, \gamma : [0,1]^2 \mapsto \{0,1\}$ extend the thin plate model with weighting functions which can be used to define local surface discontinuities. Zhang et al. [63] discretize this minimization problem by introducing a function $g_{i,j}$ that samples values of the depth map function $f : [0,1]^2 \mapsto \mathbb{R}$ on a discrete rectangular grid, that is, $g_{i,j} = f(id, jd)$, with d being the distance between neighboring grid points. For efficiency and accuracy the grid resolution can be locally refined by the user. By stacking all values $g_{i,j}$ into a single vector $\mathbf{g}$ and by discretizing the partial derivatives of g, the energy in Eq. (16.2) can be written in matrix form as

$$\mathbf{g}^T \mathbf{C} \mathbf{g} \qquad \text{subject to} \quad \mathbf{A}\mathbf{g} = \mathbf{b} \ , \tag{16.3}$$

where the condition $\mathbf{A}\mathbf{g} = \mathbf{b}$ may contain any constraints on the surface that can be expressed in linear form. For a detailed description on how the constraints are incorporated into this quadratic optimization problem we refer to [63]. A description of these constraints from the user's point of view is given later together with the experimental comparison (Sect. 16.5.2).

16.3.1.2 Prasad et al.

The works [45] and [44] of Prasad et al. introduce a framework for single view reconstruction of curved surfaces. The method is related to the one by Zhang et al. [63] but aims at reconstructing closed surfaces.

The main idea involves computing a parametric minimal surface by globally minimizing the same thin plate energy (Eq. (16.2)) as Zhang et al. [63] the difference being, however, that they minimize with respect to a parametrized 3D surface $f : [0,1]^2 \mapsto \mathbb{R}^3$ instead of a depth map. As a result, function domain and image domain are no longer equivalent. For implementation purposes, the discretization of the optimization problem with constraints is done similar to Zhang et al. (see Eq. (16.3)).

The choice of constraints is mostly different from Zhang et al. [63]. The main source of reconstruction information is the silhouette: Prasad et al. [45] use the fact that normals along the contour generator $c(t)$ can be inferred from the 2D silhouette as by definition they are parallel to the viewing plane for a smooth surface. This leads to the constraints

$$\pi(f(u(t), v(t))) = c(t) \tag{16.4}$$

$$n(c(t))\, f(u(t), v(t)) = 0 \qquad \forall t \in [0,1] \;, \tag{16.5}$$

where $n(c(t))$ is the normal at the point $c(t)$ in $\mathbb{R}^3$ and π the orthographic projection function. The user has to determine the coordinates $(u(t), v(t))$ of the contour generator in parameter space. This is done by placing lines within the grid of the parameter space and setting them in correspondence with the parts of the contour generator. Similar to Zhang et al. [63] the user can employ position constraints to define the object inflation locally. Also, surface discontinuities can be optionally specified to relax the surface smoothness along curves in the parameter space.

Important to note is that in order to define the topology of the object, the user has to define which parts of the parameter space boundary are connected. For example, the connection of the left and right boundary defines a cylindrical shape of the function domain.

16.3.1.3 Oswald et al.

The variational single view approach by Oswald et al. [41] computes closed curved minimal surface reconstructions. The input comprises of a side view of an object and its silhouette which is obtained by interactive segmentation of the input image.

An object reconstruction is achieved by searching for a smooth minimal surface that complies with a shape based on the silhouette distance function.

The minimal object surface is represented by indicator function $u : V \mapsto \{0, 1\}$ defining object interior ($u = 1$) and exterior ($u = 0$) and is found by minimizing the following convex energy

$$E(u) = \underbrace{\int_V g(x)|\nabla u(x)|\, d^3x}_{\text{smoothness term}} + \underbrace{\int_V u(x)\, \left(\phi_{\text{vol}}(x) + \phi_{\text{sil}}(x)\right) d^3x}_{\text{data term}} \;, \qquad (16.6)$$

where $V \subset \mathbb{R}^3$ is the reconstruction domain and the data term promotes volume inflation via ϕ_{vol} and silhouette consistency via ϕ_{sil}. Function $g : V \mapsto \mathbb{R}_+$ is used to locally adapt the smoothness of the surface if desired by the user – its default value is $g \equiv 1$. Given the input silhouette $S \subset \Omega \subset V$ as part of the image plane Ω a height map $h : \Omega \mapsto \mathbb{R}$ is defined

$$h(p) = \min\left\{\lambda_{\text{cutoff}}\, ,\, \lambda_{\text{offset}} + \lambda_{\text{factor}} * dist(p, \partial S)^k\right\} \qquad (16.7)$$

with the help of the silhouette distance function $dist(p, \partial S) = \min_{s \in \partial S} \|p - s\|$, which returns the distance of any point $p \in \Omega$ to the silhouette boundary ∂S. The parameters $k, \lambda_{\text{cutoff}}, \lambda_{\text{offset}}, \lambda_{\text{factor}}$ modify the shape of h.

Now, the data term can be expressed in terms of h so that the object locally has a depth proportional to the silhouette distance. Using an implicit surface representation the data term can thus be written as

$$\phi_{\text{vol}}(x) = \begin{cases} -1 & \text{if } dist(x, \Omega) \leq h(\pi(x)) \\ +1 & \text{otherwise}\;, \end{cases} \qquad (16.8)$$

where $\pi : V \mapsto \Omega$ is the orthographic projections of 3D points onto Ω. Silhouette consistency is brought forward by

$$\phi_{\text{sil}}(x) = \begin{cases} -\infty & \text{if } \chi_S(\pi(x)) = 1 \text{ and } x \in \Omega \\ +\infty & \text{if } \chi_S(\pi(x)) = 0 \\ 0 & \text{otherwise}\;, \end{cases} \qquad (16.9)$$

where characteristic function $\chi_S : \Omega \mapsto \{0, 1\}$ indicates exterior or interior of the silhouette, respectively. The convex energy in Eq. (16.6) is minimized with the help of convex relaxation techniques and the global optimum defines the reconstructed object.

16.3.1.4 Töppe et al.

Very similar to Oswald et al. [41], the variational single view approach by Töppe et al. [52] also computes closed curved minimal surface reconstructions. It follows the same work flow and also requires an object silhouette as input.

The same regularizer as in Oswald et al. [41] makes for the smooth surface, however, instead of using a heuristic data term for surface inflation, the authors suggest that the surface volume complies with a user specified volume V_t. This is achieved by adding a constraint to the minimization of the following convex energy:

$$E(u) = \int_V g(x)|\nabla u(x)|\, d^3x \qquad \text{s.t.} \quad \int_V u(x)\, d^3x = V_t \ . \tag{16.10}$$

Again, the surface is represented implicitly by the indicator function $u : V \mapsto \{0, 1\}$ where $V \subset \mathbb{R}^3$ is the reconstruction domain. Similarly, the purpose of function $g : V \mapsto \mathbb{R}_+$ is to account for optional local smoothness changes (default: $g \equiv 1$). Silhouette consistency is enforced by another constraint

$$u(x) = \begin{cases} 0, & \pi(x) \notin S \\ 1, & x \in S \end{cases} , \tag{16.11}$$

where $\pi : V \mapsto \Omega$ is the orthographic projection of 3D points onto the image plane Ω and $S \subset \Omega$ is the input silhouette. Convex relaxation techniques are used to minimize energy in Eq. (16.10) and a GPU implementation allows for interactive rates on standard PC hardware.

The benefit of the implicit surface representation used by Oswald et al. [41] and by Töppe et al. [52] is the topological freedom of the object's surface. In contrast to Prasad et al. [45] and Zhang et al. [63] the surface can contain any number of holes that are induced by the input silhouette. As opposed to [45], however, both reconstruction approaches by Oswald et al. [41] and Töppe et al. [52] assume the contour generator to be planar, which again limits the set of possible reconstructions.

16.3.1.5 Colombo et al.

Another approach to 3D reconstruction are surfaces of revolution (SORs) [6,54,59]. They are common in man-made objects and represent a subclass of Straight Homogeneous Generalized Cylinders. SOR approaches strongly rely on the assumption of rotational symmetry of the objects. A surface of revolution is obtained by revolving a planar curve, referred to as scaling function, around a straight axis, the revolution axis. SORs can be parametrized in polar coordinates:

$$S(\theta, z) = (\rho(z)cos(\theta), \rho(z)sin(\theta), z) \ . \tag{16.12}$$

Important for reconstruction approaches are cross sections and meridians. Cross sections are intersections of planes orthogonal to the revolution axis with the object, which leads to circles. Meridians are intersections of planes containing the revolution axis with the object. This leads to curves which all have the same shape and coincide with the SOR scaling function. Reconstruction approaches based on SORs try to infer the scaling function and the axis of revolution from the image.

In Colombo et al. [6], the task of 3D reconstruction is formulated as the problem of determining the meridian curve from the imaged object silhouette and two given imaged cross sections. Based on the computation of fixed entities such as the vanishing line or the SOR symmetry axis, camera calibration can be done and the SOR is inferred. Texture acquisition is obtained by inverse normal cylindrical projection.

16.3.1.6 Other Approaches

Francois and Medioni [13] present an interactive 3D reconstruction method based on user labeled edges and curves, which are represented by non-uniform rational basis splines (NURBS). The reconstructed objects are either modeled as generalized cylinders or as a set of 3D surfaces. Terzopoulos et al. [51] propose deformable elastic 3D shape models, which evolve around a symmetry axis and whose projection into the image is attracted by strong image gradients. Cohen and Cohen [5] propose a generalization of snakes to 3D objects based on a sequence of 2D contour models for medical images.

16.3.2 Piecewise Planar Objects and Scenes

16.3.2.1 Criminisi et al.

In [8], Criminisi et al. describe how 3D affine measurements can be obtained from a single perspective view. The authors concentrate on scenes containing planes and parallel lines which are obtained by perspective projection. The camera is assumed to be uncalibrated with unknown internal parameters.

The authors assume that a vanishing line of a reference plane as well as a vanishing point can be computed from the image. Given the vanishing line of a reference plane, a vanishing point for a direction not parallel to the plane and a known reference length in the image, the authors derive affine scene structure from the image. In this context, three different types of measurements are computed from the image: distances between planes parallel to the reference plane, measurements on these planes (e.g. length and area ratios) and the camera position. These measurements are obtained from cross-ratios (special numbers associated with an ordered quadruple of collinear points, which is invariant under specific transformations in projective geometry) and specific image mappings.

To simplify computations, Criminisi et al. [8] developed an algebraic approach based on a parametrization of the projection matrix. As the approach relies on feature detection in the image (e.g. for reference points), the resulting measurements can only be of limited accuracy, which is assessed by a confidence measure.

The approach is applied among others to 3D reconstruction. Based on given reference heights (e.g. the true height of a window and a pillar of a house) and a precomputed vanishing line and point in the image the complete 3D reconstruction can be obtained. The position of the camera is also estimated.

16.3.2.2 Delage et al.

Delage et al. [10] describe an approach for the automatic reconstruction of 3D indoor scenes, which mainly consist of orthogonal planes. The following assumptions are made: (1) The image is obtained by perspective projection with known camera calibration. (2) The objects in the scene are composed of planes and edges with only three mutually orthogonal orientations ('Manhattan world assumption'). (3) The camera's axis is vertical to the floor with known height.

To distinguish planes from edges with their orientations, the authors devise a Markov Random Field (MRF) consisting of vertices V and edges E with six labels $y_v = \{p_1, p_2, p_3, e_1, e_2, e_3\}$ for $v \in V$. $\{p_1, p_2, p_3\}$ encodes plane orientation and $\{e_1, e_2, e_3\}$ encodes edge orientation. Let x_v and $x_{u,v}$ denote the feature vector at node v and at node pair (u, v) respectively. The following joint probability distribution is defined over all labels y_v:

$$P_\theta(y|x) = \frac{1}{Z_\theta(x)} \exp\left(- \sum_{v \in V} \psi_1(y_v, x_v, \theta_1) - \sum_{(u,v) \in E} \psi_2(y_u, y_v, x_{u,v}, \theta_2) \right) . \tag{16.13}$$

Here, $Z_\theta(x)$ stands for the partition function and $\theta = [\theta_1, \theta_2]$ indicates the model parameters. The unary term $\psi_1(y_v, x_v, \theta_1) = \theta_1^T \phi(y_v, x_v)$ is a linear combination of image features $\phi(y_v, x_v)$ such as edge orientation or floor affiliation, whereas the pairwise term $\psi_2(y_u, y_v, x_{u,v}, \theta_2) = \theta_2^T \phi(y_u, y_v, x_{u,v})$ encodes features indicating label consistency or if two pixels were members of the same partition after a segmentation of the image.

To obtain a 3D reconstruction from the MRF labeling, a constrained iterative optimization problem is formulated to obtain the location and orientation of planes and edges in 3D space. The constraints yielding 3D points for each image pixel can be formulated from the Manhattan world assumption which ensures a unique solution. To obtain a correct result, occlusion vs. non-occlusion of edges is inferred. Delage et al. [10] can be viewed as a modification and generalization of Sturm and Maybank [49].

16.3.2.3 Koutsourakis et al.

In [30], Koutsourakis et al. generate urban 3D reconstructions from images by estimating the parameters of a 3D shape grammar in a MRF approach, so that the generated building best matches the image.

A shape grammar describes how basic shapes interact together through a set of replacement rules to produce complex structured geometries. It contains basic or atomic shapes, which are modified by operators. The operators either transform the object by means of translation, rotation or scaling, or they perform structural operations such as splitting or mirroring. The main advantages of using a shape grammar are that it always produces well-defined buildings and that the complexity of the optimization as well as the dimensionality of the problem is strongly reduced.

For optimization the authors formulate a MRF approach. The unary terms ensure that object boundaries coincide with image boundaries, whereas the pairwise terms measure the appropriateness of the configuration of atomic shapes and ensure the consistency between the operator and the image.

16.3.2.4 Other Approaches

Kanade [26] recovers shape from geometric assumptions. The world is modeled as a collection of plane surfaces, which allows for a qualitative object recovery. Quantitative recovery is achieved by mapping image regularities into shape constraints. Piecewise planar scenes are computed in Liebowitz et al. [35] based on camera and geometric constraints such as parallelism and orthogonality, e.g. for the reconstruction of buildings. Apart from symmetry and planarity, two additional shape constraints are introduced by Li et al. [33] for object reconstruction: maximum compactness and minimum surface. Instead of computing vanishing lines, Kushal et al. [31] perform 3D reconstruction of structured scenes by registering two user indicated world planes. Hong et al. [18] study the relation between symmetry of objects and the viewer's relative pose to the object. An important principle for the reconstruction of symmetric objects is that one image of a symmetric object is equivalent to multiple images. Li et al. [34] describe a method for reconstructing piecewise planar objects by using connectivity and perspective symmetry of objects.

16.3.3 Learning Specific Objects

16.3.3.1 Han and Zhu

Han and Zhu [14] propose a 3D reconstruction approach based on manually defined shape priors, which can on the one hand be applied to polyhedral objects and on the other hand to grass and tree-like objects. They argue that learning priors is hard in practice, because there is not enough real world training data available. Hence, they revert to manually defined prior models.

The image is represented by two layers, one containing image regions such as sky or planar objects such as polyhedra, the other containing this curved structures such as grass or trees. The full 3D scene is represented by two graphs, one consisting of 3D objects, the other representing the relations between the objects in the scene. The objective of the paper is then to optimally reconstruct the 3D scene given the layer representation of the image in a Bayesian approach. To this end, the authors manually introduce prior knowledge. For polyhedral objects they assume planarity of faces as well as similar angles and edge lengths. For grass and tree-like objects they assume smooth and evenly spread curves. The relation graph is used to impose regularity constraints on touching objects, e.g. on angles. Furthermore, hidden information is inferred based on human geometric reasoning on missing vertices, parallelism, rotational or mirror symmetry and stability assumptions.

Optimization is done by Markov Chain Monte Carlo Methods with reversible jumps which define rules for altering sub-graph structures.

16.3.3.2 Rother and Sapiro

Rother and Sapiro [46] present a general framework for pose estimation, 2D segmentation, object recognition and 3D reconstruction from a single image. The approach is well-suited to reconstruct bounded objects, but not for elaborate scenes.

The 3D object to be reconstructed is represented by voxels having either state full or empty: $V = \{V_i\}_{i=1}^{M},\ V_i \in \{\mathit{empty}, \mathit{full}\}$. The corresponding segmentation of the image is given by the pixel states $Q = \{Q_j\}_{j=1}^{N},\ Q_j \in \{\mathit{background}, \mathit{foreground}\}$. It is assumed that the camera matrix is given, which relates 3D voxels to 2D image pixels.

To obtain the most likely 3D reconstruction from the image, the authors formulate a graphical model based on two fidelity terms: (1) The object fidelity $P(V|K)$, which means its conformity with a learned 3D shape prior for a given shape class K. The shape prior $P(V|K)$ is learned for each class separately from sample images as the relative frequency of the voxel for being part of the object. (2) The data fidelity, which defines the probability that the given input image is generated by the occupied voxels under projection onto the image plane. This term is again composed of two probabilities for each of the N pixels in the image plane: (a) The color probability given by the user specified foreground and background color model for each pixel with color C_j, $P(C_j|Q_j)$, and (b) the probability of obtaining a particular pixel state (foreground or background) based on the number of full voxels projected onto pixel j along its corresponding projection ray $R(j)$, $P(Q_j|V_{R(j)})$.

The likelihood for a given hypothesis (shape class and affine transformation or pose) is then defined as follows:

$$L(Q, V) = \left(\prod_{j=1}^{N} P(C_j|Q_j) P(Q_j|V_{R(j)}) \right) P(V|K) \ . \tag{16.14}$$

To select the best hypothesis for each shape class and pose, their likelihood must be compared. In order to reduce computational complexity, an efficient branch and bound algorithm is proposed to discard suboptimal hypotheses and refine only plausible ones.

16.3.3.3 Chen and Cipolla

Chen and Cipolla [4] propose to infer 3D information directly from learned shape priors. They assume a number of given training shapes each consisting of the silhouette and the corresponding depth map. In a first step, the training silhouettes are registered. To ensure the independence of different data dimensions and to reduce the complexity of the approach, Principal Component Analysis is applied to the silhouettes and the depth maps separately in order to find lower-dimensional subspaces of the training shapes. In this way, the authors obtain 2D lower-dimensional training feature pairs consisting of position and depth information. To model the low-dimensional sub-manifold structure underlying the feature pair space, Gaussian Process Latent Variable Models (GPLVM) are used. They assume Gaussian processes for the position and depth information as well as for the latent variables (the PCA coefficients). Then they estimate the underlying parameters in an optimization problem, which leads to a shape model learned from sample data.

Given an unknown silhouette, 3D information can now be inferred from the learned shape model. First, the silhouette is registered and projected into the lower-dimensional PCA subspace yielding a position feature vector. Given this feature vector and the shape model, we ask for the most likely depth estimate at each point. Since there is no closed-form solution to this problem, the authors revert to a two-stage approach: (1) They find the most likely latent variables generating the given silhouette feature. (2) From these latent variables the most likely depth estimate is inferred as the mean value of a Gaussian distribution. The final 3D reconstruction at each point is the sum of the most likely depth value and the linear combination of the PCA eigenvectors determined by the latent variables.

16.3.3.4 Hassner and Basri

Hassner and Basri [15] aim at depth reconstruction from a single image based on examples. The samples are given in a database S containing mappings of images to their corresponding depth maps $S = \{M_i\}_{i=1}^n = \{(I_i, D_i)\}_{i=1}^n$. For a given image patch W_p centered on p its depth is inferred from known depth values of the most similar patches V in the database by maximizing its plausibility: $Plaus(D|I, S) = \sum_{p \in I} maxSim(W_p, V)$. Similarity Sim between patches is measured in the least squares sense. The image patches overlap leading to several depth estimates for each pixel. These are combined by averaging. To ensure consistency of neighboring patches a global optimization procedure is proposed which iteratively refines depth estimates.

16.3.3.5 Other Approaches

Vetter [55] learned a parametric model for the reconstruction of faces by applying PCA to a database of registered 3D faces. Then the model parameters can be found, which best explain the given image of a face. In Nagai et al. [38], objects are learned from a sample database. A Hidden Markov Model is used to model the correspondence between intensity and depth.

16.3.4 3D Impression from Scenes

16.3.4.1 Hoiem et al.

In [17], Hoiem et al. propose a fully automatic approach for creating 3D models from single photographs, which is similar to the creating of pop-up illustrations in children's books. They divide the world into ground, sky and vertical objects, which they call geometric classes. The appearance of these classes is described by image cues, which are learned from sample images.

In a first step the image is automatically segmented into superpixels, which are grouped to constellations. Constellations consist of superpixels, which are likely to have the same label. This probability is expressed by

$$S(C) = \sum_{k=1}^{N_c} \frac{1}{n_k(1-n_k)} \sum_{i,j \in C_k} \log P(y_i = y_j|\, |z_i - z_j|) \ , \tag{16.15}$$

where N_c is the number of constellations, n_k the number of superpixels in constellation k and $P(y_i = y_j|\, |z_i - z_j|)$ is the estimated probability that two superpixels have the same label y given the absolute difference of their feature vectors z.

The likelihood of a superpixel label given its feature vector x is estimated by marginalizing over the constellation likelihoods

$$P(y_i = t|x) = \sum_{k:s_i \in C_k} P(y_k = t|x_k, C_k) P(C_k|x_k) \ , \tag{16.16}$$

where both probabilities are learned from training data. By means of a machine learning approach the constellations are labeled as ground, sky or vertical objects. To reconstruct a 3D model from the data, the boundary of the bottom of the vertical regions is fit with the ground, the sky is removed from the model, and vertical pixels are assumed to belong to objects sticking out of the ground. Line segments are then fit to the ground-vertical label boundary and the segments are formed into poly-lines. The image is finally 'folded' along the poly-lines and 'cut' upward at the endpoints of these lines as well as at ground-sky and vertical-sky boundaries. A reasonably scaled camera model can be obtained by estimating the horizon line and setting the remaining parameters to constants.

16.3.4.2 Saxena et al.

In [48], Saxena et al. propose another approach for obtaining 3D structure from a single image of an unstructured environment. The only assumption the authors make is that the world consists of small planes, whose 3D position and orientation is to be estimated. Similar to Hoiem et al. [17], the authors start out from a superpixel segmentation of the image. In this way, they obtain N superpixels (miniature planes) each containing S_i image pixels. But instead of grouping superpixels into constellations defining ground, sky and vertical object classes, for each single superpixel the depth and orientation is inferred. This is done by a Markov Random Field (MRF) model. Each plane or superpixel is represented by a vector α such that all plane points q fulfill the equation $\alpha^T q = 1$. For a plane pixel lying on a normalized camera ray vector R at depth d we thus have $\alpha^T(dR) = 1$. The authors use a learning approach to obtain parameters θ which identify feature vectors X with depth estimates $\hat{d} = X^T\theta$. The authors maximize the following probability in an MRF approach

$$P(\alpha|X,Y,R,\theta) = \frac{1}{Z}\prod_{i=1}^{N} f_\theta(\alpha_i, X_i, Y_i, R_i) \prod_{i,j\in\{1,\ldots,N\}} g(\alpha_i,\alpha_j,Y_{ij},R_i,R_j) \;, \tag{16.17}$$

where α_i is the plane parameter of the superpixel i, X_i its corresponding feature vector, Y_i indicates feature confidence and R_i is the set of rays connecting the camera viewpoint to the pixels of the superpixel. Then f_θ relates image features to estimated depth by

$$f_\theta(\alpha_i, X_i, Y_i, R_i) = \exp\left(-\sum_{s_i=1}^{S_i} y_{i,s_i} |\alpha_i^T (X_{i,s_i}^T \theta R_{i,s_i}) - 1|\right) \;. \tag{16.18}$$

The pairwise term g can contain different definitions which capture features such as coplanarity, connectedness and colinearity.

The MRF represents two important aspects of depth estimation: it learns, (1) how visual cues such as color and texture are associated with depth, and (2) the relative depth of nearby superpixels based on geometric relations such as connectedness, coplanarity and collinearity. They also give estimates for occlusions and folds of the object.

Learning can be done only approximately due to the complexity of the model. To this end, the graphical model is approximated by a product of several marginal conditional likelihoods. MAP inference to infer the plane position and orientation of each superpixel is done by solving a linear program. The result is a polygonal mesh representation of the 3D model.

16.3.4.3 Other Approaches

In Horry et al. [20], simple 3D scenes are reconstructed based on user input such as vanishing points and foreground objects. The background of the scene is then modeled by rectangles, the foreground by hierarchical polygons. Barinova et al. [2] propose a reconstruction approach for urban scenes yielding visually pleasant results. The method is based on fitting 3D models containing vertical walls and ground plane to the scene.

16.4 General Comparison of High-Level Approaches

In the previous section we have presented important high-level single view reconstruction approaches. In this section, these approaches will be compared with respect to several categories, which we find important for their successful application. Table 16.1 compares the presented approaches with respect to these categories. It also indicates related image cues and shape priors for each approach, which were described in the previous section.

Applicability The applicability of a reconstruction approach depends on several aspects. First, methods are usually limited to some **reconstruction domain** which refers to the set of objects or scenes which can be reconstructed "successfully". These domains range from architectural such as buildings over man-made objects and piecewise planar environments to natural scenes. The examined single view approaches are therefore listed with respect to the categories introduced in Sect. 16.3 (first column of Table 16.1).

The applicability of an approach is also characterized by the **assumptions** made by the method. If specific assumptions are not met, the reconstruction process easily fails. Assumptions for each method are given in column five of Table 16.1. Typical assumptions are a calibrated camera [10], a simplified scene composition [10, 17], an object database containing samples for learning shape priors [4, 46], a specific viewpoint [41, 45, 52] or given geometric properties such as vanishing lines of reference planes [8].

Another aspect which determines the applicability of an approach to a special problem is its envisaged **reconstruction precision**. The precision of a method describes the consistency of the reconstructed 3D model with the actual real-world scene. There is a trade-off between precision and reconstruction feasibility. One can witness a correlation between reconstruction precision and requirements: the higher the envisaged reconstruction precision, the more assumptions and priors have to be made on the reconstruction domain.

Reconstructions can be exact, if the computed lengths and orientations of the inferred 3D objects accurately correspond to the true object. This is usually only possible from a single image if strong assumptions are made, e.g. piecewise planarity with only three orientations (Manhattan assumption) [10] or known reference heights and a calibrated camera [8]. Since such strict assumptions strongly limit the applicability of the approach, most approaches revert to computing the most likely solution to the ill-posed reconstruction problem without guaranteeing accuracy. The probability of a solution is usually measured by means of manually defined [14] or learned shape priors [4, 46]. We call this a *plausible* precision. Finally, there are approaches, which do not aim at reconstructing the real object. Instead, they find solutions which look good to the viewer when animated [17, 20, 48] or can be used to synthesize approximate new views of a scene. We call these reconstructions *pleasing*. The reconstruction precision is indicated in the third column of Table 16.1.

Table 16.1 Detailed overview of single view methods: for each approach the most important characteristics are indicated: Precision of the method (exact '$=$', plausible '$\simeq$', pleasing '$\approx$'), the representation of the 3D object, important assumptions made by the approach, the necessary user input and image cues as well as priors which are used in the reconstruction process. The 'L' indicates a prior which is not assumed but learned by the approach. Terms in brackets are optional.

						Image Cues				Priors				
	Method	Precision	Surface Representation	Assumptions	User Input	Silhouette	Edges	Location	Texture	Smoothness	Volume / Area	Semantic Relation	Geom. Relation	Shape
Curved Objects	Prasad et al. [45]	$\simeq$	[closed] parametric	charateristic sideview, max. genus 2	contours, [creases]	x			x	x				
Curved Objects	Zhang et al. [63]	$\simeq$	depth map	none	constraints					x				
Curved Objects	Oswald et al. [41]	$\simeq$	closed implicit	sideview, symmetry	silhouette, [creases], [data term]	x				x				
Curved Objects	Töppe et al. [52]	$\simeq$	closed implicit	sideview, symmetry	silhouette, [creases], [volume]	x				x	x			
Curved Objects	Colombo et al. [6]	$\simeq$	closed parametric	rotational symmetry	silhouette, cross sec.	x							x	
Piecewise Planar	Criminisi et al. [8]	$=$	non-closed polygonal	vanishing line, refer. height, ground plane	all edges to be measured		x						x	
Piecewise Planar	Delage et al. [10]	$\simeq$	non-closed polygonal	calibration, Manhattan	none		x	x	x	x		x	x	
Piecewise Planar	Koutsourakis et al. [30]	$=$	closed polygonal	rectified image, buildings	none		x		x			x		shape grammar
Learning Specific Objects	Han & Zhu [14]	$\simeq$	closed polygonal	polyhedra, plants	none		x			x		x	L	proba-bilistic
Learning Specific Objects	Rother & Sapiro [46]	$\simeq$	closed implicit	calibration, color models, database	none			x						learned voxel model
Learning Specific Objects	Chen & Cipolla [4]	$\simeq$	depth map	database	silhouette	x								learned PCA model
Learning Specific Objects	Hassner & Basri [15]	$\simeq$	depth map	fixed view, spec. object database	none				x	x				learned example based
Scenes	Hoiem et al. [17]	$\approx$	pw. planar depth map	simple scene: sky, vertical walls&ground	none		x	x	x	L		x		
Scenes	Saxena et al. [48]	$\approx$	pw. planar depth map	world consists of planes	none		x	x	x	L			x	

'$=$' indicates exact precision, '$\simeq$' plausible precision and '$\approx$' a pleasing approach. Surely there are smooth transitions between these classes.

Representation The form of representation is closely connected to the reconstruction algorithm. Firstly, only those objects are reconstructable that can be adequately represented. Seen the other way, the representation has to reflect the reconstruction domain well. And secondly, the representation model has to conform to the reconstruction process.

Different representations have been used for single view reconstruction approaches. We distinguish between parametric and implicit representations. Each point on a *parametric surface* can be uniquely described by a coordinate. Finding a good parametrization for an object is not straightforward and generally does not allow for arbitrary topology. *Implicit surfaces* are a remedy to this problem. In this case, the surface is a level set of a function defined on $\mathbb{R}^3$. In contrast to parametric surfaces, single points on the surface are not easily addressed. *Polygonal* surface representations are neither parametric nor implicit and can be described as a planar graph with nodes, edges and faces. Note that polygonal surfaces often describe piecewise planar objects but are also used for approximating curved parametric surfaces. Finally, representations can describe *closed* and *non-closed* 3D surfaces. As a special case we also regard *depth maps*, which assign a depth to each pixel.

User Input and Runtime Completely automatic reconstruction on a single input image is often not feasible. Therefore, the user may be required to give cues on important image features. Most single view approaches aim to keep user input simple. User input can convey low-level and high-level information. High-level input is of semantic quality which helps to dissolve ambiguities, e.g. the object silhouette.

This stands in contrast to tools, where the user *models* the reconstruction with the help of low-level operations, e.g. by specifying surface normals or cutting away object parts. Many of these *modeling tools* [3, 25, 58] are not image-based and therefore only remotely related to single view reconstruction. In *Sketch-based modeling tools* [23,27,40,61] such modeling operations are driven by user indicated lines. The Teddy tool will be examined in more detail in Sect. 16.5. A pioneering work on free-form modeling was done by Welch and Witkin [57].

There is 2D and 3D user input. Most approaches use 2D input which in most cases is directly applied to the input image [52]. This involves tracing contour edges such as creases or vanishing lines. 3D input is directly applied to the reconstruction surface and is often more involved for the user as he needs to navigate in 3D space (e.g. specifying normals).

For some approaches the user input stage is separate from the reconstruction stage [4, 8]. Other methods compute a first reconstruction, then the user can add further input and the process is continued [41, 45, 52, 63]. For approaches of the latter kind, runtime is obviously an important factor of the reconstruction approach.

16.5 Comparison of Approaches for Curved Surface Reconstruction

In this section we concentrate on methods that aim for the reconstruction of curved surfaces. In particular, we discuss the methods by Zhang et al. [63], Prasad et al. [45], Oswald et al. [41], Töppe et al. [52], and Igarashi et al. [23]. Although Igarashi et al. is a pure modeling tool it inflates object contours and is thus related to silhouette based single view methods. Comparison is done theoretically and experimentally. The user input is discussed separately.

16.5.1 Theoretical Comparison

In the following we will compare the aforementioned approaches with respect to four topics which are important in surface reconstruction.

The Inflation Problem A common problem of approaches for curved surface reconstruction is that reconstructions tend to be flat since – by default – there are no inflation forces present due to a lack of depth information. A remedy is to let the user specify the depth of certain constraint points of the reconstruction which are then interpolated by the minimal surface [41, 45, 63]. This is tedious for the user. The depth constraints can be estimated fully automatically from the silhouette only for cylindrical objects as is done in some examples by Prasad et al. [43]. Several heuristics are conceived for more complicated cases. Igarashi et al. [23] automatically set the depth by a heuristic based on a signed distance function of the silhouette outline. A similar heuristic is used by Oswald et al. [41] in order to define a data term for their variational minimal surface approach. However, in contrast to Igarashi et al. [23] the user is able to adapt the parameters of this data term and thus the final surface. Töppe et al. [52] use a constraint on the volume of the reconstruction, which in many cases leads to natural inflation behavior.

Surface Representation and Topology The reconstructability of curved surfaces with arbitrary topology depends on the surface representation. Implicit representations [41, 52] are generally better suited for this task than parametric ones [45, 63], since the parameter space has to reflect the topology. The same holds for mesh-based approaches such as the one by Igarashi et al. [23]: during modeling operations it can be tedious to keep the mesh consistent, especially during topology changes.

The parametric representation by Prasad et al. [45] has other implications. Firstly, uniformly distributed points in the parameter space are not uniformly distributed on the surface. This property may lead to oscillations, especially in the case of higher genus. Further, the relation between points in parameter space and points on the surface is non-trivial for inexperienced users.

Silhouettes Silhouettes are used by Prasad et al. [45], Oswald et al. [41], and Töppe et al. [52] for surface inference. Full silhouette consistency of the reconstruction, however, is only enforced in the latter approaches as Prasad et al. [45] derive merely local constraints from the silhouette.

View Constraints Finally, view constraints are of practical importance. Oswald et al. [41] as well as Töppe et al. [52] assume symmetric objects. Reconstructions work best if symmetry and viewing plane are parallel. This implies that the contour generator is planar. The approach by Prasad et al. [45] allows for non-planar contour generators and, thus, in some cases for slightly more general view points than just a side-view.

16.5.2 Experimental Comparison

In this section we experimentally compare the five methods discussed in the previous subsection. For all experiments, we used the single view modeling tool by Zhang et al. [63] and the software called SmoothTeddy which incorporates results of several works by Igarashi et al. [21–23]. Both of them are publicly available for download. The reconstruction results by Prasad et al. are taken from the works [43–45]. For the reconstruction results of the method by Oswald et al. [41] and Töppe et al. [52] we used our own C++ and CUDA-based implementations.

In Figs. 16.1–16.3 we compare the reconstruction results of all five methods on ten different examples, covering various issues such as object shape, topology, viewing angle and image type. Instead of explaining every example, we rather concentrate on the examples which demonstrate properties, advantages or drawbacks discussed in the theoretical comparison as well as issues we identified during the experiments. Since the necessary amount of user input and thus the simplicity of the modeling process is of particular interest for practical purposes, we also explain and compare the user input for each method. The user inputs for each method is summarized in Table 16.2. Necessary user input is printed in bold. The other inputs are either optional or the program provides a heuristic to initialize these values reasonably well.

For each method we discuss the user input separately and illustrate it in several Figures for the teapot example, which is the most sophisticated one. Since we identified significant differences in the necessary amount of time a medium experienced user needs to generate the above examples, we quantitatively compare the amount of user input by listing the modeling times for every method and each example in Table 16.3.

The modeling times for Zhang et al. [63] include all user defined constraints for a medium experienced user. The respective times for Igarashi et al. [23] only contain the modeling time without model coloring as an automated model texturing could easily be integrated. The modeling times for Oswald et al. [41] and Töppe et al. [52] include the user scribbles for the silhouette extraction, which is rather simple

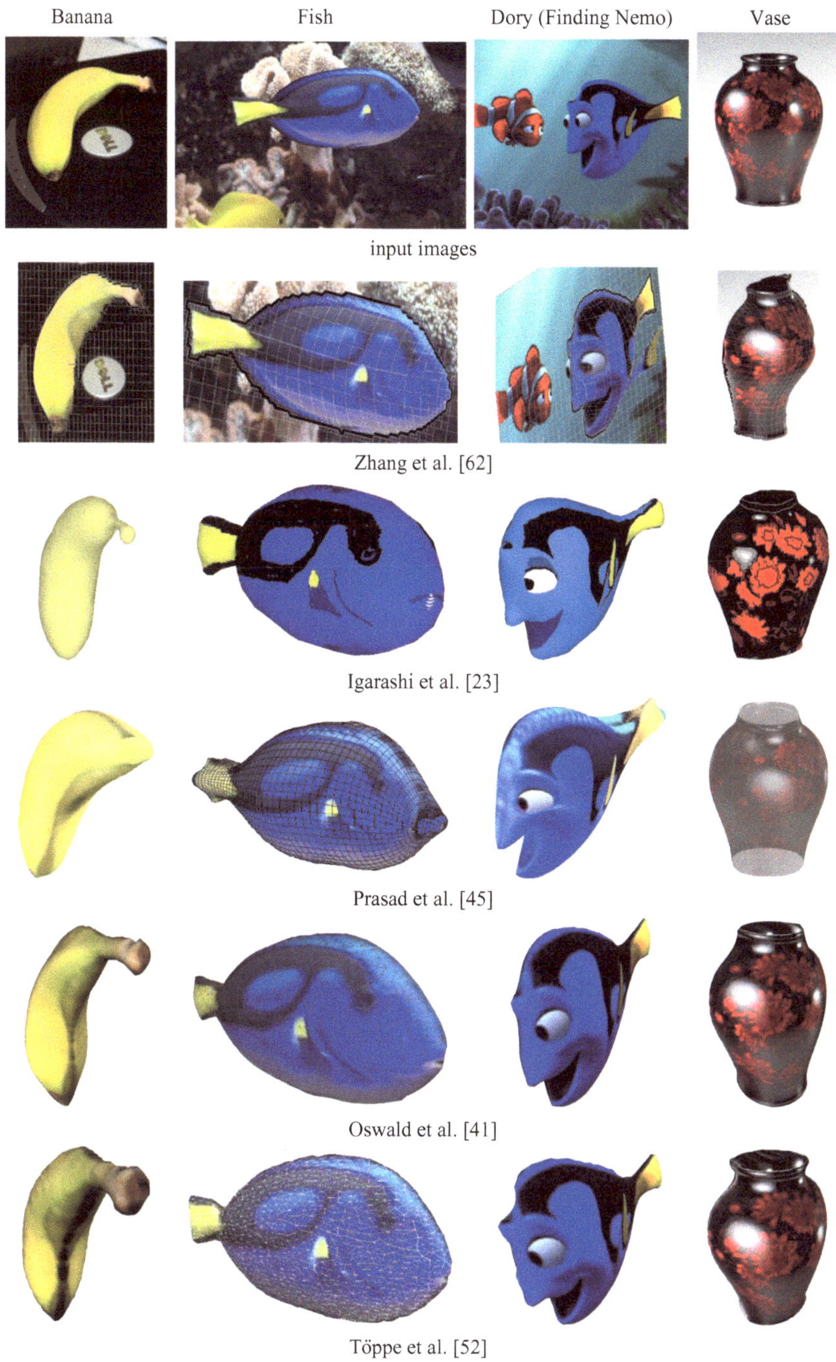

Fig. 16.1 Experimental comparison of several methods for curved object reconstruction (The figures for Prasad et al. are taken from [43])

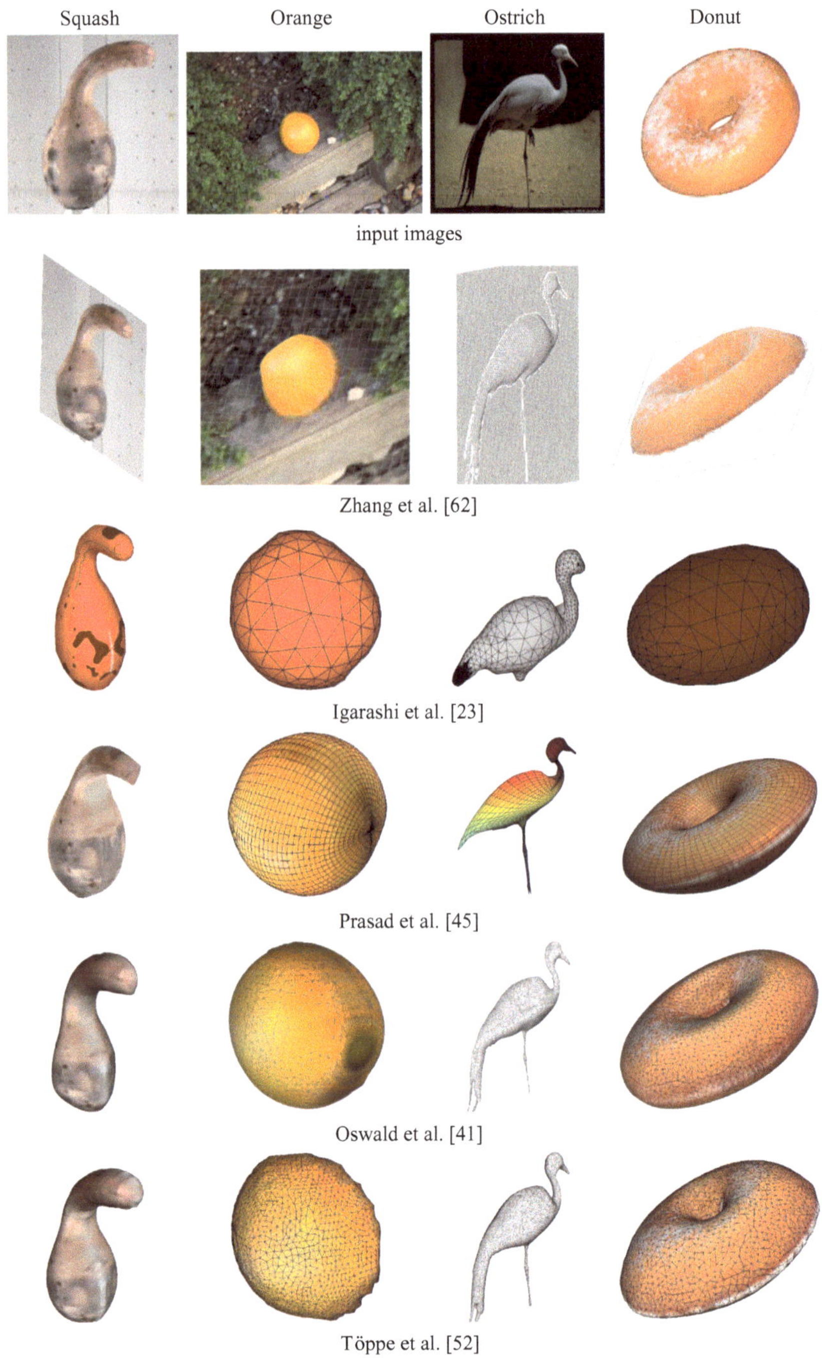

Fig. 16.2 Continuation of Fig. 16.1: experimental comparison of several methods for curved object reconstruction (The figures for Prasad et al. are taken from [43])

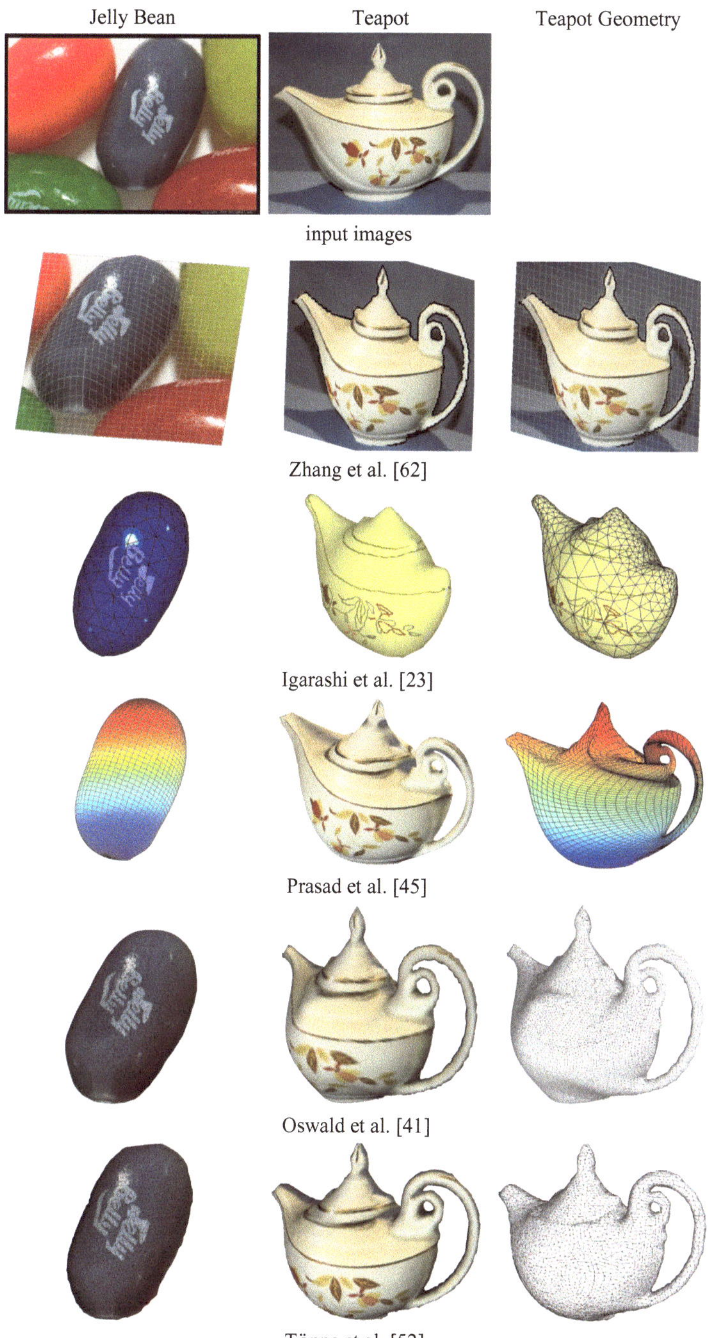

Fig. 16.3 Continuation of Fig. 16.2: experimental comparison of several methods for curved object reconstruction (The figures for Prasad et al. are taken from [43])

Table 16.2 Necessary (**bold**) and optional user inputs and modeling steps for several methods in comparison. Optional user inputs are still required algorithm inputs but they can be predefined by default values or simple heuristics and later on changed by the user if desired. Note that the variety of user input shown in this table does not reflect the amount or complexity of the input that is necessary for a reconstruction

Method	User input (optional and **required**)
Zhang et al. [63]	• Depth map dimensions • Normal/position constraints • Discontinuity lines (position/normal discontinuity) • Planar region constraint • Curvature/torsion minimizing fairing curve constraints • Manual mesh-subdivision
Igarashi et al. [23]	• **Rough contour lines** • Union or cut operations between objects • Object coloring
Prasad et al. [45]	• Mesh resolution • **Silhouette extraction** • **Define corresponding parameter space boundaries** (defines topology) • **Assign parts of the contour to lines in the parameter space** • Choose **inflation heuristic** (cylindrical, cylindrical by parts, distance transform, approximation constraints) + further inflation input • Spillage correction (correct silhouette consistency violated through optimization) • Surface creases
Oswald et al. [41]	• Volume dimensions • **Silhouette extraction** • Define data term shape interactively (4 parameters) • Surface creases
Töppe et al. [52]	• Volume dimensions • **Silhouette extraction** • Define target volume interactively • Surface creases

Table 16.3 Approximate modeling times for all five methods and all examples in Figs. 16.1–16.3 for a medium experienced user. Considering the very similar reconstruction results, this table reveals significant differences for the time of their modeling

Example	Zhang et al. [63]	Igarashi et al. [23]	Prasad et al. [45]	Oswald et al. [41]	Töppe et al. [52]
Banana	20 min	<1 min	10 min	5 min	<1 min
Fish	15 min	<1 min	2 min	8 min	1 min
Dory	40 min	<1 min	5 min	7 min	1 min
Vase	20 min	<1 min	2 min	13 min	4 min
Squash	12 min	<1 min	2 min	2 min	1 min
Orange	14 min	<1 min	<1 min	3 min	<1 min
Ostrich	30 min	<1 min	15 min	7 min	2 min
Donut	55 min	<1 min	10 min	3 min	1 min
Jelly Bean	15 min	<1 min	2 min	4 min	1 min
Teapot	35 min	<1 min	20 min	15 min	4 min

with any graph-cut based method for most of the examples. The reconstructions by Prasad et al. [45] are taken from their respective publications and the authors provided modeling times which include the time taken to annotate contours and all other constraints.

16.5.2.1 Zhang et al.

Evaluation of Experiments This method is more a single view modeling tool rather than a reconstruction tool. Every detail, every extrusion or inflation has to be modeled by the user. Due to the depth map approach it is easy to ensure silhouette consistency when using discontinuity constraints at the silhouette boundary because points are only allowed to change their elevation. These discontinuity constraints have been used for all examples in Figs. 16.1–16.3. The modeling process with this tool can be cumbersome because most constraints only have a local influence on the surface shape. Usually many constraints are necessary for modeling 3D structures from a single image. Further, the oblique position of the donut with respect to the image plane (Fig. 16.2) is difficult to model with local constraints only. The same difficulties emerged with the banana example in Fig. 16.1. Fine scale structures can hardly be modeled and elongated structures such as the handle of the teapot (Fig. 16.3) are often tedious to be modeled properly. The leg of the ostrich (Fig. 16.3) could not be modeled at all due to the limited mesh resolution.

An advantage of this method is the variety of modeling possibilities. The user can fully control each part of the mesh and can even disconnect surface parts. For that reason, the user is able to model details like the round shaped eye of Dory in Fig. 16.1 or the side fin of Dory which bends away from the fish body. Such details cannot be modeled with the other four methods in this comparison. However, the freedom in modeling incurs a larger amount of user input.

User Input Figure 16.4 illustrates the variety of different constraints listed in Table 16.2 and their general effects on the surface shape. None of these constraints is required by the method, but for most reconstructions a reasonable amount of constraints will be necessary. An example input of constraints for the teapot can also be seen in Fig. 16.4. Yellow crosses depict position constraints and the red curve is a position discontinuity constraint.

Although we observed a decrease of user input and modeling time with increasing user experience the amount of modeling time remains comparatively high even for simple reconstructions such as the banana and vase (Fig. 16.1), squash and orange (Fig. 16.2) and jelly bean (Fig. 16.3). The large amount of user input results in higher modeling times which are shown in Table 16.3. The difficulty of modeling a non-side view considerably increased the modeling time for the donut example (Fig. 16.2).

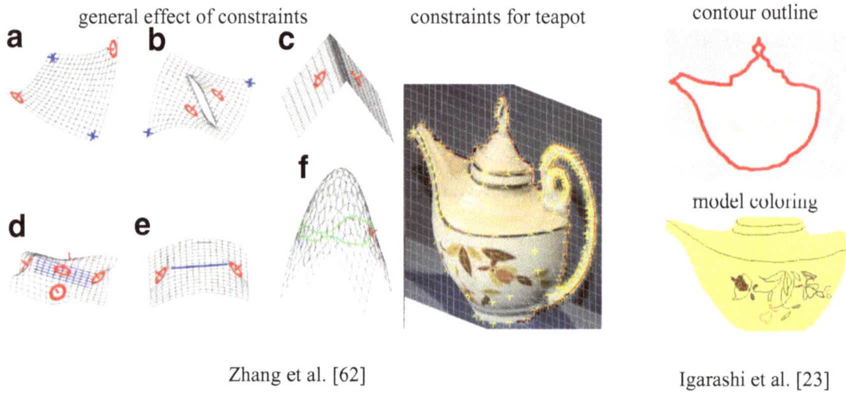

Fig. 16.4 User input for the methods of Zhang et al. [63] and Igarashi et al. [23]. The first image shows general effects of different constraints and is taken from [63]. In particular, the constraints are: (**a**) position (*blue*) and normal (*red*) constraints, (**b**) depth discontinuity constraint, (**c**) crease constraint, (**d**) planar region constraint, (**e**) curvature minimizing fairing curve and (**f**) torsion minimizing fairing curve (see [63] for further details)

16.5.2.2 Igarashi et al.

Evaluation of Experiments This method generally over-smoothes the input silhouette which can be seen in many examples but especially at the peak of the bird in the ostrich example in Fig. 16.2. Although part of the input silhouette the leg of the ostrich is totally missing. Similarly, the grip of teapot lid in Fig. 16.3 is not reconstructed although it is part of the input silhouette which is shown in Fig. 16.4 (top right). Consequently, the reconstructions resulting from this method are not silhouette consistent. Further, the mesh generation in this approach is not straightforward and may lead to mesh inconsistencies which can be observed in the reconstruction of the ostrich example in the lower neck area (Fig. 16.2).

The main advantage of this approach is the fast and intuitive modeling of geometrically simple objects. One of the drawbacks is the restricted topology, the hole in the donut example in Fig. 16.2 cannot be reconstructed. A further disadvantage is the limited influence of the user during the modeling process. The fact that surface discontinuities like sharp edges are not allowed largely decreases the class of reconstructable objects. For instance, the tail fin of Dory in Fig. 16.1 cannot be modeled to end with a sharp edge. The same holds for the top and bottom parts of the vase example in the same figure and for the bottom part of the teapot in Fig. 16.3. Only very simple roundish objects like the banana (Fig. 16.1), squash and orange (Fig. 16.2) or the jelly bean (Fig. 16.3) can be easily and reliably reconstructed.

User Input None of the user input in Table 16.2 needs much experience or even expert knowledge: From a given closed contour line the method instantly inflates a 3D mesh. For better user experience the user interface of the tool provides simple paint operations to color the model. See Fig. 16.4 for the exemplary user input of the teapot example. In all experiments this method needed the least user input (cf. Table 16.3) at the price of producing the least accurate reconstructions with respect to the given input silhouette (see Figs. 16.1–16.3).

16.5.2.3 Prasad et al.

Evaluation of Experiments The main benefit of this method stems from a few extensions over the method by Zhang et al. [63] which facilitate the object inflation and allow the reconstruction of closed surfaces. Apart from the silhouette outline the user can assign any contour line to a line in the parameter space. For instance, this has been used to relate a contour edge of the banana (Fig. 16.1) to a crease in the reconstructed geometry (see [45] for further details).

For closed surfaces a correspondence of parameter space boundaries has to be defined. One the one hand, this facilitates the reconstruction of objects with a simple geometry such as the orange example in Fig. 16.2 or the jelly bean in Fig. 16.3. The orange and the jelly bean get naturally inflated only be defining the silhouette contour and the correspondence of parameter space boundaries. On the other hand, the definition of the parametrization can be sophisticated for objects of higher genus such as in the teapot example (Fig. 16.3) which has genus 2. However, not only the object topology limits the applicability of this method: it is also difficult to define a proper parametrization for objects which have many elongated structures such as legs, arms, tails or spikes. A limiting example may be the ostrich in Fig. 16.2 which is parametrized as a cylinder starting from the peak, going through the neck down to the leg. The body of the ostrich is modeled as a 'bump' in this cylinder which already leads to very sparsely sampled regions at the back of the ostrich. A further modeling of the tail feathers is done with the other methods would need an enormous increase of the mesh resolution in the tail part of the ostrich in order to have enough quads that can align to the silhouette. Thus, finding a good parametrization is a demanding task despite a simple object genus of 0 as for example when modeling a cow with 4 legs and a tail. Moreover, an undesired effect of such a parametrization is the non-uniform distribution of points in the parameter space on the surface as shown in the ostrich (Fig. 16.2) and teapot (Fig. 16.3) examples. Together with further constraints this may lead to non-controllable surface oscillations of the object surface. This can be observed in the top part of the teapot which is connected to the teapot handle (Fig. 16.3).

Another disadvantage is the fact that silhouette consistency is only incorporated with boundary conditions which may get violated during the optimization process. The user may need to add further constraints to ensure a silhouette consistent reconstruction. This process is called spillage correction in Table 16.2 and Fig. 16.5f. Nonetheless, this method generated the most accurate results for the non-side-view examples (banana and teapot).

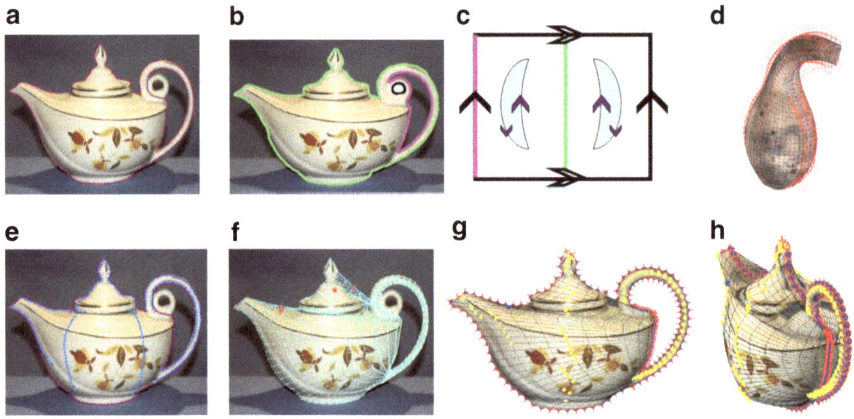

Fig. 16.5 Necessary and optional steps and user input for Prasad et al. [45]: (**a**) contour extraction; (**b**) lines of the contour have to be related to lines in the parameter space (**c**); (**d**) and (**e**) demonstrate different inflation heuristics; (**f**) during the optional spillage correction, the user can account for silhouette inconsistencies by adding further constraints; (**g**) and (**h**) show the final model and generated interpolation constraints as *yellow dots*. Note that (**b**) and (**c**) show a genus 2 reconstruction, while the other teapot images show a genus 1 reconstruction (All figures are taken from [43])

User Input Most of the user input listed in Table 16.2 is illustrated in Fig. 16.5. After the *contour extraction* (Fig. 16.5a) normal constraints are generated along the contour shown as red discs with needles in Fig. 16.5d, g, and h. The definition of corresponding parameter space boundaries and the assignment of contour line parts to lines in the parameter space is shown in Fig. 16.5b and c.

Figure 16.5d shows generated constraints as red dots for a *cylindrical inflation*. Objects with cylindrical shape about a virtual free-form 3D curve or spine can be inflated by generating interpolation constraints along the spine with a depth value equal to the minimal distance between the point of the spine and the contour line. This inflation heuristic is generalized for more complex objects as object parts can be independently inflated with the same technique. To this end the user defines pairs of *inflation curves* (Fig. 16.5e) for which interpolation constraints are generated along the (virtual) medial spine (Fig. 16.5g and h).

The necessity and complexity of each single user input step depends on the object to be reconstructed leading to very different modeling times for the presented experiments (see Table 16.3).

16.5.2.4 Oswald et al.

Evaluation of Experiments Single view modeling with this method is mostly intuitive and many examples did not need much effort. For instance, the banana, fish, dory (Fig. 16.1), squash, orange, ostrich and donut examples (Fig. 16.2) or the

Fig. 16.6 Both methods, Oswald et al. [41] and Töppe et al. [52] have a similar workflow: a silhouette can be estimated with a few user scribbles (*red* = foreground, *blue* = background). Optionally, the user can add creases (*green line*) to locally change the smoothness properties of the final surface. The final geometry shows the discontinuity around the lid and is a result of the method by Töppe et al.

jelly bean (Fig. 16.3) example are easy to accomplish, especially in comparison to the method by Zhang et al. [63]. The shape of the data term mainly defines the shape of the final reconstruction and a proper modeling needs a little user experience. However, the edged shape of the data term which is governed by the silhouette distance function is sometimes hard to conceal. This can be observed in the squash (Fig. 16.2) and the teapot (Fig. 16.3) examples in comparison to Töppe et al. [52]. This characteristic shape, however, can also be beneficial as shown in the banana (Fig. 16.1) and ostrich examples (Fig. 16.2). In contrast to the methods discussed previously, this one assumes objects to be symmetric to the image plane. That is, it assumes a side view of the object, which in turn restricts the applicability of this method to a smaller class of objects that can be reconstructed. This can be observed in the donut example in Fig. 16.2. The input image is not a side view and the resulting reconstruction underestimates the size of the hole.

User Input Although the silhouette extraction is the only necessary user input an adaption of the data term shape by changing parameters $k, \lambda_{\text{offset}}, \lambda_{\text{factor}}, \lambda_{\text{cutoff}}$ from Eq. (16.7) is necessary in most cases. The effect of optional creases to locally relax the smoothness regularization and the general workflow of this method is depicted in Fig. 16.6.

16.5.2.5 Töppe et al.

Evaluation of Experiments Reconstructions with this method are easily obtained with even less user input in comparison to Oswald et al. [41], that is, by an adaption of a single parameter: the object volume. Only the teapot (Fig. 16.3) and vase (Fig. 16.1) examples needed additional user input for surface creases. Similar to Oswald et al. [41] this method assumes symmetry of objects and requires a representative silhouette from a side view. Therefore, the hole of the donut (Fig. 16.2) is underestimated in the same way. For the banana example (Fig. 16.1), the method still yields good results because the oblique object view did not change much on the characteristic silhouette properties.

Table 16.4 Overview of advantages and disadvantages for each method. Note that the number of advantages and disadvantages is not important in this listing since each point weights differently depending on the desired application of each method

Method	Advantages (+) and disadvantages (−)
Zhang	+ Large variety of constraints allows for flexible user modeling
et al. [63]	+ User has full control of every surface detail
	Reconstructions are restricted to a depth map
	− Occluded object parts cannot be modeled,
	synthesized views from different angles will reveal those areas
	− Large amount of user input is often necessary
Igarashi	+ Very easy to use and fast interactive modeling
et al. [23]	− Over-smoothes the input silhouette
	− Smoothness properties cannot be changed by the user
	− Not silhouette consistent
	− Topology is limited to genus 0
Prasad	+ Objects can also be modeled from oblique view points
et al. [45]	+ Apart from the silhouette the user can also use contour edges for modeling
	− Parametric surface representation limits topology and
	object shape (many long elongated structures are difficult to model)
	− Higher complexity of user input (requires expert knowledge)
	− Silhouette consistency is not guaranteed and may require additional user input
Oswald	+ Moderately fast modeling
et al. [41]	+ Reconstructions are silhouette consistent
	− Objects need to be symmetric, a side view is required
Töppe	+ Fast modeling
et al. [52]	+ Very little user input
	+ Reconstructions are silhouette consistent
	− Objects need to be symmetric, a side view is required
	− User can barely influence the surface shape
	− Limited possibilities to add surface creases

User Input Similar to Oswald et al. [41] the object silhouette is the only necessary user input (see also Fig. 16.6). Apart from the silhouette the object shape is mainly defined by the adapting the object volume. Also similar to Oswald et al. [41] surface creases can be added to create non-smooth surface parts. However, changing the smoothness locally may behave differently at different locations within the object silhouette, because there is no data term that governs the object shape in non-smooth surface areas. In these areas the surface shape is more defined by the distribution of volume which minimizes the overall surface area for given smoothness properties.

16.5.2.6 Summary

The theoretical and experimental comparison of the five methods for curved object reconstruction identified several advantages and disadvantages of the presented approaches which are listed in Table 16.4. In general, the performance of a method highly depends on the application. Each method has its strengths and weaknesses when applied to a specific class of single view reconstruction problems.

The results of our experiments support the hypothesis that *generality and flexibility* of a reconstruction method is traded for the *amount of user input* or *expert knowledge*. Expert knowledge refers to the *variety* or *complexity* of the user input. The flexibility of modeling fine details with the method by Zhang et al. [63] either requires the user to know and understand a variety of modeling constraints or it needs a large amount of user input. On the other hand, the method by Prasad et al. [45] needs less user input, but increases its complexity such as the definition of a suitable surface parametrization. The comparatively simple and small amount of user input for the methods by Igarashi et al. [23], Oswald et al. [41], and Töppe et al. [52] comes along with the limited generality and flexibility of these methods.

16.6 Conclusion

Single view reconstruction approaches infer the structure of 3D objects from 2D images. In this chapter, we discussed a representative set of existing algorithms and grouped them into four classes: curved objects, piecewise planar objects, learning specific objects and 3D impression from scenes. These groupings are based on the different reconstruction objectives of the algorithms. We identified several important categories for comparing the approaches: the reconstruction precision, the 3D object representation, the assumptions made by the algorithm, the required user interaction, as well as image cues and priors used by the algorithms. Moreover, we have concentrated on five methods aiming for curved object reconstruction and provided an extensive experimental comparison of established and recent methods.

Acknowledgements We thank Li Zhang and Steve Seitz for sharing their data and providing their single view modeling tool. Further, we are grateful to Mukta Prasad and co-authors for providing test data and results of their single view reconstruction method.

References

1. Bae, S., Durand, F.: Defocus magnification. In: Proceedings of the Eurographics. Blackwell, Oxford (2007)
2. Barinova, O., Konushin, V., Yakubenko, A., Lee, K., Lim, H., Konushin, A.: Fast automatic single-view 3-d reconstruction of urban scenes. In: Proceedings of the European Conference on Computer Vision, pp. 100–113. Springer, Berlin/Heidelberg (2008)
3. Botsch, M., Kobbelt, L.: An intuitive framework for real-time freeform modeling. In: ACM Transactions on Graphics (Proc. SIGGRAPH), vol. 23, pp. 630–634. ACM/Addison-Wesley Publishing Co., New York (2004)
4. Chen, Y., Cipolla, R.: Single and sparse view 3d reconstruction by learning shape priors. Comput. Vis. Image Underst. **115**, 586–602 (2011)

5. Cohen, L.D., Cohen, I.: Finite-element methods for active contour models and balloons for 2-d and 3-d images. IEEE Trans. Pattern Anal. Mach. Intell. **15**(11), 1131–1147 (1993)
6. Colombo, C., Del Bimbo, A., Pernici, F.: Metric 3d reconstruction and texture acquisition of surfaces of revolution from a single uncalibrated view. IEEE Trans. Pattern Anal. Mach. Intell. **27**, 99–114 (2005)
7. Courteille, F., Crouzil, A., Durou, J.-D., Gurdjos, P.: Towards shape from shading under realistic photographic conditions. pp. 277–280 (2004)
8. Criminisi, A., Reid, I., Zisserman, A.: Single view metrology. Int. J. Comput. Vis. **40**(2), 123–148 (2000)
9. Daum, M., Dudek, G.: On 3-d surface reconstruction using shape from shadows. In: Proceedings of the International Conference on Computer Vision and Pattern Recognition, Santa Barbara, CA, USA, pp. 461–468 (1998)
10. Delage, E., Lee, H., Ng, A.Y.: Automatic single-image 3d reconstructions of indoor Manhattan world scenes (2005)
11. Duchon, J.: Splines Minimizing Rotation-Invariant Semi-norms in Sobolev Spaces. In: Schempp, W., Zeller, K. (eds.) Constructive Theory of Functions of Several Variables, vol. 571/1977, pp. 85–100. Springer, Berlin (1977)
12. Durou, J.-D., Falcone, M., Sagona, M.: Numerical methods for shape-from-shading: a new survey with benchmarks. Comput. Vis. Image Underst. **109**, 22–43 (2008)
13. Francois, A.R.J., Medioni, G.G.: Interactive 3d model extraction from a single image. Image Vis. Comput. **19**(6), 317–328 (2001)
14. Han, F., Zhu, S.-C.: Bayesian reconstruction of 3d shapes and scenes from a single image. In: Proceedings of the Higher-Level Knowledge in 3D Modeling and Motion Analysis, Nice, France, pp. 12–20. (2003)
15. Hassner, T., Basri, R.: Example based 3d reconstruction from single 2d images. In: Beyond Patches Workshop at IEEE Conference on Computer Vision and Pattern Recognition, New York, NY, USA, p. 15. (2006)
16. Hatzitheodorou, M.: The derivation of 3-d surface shape from shadows. In: Proceedings of a Workshop on Image Understanding Workshop, pp. 1012–1020. Morgan Kaufmann, San Francisco (1989)
17. Hoiem, D., Efros, A.A., Hebert, M.: Automatic photo pop-up. ACM Trans. Graph. **24**(3), 577–584 (2005)
18. Hong, W., Yang, A.-Y., Huang, K., Ma, Y.: On symmetry and multiple-view geometry: structure, pose, and calibration from a single image. Int. J. Comput. Vis. **60**, 241–265 (2004). doi:10.1023/B:VISI.0000036837.76476.10
19. Horaud, R.P., Brady, M.: On the geometric interpretation of image contours. Artif. Intell. **37**(1–3), 333–353 (1988). Special Issue on Geometric Reasoning
20. Horry, Y., Anjyo, K.-I., Arai, K.: Tour into the picture: using a spidery mesh interface to make animation from a single image. In: SIGGRAPH '97: Proceedings of the 24th Annual Conference on Computer Graphics and Interactive Techniques, pp. 225–232. ACM/Addison-Wesley, New York (1997)
21. Igarashi, T.: Adaptive unwrapping for interactive texture painting. In: Proceedings of the ACM Symposium on Interactive 3D Graphics, pp. 209–216. ACM, New York (2001)
22. Igarashi, T.: Smooth meshes for sketch-based freeform modeling. In: Proceedings of the 2003 Symposium on Interactive 3D Graphics, pp. 139–142. ACM, New York (2003)
23. Igarashi, T., Matsuoka, S., Tanaka, H.: Teddy: a sketching interface for 3d freeform design. In: ACM Transactions on Graphics (Proceedings of the SIGGRAPH), pp. 409–416. ACM/Addison-Wesley, New York (1999)
24. Ikeuchi, K., Horn, B.K.P.: Numerical shape from shading and occluding boundaries. Artif. Intell. **17**, 141–185 (1981)
25. Joshi, P., Carr, N.: Repoussé: automatic inflation of 2d art. In: Proceedings of the Eurographics Workshop on Sketch-Based Modeling, Annecy, France (2008)
26. Kanade, T.: Recovery of the three-dimensional shape of an object from a single view. Artif. Intell. **17**, 409–460 (1981)

27. Karpenko, O.A., Hughes, J.F.: Smoothsketch: 3d free-form shapes from complex sketches. ACM Trans. Graph. **25**(3), 589–598 (2006)
28. Karpenko, O., Hughes, J.F., Raskar, R.: Free-form sketching with variational implicit surfaces. Comput. Graph. Forum **21**(3), 585–594 (2002)
29. Kender, J., Smith, E.: Shape from darkness. In: Proceedings of the International Conference on Computer Vision, London, pp. 539–546 (1987)
30. Koutsourakis, P., Simon, L., Teboul, O., Tziritas, G., Paragios, N.: Single view reconstruction using shape grammars for urban environments. In: Proceedings of the International Conference on Computer Vision, Kyoto. IEEE (2009)
31. Kushal, A.M., Sanyal, S., Bansal, V., Banerjee, S.: A simple method for interactive 3d reconstruction and camera calibration from a single view. In: Proceedings Indian Conference in Computer Vision, Graphics and Image Processing, Ahmedabad, India (2002)
32. Levin, A.: Analyzing depth from coded aperture sets. In: Proceedings of the European Conference on Computer Vision, pp. 214–227. Springer, Berlin (2010)
33. Li, Y., Pizlo, Z., Steinman, R.M.: A computational model that recovers the 3d shape of an object from a single 2d retinal representation. Vis. Res. **49**, 979–991 (2009)
34. Li, Z., Liu, J., Tang, X.: A closed-form solution to 3d reconstruction of piecewise planar objects from single images. In: Proceedings of the International Conference on Computer Vision and Pattern Recognition, Minneapolis, MN, USA, pp. 1–6 (2007)
35. Liebowitz, D., Criminisi, A., Zisserman, A.: Creating architectural models from images. In: Proceedings of the EuroGraphics, vol. 18, pp. 39–50 (1999)
36. Lowe, D.G.: Three-dimensional object recognition from single two-dimensional images. Artif. Intell. **31**, 355–395 (1987)
37. Malik, J., Rosenholtz, R.: Computing local surface orientation and shape from texture for curved surfaces. Int. J. Comput. Vis. **23**(2), 149–168 (1997)
38. Nagai, T., Naruse, T., Ikehara, M., Kurematsu, A.: Hmm-based surface reconstruction from single images. In: Proceedings of the International Conference on Image Processing, Rochester, NY, USA, vol. 2, pp. II–561–II–564 (2002)
39. Nayar, S.K., Ikeuchi, K., Kanade, T.: Shape from interreflections. In: Proceedings of the International Conference on Computer Vision, Osaka, Japan, pp. 2–11 (1990)
40. Nealen, A., Igarashi, T., Sorkine, O., Alexa, M.: Fibermesh: designing freeform surfaces with 3d curves. ACM Trans. Graph. **26**(3), 41 (2007)
41. Oswald, M.R., Töppe, E., Kolev, K., Cremers, D.: Non-parametric single view reconstruction of curved objects using convex optimization. In: Pattern Recognition (Proceedings of the DAGM), Jena (2009)
42. Prados, E., Faugeras, O.: Shape from shading: a well-posed problem? In: Proceedings of the International Conference on Computer Vision and Pattern Recognition, vol. 2, pp. 870–877. IEEE, San Diego/Etats-Unis (2005)
43. Prasad, M.: Class-based single view reconstruction. Ph.D. thesis, University of Oxford (2009)
44. Prasad, M., Zisserman, A., Fitzgibbon, A.W.: Fast and controllable 3D modelling from silhouettes. In: Eurographics, Short Papers, pp. 9–12 (2005)
45. Prasad, M., Zisserman, A., Fitzgibbon, A.W.: Single view reconstruction of curved surfaces. In: Proceedings of the International Conference on Computer Vision and Pattern Recognition, New York, pp. 1345–1354 (2006)
46. Rother, D., Sapiro, G.: Seeing 3d objects in a single 2d image. In: Proceedings of the International Conference on Computer Vision, Kyoto, pp. 1819–1826 (2009)
47. Saxena, A., Chung, S.H., Ng, A.Y.: 3-d depth reconstruction from a single still image. Int. J. Comput. Vis. **76**, 53–69 (2007)
48. Saxena, A., Sun, M., Ng, A.Y.: Make3d: learning 3d scene structure from a single still image. IEEE Trans. Pattern Anal. Mach. Intell. **31**(5), 824–840 (2009)
49. Sturm, P.F., Maybank, S.J.: A method for interactive 3d reconstruction of piecewise planar objects from single images. In: Proceedings of the BMVC, pp. 265–274. British Machine Vision Association, England (1999)

50. Super, B.J., Bovik, A.C.: Shape from texture using local spectral moments. IEEE Trans. Pattern Anal. Mach. Intell. **17**, 333–343 (1995)
51. Terzopoulos, D., Witkin, A., Kass, M.: Symmetry-seeking models and 3d object reconstruction. Int. J. Comput. Vis. **1**, 211–221 (1987)
52. Töppe, E., Oswald, M.R., Cremers, D., Rother, C.: Image-based 3d modeling via cheeger sets. In: Proceedings of the Asian Conference on Computer Vision, Queenstown (2010)
53. Ulupinar, F., Nevatia, R.: Shape from contour: straight homogeneous generalized cylinders and constant cross section generalized cylinders. IEEE Trans. Pattern Anal. Mach. Intell. **17**, 120–135 (1995)
54. Utcke, S., Zisserman, A.: Projective reconstruction of surfaces of revolution. In: Pattern Recognition (Proceedings of the DAGM), pp. 93–102. Springer, Berlin/New York (2003)
55. Vetter, T.: Synthesis of novel views from a single face image. Int. J. Comput. Vis. **28**, 103 (1998)
56. Wang, G., Su, W., Song, Y.: A new shape from shading approach for specular surfaces. In: Proceedings of the Third International Conference on Artificial Intelligence and Computational Intelligence – Volume Part III. Lecture Notes in Computer Science, pp. 71–78. Springer, Berlin/Heidelberg (2011)
57. Welch, W., Witkin, A.P.: Free-form shape design using triangulated surfaces. In: ACM Transactions on Graphics (Proceedings of the SIGGRAPH), pp. 247–256 (1994)
58. Williams, L.: 3d paint. In: Proceedings of the 1990 Symposium on Interactive 3D Graphics, Snowbird, pp. 225–233 (1990)
59. Wong, K.-Y.K., Mendonça, P.R.S., Cipolla, R.: Reconstruction of surfaces of revolution from single uncalibrated views. In: Proceedings of the British Machine Vision Conference, Cardiff, pp. 265–272 (2002)
60. Yu, Y., Chang, J.: Shadow graphs and surface reconstruction. In: Proceedings of the European Conference on Computer Vision, Copenhagen, pp. 31–45 (2002)
61. Zeleznik, R.C., Herndon, K.P., Hughes, J.F.: Sketch: an interface for sketching 3d scenes. In: ACM Transactions on Graphics (Proceedings of the SIGGRAPH), pp. 163–170. ACM, New York (1996)
62. Zhang, R., Tsai, P.-S., Cryer, J.E., Shah, M.: Shape from shading: a survey. IEEE Trans. Pattern Anal. Mach. Intell. **21**(8), 690–706 (1999)
63. Zhang, L., Dugas-Phocion, G., Samson, J.-S., Seitz, S.M.: Single view modeling of free-form scenes. In: Proceedings of the International Conference on Computer Vision and Pattern Recognition, Kauai, HI USA, pp. 990–997 (2001)

Chapter 17
On Globally Optimal Local Modeling: From Moving Least Squares to Over-parametrization

Shachar Shem-Tov, Guy Rosman, Gilad Adiv, Ron Kimmel, and Alfred M. Bruckstein

Abstract This paper discusses a variational methodology, which involves locally modeling of data from noisy samples, combined with global model parameter regularization. We show that this methodology encompasses many previously proposed algorithms, from the celebrated moving least squares methods to the globally optimal over-parametrization methods recently published for smoothing and optic flow estimation. However, the unified look at the range of problems and methods previously considered also suggests a wealth of novel global functionals and local modeling possibilities. Specifically, we show that a new non-local variational functional provided by this methodology greatly improves robustness and accuracy in local model recovery compared to previous methods. The proposed methodology may be viewed as a basis for a general framework for addressing a variety of common problem domains in signal and image processing and analysis, such as denoising, adaptive smoothing, reconstruction and segmentation.

17.1 Introduction

A fundamental problem in both image and signal processing is that of recovering a function, a curve or a surface (i.e., a signal or an image) from its noisy and distorted samples. Significant research effort was invested in this problem and the results obtained so far are quite remarkable. The most important ingredient in the success of any method that extracts signals from noise is, of course, the set of assumptions that

S. Shem-Tov (✉) · G. Rosman · R. Kimmel · A.M. Bruckstein
Technion - Israel Institute of Technology, 2000, Haifa, Israel
e-mail: shemtov@cs.technion.ac.il; rosman@cs.technion.ac.il; ron@cs.technion.ac.il; freddy@cs.technion.ac.il

G. Adiv
Rafael, Haifa, Israel
e-mail: gilad3a@gmail.com

M. Breuß et al. (eds.), *Innovations for Shape Analysis*, Mathematics and Visualization,
DOI 10.1007/978-3-642-34141-0_17,

summarizes our prior knowledge about the properties of the signal that effectively differentiates it from the noise. These assumptions range from some vague general requirements of smoothness on the signals, to quite detailed information on the structure or functional form of the signals that might be available due to prior knowledge on their sources.

The prior information on signal/image is often expressed in the form of a parameterized model. For instance, in speech recognition [25] slowly varying coefficients of the short-time Fourier transform (STFT) are used to locally describe and model highly fluctuating spectral characteristics over time. In object recognition the choice of the correct spatial support for objects i.e., the segmentation, is a fundamental issue [19], hence, in general scene understanding, support maps are used to represent segmentation of images into homogeneous chunks, enabling the representation of objects as disjoint regions with different local modeling parameters [9]. This concept of spatial support selection for estimation problems is closely related to layered models of scenes which offer significant benefits for motion estimation producing state of the art results [33]. In geometric modeling, B-splines (which are essentially a continuous set of piecewise polynomials), are used for local curve and surface approximation, interpolation and fitting from noisy samples [8]. In model-based texture segmentation [13], the selection of an appropriate support set for the local model, is important to obtain a good local texture representation, which then serves as a basis for the segmentation process. In sparse coding [1, 23], the main goal is to model data vectors (signals) as a linear combination of a few elements (support set) from a known dictionary. Sparse coding has proven to be very effective for many signal or image processing tasks, as well as advances in computer vision tasks such as object recognition.

One of the widespread and successful methods for local signal modeling, is the celebrated *moving least squares* local fitting method (MLS), which in recent years has evolved to become an important tool in both image and signal processing and in computational graphics. In [17], Levin explored the moving least-squares method and applied it to scattered-data interpolation, smoothing and gradient approximation. In [2, 4, 18] the moving least squares technique was employed for modeling surfaces from point-sampled data, and proved to be a powerful approach. This was followed by the work of Fleishman et al. [2, 11], incorporating robust statistics mechanisms for outlier removal. Common to these works is the locality of the fitting procedure and the lack of global assumptions expressing prior knowledge on the variations of local parameters.

The aim of this paper is to show how one can design variational functionals that exploit local fitting of models and global smoothness assumptions on the variations of model parameters, that are natural for various types of signals. A first attempt, at such a variational methodology, was made by Nir et al. [21, 22]. This work was subsequently broadened and generalized by Bruckstein in [6], by the realization that over-parametrization methods naturally follow from combining moving least squares, or other local fitting methods, with global priors on parameter variations. The local modeling relates to a wealth of classical methods, such as Haralick's and Watson's *facet model* for images [14] and extends them in many ways.

More importantly, the discussion and experimental results reported in this paper point at a rather general methodology for designing functionals for variational model estimation and signal reconstruction and focusing on the denoising problem is merely an illustrative test case.

The importance of the proposed variational framework, lies in the fact that it allows for directly incorporating knowledge of the problem domain, hence it is easily extendable to address numerous problem areas, such as denoising, deconvolution and optical flow in image and signal processing and various other fields of research. Moreover, due to the structure of the proposed functionals, our variational framework is able, unlike many common methods, to accurately recover the underlying model of a signal while addressing its main aim.

17.2 The Local Modeling of Data

For the sake of simplicity, we shall here limit our discussion to one dimensional signals and address the generic problem of denoising. Let $f(x)$ be a one dimensional signal and $f_{noisy}(x) = f(x) + n(x)$ be it's noisy counterpart. Also, denote by $\{f_j = f(x_j) + n(x_j)\}$ the set of samples of the noisy signal. Suppose that f can be (locally) described by a parameterized model of the form

$$f(x) = \sum_{i=1}^{n} A_i \phi_i(x) \tag{17.1}$$

where $A = \{A_i\}_{i=1}^n$ is a set of parameters, and $\phi = \{\phi_i\}_{i=1}^n$ is a set of *'basis'* signals. *Local modeling* is the process of estimating $A = \{A_i\}$ from the noisy signal f_{noisy} or it's samples $\{f_j\}$, in the neighborhood of a point x. As a simple example, we can consider the Taylor approximation as a parameterized model with polynomial basis functions $\phi_i = x^{i-1}$.

Suppose we would like to use this local modeling for denoising our noisy data $f_{noisy}(x)$ or $\{f_j\}$. Then, around $x = x_0$, we want to estimate $A_i(x_0)$, i.e., the parameters of the model (17.1), by solving:

$$\arg \min_{[A_1, A_2, \ldots, A_n]} \left\| f_{noisy}(x_0) - \sum_{i=1}^{n} A_i \phi_i(x_0) \right\|, \tag{17.2}$$

in some local neighborhood of x_0 and a distance norm $\|\cdot\|$. This minimization gives us the best local estimate of $\hat{f}(x_0)$

$$\hat{f}(x_0) = \sum_{i=1}^{n} A_i(x_0) \phi_i(x_0). \tag{17.3}$$

Repeating this process for every location x give us the "moving" best estimate of f.

The choice of the distance or measure of error is of a great importance. One common choice is the weighted least squares distance, as considered, for example, by Farnebäck [10], as a generalization to the facet model:

$$\left\| f_{noisy}(x) - \sum_{i=1}^{n} A_i(x)\phi_i(x) \right\|_w = \int \left(f_{noisy}(y) - \sum_{i=1}^{n} A_i(x)\phi_i(y) \right)^2 w(y-x)\,dy. \tag{17.4}$$

This is essentially a weighted L_2 norm, where $w(\cdot)$ is a weight function which localizes the estimation of the parameters $A_i(x_0)$ (This error measure can also be applied to a sampled signal). Both the continuous and the discrete cases, described above, yield a process to compute the local model parameters $A_i(x)$ and therefore for estimating the signal $\hat{f}(x)$, not only at x_0 but at all x. This process is the well-known moving least squares estimation process.

We wish to emphasize that although our proposed methodology, discussed in the next section, focuses on the L_2 norm, this is by no means the only choice of error measure. The choice of the error measure may be adapted to each particular problem of interest. Other measures, such as the *least sum of absolute values* distance [34] or L_1 norm, can readily be substituted into the cost functionals.

17.3 Global Priors on Local Model Parameter Variations

In the previous discussion we did not impose any conditions on the model parameters A_i, and in the definition of the neighborhood around x_0, via the weight functions, we did not use any idea of adaptation to the data.

Suppose now that we are given some structural knowledge of the signal $f(x)$. We would like to use this knowledge to improve the estimation process. For example, suppose we a priori know that $f(x)$ is a piecewise polynomial signal over a set of intervals, i.e., we know that:

$$f(x) = \sum_{i=1}^{n} A_i^r x^{i-1} \quad for \quad x \in [x_r, x_{r+1}] \tag{17.5}$$

but we do not know the sequence of breakpoints $\{x_r\}$. Using polynomial basis functions $\left[1, x, x^2, \ldots, x^{n-1}\right]$, we know a priori that a good estimate for $f(x)$ may be provided by piecewise constant sets of parameters $\{A\}$, over $[x_r, x_{r+1}]$ segments, and changes in the parameters occur only at the breakpoints $\{x_r\}$. Such knowledge provides us the incentive to impose some global prior on the parameters, such that the minimum achieved by the optimization process, will indeed favor them to be piecewise constant. This may be achieved by supplementing the moving least squares local fitting process with constraints on the variations of the parameters in the minimization process. Thus, we shall force the estimated parameters not only to provide the best weighted local fit to the data, but also to be consistent with the local fitting over adjacent neighborhoods. This is where we diverge and extend the facet model, which assigns the basis function's parameters at each point, using solely a local least square fit process.

In this paper, the assumed structural knowledge of the signal implies that good estimates can be achieved by a piecewise constant model i.e., a model whose parameters are piecewise constant. Therefore the focus of this work will be to design functionals which also impose global priors on the model parameters. We shall demonstrate that one can design a functional that does indeed fulfil this requirement.

The mix of local fitting and global regularity is the main idea and power behind over-parameterized variational methods and is what makes them a versatile problem solving tool. By adapting the local fitting process and incorporating the global prior (in a way that will be described in the following sections), this methodology can be readily applied to address problems in various domains.

17.4 The Over-parameterized Functional

In [21] Nir and Bruckstein presented a first attempt at noise removal based on an over-parameterized functional. This functional was, similarly to many well known functionals, a combination of two terms, as follows:

$$E\left(f, A\right) = E_D\left(f, A\right) + \alpha E_S\left(A\right). \tag{17.6}$$

Here α, is some fixed relative weight parameter, E_D is a data or fidelity term, and E_S is a regularization term. The data term E_D was chosen to:

$$E_D(f, A) = \int \left(f_{noisy}(x) - \sum_{i=1}^{n} A_i(x)\phi_i(x) \right)^2 dx. \tag{17.7}$$

Note that this functional implies a neighborhood of size 0 in the distance measure (17.4), which means that this data term only penalizes for *point-wise* deviations of the estimated signal via $\sum_{i=1}^{n} A_i(x)\phi_i(x)$ from f_{noisy}.

The smoothness or regularization term, E_S, was defined to penalize variations of the parameters A_i as follows

$$E_S(A) = \int \Psi \left(\sum_{i=1}^{n} A_i'(x)^2 \right) dx \tag{17.8}$$

where $\Psi\left(s^2\right) = \sqrt{s^2 + \epsilon^2}$.

The resulting functional yields a channel-coupled *total variation* (TV) regularization process for the estimation of the model parameters. Note that (17.8) is an approximated L_1 type of regularizer (sometimes referred to as the Charbonnier penalty function). This regularizer causes the functional to be more robust to outliers, and allows for smaller penalties for high data differences (compared to a quadratic regularizer), while maintaining convexity and continuity [5, 20]. The regularization term was designed to impose the global prior on the parameters. It is channel-coupled to "encourage" the parameters to change simultaneously, thus preferring a piecewise constant solution as described in Sect. 17.3.

In our experiments (which are discussed in Sect. 17.7), as well as in [21] , this functional displayed good performance for noise removal compared with Rudin Osher and Fatemi's [28] classical total variation noise removal functional. A similar functional, with data term modifications, was used by Nir et al. in [22] for optical flow estimation, producing state of the art results.

17.4.1 The Over-parameterized Functional Weaknesses

Despite the good performance displayed by the over-parameterized functional, it still lacks with regard to the following shortcomings, that were clear in our experiments:

Discontinuities smearing: As mentioned, the regularization term is an approximate L_1 regularizer. A precise L_1 regularizer is indifferent to the way signal discontinuities appear, i.e. the same penalty is given to a smooth gradual signal change, and to sharp discontinuities (as long as the total signal difference is the same). See for example Pock's Ph.D. work [24] for a detailed example. We consider this property as a shortcoming, because we expect the reconstructed parameters to be piecewise constant, where discontinuities appear as relatively few and sharp changes, hence this regularizer does not convey the intended global prior on the parameters, and does not prefer a "truly" piecewise constant solution. In practice the problem is even more severe: first the selection of ϵ constant, in the Charbonnier penalty function, proves to be problematic. Choosing a bigger ϵ value causes the functional to lose the ability to preserve sharp discontinuities and actually prefers to smooth out discontinuities. On the other hand, choosing a smaller ϵ value degenerates the penalty function. In fact, for any choice of ϵ, this penalty function will tend to smooth sharp discontinuities. Second, as discussed above the TV-L_1 model suffers from the so called *staircasing* effect, where smooth regions are recovered as piecewise constant staircases in the reconstruction. See the work of Savage et al. [29] and references therein, for a detailed review of such effects.

Origin biasing: The over-parameterized functional's global minimum may depend on the selected origin of the model. In the over-parameterized methodology we consider basis functions which are defined globally across the signal domain. This definition requires us to fix an arbitrary origin for the basis functions. As the true parameters value may vary with a change of the origin choice, thus the value of the regularization term may also vary. For instance, consider the value of the constant term in a linear function, which determines the point at which the line crosses the y-axis. Change of the y-axis location i.e., the origin location, will incur achange in the value the constant term. This dependency on the choice of the basis function origin is termed Origin biasing. A detailed example, regarding the optical flow over-parameterized functional with affine flow basis functions, is given in the work of Trobin et al. [35].

We also note that the data term presented above only penalizes for point-wise deviation form the model, hence it imposes only a point-wise constraint on the functional's minimum, relying only on the regularization term to impose the global constraint. A discussion why this is a problematic issue is given in Sect. 17.5.3.

Overall, it is evident, that despite producing good results when applied to various applications, the over-parameterized functional model is fundamentally flawed when attempting to accomplish parameter reconstructions. On one hand the over-parameterized model provides a solution domain wider than the TV model, for the functional to "choose" from, thus often enabling convergence to excellent denoising solutions, on the other hand the constraints applied to the solution domain, through the functional, are not strong enough as to impose convergence to piecewise-constant-parameters solution, as demonstrated in Sect. 17.7.

17.5 The Non-local Over-parameterized Functional

To overcome the shortcomings described in Sect. 17.4.1, we shall modify the functional, both in the data term and in the regularization term, as described below

17.5.1 The Modified Data Term: A Non-local Functional Implementing MLS

In order to overcome the point-wise character of the data term, and to impose a neighborhood constraint in the spirit of (17.4) in the data term, we extend it to what is commonly referred to as a *non-local functional* [6, 12, 16]. This is done simply by means of defining a weighting function which considers more then point-wise differences.

Making the data term a non-local functional, requires the parameters to model the signal well over a neighborhood of each point. We note that the robustness of the parameters estimate increases with the size of the support set. On the other hand, increasing the size of the support set too much, may reduce the functional's ability to detect discontinuities and to preserve them.

The non-local data term functional is

$$E_D = \int_x \int_y \left(f_{noisy}(y) - \sum_{i=1}^{n} A_i(x)\,\phi_i(y) \right)^2 w(x,y)\,dydx \tag{17.9}$$

and it conforms to the functional form of the weighted least squares fit distance defined in Sect. 17.2. Note that there is yet no restriction on the size or shape of the support set around each point that is induced by the weighting function $w(x,y)$.

For the 1D case, which is thoroughly explored in this work, we defined a simple yet powerful sliding window weighting function, as described in Sect. 17.5.1.1. The generalization of the functional to the 2D case may require a more sophisticated optimal data dependent weighting function, which will be the subject of future research.

17.5.1.1 Weighting Scheme

The choice of the weighting scheme is of great importance. For each location x_0, the weighting scheme expresses the support set from which the parameters are to be reconstructed from i.e. the set of samples which are used for the implicit MLS fitting process calculated in the minimization process. Thus the weighting scheme should be designed in such a manner which, at points near a discontinuity, will prevent the combination of samples from different sides of the discontinuity, thus enables the preservation of signal discontinuities in the reconstruction process. This implies that a simple box or gaussian window around each location x_0 will not suffice, as it may spread a discontinuity across several samples around the actual location of the discontinuity (which was evident in various experiments we performed).

In order to avoid this behavior, we defined an adaptive sliding window weighting function, which is closely related to concepts of ENO schemes used for the approximation of hyperbolic conservation laws first introduced in the fundamental work of Harten et al. [15] (a detailed description was presented by Shu in [32]). For each point x of the signal, we choose a window W of length N such that, $x \in W$ and $r(x) = \sum_{j=1}^{N} r(x, y_j)$ is minimal, where $y_j \in W$ and $r(x, y_j)$ is the non-weighted least squares fit distance at x:

$$r(x, y_j) = \left(f_{noisy}(y_j) - \sum_{i=1}^{n} A_i(x)\,\phi_i(y_j) \right)^2. \tag{17.10}$$

Specifically, we chose a successive set of points of size N which include x and also minimizes $r(x)$. For example, for a window of size 5, there are 5 window

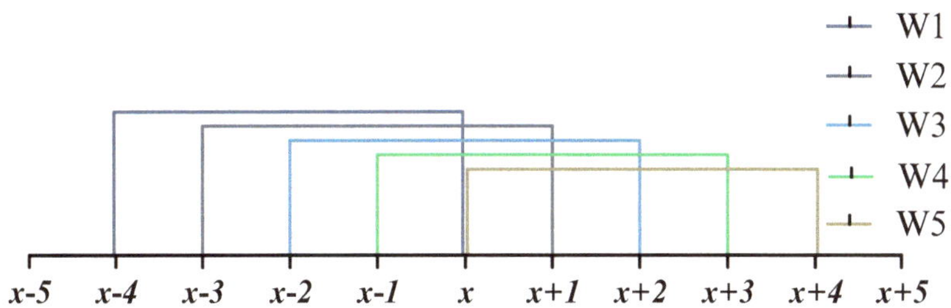

possibilities as described below: We mark the chosen window by w^x, and the selected weight function will then be:

$$w(y-x) = \begin{cases} \frac{1}{N} & if \quad y \in w^x \\ 0 & otherwise \end{cases} \tag{17.11}$$

By no means do we claim that this is the best choice of the weighting function. This is but one possible, adaptive weighting function selection process, that enhances the data term to impose a non point-wise constraint, while allowing for the preservation and sharpening of discontinuities in the signal and that was found to yield good results. Note that the choice of the window is sensitive to noise, therefore for achieving better results we updated the selected window at each location throughout the minimization process as described in Sect. 17.6.

17.5.2 The Modified Regularization Term

We extend the over-parameterized regularization term using the *Ambrosio-Tortorelli (AT)* scheme [3] in a similar manner as done in [7] for the TV functional, while retaining the L_1 penalty function as proposed by Shah [30, 31], and applied to the over-parameterization functional by Rosman et al. [26]. This functional transforms the regularization into an implicit segmentation process, each segment with it's own set of parameters. In effect, the AT scheme allows the regularization term to prefer few parameter discontinuities and to prevent discontinuities from smearing into neighboring pixels via the diffusion process, thus allowing piecewise smooth solution. This directly addresses the discontinuities smearing effect described in Sect. 17.4.1. The choice of L_1 regularizer, as opposed to the L_2 regularizer in original AT scheme, is due to the fact that a L_1 regularizer better encourages a piecewise constant solution, which is the intended global prior we wish to impose on the solution. Exploration of different sub-L_1 regularizer functions, is beyond the scope of this paper.

The chosen AT regularization term is:

$$E_{S,AT} = \int (1 - v_{AT})^2 \Psi\left(\sum_{i=1}^{n} ||A_i'||^2\right) + \rho_1 (v_{AT})^2 + \rho_2 \left\|v_{AT}'\right\|^2 \tag{17.12}$$

where v_{AT} is a diffusivity function, ideally serving as an indicator of the parameters discontinuities set in the signal model. Suppose we have a piecewise linear signal, and an ideal solution (A^*, v^*_{AT}) where A^* is piecewise constant, and the diffusivity function v^*_{AT} is 1 at the linear regions boundaries and 0 elsewhere. With such a solution, we expect two neighboring points, belonging to different regions, to have a very small, in fact negligible, diffusivity interaction between them. This is controlled by the value of v^*_{AT} at those points, which effectively cancels the diffusivity interaction between the different sets of linear parameters. Furthermore, the cost associated with this solution is directly due to the discontinuity set measure in the signal i.e. to

$$\int \rho_1 (v_{AT})^2 + \rho_2 \left\| v'_{AT} \right\|^2 \tag{17.13}$$

hence the penalty no longer depends on the size of the signal difference at discontinuities. Moreover, the AT regularization addresses the origin biasing effect, described in Sect. 17.4.1, by making the functional much less sensitive to the selected origin. This is due to the fact that ideally we consider only piecewise constant parameters solutions. These solutions nullifies the regularization term energy at every location except for discontinuities where the energy depends solely on energy (17.13). Therefore the ideal piecewise constant solution becomes a global minimizer of the functional.

17.5.3 *Effects of the Proposed Functional Modifications*

An obvious question arises: why do we need to modify both data and regularization terms? To answer this, we first notice that using only the non-local data term improves the local parameter estimates, but cannot prevent the discontinuities smearing effect. A moving least squares process, with small window, will yield excellent parameter estimates, but is unable to prevent the diffusion process from combining data from neighboring segments, thus smoothing and blurring the estimates at boundaries locations. Therefore we need to add the AT scheme to the functional, which has the ability to prohibit the diffusion process at discontinuities.

But then one might expect that the AT scheme without the non-local data term would suffice, by segmenting the regularization into piecewise constant regions, and relying on the data term and the global regularization to recover the correct parameters for each segment. In practice this is not the case. Consider the following illustrative test case: y_s a discontinuous signal, depicted in Fig. 17.1, and suppose we initialize the model parameters to a smoothed version of y_s achieved by calculating the moving least squares fit solution, with a centralized window, on the clean signal. We will now discuss applying different functionals for reconstructing the model parameters from clean signal y_s and the initialized smooth parameters.

Figures 17.2 and 17.3, depict snapshots of the reconstructed signal, parameters, and the resulting AT indicator function v_{AT} at different time steps of the minimization process. We used $\alpha = 0.05$ and $\alpha = 10$ respectively. These are snapshots of the

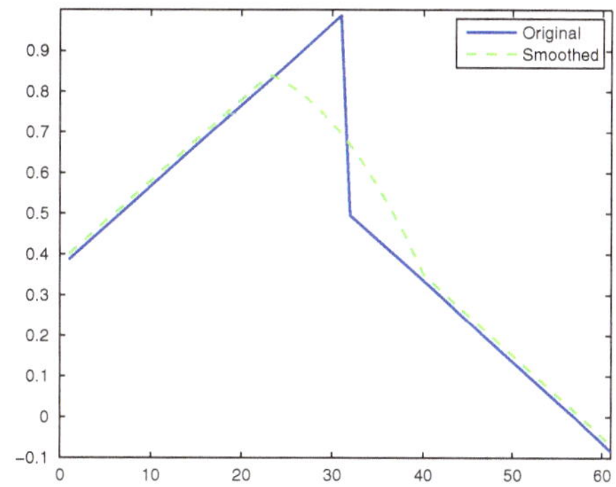

Fig. 17.1 *Solid line* – y_s a piecewise linear signal with one discontinuity. *Dashed line* – a smooth version of y_s with which we initialized the model parameters

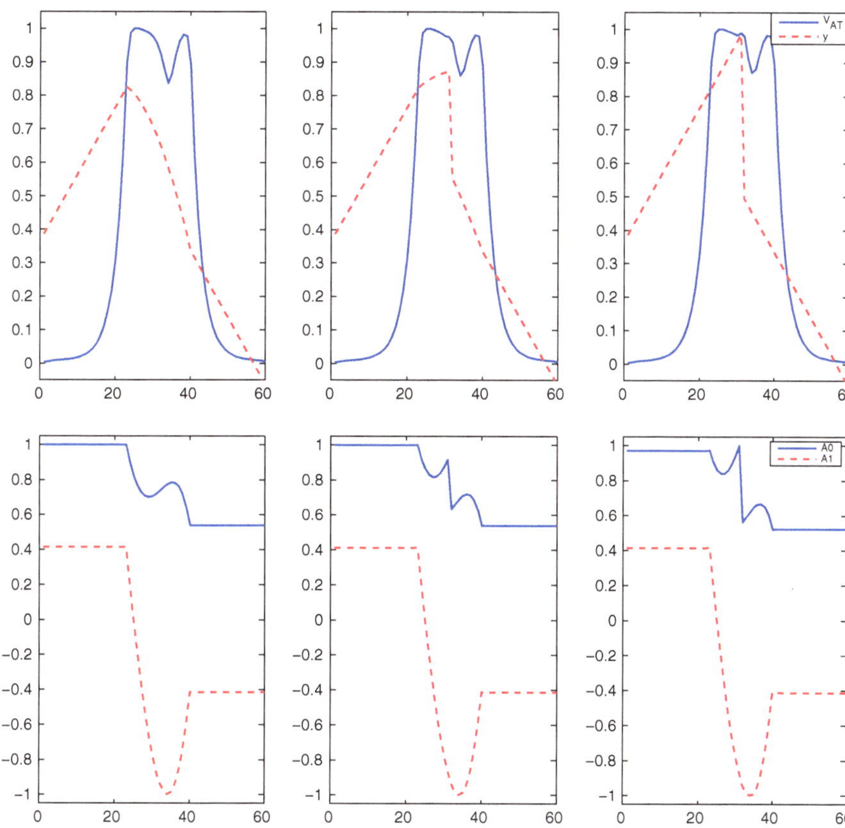

Fig. 17.2 From *left* to *right*, snapshots at various times, of the point-wise over-parameterized functional modified with the AT regularization term (MOP), with relative weight of E_s $\alpha = 0.05$, $\rho_1 = 7.5$ and $\rho_2 = 5$. The *top image* displays the reconstructed signal and the v_{AT} indicator function. The *bottom image* displays the reconstructed parameters

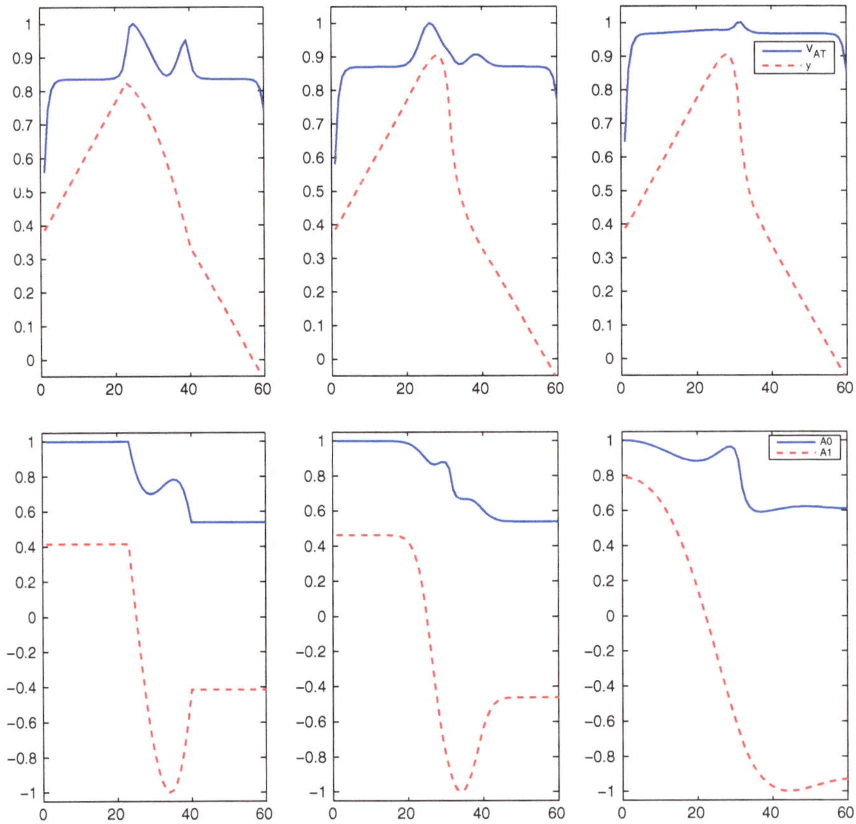

Fig. 17.3 From *left* to *right*, snapshots at various times, of the Mpoint-wise over-parameterized functional modified with the AT regularization term (MOP), with relative weight of E_s $\alpha = 10$, $\rho_1 = 7.5$ and $\rho_2 = 5$. The *top image* displays the reconstructed signal and the v_{AT} indicator function. The *bottom image* displays the reconstructed parameters

minimization process of the point-wise over-parameterized functional, modified with the AT regularization term (MOP). Both minimizations were carried out until convergence. It is evident that the reconstructed parameters are not piecewise constant, as we would like to have from our prior information on the signal. Also, it is important to note that in the first series (Fig. 17.2) the signal is perfectly reconstructed, effectively nullifying the data term, and in the second series (Fig. 17.3) the signal is smoothed, however the data term energy is still very low.

In contrast, Fig. 17.4 depicts a series of snapshots of the minimization process of the non-local over-parameterized functional (NLOP). Note that here the parameters are perfectly reconstructed, the AT indicator function v_{AT} receives a value close to one only in the vicinity of the discontinuity and also the signal is perfectly reconstructed.

We wanted to try and understand why did the MOP functional converged to such a solution, is this solution a local/global minimum of the functional? We addressed

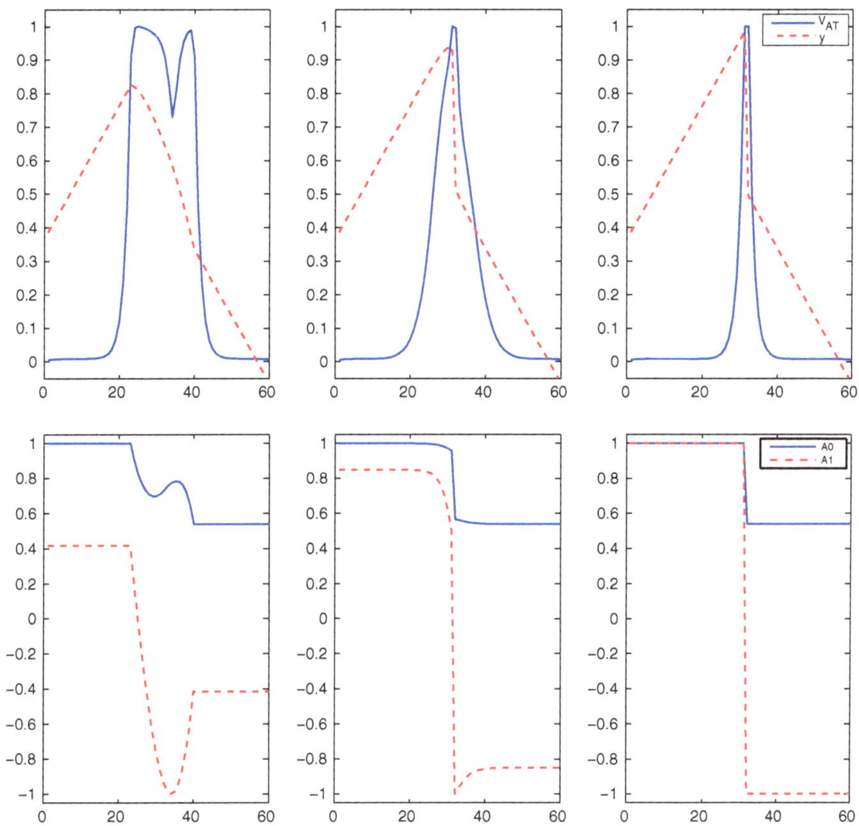

Fig. 17.4 From *left* to *right*, snapshots at various times, of the NLOP functional, with relative weight of E_s $\alpha = 0.1$, $\rho_1 = 0.005$ and $\rho_2 = 0.0025$. The *top image* displays the reconstructed signal and the v_{AT} indicator function. The *bottom image* displays the reconstructed parameters

the question in hand through functional energy consideration. Using the expected global minimum of both functionals (which was achieved by the minimization process of the NLOP functional), we calculated using interpolation, hypothetical steps of a minimization from the solution achieved by the MOP functional to the global minimum. We then calculated the energy of the MOP functional on each hypothetical step. It becomes obvious that the energy of the MOP functional is rising before dropping to the energy level of the expected global minimum, indicating that indeed the MOP functional converged into a local minimum. Separate calculation of the energy of the data and the regularization terms achieved in the minimization of the MOP functional, indicates that most of the functional energy is concentrated in the regularization term. In the transition between the local and global minimum solutions, the regularization term energy rises and dictates the total energy change, while the data term contribution is negligible.

The solutions to which the MOP functional converges and energy considerations, lead us to the conclusion that the MOP functional is converging into a local minimum solution. This "trapping" effect is alleviated in the NLOP functional, where the presumed local minimum, achieved by the MOP, is no longer a local minimum. This is due to the fact that it is much more difficult to drive the energy of the non-local data term close to zero and it contributes significantly to drive the parameters toward their correct values. Thus, the minimization process does not halt and continues toward the global minimum.

17.5.4 Euler-Lagrange Equations

Once we designed the functionals to be minimized, by interchangeably fixing the $A_i(x), i = 1 \ldots n$ and $v_{AT}(x)$ functions, we readily obtain the Euler-Lagrange equations which characterize the minimizers of the functional.

17.5.4.1 Minimization with Respect to $A_q(x), q = 1 \ldots n$ (Parameter Minimization Step)

Fixing $v_{AT}(x)$, we obtain

$$\forall q = 1 \ldots n, \quad \nabla_{A_q} E_D - \frac{d}{dx}\left(\nabla_{A'_q} \alpha E_{S,AT}\right) = 0. \tag{17.14}$$

the variation of the data term with respect to the model parameter functions $A_q(x)$ is given by

$$\nabla_{A_q} E_D = 2 \int_y \left(f_{noisy}(y) - \sum_{i=1}^{n} A_i(x)\, \phi_i(y) \right) \phi_q(y) w(x, y)\, dy. \tag{17.15}$$

For the smoothness term, the Euler-Lagrange equations are

$$\begin{aligned}\frac{d}{dx}\left(\nabla_{A'_q} \alpha E_{S,AT}\right) = {} & 4\alpha\left(1 - v_{AT}\right) v'_{AT} \Psi' \left(\sum_{i=1}^{n} \|A'_i\|^2 \right) A'_q \\ & + 2\alpha(1 - v_{AT})^2 \frac{d}{dx}\left(\Psi' \left(\sum_{j} \|A'_i\|^2 \right) A'_q \right)\end{aligned} \tag{17.16}$$

thus, the energy is minimized by solving the following nonlinear system of equations at each point x, $\forall q = 1 \ldots n$

$$
\begin{aligned}
&2\int_{y}\left(f_{noisy}(y)-\sum_{i=1}^{n}A_i\,(x)\,\phi_i(y)\right)\phi_q(y)w\,(x,y)\,dy \\
&-4\alpha\,(1-v_{AT})\,v'_{AT}\Psi'\left(\sum_{i=1}^{n}||A_i'||^2\right)A_q' \\
&-2\alpha(1-v_{AT})^2\frac{d}{dx}\left(\Psi'\left(\sum_{j}||A_i'||^2\right)A_q'\right)=0.
\end{aligned}
\tag{17.17}
$$

17.5.4.2 Minimization with Respect to $v_{AT}(x)$ (AT Minimization Step)

Fixing the functions $A_i(x), i = 1\ldots n$, we obtain

$$
-2\alpha\,(1-v_{AT})\,E_s+2\rho_1\,(v_{AT})-\rho_2(v''_{AT})=0. \tag{17.18}
$$

17.6 Implementation

We used central first and second derivatives and reflecting boundary conditions. In all the methods, we used various α, ρ_1 and ρ_2 constants, depending on the noise level, as is common in noise reduction methods (this was done for all the considered algorithms, with the appropriate parameters). In all the examples, we assumed a sampling interval of $dx = 1$. For the minimization process we used gradient descent with 200,000 iterations. This minimization method is a notoriously slowly converging method, but it is fast enough for our 1D example, and we intend to pursue a faster implementations in future work.

We performed an AT minimization step every 100 parameter minimization steps, and updated the weighting function every 1,000 parameter minimization steps. We used a window size of ten sample points both for the NLOP functional, and for the weighted least square fit method (used for initialization and reconstruction). We note that using a bigger window size resulted in significantly superior results on all tests, but this may not be the case in other signals. Choosing too big a window may cause an overlap between adjacent discontinuities and prevent the functional from correctly recovering them.

17.6.1 Initialization

In order to prevent trapping into local minima, we initialize the model parameters from the noisy data by means of a robust MLS fitting i.e. we compute each point's parameters by choosing the best least square fitting approximation via a sliding window least square calculation. This was done in exactly the same manner as the sliding window weighting function described in Sect. 17.5.1.1. The parameters chosen are those which generated the minimal reconstruction error. This computation provided a robust initialization, which already preserves, to some extent, discontinuities in the signal. This claim is strengthened by experiments we performed comparing with a regular moving least square fitting initialization. We found that with the robust initialization, the functional converges to a better solution, and nicely exposes the true discontinuities in the signal models.

17.7 Experiments and Results

We conducted various experiments in order to verify the performance of the proposed functional. In this paper we focus mainly on 1D examples leaving 2D extensions to images for future publications, nevertheless we exhibit some initial experiments conducted on 2D synthetic images which yield a good idea of the performance to be expected in the 2D case.

17.7.1 1D Experiments

We begin with the selection of the basis functions. We consider linear basis functions of the form:

$$\begin{cases} \phi_1 = 1 \\ \phi_2 = x \end{cases} \tag{17.19}$$

This seemingly simple choice of functions, enables us to make comprehensive tests of the functional performance. Under this choice, the functional is expected to have the best performance on piecewise linear signals. We note that this is an arbitrary choice of basis functions, and one should choose other basis functions appropriate for the signal domain.

To perform the tests we devised a set of synthetic 1D signals, and added white Gaussian noise with standard deviation ranging from $STD = 0.01$ up to $STD = 0.25$. These signals can be separated to two main groups:

- The first group is comprised of noisy piecewise linear signals. This group is interesting because it enables us to test the parameter reconstruction performance as well as noise reduction capabilities, under the optimal basis functions. The piecewise linear signals are depicted in Fig. 17.5.
- The second group is comprised of noisy nonlinear signals, such as higher degree polynomials. This group is interesting only with regard to the noise removal performance, because generating the ground truth parameters for linear basis

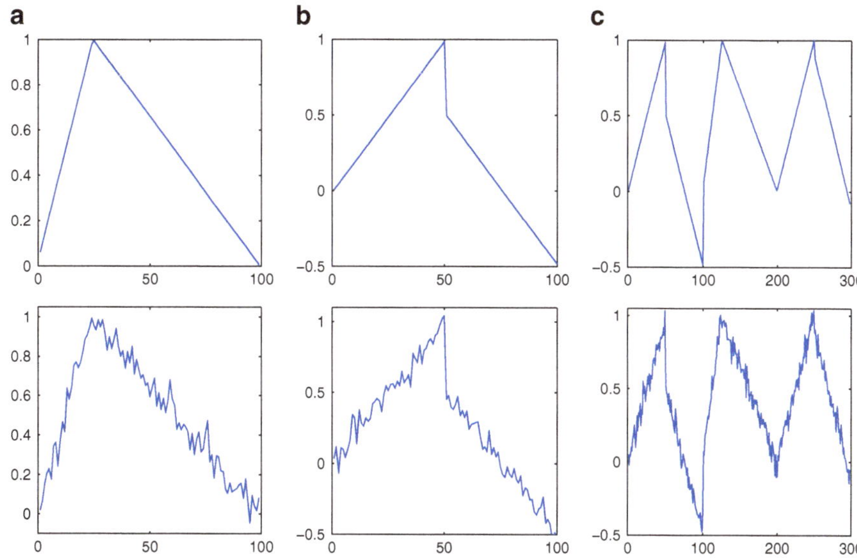

Fig. 17.5 The piecewise Linear test signals and their STD 0.05 noisy counterparts. (**a**) One jump, (**b**) one discontinuity, and (**c**) many jumps & discontinuities

functions is problematic and we do not expect the linear parameters to be piecewise constant. Naturally, we could choose appropriate basis functions for these signals too, but we wanted to demonstrate that our functional preforms surprisingly well, even when applied with suboptimal basis functions. The nonlinear signals are depicted in Fig. 17.6.

In order to the check the performance of NLOP functionals, and to test them against other denoising algorithms, we also implemented the following noise removal algorithms. The first two are the classic TV functional and the original over-parameterized functional (*OP*). We chose these functionals due to their relation to the NLOP functional, enabling us to show the improvement that the NLOP has on predecessor functionals. The third and final algorithm, is the state of the art *K-SVD* noise removal algorithm, firstly proposed by Aharon et al. [1]. We used implementation published by Rubinstein et al. [27].

We compared the various algorithms noise reduction performance and, more importantly, we compared their parameter reconstruction capability. For the latter comparison, we reconstructed parameters from the K-SVD denoised signal, using our robust least square fitting method (see Sect. 17.6.1), and compared the results with both our NLOP functional and the original OP functional.

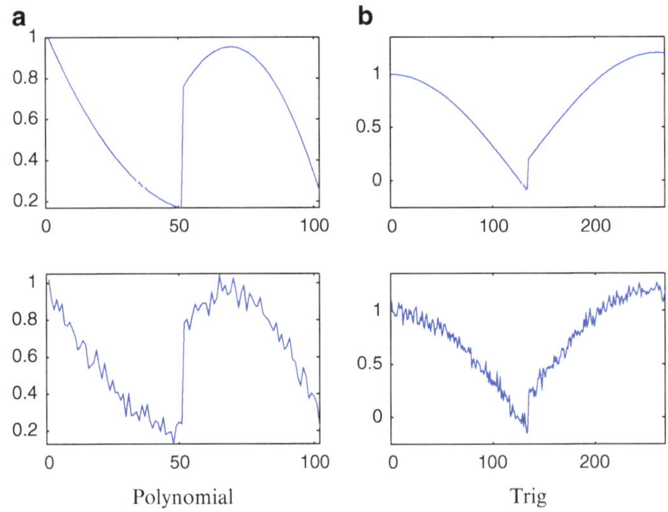

Fig. 17.6 The nonlinear test signals and their STD 0.05 noisy counterparts. *Left*: a polynomials signal of degree 2. *Right*: a signal of combined sine and cosine functions

17.7.2 1D Results

Noise removal performance testing, was done by comparing the L_2 norm of the *residual noise*, that is the difference between the cleaned reconstructed signal and the original signal:

$$E_{L_2} = \left\| f(x) - \sum_{i=1}^{n} A_i(x)\phi_i(x) \right\|_2 . \tag{17.20}$$

Parameters reconstruction performance testing (which with the linear basis function, is only relevant on the piecewise linear signals), was done by calculating the L_2 norm of the difference between the reconstructed parameters and the original parameters, with whom we generated the signal. Figure 17.7 depicts an example of the recovered parameters and the v_{AT} indicator function obtained for the "Many jumps & discontinuities" signal.

Figure 17.8, displays graphs comparing performance of the various algorithms on piecewise linear signals. The left graphs display the noise removal error norms, while the right graphs display parameters reconstruction error norms.

On various noise removal performance graphs we can see the excellent performance of both OP and NLOP functionals, reinforcing the claims of outstanding noise removal performance obtained by the OP functional and maintained by our new functional. Also, we can see that when the signal contains a discontinuity, such as in the "One discontinuity" signal (as apposed to a continuous signal with only parameters discontinuity such as the "One jump" signal), the NLOP functional has greater ability to cope with the discontinuity, thus generating better results than

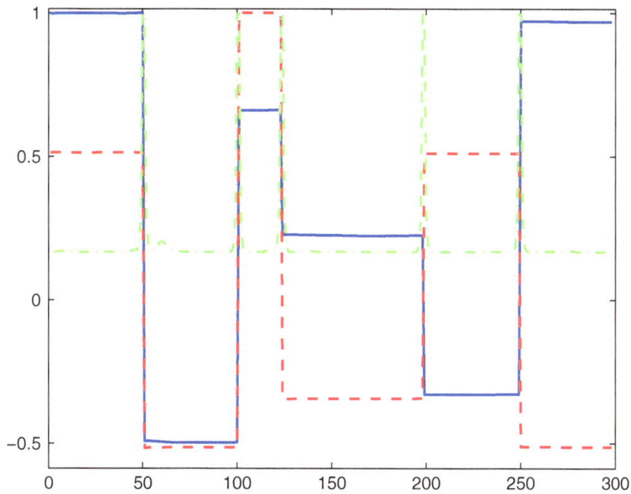

Fig. 17.7 Example of parameters reconstruction. Note that the indicator function tends to 1 where the signal has parameters discontinuities and tends to 0 in almost any other location

the OP functional. In Fig. 17.9 we display comparison of the residual noise of all the algorithms on the "One discontinuity" and the "Many jumps & discontinuities" signals.

When considering parameters reconstruction performance, we see a totally different picture. We see that on most cases, particularly in signals which contains signal discontinuity, the NLOP functional and the K-SVD algorithms both outperform the OP functional. This result demonstrates our claim, in Sect. 17.4.1, that the OP functional lacks the possibility to well enforce a global prior on the reconstructed parameters. Figure 17.10 compares the reconstruction results of NLOP functional, OP functional and the reconstruction from the denoised K-SVD signal, on the "One discontinuity" signal. Note the reconstruction of NLOP functional is close to a piecewise constant solution, while the OP reconstruction is seemingly a smoothly changing function. In the K-SVD reconstruction, where at each point a set of parameters is chosen regardless of the choice made for it's adjacent neighbors, the lack of influence of enforcement of a global constraint is evident.

In Fig. 17.11, we compare the noise removal performance on the nonlinear signals. We can see that the NLOP functional still exhibits the best performance but it is not unchallenged by the OP functional. This is due to the strong constraints the NLOP functional has in trying to enforce the linear basis functions, i.e. trying to find a piecewise linear solution suitable for the given signal.

Indeed, in order to see that our functional performance is not restricted by the basis function, and to verify that indeed better performance is achieved if we choose a better set of basis functions to model the signal domain, we performed several tests with higher degree polynomials. We display in Fig. 17.12, results achieved by denoising the polynomial signal displayed in Fig. 17.6, while changing the NLOP functional basis functions to a second degree polynomial. In general we expect

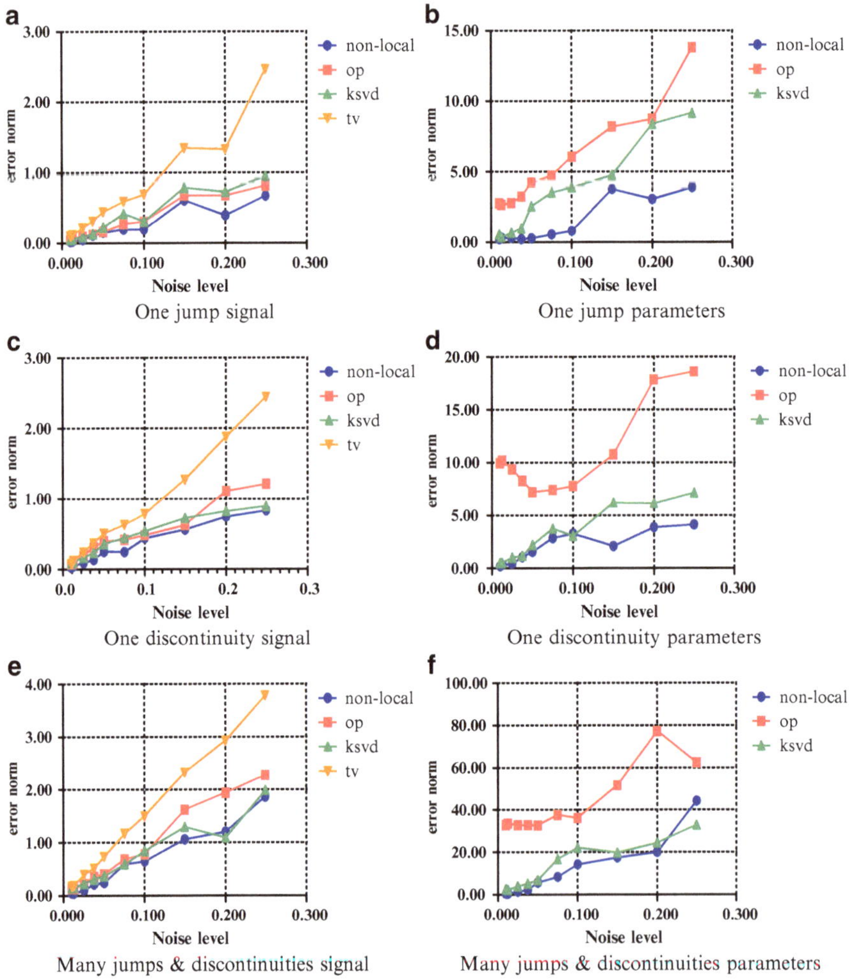

Fig. 17.8 Signal noise removal and parameters reconstruction comparison for the various piece-wise Linear signals. On the *left column* depicted are graphs compering the L_2 norm of the residual noise of the various algorithms. On the *right column* depicted are graphs compering the L_2 norm of the various algorithms reconstructed parameters compared to the expected parameters

a polynomial signal to be best recovered by polynomial basis functions of the same degree. This is clearly depicted in the graph displayed in Fig. 17.12, where we see better performance by NLOP with polynomial basis function compared to NLOP with linear basis functions.

Another test we performed with polynomial basis functions, was on a C_1 continuous second degree polynomial displayed in Fig. 17.13. This is an interesting case, as both this signal and it's first derivative are continuous, and only the second derivative is discontinuous. We found that this signal proved challenging for the MLS initialization method, causing it to misplace the point of discontinuity

a

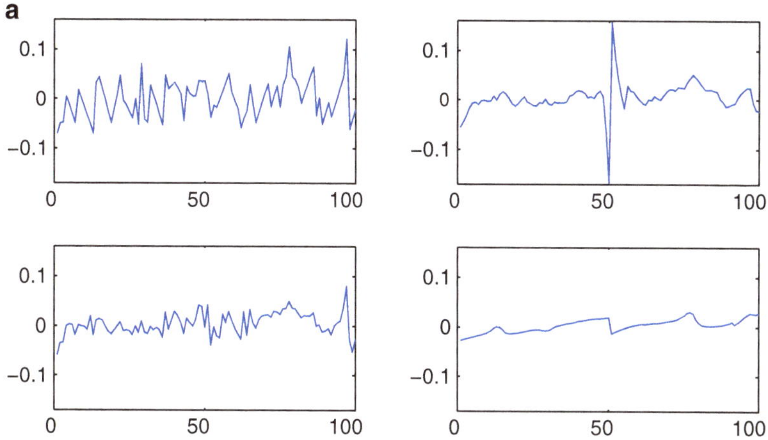

Top left: TV. Top right: OP. Bottom left: K-SVD. Bottom right: NLOP.

b

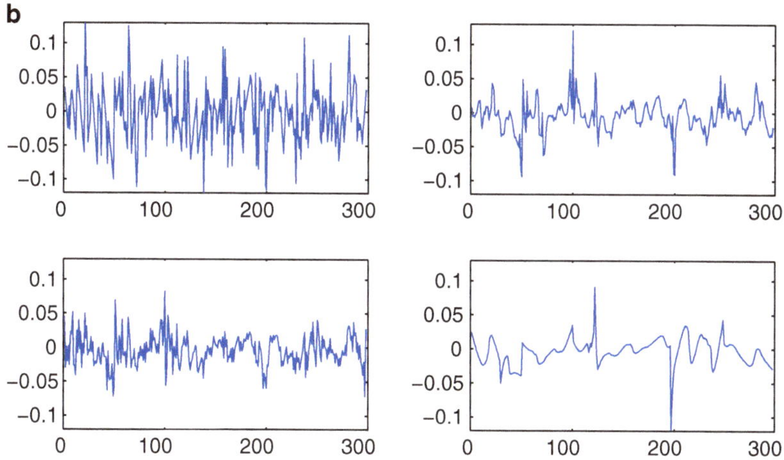

Top left: TV. Top right: OP. Bottom left: K-SVD. Bottom right: NLOP.

Fig. 17.9 Residual noise comparison of noise removal. (**a**) One discontinuity signal with noise $STD = 0.0375$. (**b**) "Many jumps & discontinuities" signal with noise $STD = 0.05$

by several points. This initialization error was not detected by the NLOP functional, which maintained it throughout the minimization process, as displayed for example in Fig. 17.13. The location of the discontinuity point depends on the random noise. We wish to emphasize that the reconstructed solutions achieved by the minimization of the NLOP functional have piecewise constant reconstructed parameters, which generate a piecewise smooth polynomial solution. This reconstructed signal may as well be the signal from which the noisy signal was generated. Also, the solution achieved by the NLOP functional outperformed the K-SVD method, as displayed in the graphs in Fig. 17.14.

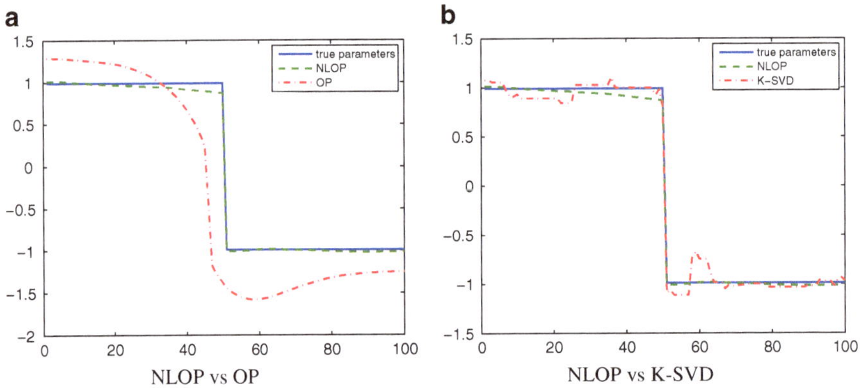

Fig. 17.10 This figure compares the reconstructed second parameters on the various algorithms, when denoising the "One discontinuity" signal. *Left image*: comparison of parameter reconstruction between NLOP functional and OP functional on one discontinuity signal. Note how far the OP reconstruction is from a piecewise solution, while generating an excellent denoising result (seen in the relevant graph). *Right image*: comparison of parameter reconstruction between NLOP functional and K-SVD algorithm on one discontinuity signal. Note the apparent lack of global constraint on the parameters in the K-SVD reconstruction

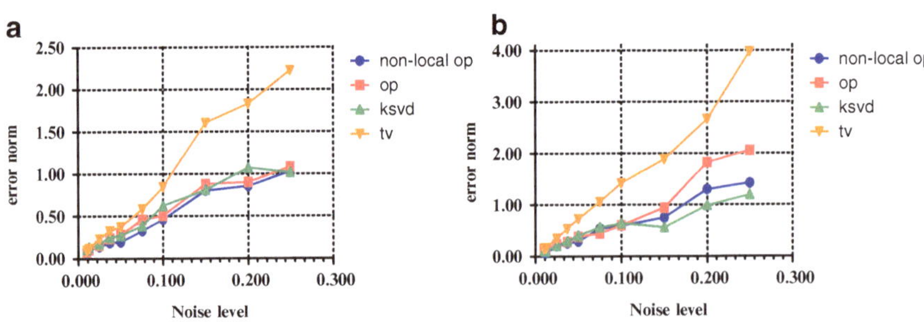

Fig. 17.11 Signal noise removal comparison of the nonlinear signals. (**a**) Polynomial signal and (**b**) trig signal

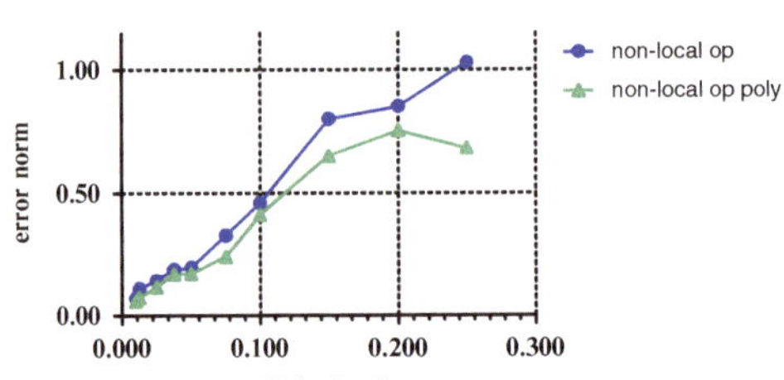

Fig. 17.12 Comparison of noise removal performance on the polynomial signal. In this figure we compare performance of the NLOP functional with linear basis functions (marked by non-local OP), and NLOP functional with polynomial basis functions (marked by non-local OP poly)

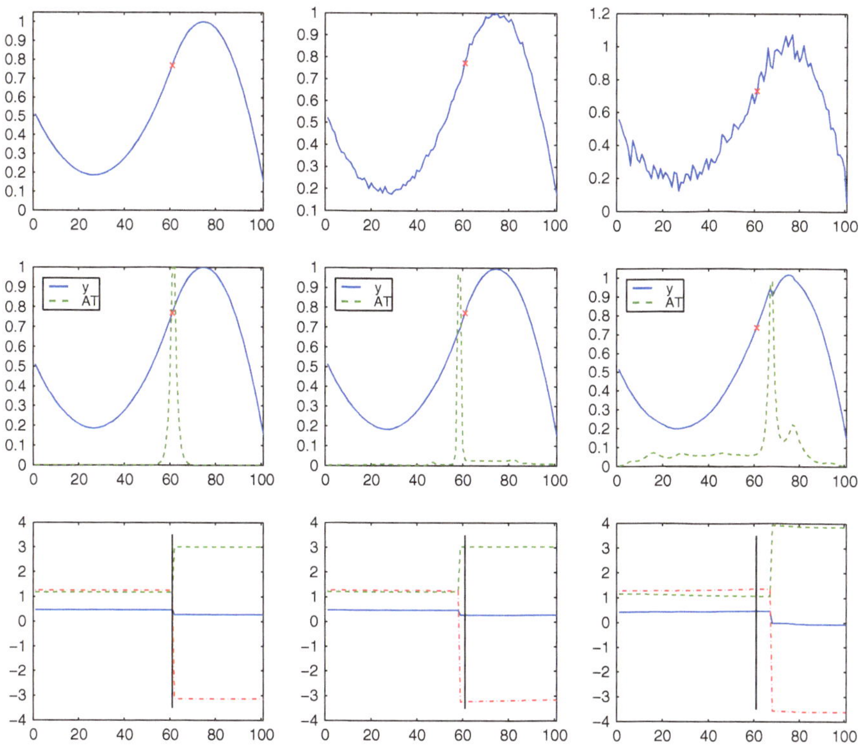

Fig. 17.13 C_1 continuous second degree polynomial. Point of the second derivative discontinuity is marked by a *red cross* (or *vertical black line*). *Top row*: images of the signals. *Middle row*: images of the reconstructed signals and the v_{AT} indicator functions. *Bottom row*: the reconstructed parameters. From *left* to *right*: clean signal, 0.01 STD noise, 0.0375 STD noise

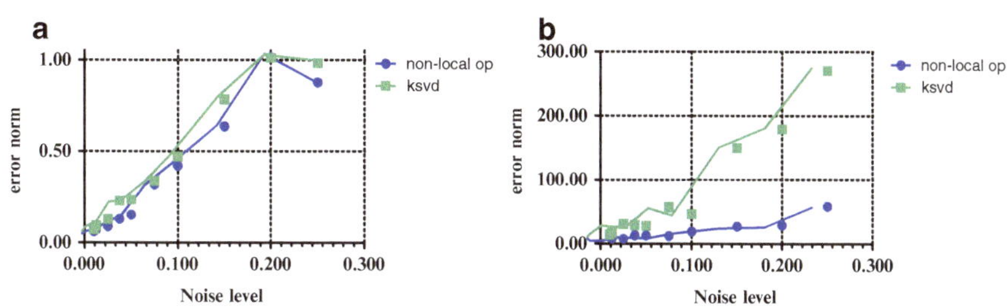

Fig. 17.14 Signal noise removal and parameters reconstruction comparison for the C_1 continuous second degree polynomial. (**a**) Signal and (**b**) parameters

17.7.3 2D Example

We also ran initial tests on a 2D example. The non-local over-parameterized method extends naturally to higher dimensions. We implemented the 2D case in a similar manner to the 1D case, and chose a linear basis functions of the form

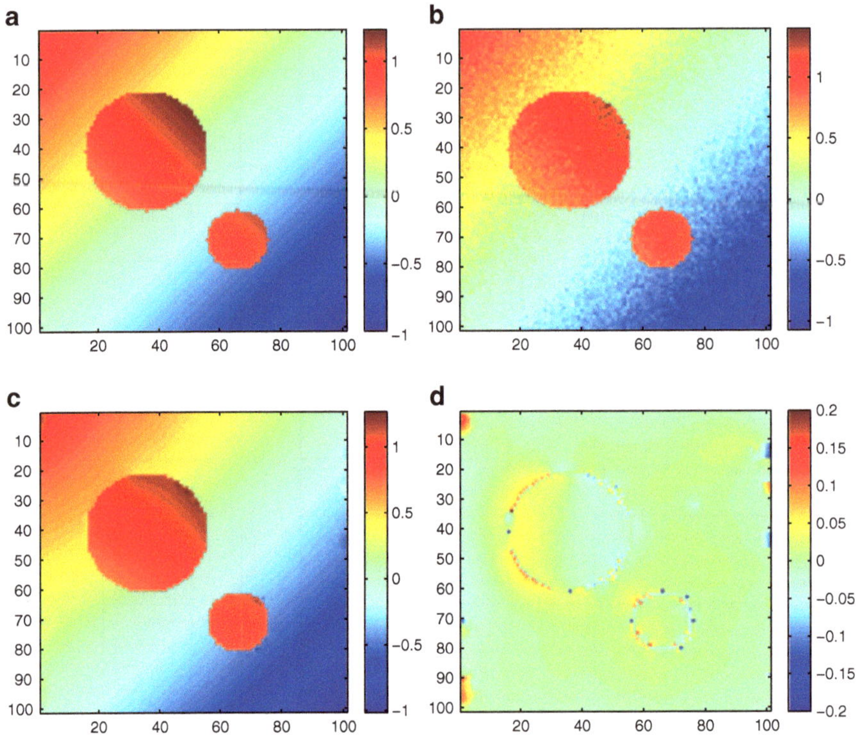

Fig. 17.15 A 2D noise removal example of the 2D NLOP functional. (**a**) Original image, (**b**) noisy image, (**c**) denoised image, and (**d**) residual noise

$$\begin{cases} \phi_1 = 1 \\ \phi_2 = x \\ \phi_3 = y \end{cases} \tag{17.21}$$

In Fig. 17.15, we show the 2D example, the NLOP functional noise removal result. In Fig. 17.16, we display the parameters which were reconstructed in the minimization process and the generated v_{AT} indicator function. We can see that the v_{AT} indicator function managed to segment the signal in a good manner, although still delineated some ghost segments especially near the image edges. Note that the recovered parameters are almost piecewise constant as expected.

A more thorough discussion of the 2D case is out of the scope of this paper. We intend to explore it extensively in the near future.

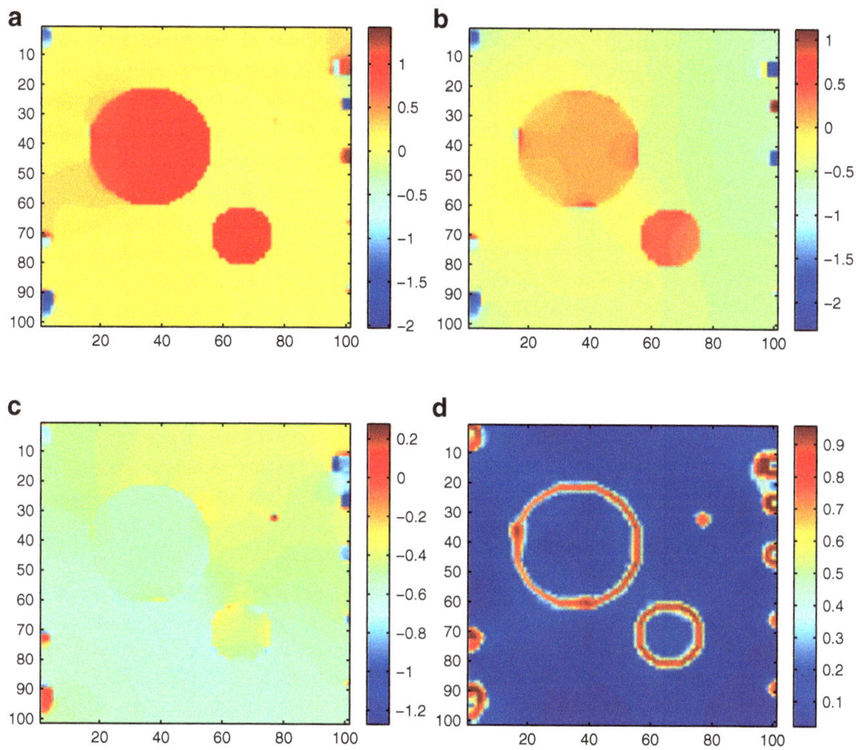

Fig. 17.16 Example of the reconstructed parameters from the noisy image and v_{AT} AT scheme indicator function. (**a**) Parameter 1, (**b**) parameter 2, (**c**) parameter 3, and (**d**) v_{AT}

17.8 Conclusion

A general over-parameterized variational framework for signal and image analysis was presented. This framework can be applied to various image processing and vision problems, such as noise removal, segmentation and optical flow computation. This framework is closely related to the powerful moving least squares method, enhancing it by globally constraining the parameter variations in space based on knowledge available on the problem domain. This knowledge enables a model based reconstruction of the considered signal, by effectively recovering parameters of an a priori, assumed to be known, set of "basis" or "dictionary functions".

The new variational framework relies on the successful over-parameterized functional, and significantly improves it by making it non-local and giving it the power not only to generate excellent results for the problem domain (such as noise removal), but also to reconstruct the underling model parameters that might capture prior information on the problem.

This paper may be viewed as the basis of extensive future research on the subject of over-parametrization. In future work we intend to thoroughly explore the extension of the non-local over-parametrization denoising functional into $2D$ settings, where the choice of the support set, expressed by the weighting function, becomes a challenging matter. Also we intend to explore a variety of different basis functions, such has higher degree polynomials and non-polynomial basis functions, which may be relevent for other applications. Finally we would like to extend the proposed framework and to apply it to various other problem domains in computer vision.

Acknowledgements This research was supported by the Israel Science foundation (ISF) grant no. 1551/09.

References

1. Aharon, M., Elad, M., Bruckstein, A.: K-SVD: design of dictionaries for sparse representation. In: Proceedings of SPARS'05, Rennes, pp. 9–12 (2005)
2. Alexa, M., Behr, J., Cohen-Or, D., Fleishman, S., Levin, D., Silva, C.T.: Point set surfaces. In: IEEE Visualization 2001, pp. 21–28. IEEE Computer Society, Piscataway (2001)
3. Ambrosio, L., Tortorelli, V.M.: Approximation of functional depending on jumps by elliptic functional via Γ-convergence. Commun. Pure Appl. Math. **43**(8), 999–1036 (1990)
4. Amenta, N., Kil, Y.J.: Defining point-set surfaces. ACM Trans. Graph. **23**, 264–270 (2004)
5. Black, M.J., Anandan, P.: The robust estimation of multiple motions: parametric and piecewise-smooth flow fields. Comput. Vis. Image Underst. **63**, 75–104 (1996)
6. Bruckstein, A.M.: On globally optimal local modeling: from moving least squares to over-parametrization. In: Workshop on Sparse Representation of Multiscale Data and Images: Theory and Applications. Institute of Advanced Study, Nanyang, Technological University, Singapore (2009)
7. Chan, T.F., Shen, J.: On the role of the bv image model in image restoration. AMS Contemp. Math. **330**, 25–41 (2003). accepted (by two referees)
8. Cohen, E., Riesenfeld, R.F., Elber, G.: Geometric Modeling with Splines: an Introduction. AK Peters, Natick (2001)
9. Darrell, T.J., Pentland, A.P.: Cooperative robust estimation using layers of support. IEEE Trans. Pattern Anal. Mach. Intell. **17**, 474–487 (1991)
10. Farnebäck, G.: Spatial domain methods for orientation and velocity estimation. Lic. Thesis LiU-Tek-Lic-1999:13, Dept. EE, Linköping University, SE-581 83 Linköping, (1999). Thesis No. 755, ISBN 91-7219-441-3
11. Fleishman, S., Cohen-Or, D., Silva, C.T.: Robust moving least-squares fitting with sharp features. ACM Trans. Graph. **24**, 544–552 (2005)
12. Gilboa, G., Osher, S.: Nonlocal operators with applications to image processing. Multiscale Model. Simul. **7**(3), 1005–1028 (2008)
13. Haindl, M., Mikes, S.: Model-based texture segmentation. In: Proceedings of the International Conference on Image Analysis and Recognition, pp. 306–313. Springer, Berlin (2004)
14. Haralick, R.M., Watson, L.: A facet model for image data. Comput. Graph. Image Process. **15**(2), 113–129 (1981)
15. Harten, A., Osher, S.: Uniformly high-order accurate nonoscillatory schemes. SIAM J. Numer. Anal. **24**, 279–309 (1987)
16. Kindermann, S., Osher, S., Jones, P.W.: Deblurring and denoising of images by nonlocal functionals. Multiscale Model. Simul. **4**(4), 1091–1115 (2005)

17. Levin, D.: The approximation power of moving least-squares. Math. Comput. **67**, 1517–1531 (1998)
18. Levin, D.: Mesh-independent surface interpolation. In: Brunnett, G. et al. (eds.) Geometric Modeling for Scientific Visualization, pp. 37–49. Springer, Berlin/London (2003)
19. Malisiewicz, T., Efros, A.A.: Improving spatial support for objects via multiple segmentations. In: Proceedings of the British Machine Vision Conference (BMVC), Warwich (2007)
20. Memin, E., Perez, P.: A multigrid approach for hierarchical motion estimation. In: Proceedings of the Sixth International Conference on Computer Vision, 1998, Bombay, pp. 933–938 (1998)
21. Nir, T., Bruckstein, A.M.: On over-parameterized model based TV-denoising. In: International Symposium on Signals, Circuits and Systems (ISSCS 2007), Iasi (2007)
22. Nir, T., Bruckstein, A.M., Kimmel, R.: Over-parameterized variational optical flow. Int. J. Comput. Vis. **76**(2), 205–216 (2008)
23. Olshausen, B.A., Fieldt, D.J.: Sparse coding with an overcomplete basis set: a strategy employed by v1. Vis. Res. **37**, 3311–3325 (1997)
24. Pock, T.: Fast total variation for computer vision. Ph.D. thesis, Graz University of Technology (2008)
25. Quatieri, T.: Discrete-Time Speech Signal Processing: Principles and Practice, 1st edn. Prentice Hall, Upper Saddle River (2001)
26. Rosman, G., Shem-Tov, S., Bitton, D., Nir, T., Adiv, G., Kimmel, R., Feuer, A., Bruckstein, A.M.: Over-parameterized optical flow using a stereoscopic constraint. In: Proceedings of the Scale Space and Variational Methods in Computer Vision. Springer, Berlin/London (2011)
27. Rubinstein, R., Zibulevsky, M., Elad, M.: Efficient implementation of the k-svd algorithm using batch orthogonal matching pursuit. Technical report, Technion – Israel Institute of Technology (2008)
28. Rudin, L.I., Osher, S., Fatemi, E.: Nonlinear total variation based noise removal algorithms. Phys. D **60**, 259–268 (1992)
29. Savage, J., Chen, K.: On multigrids for solving a class of improved total variation based staircasing reduction models. In: Tai, X.-C., Lie, K.-A., Chan, T.F., Osher, S. (eds.) Proceedings of the International Conference on PDE-Based Image Processing and Related Inverse Problems, CMA, Oslo, 8–12 Aug 2005. Mathematics and Visualization, Image Processing Based on Partial Differential Equations, pp. 69–94 (2006)
30. Shah, J.: A common framework for curve evolution, segmentation and anisotropic diffusion. In: Proceedings of the 1996 Conference on Computer Vision and Pattern Recognition (CVPR '96), p. 136. IEEE Computer Society, Washington, DC (1996)
31. Shah, J.: Curve evolution and segmentation functionals: application to color images. In: Proceedings of the IEEE ICIP'96, Lausanne, pp. 461–464 (1996)
32. Shu, C.W.: Essentially non-oscillatory and weighted essentially non-oscillatory schemes for hyperbolic conservation laws. In: Cockburn, B., Shu, C.-W., Johnson, C., Tadmor, E. (eds.) Advanced Numerical Approximation of Nonlinear Hyperbolic Equations. Lecture Notes in Mathematics, vol. 1697, pp. 325–432. Springer (1997)
33. Sun, D., Sudderth, E., Black, M.: Layered image motion with explicit occlusions, temporal consistency, and depth ordering. In: Lafferty, J., Williams, C.K.I., Shawe-Taylor, J., Zemel, R.S., Culotta, A. (eds.) Proceedings of the Advances in Neural Information Processing Systems, Vancouver, vol. 23, pp. 2226–2234 (2010)
34. Tarantola, A.: Inverse Problem Theory and Methods for Model Parameter Estimation. Society for Industrial and Applied Mathematics, Philadelphia (2004)
35. Trobin, W., Pock, T., Cremers, D., Bischof, H.: An unbiased second-order prior for high-accuracy motion estimation. In: Proceedings of the DAGM Symposium, pp. 396–405. Springer, Berlin/New York (2008)

Chapter 18
Incremental Level Set Tracking

Shay Dekel, Nir Sochen, and Shai Avidan

Abstract We consider the problem of contour tracking in the level set framework. Level set methods rely on low level image features, and very mild assumptions on the shape of the object to be tracked. To improve their robustness to noise and occlusion, one might consider adding shape priors that give additional weight to contours that are more likely than others. This works well in practice, but assumes that the class of object to be tracked is known in advance so that the proper shape prior is learned. In this work we propose to learn the shape priors on the fly. That is, during tracking we learn an eigenspace of the shape contour and use it to detect and handle occlusions and noise. Experiments on a number of sequences reveal the advantages of our method.

18.1 Introduction

Contour tracking is a task in which the contour of the object(s) of interest has to be extracted for each frame in a video in a way which is robust to noise and clutter. This task is different from object tracing in which a bounding box that contains the object is sought or segmentation in which the contour in a given image is extracted. The main issues that we address in this paper is the robustness to noise, clutter and occlusions on one hand and the ability to deal with shape/behavior change. Accordingly, this paper introduces a way to achieve such a robustness as a matter of principle without paying too much attention to the quality of segmentation. The way it is done is via variational and level-set methods.

S. Dekel (✉) · S. Avidan
Department of Electrical Engineering, Tel Aviv University, Tel Aviv, Israel

N. Sochen
Department of Applied Mathematics, Tel Aviv University, Tel Aviv, Israel
e-mail: sochen@math.tau.ac.il

M. Breuß et al. (eds.), *Innovations for Shape Analysis*, Mathematics and Visualization,
DOI 10.1007/978-3-642-34141-0_18, © Springer-Verlag Berlin Heidelberg 2013

Level set methods are a convenient way to parameterize and track object contours. They work by evolving a contour which tightly encloses the deformable object to be tracked. However this method cannot handle missing or misleading information due to noise, clutter or occlusions in the input images. In order to overcome these problems one can derive a parametric model for implicit representation of the segmentation curve by applying low dimensional subspace representation, such as Principle Component Analysis (PCA) to a specific collection of training set images before the tracking begins. In this case the evolving curve has limited degrees of freedom since the curve lies on the PCA subspace. This property enables the segmentation to be robust to noise and partial occlusions. However this model relies on a fixed training set and assumes that the object class to be tracked is known in advance.

We present an extension that deals with these shortcomings. Our approach learns, on line, a shape prior that is then used during tracking. This enables our tracker to overcome occlusions and, unlike previous algorithms, it does not demand a specific predefined training set.

We formulate the tracking process as a mixture of two major models: In the **On-line Learning Model**, we perform region-based segmentation on each frame using the Chan-Vese approach [3] together with the edge-based approach which is based on the Geodesic Active Contour [2]. The segmentation results are then used to incrementally learn an on-line low dimensional subspace representation of the objects' contour, efficiently adapting to changes in the appearance of the target.

In case of occlusion, the eigencoordinates of the segmented shape will differ considerably from those obtained so far and, in that case, we switch to the **PCA Representation Model** that tracks the object using the currently available PCA space. This PCA eigenbase representation together with the temporal prior representation allows us to limit the degree of freedom of the evolving contour which enables it to cope with missing or misleading information due to occlusions, partial occlusions and noise. Once the object reappears we switch back to the online learning model and keep updating our representation model. Hence, we can properly track deformable objects through occlusion and noise.

We provide experimental results that present several properties of our method: We show that our method can cope with partial or total occlusions, as well as examples in which the images are severely contaminated with strong Gaussian noise. In addition, we show that our algorithm can adapt to considerable deformations in shape.

18.2 Background

Contour tracking via variational methods and level-sets is based on the seminal works [1, 3, 18] and many more; see [6] for a very nice overview and for further references on level-set based tracking.

Several authors combined prior shape information into level-set-based segmentation. Leventon et al. [11] incorporated a training set information as a prior model to restrict the flow of the geodesic active contour using Principle Component Analysis (PCA). Tsai et al. [25] used only the first few eigenmodes by performing optimization. Rousson et al. [20, 21] introduced shape information on the variational level. Chen et al. [4] imposed shape constraints directly on the contour. However these authors ignored the temporal coherence of the shapes which leads to degredation in performance when dealing with occlusions.

Cremers [5] proposes to model the embedding functions by a Principle Component Analysis (PCA) and to use dynamical shape prior. He learns a specific set of training shapes before the tracking begins and also exploits the temporal correlations between consecutive shapes. This enables him to handle occlusions and large amounts of noise. His method is well-suited for specific tracking missions where a pre-defined training set can be performed off-line.

Another approach is presented in the work of Fussenegger et al. [7]. In that work a level-set method is combined with PCA decomposition of shape space. This is very similar and relevant to this paper. The difference is in the aim and type of video treated. In Fussengger et al. there are many, mainly rigid, objects to segment such that each individual shape doesn't change too much from frame to frame. Our paper deals with mainly one object with **changing shape** and the main focus is on the way to deal with occlusions and changes of shape along the video.

Our work is motivated in part by the power of subspace representation and exploits the temporal correlations between consecutive shapes following the work of Cremers [5]. But in contrast to eigentracking algorithms, our algorithm does not require a specific training set before tracking begins. It learns the eigenbase on-line during the object tracking process, thus eliminating the need to collect the training images prior to tracking.

18.2.1 Integrated Active Contours

We start with a generic algorithm for data-based segmentation. The model is formulated in variational way and we use the Integrated Active Contour model [19, 22] that combines region-based and edge-based segmentation via the level-set formulation. In order to perform region-based segmentation in each frame we use the Chan-Vese algorithm which attempts to partition the image into two regions according to common image properties. Then we add to the functional an edge-based term which is based on the Geodesic Active Contour (GAC).

Let $I_t : \Omega \to \mathbb{R}$ be the image at time t that assigns for each pixel $x \in \Omega \subset \mathbb{R}^2$ a real value grey level. A contour that separates the object (or objects) from the background is encoded as a zero level-set of a function $\phi_t : \Omega \to \mathbb{R}$. The contour at frame t is $C_t = \{(x, y)|\phi_t(x, y) = 0\}$. We denote the region inside the zero-level set by $\Omega_+ = \{(x, y)|\phi_t(x, y) > 0\}$ and similarly the region outside the zero level-set $\Omega_- = \{(x,y)|\phi_t(x,y) < 0\}$. The probability of the contour ϕ_t, given the previous contours and all the measurements $[I_0(\mathbf{x})\dots I_t(\mathbf{x})]$ using the Bayes rule is:

$$\mathscr{P}(\phi_t|I_{0:t},\phi_{0:t-1}) \propto \underbrace{\mathscr{P}_+(I_t|\phi_t)}_{Target}\cdot\underbrace{\mathscr{P}_-(I_t|\phi_t)}_{Background}\cdot\underbrace{\mathscr{P}(\phi_{0:t-1}|\phi_t)}_{Shape}\cdot\underbrace{\mathscr{P}(\phi_t)}_{Smoothness} \tag{18.1}$$

Here $P_\pm$ are the probability distributions of the grey value intensities inside and outside of the zero level-set of ϕ_t.

While $P_\pm$ can be quite involved in real-life applications we choose here to stick to the simple Gaussian model in order to concentrate on the tracking part. This choice leads to the Chan-Vese model. In this approach we find a contour, represented by $\phi(x)$, that partitions the image into two regions Ω_+ and Ω_-, that describe an optimal piecewise constant approximation of the image. We also assume that the intensities of the shape and the background are independent samples from two Gaussian probabilities, therefore:

$$\mathscr{P}_\pm(I_t|\phi_t) = \prod_{x\in\Omega_\pm} \frac{1}{\sqrt{2\Pi\sigma_\pm^2}} \exp^{-\frac{(I_t-C_\pm)^2}{2\sigma_\pm^2}} \tag{18.2}$$

Thereby $C_\pm$ and $\sigma_\pm$ are the mean and standard deviation of the intensities inside and outside of the zero level-set of ϕ_t.

The region-based energy is defined as

$$E_{RB}(\phi_t, C_+, C_-) = -\log\{\mathscr{P}_+(I_t|\phi_t)\cdot\mathscr{P}_-(I_t|\phi_t)\} \tag{18.3}$$

The contour with the highest probability is the one that minimizes the following region-based energy functional:

$$\begin{aligned} E_{RB}(\phi_t, C_+, C_-) = &\int_\Omega \{\frac{1}{2\sigma_+^2}(I_t(x) - C_+)^2 + \frac{1}{2}\log(2\pi\sigma_+^2)\}H(\phi_t(x))dx \\ &+ \int_\Omega \{\frac{1}{2\sigma_-^2}(I_t(x) - C_-)^2 + \frac{1}{2}\log(2\pi\sigma_-^2)\}\{1 - H(\phi_t(x))\}dx \end{aligned} \tag{18.4}$$

where $H(\phi_t(x))$ is the Heaviside step function:

$$H(\phi_t(x)) = \begin{cases} 1 \text{ if } \phi_t(x) > 0 \\ 0 \text{ if } \phi_t(x) \leqslant 0 \end{cases} \tag{18.5}$$

The smoothness prior is taken to be the Geodesic Active Contour (GAC) term [2]. This term defines the object boundaries as a (locally) minimal length weighted by the local gradients. In other words it is a geodesic over a Riemannian manifold whose metric is defined via the gradients of the image. It leads to the following functional:

$$E_{GAC}(\phi_t) = \int_{\Omega} g_{GAC}(|\nabla I|)|\nabla H(\phi_t(x))|dx \tag{18.6}$$

where $g_{GAC} = 1/(1 + |\nabla I|^2)$.

Finally, the Integrated Active Contour functional E_{IAC} is obtained by the summation of the region-based energy E_{RB} (18.4) and the edge-based geodesic active contour energy E_{GAC} (18.6) as:

$$E_{IAC}(\phi_t, C_+, C_-) = E_{RB}(\phi_t, C_+, C_-) + \mu E_{GAC}(\phi_t) \tag{18.7}$$

with the Euler-Lagrange equation:

$$\frac{\delta E_{IAC}}{\delta \phi_t} = \Big[\log(\frac{\sigma_+}{\sigma_-}) + \frac{1}{2\sigma_+^2}(I_t(x) - C_+)^2$$

$$-\frac{1}{2\sigma_-^2}(I_t(x) - C_-)^2 - \mu \mathrm{div}(g_{GAC}(|\nabla I|)\frac{\nabla \phi_t(x)}{|\nabla \phi_t(x)|})\Big]\delta(\phi_t(x)) \tag{18.8}$$

18.2.2 *Building the PCA Eigenbase*

The shape term, as explained earlier, is an on-line learning model that produces and updates an eigenbase representation during tracking. We work here with a PCA decomposition, where we first build a PCA eigenbase from the first n frames of the sequence and then incrementally update it as new m observations arrive. For efficiency we use an incremental PCA algorithm and only keep the top k eigenvalues. The corresponding eigenvectors are denoted by ψ_i. This PCA eigenbase, ψ_i, will help us cope with occlusions in the PCA representation model. Each shape is represented as:

$$\phi_i(x) = \bar{\phi}_0(x) + \sum_{j=1}^{k} \alpha_{ij}\psi_j \tag{18.9}$$

where $\phi_i(x)$ represents the i-th shape from the PCA subspace model, $\bar{\phi}_0(x)$ is the mean shape and α_{ij} is the PCA coefficient of the i-th shape.

18.2.2.1 First PCA Eigenbase

We produce the first PCA eigenbase from the previous n segmentation results of the On-Line Learning model. Let $A = \{\phi_1(x), \phi_2(x) \ldots \phi_n(x)\}$ be the previous n level set function segmentations. Each data element, $\phi_i(x)$, is a $d \times 1$ vector that contains the level set function of the i-th shape. We calculate the mean shape $\phi_A(x)$ as:

$$\bar{\phi}_A(x) = \frac{1}{n}\sum_{i=1}^{n}\phi_i(x) \tag{18.10}$$

Then we apply singular value decomposition (SVD) on the centered n previous level set functions

$$U_A \Sigma_A V_A^T = SVD\{\phi_1 - \bar{\phi}_A, \phi_2 - \bar{\phi}_A, \ldots, \phi_n - \bar{\phi}_A\} \tag{18.11}$$

Here U_A and Σ_A contain the eigenvectors and the eigenvalues, respectively. Then the first PCA eigenbase denoted as $\psi_A(x) = U_A^{(1:k)}$ contains the eigenvectors corresponding to the k largest eigenvalues, i.e. vectors of U_A. These terms $\psi_A(x)$ and Σ_A will serve as initialization to the incremental PCA algorithm.

18.2.2.2 Updating the Eigenspace

Incremental PCA combines the current eigenbase with new observations without recalculating the entire SVD. Numerous algorithms have been developed to efficiently update an eigenbase as more data arrive [8, 9]. However, most methods assume a fixed mean when updating the eigenbase. We use the Sequential Karhunen-Loeve (SKL) algorithm of Levy and Lindenbaum [12]. They present an efficient method that incrementally updates the eigenbase as well as the mean when new observations arrive. They also add a forgetting factor $f \in [0, 1]$ that down-weights the contribution of the earlier observations. This property plays an important role in the on-line learning. As time progresses the observation history can become very large and the object may change its appearance, and the forgetting factor allows us to strengthen the contribution of the current data such that the updated PCA eigenbase will be able to cope with that change. This algorithm allows us to update the PCA eigenbase online while tracking using the segmentation results during the on-line learning model.

18.2.2.3 Detection of Occlusion

In the on-line learning model we performed region-based segmentation in each frame and incrementally updated the PCA eigenbase. We want to know when the current contour encounters an occlusion in order to switch to the PCA representation model. Then, after the occlusion ends we switch back to the on-line learning model to keep updating our representation model.

For this purpose, we rely on the PCA coefficients that represent the current shape and observe that under occlusions these coefficients are farther away from the mean PCA coefficients.

During the on-line learning model we project each contour segmentation ϕ_t on the current PCA subspace $\Psi(x)$ to obtain its current PCA coefficient α_t. Then we measure the Mahalanobis distance between the current PCA coefficient α_t and the mean PCA coefficient $\bar{\alpha}$:

$$D_t(\alpha_t) = \sqrt{(\alpha_t - \bar{\alpha})^T S^{-1} (\alpha_t - \bar{\alpha})} \tag{18.12}$$

Here $\bar{\alpha}$ is the mean PCA coefficient and S is the covariance matrix. These two terms were obtained by collecting only the good PCA coefficients every frame during the on-line learning model. From (18.12), if $D_t(\alpha_t) > Th$ our method switches to the PCA representation model and when $D_t(\alpha_t) < Th$ we return back to the on-line learning model.

Experimental results show that if the scene is free of occlusions, the Mahalanobis distance is usually $D_t(\alpha_t) < Th$. Figure 18.1 shows the Mahalanobis distance in each frame during the two models. We can see that during the on-line learning model the contour encounters an occlusion and its appropriate Mahalanobis distance is above Th (the peaks in the blue bars in the middle), in that moment we switch to the PCA representation model and see the improvement in the Mahalanobis distances during the occlusion (the red bars in the middle). Then we switch back to the on-line learning model when the coefficients are below Th.

18.2.3 Dynamical Statistical Shape Model

Once the algorithm detects an occlusion, it switches to the PCA representation model with the updated PCA eigenbase. But before we switch to the PCA representation model, we want to exploit the temporal correlations between the shapes. As explained in Eq. (18.9) we can represent each shape using the PCA eigenbase and the mean shape. Therefore the segmentation in the current frame $\phi_t(x)$ can be represented by the PCA coefficient vector α_t. This will lead us to write the probability of the shape prior from (18.1) as: $\mathscr{P}(\alpha_t|\alpha_{0:t-1})$ instead of $\mathscr{P}(\phi_t|\phi_{0:t-1})$. In addition during the on-line learning model, we ignored the correlation between the frames since we assumed that the object wasn't occluded and therefore no temporal prior information was needed. In the PCA representation model, where the deformable object may be occluded, we have to obtain more powerful prior which relies on the correlation between consecutive frames. For this reason we can represent each shape by a Markov chain of order q in a manner similar to Cremers [5]. More formally, the current shape at time t can be represented by the previous shapes using an Auto Regressive (AR) model as follows:

$$\alpha_t = \vartheta + A_1\alpha_{t-1} + A_2\alpha_{t-2} \ldots + A_q\alpha_{t-q} + \eta \tag{18.13}$$

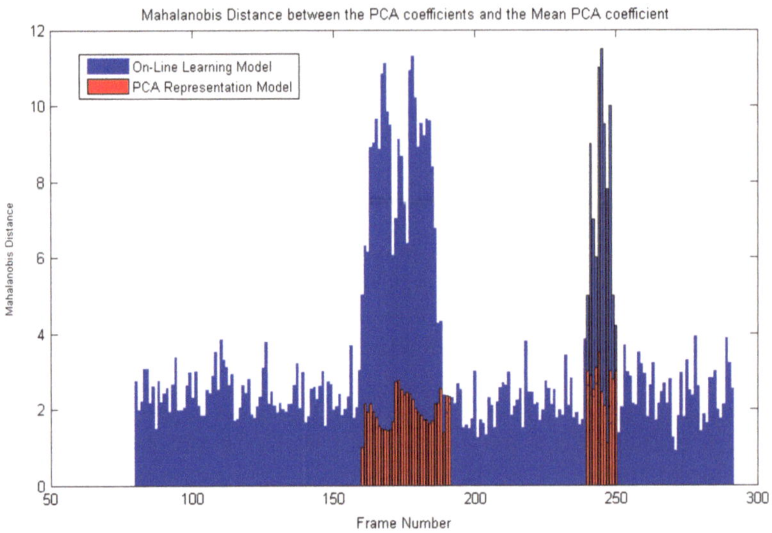

Fig. 18.1 The Mahalanobis distance between the current PCA coefficient α_t and the mean PCA coefficient $\bar{\alpha}$ in each frame during the two models on-line learning model (*blue*) and PCA representation model (*red*)

Here η is Gaussian noise with zero mean and covariance matrix Λ, A_i are the transition matrices of the AR model. and ϑ is the mean of the process. With this AR model we can determine the probability $\mathscr{P}(\alpha_t|\alpha_{0:t-1})$ for observing a particular shape α_t at time t given the shapes estimated on the previous frames as follows:

$$\mathscr{P}(\alpha_t|\alpha_{0:t-1}) \propto \exp(-\frac{1}{2}\omega^T \Lambda^{-1}\omega) \tag{18.14}$$

Where:

$$\omega = \alpha_t - \vartheta - A_1\alpha_{t-1} - A_2\alpha_{t-2} \ldots - A_q\alpha_{t-q} \tag{18.15}$$

Various methods have been proposed in the literature to estimate the model parameters: η, Λ and A_i. We applied a Stepwise Least Squares algorithm as proposed in [16]. The order q determines the accuracy of the AR model approximation and its value depends on the input sequence. In order to estimate its value we use the Schwarz Bayesian Criterion [23].

18.3 PCA Representation Model

The algorithm switches to that model only if it detects an occlusion. Once detecting an occlusion, it continually tracks the same target using the PCA eigenbase and the AR parameters which were obtained in the on-line learning model. As explained in Sect. 18.2.2, according to (18.9) the segmentation in the current frame $\phi_t(x)$ can be represented by the PCA coefficient vector α_t. Therefore we actually exchange in (18.1) each level set function ϕ by the appropriate shape vector representation α:

$$\mathscr{P}(\alpha_t|I_{0:t},\alpha_{0:t-1}) \propto \underbrace{\mathscr{P}_+(I_t|\alpha_t)}_{Target}\cdot\underbrace{\mathscr{P}_-(I_t|\alpha_t)}_{Background}\cdot\underbrace{\mathscr{P}(\alpha_t|\alpha_{0:t-1})}_{Shape\ Prior} \tag{18.16}$$

This will lead us in this model to focus on estimating the shape vector representation α_t by minimizing the following energy functional:

$$\begin{aligned} E_{PCA}(\alpha_t, C_+, C_-) &= -log(\mathscr{P}(\alpha_t|I_{0:t},\alpha_{0:t-1})) \\ &= \underbrace{-\log \mathscr{P}_+(I_t|\alpha_t)\cdot\mathscr{P}_-(I_t|\alpha_t)}_{Data} - \log \underbrace{\mathscr{P}(\alpha_t|\alpha_{0:t-1})}_{Shape\ Prior} \end{aligned} \tag{18.17}$$

Thereby the probabilities $\mathscr{P}_\pm(I_t|\alpha_t)$ are similar to the Chan-Vese probabilities (18.2), except that $\phi_t(x)$ is determined by α_t as (18.9).

Applying these probabilities and (18.14)–(18.17) leads to the following energy functional:

$$\begin{aligned} E_{PCA}(\alpha_t, C_+, C_-) &= \frac{1}{2\sigma_+^2}\int_\Omega (I_t(x)-C_+)^2 H(\phi(\alpha_t))dx + \frac{1}{2}\log(2\pi\sigma_+^2)\int_\Omega H(\phi(\alpha_t))dx \\ &+\frac{1}{2\sigma_-^2}\int_\Omega (I_t(x)-C_-)^2(1-H(\phi(\alpha_t)))dx + \frac{1}{2}\log(2\pi\sigma_-^2)\int_\Omega (1-H(\phi(\alpha_t)))dx + \lambda\frac{1}{2}\omega^T\Lambda^{-1}\omega \end{aligned} \tag{18.18}$$

Here λ is an additional parameter that allows relative weighting between data and prior and $\phi(\alpha_t)$ is the level set estimation which is determined by α_t. The segmentation in each frame requires the estimation of the shape vector α_t, which is done by minimizing (18.18) with respect to α_t using gradient descent strategy.

18.4 Motion Estimation

In each frame we estimate the translation positions $(u,v)_t$ and use this to translate the previous contour ϕ_{t-1} as initialization to estimate the current contour. In the On-Line Learning model the segmentations aren't sensitive to the initial contour since we assume that the object isn't occluded. Therefore we use Lucas-Kanade approach [13] to estimate the translation position $(u,v)_t$ by minimizing

$$E_{LK}(u,v)_t = K_W * (I_x u + I_y v + I_t)^2 \tag{18.19}$$

where $K_W * (\cdot)$ denotes the convolution with an integration window of size W. I_x, I_y are the x, y derivatives of the image in each axis and I_t is the derivative

Fig. 18.2 Comparison between our algorithm (*Green*) and Chan-Vese (*Red*) on walking man sequence (319 frames) with full occlusion. In the on-line learning model (frame 22) and in the PCA representation model (frames: 153,161,165,176,266). As can be seen, the Chan-Vese model cannot handle the case of occlusion

between two consecutive frames. In the on-line learning model we also learn the temporal translations between consecutive frames to build a motion prior. This is done by collecting all the translations seen so far and build it into a AR model in the same way as we build the shape prior (18.13)

$$\begin{bmatrix} u \\ v \end{bmatrix}_t = \begin{bmatrix} \bar{u} \\ \bar{v} \end{bmatrix} + B_1 \begin{bmatrix} u \\ v \end{bmatrix}_{t-1} + B_2 \begin{bmatrix} u \\ v \end{bmatrix}_{t-2} + \dots B_q \begin{bmatrix} u \\ v \end{bmatrix}_{t-q} + \eta_{pos}$$

Here η_{pos} is Gaussian noise with zero mean with covariance matrix Λ_{pos}, B_i are the transitions of the AR model, and $(\bar{u}, \bar{v})$ are the mean values of u and v.

In the PCA representation model, when the object is occluded we use the learned AR motion parameters B_i, Λ_{pos} and $(\bar{u}, \bar{v})$ to estimate u and v in each frame, (u_p, v_p), as a prior and combine this to the LK functional (18.19):

$$E_{LK-PRIOR}(u, v)_t = E_{LK}(u, v) + \xi((u - u_p)^2 + (v - v_p)^2) \qquad (18.20)$$

This addition prevents the estimation of (u, v) during occlusion from being too far from their prior estimations.

18.5 Results

We tested our algorithm on different sequences with a deformable shape that are partially or fully occluded.

In each example the on-line learning model provides the contour based segmentations of the deformable shape and incrementally constructs the PCA eigenbase.

Fig. 18.3 Results of our algorithm on walking man sequence (200 frames) with partial traffic occlusions (*yellow* and *silver car*). The algorithm automatically switches from on-line learning (frame 60) to PCA representation as soon as occlusion is detected (frames: 11,66,68,70,77,84)

Fig. 18.4 Results of our algorithm on walking man sequence (173 frames) with full occlusion (walking woman hides the target). The algorithm automatically switches from on-line learning (frame 95) to PCA representation as soon as occlusion is detected (frames: 35,100,101, 103,104,109)

When it detects an occlusion, it estimates the AR parameters that capture the temporal dynamics of the shapes evolution seen so far and switches to the PCA representation model. The PCA model uses the current PCA eigenbase and the estimated AR prior parameters to keep segmenting the deformable shape during occlusion. Finally, when the target reappears it switches back to the on-line learning model and keeps tracking the target and updating the PCA eigenbase. We can see that it maintains the appropriate contours when the shape is totally or partially occluded.

Fig. 18.5 Results of our algorithm on jumping man sequence (225 frames) with full occlusion (walking man hides the target). Frames: 103,120,121,122,124,125

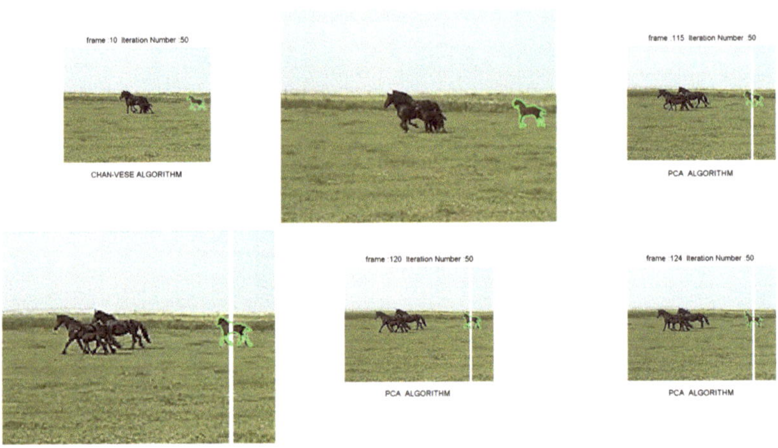

Fig. 18.6 Results of our algorithm on running horse sequence (290 frames) with one partially long period (synthetic) occlusion (20 frames with white label hides partially the target). We showed that our method was able to remain locked onto the target and illustrate the appropriate contours. Frames: 10,89,115,119,120,124

First, we compared our method to a stand-alone Chan-Vese algorithm on a sequence of walking man with one occlusion (left column). As can be seen in Fig. 18.2, the Chan-Vese model could not handle the occlusion properly, while our method kept tracking the person through the entire sequence and was able to illustrate the appropriate shapes when the man was totally occluded by the left column (Figs. 18.3–18.6).

Finally, in Fig. 18.7 we examined our method on a noisy sequence of a jumping man with additive Gaussian noise (SNR =15), and, as can be seen, our algorithm is able to cope with Gaussian noise and occlusion as well.

Fig. 18.7 Results of our algorithm on jumping man sequence with additive Gassuan noise (SNR = 15) and full occlusion (walking man hides the target). Frames: 119,120,121,132,134,142

18.6 Conclusions

We have extended level-set tracking to learn an eigenbase on the fly. This was then used to handle occlusions by switching from a Chan-Vese-based algorithm to a PCA-based representation that is more robust to noise and occlusions. In addition, we have shown that the proposed incremental level-set tracking can adjust to changes in the appearance of the object. This results in a robust tracker that can handle never-seen-before objects and deal with partial or full occlusions and noise.

References

1. Caselles, V., Kimmel, R., Sapiro, G.: Geodesic active contours. In: Proceeding of IEEE International Conference on Computer Vision, Boston, USA, pp. 694–699 (1995)
2. Caselles, V., Kimmel, R., Sapiro, G.: Geodesic active contours. Int. J. Comput. Vis. **22**(1), 61–79 (1997)
3. Chan, T.F., Vese, L.A.: Active contours without edges. IEEE Trans. Image Process. **10**(2), 266–277 (2001)
4. Chen, Y., Tagare, H., Thiruvenkadam, S., Huang, F., Wilson, D., Gopinath, K.S., Briggs, R.W., Geiser, E.: Using shape priors in geometric active contours in a variational framework. Int. J. Comput. Vis. **50**(3), 315–328 (2002)
5. Cremers, D.: Dynamical statistical shape priors for level set based tracking. IEEE Trans. Pattern Anal. Mach. Intell. **28**(8), 1262–1273 (2006)
6. Cremers, D., Rousson, M., Deriche, R.: A Review of Statistical Approaches to Level Set Segmentation: Integrating Color, Texture, Motion and Shape, IJCV (2007)
7. Fussenegger, M., Roth, P., Bischof, H., Deriche, R., Pinz, A.: A level set framework using a new incremental, robust active shape model for object segmentation and tracking. Image Vis. Comput. **27**, 1157–1168 (2009)

8. Golub, G.H., Van Loan, C.F.: Matrix Computations. The Johns Hopkins University Press, Baltimore (1996)
9. Hall, P., Marshall, D., Martin, R.: Incremental eigenanalysis for classification. In: Proceedings of British Machine Vision Conference, pp. 286–295 (1998)
10. Kichenassamy, S., Kunar, A., Olver, P.J., Tannenbaum, A., Yezzi, A.J.: Gradient flows and geometric active contour model. In: IEEE International Conference on Computer Vision, pp. 810–815 (1995)
11. Leventon, M., Grimson, W., Faugeras, O.: Statistical shape influence in geodesic active contours. In: CVPR, vol. 1, Hilton Head Island, SC, pp. 316–323 (2000)
12. Levy, A., Lindenbaum, M.: Sequential Karhunen-Loeve basis extraction and its application to images. IEEE Trans. Image Process. **9**(8), 1371–1374 (2000)
13. Lucas, B.D., Kanade, T.: An iterative image registration technique with an application to stereo vision. In: Proceedings of the 7th International Joint Conference on Artificial Intelligence, pp. 674–679 (1981)
14. Malladi, R., Sethian, J.A., Vemuri, B.C.: Shape modeling with front propagation: a level set approach. IEEE Trans. Pattern Anal. Mach. Intell. **17**(2), 158–175 (1995)
15. Mumford, D., Shah, J.: Optimal approximations by piecewise smooth functions and associated variational problems. Commun. Pure Appl. Math. **42**, 577–685 (1989)
16. Neumaier, A., Schneider, T.: Estimation of parameters and eigenmodes of multivariate autoregressive models. ACM Trans. Math. Softw. **27**(1), 27–57 (2001)
17. Osher, S.J., Sethian, J.A.: Fronts propagation with curvature dependent speed: algorithms based on Hamilton-Jacobi formulations. J. Comput. Phys. **79**, 12–49 (1988)
18. Paragios, N., Derichle, R.: Geodestic active regions and level set methods for supervised texture segmentation. Int. J. Comput. Vis. **46**(3), 223–247 (2002)
19. Paragios, N., Deriche, R.: Geodesic active regions and level set methods for motion estimation and tracking. Comput. Vis. Image Underst. **97**(3), 259–282 (2005)
20. Rousson, M., Paragios, N.: Shape priors for level set representations. In: Heyden, A. et al. (eds.) European Conference on Computer Vision. Volume 2351 of Lecture Notes in Computer Science, pp. 78–92. Springer, Berlin (2002)
21. Rousson, M., Paragios, N., Deriche, R.: Implicit active shape models for 3d segmentation in MRI imaging. In: MICCAI, pp. 209–216 (2004)
22. Sagiv, C., Sochen, N., Zeevi, Y.Y.: Integrated active contours for texture segmentation. IEEE Trans. Image Process. **1**(1), 1–19 (2006)
23. Schwarz, G.: Estimating the dimension of a model. Ann. Stat. **6**, 461–464 (1978)
24. Tsai, A., Yezzi, A., Wells, W., Tempany, C., Tucker, D., Fan, A., Grimson, E., Willsky, A.: Model-based curve evolution technique for image segmentation. In: Computer Vision and Pattern Recognition, Kauai, Hawaii, pp. 463–468 (2001)
25. Tsai, A., Yezzi, A.J., Willsky, A.S.: Curve evolution implementation of the Mumford-Shah functional for image segmentation, denoising, interpolation, and magnification. IEEE Trans. Image Process. **10**(8), 1169–1186 (2001)

Chapter 19
Simultaneous Convex Optimization of Regions and Region Parameters in Image Segmentation Models

Egil Bae, Jing Yuan, and Xue-Cheng Tai

Abstract This work develops a convex optimization framework for image segmentation models, where both the unknown regions and parameters describing each region are part of the optimization process. Convex relaxations and optimization algorithms are proposed, which produce results that are independent from the initializations and closely approximate global minima. We focus especially on problems where the data fitting term depends on the mean or median image intensity within each region. We also develop a convex relaxation for the piecewise constant Mumford-Shah model, where additionally the number of regions is unknown. The approach is based on optimizing a convex energy potential over functions defined over a space of one higher dimension than the image domain.

19.1 Introduction

Image segmentation is one of the most important problems in image processing and computer vision. The task is to group the image pixels into several regions or objects based on their intensity values. Energy minimization has become an established paradigm to formulate such problem mathematically, where both data/scene consistency and the regularity of the segmentation regions are encoded in

E. Bae (✉)
Department of Mathematics, University of California, Los Angeles, CA, USA
e-mail: ebae@math.ucla.edu

J. Yuan
Computer Science Department, University of Western Ontario, London, ON, Canada
e-mail: cn.yuanjing@gmail.com

X.-C. Tai
Department of Mathematics, University of Bergen, Bergen, Norway
e-mail: tai@math.uib.no

M. Breuß et al. (eds.), *Innovations for Shape Analysis*, Mathematics and Visualization,
DOI 10.1007/978-3-642-34141-0_19,

an energy potential. A major challenge is to solve the resulting NP-hard optimization problems numerically.

Variational Models for Image Segmentation

In this work, we focus on image segmentation with Potts regularity [16], which enforces region boundaries of minimal total length. We wish to partition the image domain Ω into n regions $\{\Omega_i\}_{i=1}^n$. The image domain is usually a rectangular subset of $\mathbb{R}^2$ or $\mathbb{R}^3$. For each point $x \in \Omega$ and each $i = 1, \ldots, n$, define the data cost function $f_i(x)$ of assigning x to the region Ω_i. Image segmentation with Potts prior and predefined data cost functions can then be formulated as

$$\min_{\{\Omega_i\}_{i=1}^n} \quad \sum_{i=1}^n \int_{\Omega_i} f_i(x)\, dx + \alpha \sum_{i=1}^n \int_{\partial\Omega_i} ds \tag{19.1}$$

$$\text{s.t.} \quad \cup_{i=1}^n \Omega_i = \Omega\,, \quad \Omega_k \cap \Omega_l = \emptyset\,, \quad \forall k \neq l\,, \tag{19.2}$$

where α is a parameter which controls the impact of the boundary regularization. The model (19.1) will be referred to as Potts model in this work. When $n > 2$ the optimization problem (19.1) in a discrete setting is NP-hard, therefore it is generally too difficult to find a global optimum. Algorithms exist that can compute good approximations [3] and in some cases exact solutions for level set representations of the problem [1]. In a continuous setting, several convex methods have recently appeared that may often lead to global solutions, or otherwise produce good approximations of global solutions [2, 4, 10, 14, 22].

An important example of the data costs in (19.1) is

$$f_i(x) = |I^0(x) - \mu_i|^\beta, \tag{19.3}$$

where $I^0(x)$ is the input image function and $\mu_i \in \mathbb{R}$, $i = 1, \ldots, n$ are predefined region parameters and $\beta \geq 1$. An intuitive explanation of μ_i is the mean of the image intensity I^0 within region Ω_i in case $\beta = 2$, or the median value within Ω_i if $\beta = 1$. They are, however, unknown in advance. Therefore, the function f_i depends on the unknown segmentation region Ω_i and does not fit into the framework of (19.1). The most ideal model should not rely on a post-processing step to determine the parameters, instead the values μ_i should be part of the optimization process. In [6] and [13] such an image segmentation model was formulated as follows

$$\min_{\{\Omega_i\}_{i=1}^n} \min_{\{\mu_i\}_{i=1}^n \in X} \quad \sum_{i=1}^n \int_{\Omega_i} |I^0(x) - \mu_i|^\beta dx + \alpha \sum_{i=1}^n \int_{\partial\Omega_i} ds \tag{19.4}$$

subject to (19.2). The set X is typically the set of feasible gray values, which may be the real line $X = \mathbb{R}$ or a discrete quantization $X = \{\ell_1, \ldots, \ell_L\}$. In contrast to

(19.1) along with (19.3), the energy is minimized over both $\{\Omega_i\}_{i=1}^n$ and the region parameters $\{\mu_i\}_{i=1}^n$. The model (19.4) is often called the Chan-Vese model. If there is no regularization, i.e. $\alpha = 0$, (19.4) can be recognized as the "k-means" model, which is also an NP-hard problem. To solve (19.4), one possibility is alternating minimization with respect to Ω_i and μ_i until convergence as follows:
Find initialization $\{\mu_i^0\}_{i=1}^n$. For $k = 0, \dots$ until convergence

$$1.\ \{\Omega_i^{k+1}\}_{i=1}^n = \arg\min_{\{\Omega_i\}_{i=1}^n} \sum_{i=1}^n \int_{\Omega_i} |I^0(x) - \mu_i^k|^\beta dx + \alpha \sum_{i=1}^n \int_{\partial\Omega_i} ds \quad \text{subject to (19.2)} \tag{19.5}$$

$$2.\ \{\mu_i^{k+1}\}_{i=1}^n = \arg\min_{\{\mu_i\}_{i=1}^n \in X} \sum_{i=1}^n \int_{\Omega_i^{k+1}} |I^0(x) - \mu_i|^\beta dx + \alpha \sum_{i=1}^n \int_{\partial\Omega_i^{k+1}} ds \tag{19.6}$$

Since (19.4) is not jointly convex, such a procedure does not in general produce a global minimum, but converges to a local optimum depending on the initialization of μ. Furthermore, there is no easy way to measure the quality of the converged result.

Closely related is the piecewise constant Mumford-Shah model [13], which can be expressed as a slight variation of (19.4) as (see Sect. 19.2.2)

$$\min_n \min_{\{\Omega_i\}_{i=1}^n} \min_{\{\mu_i\}_{i=1}^n \in X} \sum_{i=1}^n \int_{\Omega_i} |I^0(x) - \mu_i|^\beta dx + \alpha \sum_{i=1}^n \int_{\partial\Omega_i} ds \tag{19.7}$$

subject to (19.2). The energy potential (19.7) is also optimized over the number of regions n. In spite of the seemingly higher complexity, we show the problem (19.7) is easier to tackle than (19.4) in the following sections.

The optimization problem (19.4) can also be extended to more general data cost functions $f_i(\xi_i, x)$, where ξ_i is some unknown parameter associated with region Ω_i.

$$\min_{\{\Omega_i\}_{i=1}^n} \min_{\{\xi_i\}_{i=1}^n} \sum_{i=1}^n \int_{\Omega_i} f(\xi_i, x)\, dx + \alpha \sum_{i=1}^n \int_{\partial\Omega_i} ds \tag{19.8}$$
$$\text{s.t.} \quad \cup_{i=1}^n \Omega_i = \Omega\,, \quad \Omega_k \cap \Omega_l = \emptyset\,, \quad \forall k \neq l\,,$$

One example is $\xi_i = (\mu_i, \sigma_i)$, where μ_i is the mean and σ_i is the standard deviation of intensities in Ω_i. The data term can then be formulated as the log of the Gaussian distribution $f_i(\mu_i, \sigma_i, x) = \log(\frac{1}{\sqrt{2\pi\sigma_i}} \exp(-\frac{(I^0(x)-\mu_i)^2}{2\sigma_i^2}))$.

An image segmentation model based on the minimum description rule (MDL) was proposed [23] which places a direct penalty to the number of appearing regions of the Potts model (19.1). Recently, various algorithms have been proposed for computing global or good approximations of global minima [9, 18, 21] of the resulting optimization problems. We will also see that there is a close relationship between such a MDL based segmentation model and (19.4) or (19.8) if f_i and the penalty parameter are chosen in a particular way, as discussed in Sect. 19.4.

Instead of optimizing over the continuous set $X = \mathbb{R}$, the parameters μ_i in (19.4) or (19.7) can be optimized over a finite set of real numbers $X = \{\ell_1, \dots, \ell_L\}$, where L is the number of elements in X. This is the case for digital images, where the image intensity is quantized and the set X consists of a finite number of gray values, for instance 256. When the L_1 data fitting term is applied (that is, $\beta = 1$ in (19.3)), we show there exists globally optimal $\{\mu_i\}_{i=1}^n$ that are also present in the input image. Therefore, optimizing over the finite set will produce an exact global optimum. This extends the result of [8] from 2 to any number of regions. A similar result can also be found in [20] for denoising problems with total variation regularization.

Contributions and Previous Work

Little work has been devoted to global optimization over the regions and parameters simultaneously in the image segmentation models (19.4) or (19.7). In order to optimize (19.4) or (19.7) over the finite set X, one simple, but very slow approach, is to minimize the energy in the Potts model (19.1) for every combination of $\{\mu_i\}_{i=1}^n \in X$, and finally select the combination of $\{\mu_i\}_{i=1}^n \in X$ which yields the lowest energy. Since there are a total of L^n such combinations when X contains L elements, a total number of L^n problems of the form (19.1) need to be solved. In case of two regions, it is known each subproblem can be solved exactly and globally [7], but L^2 subproblems need to be solved which would be rather slow as L becomes large.

Restricted to two regions and a finite set X, Darbon [8] developed an algorithm which solved a sequence of two region problems with fixed parameters μ_1 and μ_2, but avoided to check all L^2 combinations. The number of two region subproblems to be solved is still $O(L^2)$. In [5], Brown et al. cast a relaxation of the problem with two regions and quantized parameters as an optimization problem over a higher dimensional space. The size of the convex problem is $O(|\Omega|L^2)$, where $|\Omega|$ is the number of pixels, therefore the complexity of their algorithm is also $O(|\Omega|\, L^2)$. An approach based on the branch and bound method was proposed for two region problems in [12]. In worst case its complexity is $O(|\Omega|\, L^2)$, but the method appears to converge linearly in the number of parameter values in practice. An algorithm was proposed in [17] for segmentation problems with two regions, where a sequence of L total variation regularized problems could be solved, followed by L simple thresholding operations each step. The complexity is therefore effectively $O(|\Omega|\, L)$.

Contributions: This work presents a jointly convex optimization framework for minimizing energy potentials of the form (19.4) over the regions and the parameters associated with each region (such as mean intensities). We also derive a convex relaxation of the piecewise constant Mumford-Shah model (19.7), where additionally the number of regions are unknown. The convex relaxation of (19.4) can be applied for problems with any number of regions, not just $n = 2$. Furthermore, the size of the convex relaxed problems grow at most linearly in the number of potential parameter values L, i.e. as $O(|\Omega|\, L)$.

The problems are first reformulated as minimization problems over binary functions defined in a space of one higher dimension than the image domain. Convex relaxations are then derived based on the reformulated problems. The method is not guaranteed to always produce an exact solution, but some conditions are identified for when this is possible. We begin by treating the piecewise constant Mumford Shah model in Sect. 19.2.2. Next, we present convex relaxations for the problems (19.4) and (19.8), where the number of regions are upper bounded. Fast algorithms are derived in Sect. 19.4.

19.2 Convex Relaxation Models

In this section the problems (19.4) and (19.7) are first reformulated as optimization problems in terms of a binary function in a space of one higher dimension than the image domain. Convex relaxations are then derived based on the reformulated problems. The new relaxations build on recently proposed convex relaxations for Potts model (19.1) which are briefly reviewed next.

19.2.1 Convex Relaxation for Potts Model

Several convex relaxations for Potts model (19.1) have recently been proposed [2, 10, 14, 22]. Any such convex relaxation can be used as building block for the new relaxations of the more complicated models (19.4) and (19.7) proposed in this work. However, we focus particularly on a simple relaxation for Potts model [2, 10, 22], which has demonstrated to work well for practical problems. Let $u_i(x)$ be the characteristic function of the region Ω_i, defined as

$$u_i(x) := \begin{cases} 1, \; x \in \Omega_i \\ 0, \; x \notin \Omega_i \end{cases}, \quad i = 1, \dots, n\,.$$

Then, the Potts model (19.1) can be written in terms of u_i as:

$$\min_{\{u_i\}_{i=1}^n \in B} \quad \sum_{i=1}^{n} \int_{\Omega} u_i(x) f_i(x)\, dx + \alpha \sum_{i=1}^{n} \int_{\Omega} |\nabla u_i|\, dx \tag{19.9}$$

subject to

$$\sum_{i=1}^{n} u_i(x) = 1\,, \quad \forall x \in \Omega \tag{19.10}$$

where B is the set

$$B = \{u \in BV(\Omega) \;:\; u(x) \in \{0,1\},\ \forall x \in \Omega\} \tag{19.11}$$

and the total-variation of the characteristic function $u_i(x)$ encodes the length of the boundary of the region Ω_i.

A convex relaxation of (19.9), was proposed and studied in [2, 10, 19, 22] by instead minimizing over the convex set

$$u_i \in B' = \{u \in BV(\Omega) \;:\; u(x) \in [0,1],\ \forall x \in \Omega\} \tag{19.12}$$

for $i = 1, \ldots, n$. If the solution of the relaxed problem is binary at all $x \in \Omega$, it is also a global minimum of (19.1). Otherwise different schemes were proposed [2, 10, 22] to generate a binary solution $\tilde{u}$, which may either be a global minimum or close approximation to a global minimum of (19.1). The simplest such rounding scheme is just:

$$\tilde{u}(x) = e_\ell(x), \text{ where } \ell = \arg\max_i u_i(x), \tag{19.13}$$

where e_ℓ is the indicator vector which is 1 at the ℓ-th component and 0 elsewhere.

19.2.2 Convex Relaxation for Piecewise-Constant Mumford-Shah Model

In this section, we show that the piecewise constant Mumford-Shah model (19.7) can be expressed as a special case of (19.9). In its most general form, the Mumford-Shah model seeks an approximation image I and a set of curves Γ which minimizes

$$\inf_{\Gamma,I} E_\lambda(\Gamma, I) = \int_\Omega |I^0(x) - I(x)|^\beta dx + \lambda \int_{\Omega\backslash\Gamma} |\nabla I|^2 dx + \alpha \int_\Gamma ds. \tag{19.14}$$

Its piecewise constant variant can be regarded as the limit model as the penalty parameter λ goes to infinity i.e.

$$\inf_{\Gamma,I} E_\infty(\Gamma, I) \tag{19.15}$$

Due to infinite weight on the term $\int_{\Omega\backslash\Gamma} |\nabla I|^2$, (19.15) enforces solutions $I(x)$ that are constant everywhere except for the discontinuity set Γ, i.e. the function $I(x)$ is piecewise constant. The discontinuity set Γ therefore splits the domain Ω into a set of subdomains, say n in number: $\{\Omega_i\}_{i=1}^n$. The number n is unknown in advance, and is part of the optimization process. The piecewise constant Mumford-Shah model can therefore equivalently be formulated as (19.7), which is optimized over the regions Ω_i for $i = 1, \ldots, n$, the mean values μ_i of $I(x)$ within each region Ω_i for $i = 1, \ldots, n$ and the number of regions n.

Alternatively, (19.7) can be formulated in terms of the characteristic functions $u_i(x)$ as:

$$\min_{n} \min_{\{u_i\}_{i=1}^n \in B} \min_{\{\mu_i\}_{i=1}^n \in X} \quad E(u, \mu, n) = \sum_{i=1}^{n} \int_{\Omega} u_i(x)|I^0(x) - \mu_i|^\beta \, dx + \alpha \sum_{i=1}^{n} \int_{\Omega} |\nabla u_i| \, dx \tag{19.16}$$

subject to

$$\sum_{i=1}^{n} u_i(x) = 1, \quad \forall x \in \Omega.$$

Assume now that the set of feasible values μ_i is finite, i.e. $X = \{\ell_1, \ldots, \ell_L\}$. For instance X may consist of the set of quantized gray values: $X = \{1, \ldots, L\}$. For each element $\ell_i \in X$ define the corresponding characteristic function $u_i(x) \in B$. We will show the piecewise constant Mumford-Shah model (19.7) can be written as a minimization problem over such a set $\{u_i\}_{i=1}^L$. More specifically, we show that the following minimization problem is equivalent to the piecewise constant Mumford-Shah model (19.16) if the feasible intensity values are restricted to $X = \{\ell_1, \ldots, \ell_L\}$.

$$\min_{\{u_i\}_{i=1}^L \in B} \quad E^{\text{ext}}(\{u_i\}_{i=1}^L) = \sum_{i=1}^{L} \int_{\Omega} u_i(x)|I^0(x) - \ell_i|^\beta dx + \alpha \sum_{i=1}^{L} \int_{\Omega} |\nabla u_i| dx \tag{19.17}$$

subject to

$$\sum_{i=1}^{L} u_i(x) = 1, \quad u_i(x) \geq 0, \quad \forall x \in \Omega, \; i = 1, \ldots, L. \tag{19.18}$$

The above energy has the same form as (19.9).

Proposition 19.1. *Given an optimum u^* of (19.17). Let n^* be the number of indices i for which $u_i^* \not\equiv 0$. Define the set of indices $\{i_j\}_{j=1}^{n^*} \subset \{1, \ldots, L\}$ where $u_{i_j}^* \not\equiv 0$. Then $(\{\ell_{i_j}\}_{j=1}^{n^*}, \{u_{i_j}^*\}_{j=1}^{n^*}, n^*)$ is a global optimum of the piecewise constant Mumford-Shah model* (19.16) *with $X = \{\ell_1, \ldots, \ell_L\}$.*

The proof is given in the appendix.

In view of Proposition 19.1, a convex relaxation of the piecewise constant Mumford-Shah model can be defined as the minimization of (19.17) over $B' = \{u \in BV(\Omega) : u(x) \in [0, 1] \; \forall x \in \Omega\}$. It has the same form as the convex relaxed Potts model, which has already been studied in [2, 10, 19, 22].

The piecewise constant Mumford-Shah model (19.16) will naturally result in a sparse solution, where the number of 'active' regions n is relatively low in comparison to L. The regularization parameter α controls both regularity of the region boundaries and the number of regions. If $\alpha = 0$, the solution is just $I = I^0$ and the pixels are not grouped in any way, instead each pixel is regarded as a distinct region. Therefore, the Mumford-Shah model may result in more regions than desired unless α is set sufficiently high.

19.2.3 Jointly Convex Relaxation over Regions and Region Parameters

In this section we propose a convex relaxation for image segmentation models where the number of regions are fixed, e.g. (19.4) or (19.8). In many applications, the number of regions is known in advance, but the region parameters are unknown. This is for instance the case for segmentation problems with two regions where one wishes to distinguish foreground and background.

We start by writing out (19.4) in terms of the characteristic functions u_i of each region Ω_i as follows

$$\min_{\{u_i\}_{i=1}^n \in B} \min_{\{\mu_i\}_{i=1}^n \in X} \sum_{i=1}^n \int_\Omega u_i(x)|I^0(x) - \mu_i|^\beta \, dx + \alpha \sum_{i=1}^n \int_\Omega |\nabla u_i| \, dx \,. \tag{19.19}$$

s.t. $\sum_{i=1}^n u_i(x) = 1$ for all $x \in \Omega$. In order to optimize (19.4) over the set $\mu_i \in X = \{\ell_1, \dots, \ell_L\}$, $i = 1, \dots, n$, we start by proposing two equivalent alternative reformulations of (19.4):

Alternative 1: For each binary-valued function u_i, $i = 1, \dots, L$, define a binary variable $v_i \in \{0, 1\}$, with the interpretation $v_i = 1$ if $u_i(x) \neq 0$ for some $x \in \Omega$ and $v_i = 0$ else. Then (19.4) can be formulated as

$$\min_{u,v} \quad E^{\text{ext}}(\{u_i\}_{i=1}^L) = \sum_{i=1}^L \int_\Omega u_i(x)|I^0(x) - \ell_i|^\beta + \alpha |\nabla u_i| \, dx \tag{19.20}$$

subject to

$$\sum_{i=1}^L u_i(x) = 1 \,, \quad \forall x \in \Omega, \tag{19.21}$$

$$\sum_{i=1}^L v_i \leq n \,, \tag{19.22}$$

$$u_i(x) \leq v_i\,, \quad \forall x \in \Omega,\ i = 1, \ldots, L \tag{19.23}$$

$$u_i(x) \in \{0,1\}\,, \quad \forall x \in \Omega,\ i = 1, \ldots, L \tag{19.24}$$

$$v_i \in \{0,1\}\,, \quad i = 1, \ldots, L \tag{19.25}$$

Alternative 2: The problem can also be formulated without the artificial variable v. Observe that by definition, $\sup_{x\in\Omega} u_i(x) \leq v_i$, therefore the constraints (19.22) and (19.23) can be shortened by $\sum_{i=1}^{L} \sup_{x\in\Omega} u_i(x) \leq n$, which is also convex. Therefore the problem can equivalently be formulated as

$$\min_u \quad E^{\text{ext}}(\{u_i\}_{i=1}^{L}) = \sum_{i=1}^{L} \int_\Omega u_i(x)|I^0(x) - \ell_i|^\beta + \alpha\,|\nabla u_i|\,dx \tag{19.26}$$

subject to

$$\sum_{i=1}^{L} u_i(x) = 1\,, \quad \forall x \in \Omega, \tag{19.27}$$

$$\sum_{i=1}^{L} \sup_{x\in\Omega} u_i(x) \leq n\,, \tag{19.28}$$

$$u_i(x) \in \{0,1\}\,, \quad \forall x \in \Omega,\ i = 1, \ldots, L \tag{19.29}$$

The constraint (19.28) forces the solution to satisfy $u_i \equiv 0$ for all but at most n indices $i \in \{1, \ldots, L\}$. The next result shows that an optimum of (19.19), or equivalently of (19.4), can be obtained by finding an optimal solution u^* to either of the two above problems, (19.20) or (19.26), through the following proposition.

Proposition 19.2. *Given an optimum u^* of* (19.20) *or* (19.26)*. Let n^* be the number of indices i for which $u_i^* \not\equiv 0$. Define the set of indices $\{i_j\}_{j=1}^{n^*} \subset \{1, \ldots, L\}$ such that $u_{i_j}^* \not\equiv 0$. Then $\{\ell_{i_j}\}_{j=1}^{n^*}, \{u_{i_j}^*\}_{j=1}^{n^*}$ is a global optimum to* (19.19) *with $X = \{\ell_1, \ldots, \ell_L\}$.*

Clearly, $n^* \leq n$, otherwise the constraints (19.28) or (19.23) would be violated. The rest follows by an identical proof to that of Proposition 19.1.

Both formulations (19.20) and (19.26) are nonconvex due to the binary constraints on u and v. Convex relaxations can instead be derived by replacing the binary constraints (19.24), (19.25) by

$$u_i(x) \in [0,1],\ \ \forall x \in \Omega,\ i = 1, \ldots, L \tag{19.30}$$

$$v_i \in [0,1],\ \ i = 1, \ldots, L\,. \tag{19.31}$$

The convex relaxations resulting from the two alternatives (19.20) and (19.26) are equivalent as they share the same set of minimizers u. If the computed solution of the relaxed problem is binary for all $x \in \Omega$, it is also globally optimal to the original problem. If not, a close approximation can be obtained by the binarization scheme (19.13).

19.3 Some Optimality Results

In general, the convex relaxations are not guaranteed to produce an exact global minimum, but will provide close approximations. In this section we derive some conditions under which an exact solution can be obtained. First, we show that under L^1 data fidelity, the optimal gray values belong to the set of gray values which are already contained in the image. Second, we show that in case of two regions ($n = 2$) a thresholding scheme for producing exact solutions can applied under some conditions.

19.3.1 L^1 Data Fidelity

Consider the models (19.1) and (19.7) with f_i, $i = 1, \dots, n$, given by (19.3) and $\beta = 1$, i.e. the L^1 fidelity term. Assume further the input image $I^0(x)$ is quantized and takes values in the set $\{\ell_1, \dots, \ell_L\}$. The next result shows that there exists optimal parameters μ_i, $i = 1, \dots, n$, that also take values in the same set $\{\ell_1, \dots, \ell_L\}$. Hence it suffices to optimize μ over the set $X = \{\ell_1, \dots, \ell_L\}$. This result has previously been shown by the two-region problems ($n = 2$) in [8].

Proposition 19.3. *Given $I^0 : \Omega \mapsto \{\ell_1, \dots, \ell_L\}$, and consider the data term* (19.3) *with $\beta = 1$ and $X = \mathbb{R}$. There exists globally optimal $(\{\Omega_i\}_{i=1}^n, \{\mu_i^*\}_{i=1}^n, n)$ to the Mumford-Shah model (19.7) or $\{\Omega_i\}_{i=1}^n, \{\mu_i^*\}_{i=1}^n$ to* (19.4)*, where $\mu_i^* \in \{\ell_1, \dots, \ell_L\}$ for $i = 1, \dots, n$.*

Proof. The proof is by induction. When restricted to two regions, $n = 2$, the result was proved in [8], Theorem 1. Assume the result holds for $n = k$, then there exists a globally optimal solution $\{\Omega_i\}_{i=1}^{k+1}, \{\mu_i\}_{i=1}^{k+1}, k + 1$ to the Mumford-Shah model (19.7), or $\{\Omega_i\}_{i=1}^{k+1}, \{\mu_i\}_{i=1}^{k+1}$ to (19.4), where $\{\mu_i\}_{i=1, i \neq j}^{k+1} \in \{1, \dots, L\}$. We will show the result also holds for $n = k + 1$. Pick any $j \in \{1, \dots, k + 1\}$, and consider the image domain $\Omega \backslash \Omega_j$. Clearly, $\{\Omega_i\}_{i=1, i \neq j}^{k+1}, \{\mu_i\}_{i=1, i \neq j}^{k+1}, k$ is globally optimal to the Mumford Shah model in the domain $\Omega \backslash \Omega_j$. It remains to show that also $\mu_j \in \{1, \dots, L\}$. Pick any $\ell \neq j \in \{1, \dots, L\}$. Then $\{\Omega_i\}_{i=1, i \neq \ell}^{k+1}, \{\mu_i\}_{i=1, i \neq \ell}^{k+1}, k$ is globally optimal to the Mumford Shah model in the domain $\Omega \backslash \Omega_\ell$. By the induction hypotheses it is possible that $\{\mu_i\}_{i=1, i \neq \ell}^{k+1} \in \{1, \dots, L\}$, which implies there exists optimal $\mu_j \in \{1, \dots, L\}$.

19.3.2 Optimality of Relaxation for $n = 2$

The relaxations are not in general exact, but will produce solutions that are optimal or nearly optimal. In case $n = 2$, exact solutions can be generated under some conditions. It suffices that for two indices k, j, the boundary $u_k(x) = 1$ and $u_j(y) = 1$ is attained for some $x, y \in \Omega$.

Proposition 19.4. *Let u^* be a solution of (19.26), or alternatively u^*, v^* a solution of (19.20) with $n = 2$, where the binary constraints B are replaced by the convex constraint B'. Assume the variable v is binary, or equivalently, assume there exists $k, j \in \{1, \ldots, L\}$ such that $u_k(x) = 1$ for some $x \in \Omega$ and $u_k(y) = 1$ for some $y \in \Omega$ For any threshold level $t \in (0, 1)$ define the function $\tilde{u}$ such that*

$$\tilde{u}_k(x) := \begin{cases} 1\,, \text{ if } u_i^*(x) \geq t \\ 0\,, \text{ if } u_i^*(x) < t \end{cases}, \quad \tilde{u}_j(x) := \begin{cases} 1\,, \text{ if } u_i^*(x) > 1 - t \\ 0\,, \text{ if } u_i^*(x) \leq 1 - t \end{cases}.$$

and $\tilde{u}_i = u_i$ for all $i \neq k, j \in \{1, \ldots, L\}$. Then $(\tilde{u}, v^*)$ is a binary global optimum of (19.20) subject to (19.21)–(19.23) and the binary constraints (19.24) and (19.25).

Proof. Since $u_k(x) = 1$ for some $x \in \Omega$ and $u_k(y) = 1$ for some $y \in \Omega$, it follows by constraint (19.28) that $u_i(x) = 0$ for all $i \neq k, j \in \{1, \ldots, L\}$. Define $\phi = u_k$, then since $u_k + u_j = 1, \phi = 1 - u_j$. Define

$$\tilde{\phi}(x) := \begin{cases} 1\,, \phi(x) \geq t \\ 0\,, \phi(x) < t \end{cases}.$$

and observe that $\tilde{\phi} = \tilde{u}_k$ and $\tilde{\phi} = 1 - \tilde{u}_j$. Then

$$E^{\text{ext}}(u) = \int_\Omega u_k(x)|I^0(x) - k|^\beta + u_j(x)|I^0(x) - j|^\beta + \alpha \int_\Omega |\nabla u_k| + |\nabla u_j|\, dx$$

$$= \int_\Omega \phi(x)|I^0(x) - k|^\beta + (1 - \phi(x))|I^0(x) - j|^\beta + 2\alpha \int_\Omega |\nabla \phi|\, dx$$

$$= \int_\Omega \tilde{\phi}(x)|I^0(x) - k|^\beta + (1 - \tilde{\phi}(x))|I^0(x) - j|^\beta + 2\alpha \int_\Omega |\nabla \tilde{\phi}|\, dx$$

$$= \int_\Omega \tilde{u}_k(x)|I^0(x) - k|^\beta + \tilde{u}_j(x)|I^0(x) - j|^\beta + \alpha \int_\Omega |\nabla \tilde{u}_k| + |\nabla \tilde{u}_j|\, dx = E^{\text{ext}}(\tilde{u}).$$

The third equality follows by the thresholding theorem of [7] for relaxed binary segmentation problems.

19.4 Algorithms

The convex relaxation for the piecewise constant Mumford-Shah model (19.17) has the form of the convex relaxed Potts model [11, 22], and can be optimized by established algorithms. In [19] a very efficient algorithm was proposed based on the dual formulation, which can also be parallelized over each characteristic function. This algorithm is therefore well suited for optimizing (19.17), which usually contains a large number of characteristic functions.

The convex relaxation of (19.4) is a little more complicated due the extra constraints. As stated in Sect. 19.2.3, there are two equivalent formulations of the relaxation. We will build up an algorithm based on alternative 2, i.e. solve (19.26) with constraints (19.27), (19.28) and (19.30). We assume the optimal number of regions n is attained (i.e. equality in (19.28)). If the optimal number of regions is less than n, exactly the same solution would be produced by the convex relaxation of the piecewise constant Mumford-Shah model, which is simpler to optimize and could be checked by a separate calculation. Let γ be a Lagrange multiplier for the constraint

$$\sum_{i=1}^{L} \sup_{x\in\Omega} u_i(x) - n = 0. \tag{19.32}$$

The problem can then be stated as the saddle point problem

$$\max_{\gamma} \min_{u} \mathscr{L}(u,\gamma) = \sum_{i=1}^{L} \int_{\Omega} u_i(x)|I^0(x) - \ell_i|^{\beta} + \alpha\, |\nabla u_i|\, dx + \gamma\Big(\sum_{i=1}^{L} \max_{x\in\Omega} u_i(x) - n\Big) \tag{19.33}$$

$$\text{s.t.} \quad \sum_{i=1}^{L} u_i(x) = 1, \quad u_i(x) \geq 0\ \forall x \in \Omega,\ i = 1,\dots,L, \quad \gamma \geq 0$$

In order to optimize (19.33), the Lagrangian method can be applied as follows: for $k = 1, \dots$ until convergence

1. $u^{k+1} = \arg\min_u \mathscr{L}(u, \gamma^k)$, s.t. $\sum_{i=1}^{L} u_i(x) = 1$, $u_i(x) \geq 0\ \forall x \in \Omega$, $i = 1,\dots,L$

2. $\gamma^{k+1} = \max(0, \gamma^k + c(\sum_{i=1}^{L} \max_{x\Omega} u_i^{k+1}(x) - n))$.

Observe that subproblem 1. has the same form as the label cost prior problem studied in [18, 21]. A fast algorithm for solving such problems was proposed in [21]. In particular, it was shown 1. could be written as the primal-dual problem

$$\min_{u\in B',\ \sum_{i=1}^{L} u_i(x)=1,\ \forall x\in\Omega} \mathscr{L}(u,\gamma^k) = \min_{u} \max_{p_s,p_t,q,r} \int_{\Omega} p_s\, dx + \sum_{i=1}^{n} \int_{\Omega} u_i(\operatorname{div} q_i - p_s + p_i - r_i) \tag{19.34}$$

$$\text{s.t.} \quad p_i(x) \leq |I^0(x) - \ell_i|^{\beta},\ |q_i(x)| \leq \alpha,\ \int_{\Omega} |r_i(x)|\, dx \leq \gamma^k;\ i = 1\dots n.$$

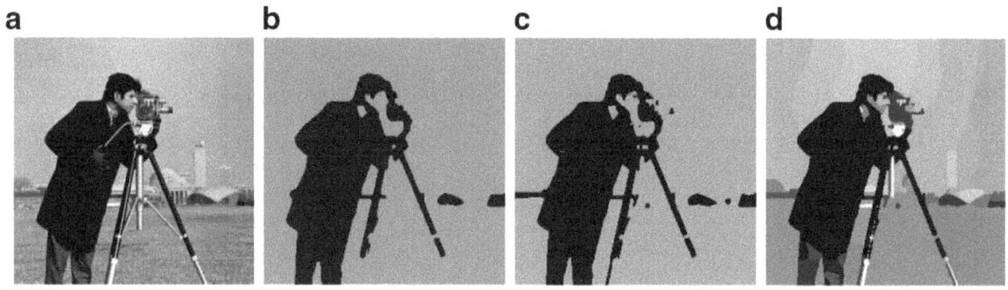

Fig. 19.1 (**a**) Input image. (**b**) and (**c**) Convex relaxation of (19.4) with $n = 2$ and $\alpha = 0.15$: (**b**) $\beta = 2$, estimated parameters $\mu_1 = 0.09, \mu_2 = 0.59$, energy $= 1.25 * 10^3$; (**c**) $\beta = 1$ estimated parameters $\mu_1 = 0.06, \mu_2 = 0.62$, energy $= 5.39 * 10^3$. (**d**) Convex relaxation of piecewise constant Mumford-Shah model (19.7) with $\beta = 1$ and $\alpha = 0.15$, energy $= 3.31 * 10^3$

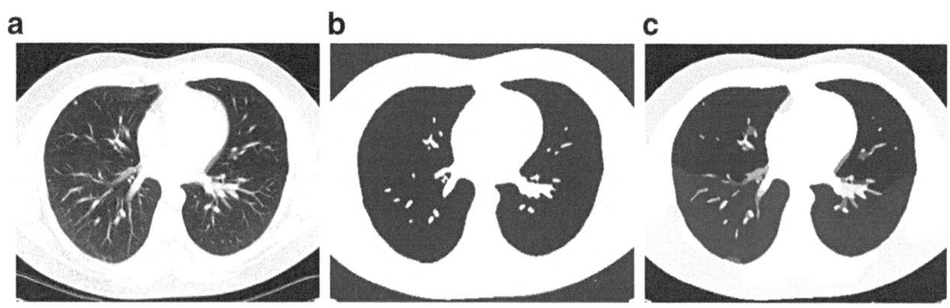

Fig. 19.2 (**a**) Input image. (**b**) Convex relaxation of (19.4) with $n = 2$, $\beta = 1$ and $\alpha = 0.15$: estimated parameters $\mu_1 = 0.23, \mu_2 = 0.93$, energy $9.23{*}10^3$. (**c**) Convex relaxation of piecewise constant Mumford-Shah model (19.7) with $\beta = 1$, energy $= 7.13 * 10^3$

where u_i works as a Lagrange multiplier. The above energy functional can be optimized separately for p_s, p_i, q_i and r_i in closed form. Therefore the augmented Lagrangian method could be applied to efficiently solve the overall problem. In practice, only a few iterations are necessary before γ is updated.

19.5 Numerical Experiments

In this section we demonstrate numerically the new convex relaxation for optimizing the energy in the image segmentation model (19.4) jointly over the regions and regions parameters, and the new convex relaxation of the piecewise constant Mumford-Shah model (19.7). In Figs. 19.1, 19.2 and 19.4, we have used 100 quantization levels for the unknown parameters, i.e. $X = \{0.01, 0.02, \ldots, 1.00\}$ and in Fig. 19.3, 255 levels have been used. In order to visualize results, we depict the function $u(x) = \ell_i$ if $x \in \Omega_i, i = 1, \ldots, n$.

Observe that the piecewise constant Mumford-Shah model may result in more regions than desired, as shown in the last subfigures. This is especially visible in Figs. 19.1 and 19.2, whereas it leads to more reasonable results in Figs. 19.3

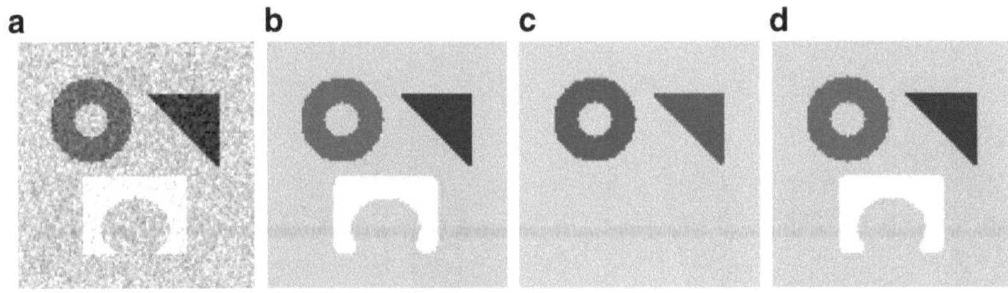

Fig. 19.3 (a) Input. (b) and (c) Convex relaxation of (19.4): (b) $n = 4$, (c) $n = 2$. (d) Convex relaxation of piecewise constant Mumford-Shah model (19.7)

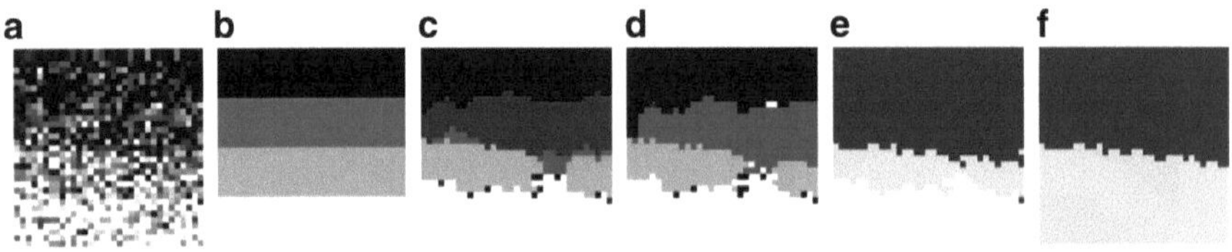

Fig. 19.4 (a) Input. (b) Ground truth. (c) Convex relaxation of Mumford-Shah functional: energy 212.02. (d)–(f) Convex relaxation of (19.4): (d) $n = 4$, (e) $n = 3$, (f) $n = 2$ energy = 219.78. In all experiments $\beta = 1$

and 19.4. By instead minimizing (19.4), with the number of regions fixed to 2, in terms of the regions and parameters μ_1 and μ_2, one is able to separate foreground and background in Figs. 19.1 and 19.2. Observe that the piecewise constant Mumford-Shah model leads to lower energy, since it is optimized over a larger feasible set.

The convex relaxations generate close approximations to a global minimum. To verify this, we have used the estimated parameters μ_1 and μ_2 from the convex relaxation as initialization of the alternating minimization algorithm (19.5) and (19.6). In all cases, the converged values of μ_1 and μ_2, after rounding to the nearest element in X, did not change. This indicates strongly that the globally optimal values of μ_1 and μ_2 within X, had been obtained by the convex relaxation method. In our experience, the alternating algorithm (19.5) and (19.6) is rather robust to initialization and converges to the same solution for many initializations of μ_1 and μ_2. However, an independent work [5] presented examples where the alternating algorithm gets stuck in poor local minima for exactly the input images in Figs. 19.1 and 19.2.

19.6 Conclusions and Future Work

We end with some discussions on future work and conclusions.

Extension to continuous label values The convex optimization framework for (19.4) and requires that the set of feasible parameter values is quantized. The relaxations can also be extended to optimization problems where the set of feasible parameter values is continuous, i.e. $X = \mathbb{R}$. Let $\ell_{\min}$ be the smallest and $\ell_{\max}$ be the largest value of μ_i. Define the one-higher dimensional binary variable $u(x,\ell)$ for each $(x,\ell) \in \Omega \times [\ell_{\min}, \ell_{max}]$, i.e. $u : \Omega \times [\ell_{\min}, \ell_{max}] \mapsto \{0,1\}$.

As a continuous generalization of (19.17), we argue the piecewise constant Mumford-Shah model (19.7) can be formulated in terms of u as

$$\min_u \int_{\ell_{\min}}^{\ell_{\max}} \int_{\Omega} u(x,\ell)|I^0(x) - \ell|^\beta + \alpha|\nabla_x u(x,\ell)|\, dx\, d\ell \tag{19.35}$$

subject to

$$\int_{\ell_{\min}}^{\ell_{\max}} u(x,\ell) * \delta(\ell)\, d\ell = 1, \quad \forall x \in \Omega \tag{19.36}$$

$$u(x,\ell) \in \{0,1\}, \quad \forall (x,\ell) \in \Omega \times [\ell_{\min}, \ell_{\max}]. \tag{19.37}$$

where $\delta(\ell)$ is the delta distribution and the convolution is defined as $u(x,\ell) * \delta(\ell) = \int_{-\infty}^{\infty} u(x,s)\delta(\ell - s)ds$.

Let $u^*(x,\ell)$ be an optimum of (19.35). We conjecture that $w^*(.) = \int_{-\infty}^{\infty} \ell u^*(.,\ell) * \delta(\ell)\, d\ell$ is a piecewise constant function that is a global minimizer of the piecewise constant Mumford-Shah model (19.15). We believe the proof can be constructed as a direct continuous generalization of the proof of Proposition 19.1, but will be more involved due to measure theoretic aspects. It would be interesting to investigate how this result relates to a recently proposed convex relaxation of the piecewise smooth Mumford-Shah model [15]. The model (19.4) with n regions can similarly be formulated by adding the constraint

$$\int_{-\infty}^{\infty} \sup_{x \in \Omega} u(x,\ell) d\ell \leq n \tag{19.38}$$

Extension to Vector-Valued Parameters The results discussed in Sect. 19.2.3 can easily be extended to more general problems of the form (19.8), where $\xi = (\xi_1, \ldots, \xi_N)$ denote the vector-valued parameter associated with each region. Let $X = \{\ell_1^1, \ldots, \ell_L^1\} \times \ldots \times \{\ell_1^N, \ldots, \ell_L^N\}$ denote the finite set of all feasible ξ. For each $i_1, i_2, \ldots, i_N \in \{1, \ldots, L\}$ define the function $u_{i_1,\ldots,i_N} : \Omega \mapsto \{0,1\}$ and variable $v_{i_1,\ldots,i_N} \in \{0,1\}$. Then the model (19.8) can be written

$$\min_u \sum_{i_1=1}^{L} \cdots \sum_{i_N=1}^{L} \int_{\Omega} u_{i_1,\ldots,i_N}(x) f(\xi_{i_1}^1, \ldots, \xi_{i_N}^N, x) + \alpha|\nabla u_{i_1,\ldots,i_N}| \tag{19.39}$$

subject to

$$\sum_{i_1=1}^{L} \cdots \sum_{i_N=1}^{L} u_{i_1,\ldots,i_N}(x) = 1, \quad \forall x \in \Omega \tag{19.40}$$

$$\sum_{i_1=1}^{L} \cdots \sum_{i_N=1}^{L} \max_{x \in \Omega} u_{i_1,\dots,i_N}(x) \leq n, \tag{19.41}$$

$$u_{i_1,\dots,i_N}(x) \in \{0,1\}, \quad \forall x \in \Omega, \; i_1,\dots,i_N \in X. \tag{19.42}$$

The equivalence between (19.8) and (19.39) follows by a straight forward generalization of Proposition 19.2.

19.6.1 Conclusions

Image segmentation problems can successfully be modeled as the minimization of an energy potential with respect to regions and parameters associated with each region. In this work, we have reformulated such problems as the optimization of binary functions in a space of one higher dimension than the image domain. Convex relaxations and optimization algorithms have been proposed which does not depend on initializations and produce close approximations to global minima. In contrast to previous work, the complexity of our algorithm grows at most linearly with the number of potential parameter values, and can be applied for segmentation problems with any number of regions.

19.7 Proofs

Proof of Proposition 19.1

Proof. Let $(\{\tilde{u}_j\}_{j=1}^{\tilde{n}}, \{\ell_{\tilde{i}_j}\}_{j=1}^{\tilde{n}}, \tilde{n})$ be any other solution of (19.16). Define the vector function

$$\bar{u}_j = 0, \text{ for } j \in \{1,\dots,L\} \backslash \{\tilde{i}_1,\dots,\tilde{i}_{\tilde{n}}\}$$

$$\bar{u}_{\tilde{i}_j} = \tilde{u}_j \text{ for } j = 1,\dots,\tilde{n}.$$

Then $\bar{u}$ belongs to the feasible set (19.18) of the problem (19.17).

$$E^{\text{ext}}(\bar{u}) = \sum_{i=1}^{L} \int_\Omega \bar{u}_i |I^0(x) - \ell_i|^2 \, dx + \sum_{i=1}^{L} \alpha \int_\Omega |\nabla \bar{u}_i| \, dx = \sum_{j=1}^{\tilde{n}} \int_\Omega \bar{u}_{\tilde{i}_j} |I^0(x) - \ell_i|^2 dx + \sum_{j=1}^{\tilde{n}} \alpha \int_\Omega |\nabla \bar{u}_{\tilde{i}_j}| dx$$

$$= \sum_{i=1}^{\tilde{n}} \int_\Omega \tilde{u}_i |I^0(x) - \ell_{\tilde{i}_j}|^2 \, dx + \sum_{i=1}^{\tilde{n}} \alpha \int_\Omega |\nabla \tilde{u}_i| \, dx = E(\{\tilde{u}_j\}_{j=1}^{\tilde{n}}, \{\ell_{\tilde{i}_j}\}_{j=1}^{\tilde{n}}, \tilde{n}).$$

But since u^* is a global minimizer of E^{ext}

$$E^{\text{ext}}(u^*) \leq E^{\text{ext}}(\bar{u}) = E(\{\tilde{u}_j\}_{j=1}^{\tilde{n}}, \{\ell_{\tilde{i}_j}\}_{j=1}^{\tilde{n}}, \tilde{n}), \tag{19.43}$$

and since

$$E^{\text{ext}}(u^*) = \sum_{i=1}^{L} \int_\Omega u_i^* |I^0(x) - \ell_i|^2\, dx + \sum_{i=1}^{L} \alpha \int_\Omega |\nabla u_i^*|\, dx$$

$$= \sum_{j=1}^{n} \int_\Omega u_{i_j}^* |I^0(x) - \ell_{i_j}|^2 dx + \sum_{j=1}^{n} \alpha \int_\Omega |\nabla u_{i_j}^*| dx = E(\{u_{i_j}^*\}_{j=1}^n, \{i_j\}_{j=1}^n, n). \tag{19.44}$$

Combining (19.43) and (19.44) it follows that

$$E(\{u_{i_j}^*\}_{j=1}^n, \{i_j\}_{j=1}^n, n) \leq E(\{\tilde{u}_j\}_{j=1}^{\tilde{n}}, \{\ell_{\tilde{i}_j}\}_{j=1}^{\tilde{n}}, \tilde{n}).$$

Hence $\{u_{i_j}^*\}_{j=1}^n, \{i_j\}_{j=1}^n, n$ must be a solution to (19.16).

References

1. Bae, E., Tai, X.-C.: Efficient global minimization for the multiphase Chan-Vese model of image segmentation. In: Cremers, D., Boykov, Y., Blake, A., Schmidt, F.R. (eds.) Energy Minimization Methods in Computer Vision and Pattern Recognition 2009. Volume 5681 of Lecture Notes in Computer Science, pp. 28–41. Springer, Berlin/Heidelberg (2009)
2. Bae, E., Yuan, J., Tai, X.-C.: Global minimization for continuous multiphase partitioning problems using a dual approach. Int. J. Comput. Vis. **92**, 112–129 (2011)
3. Boykov, Y., Veksler, O., Zabih, R.: Fast approximate energy minimization via graph cuts. IEEE Trans. Pattern Anal. Mach. Intell. **23**, 1222–1239 (2001)
4. Brown, E.S., Chan, T.F., Bresson, X.: A convex relaxation method for a class of vector-valued minimization problems with applications to mumford-shah segmentation. UCLA, Applied Mathematics, CAM-report-10-43, Department of Mathematics, UCLA, July 2010
5. Brown, E.S., Chan, T.F., Bresson, X.: Completely convex formulation of the chan-vese image segmentation model. Int. J. Comput. Vis. (2011). doi:10.1007/s11263-011-0499-y
6. Chan, T., Vese, L.A.: Active contours without edges. IEEE Image Proc. **10**, 266–277 (2001)
7. Chan, T.F., Esedoḡlu, S., Nikolova, M.: Algorithms for finding global minimizers of image segmentation and denoising models. SIAM J. Appl. Math. **66**(5), 1632–1648 (electronic) (2006)
8. Darbon, J.: A note on the discrete binary mumford-shah model. In: Proceedings of Computer Vision/Computer Graphics Collaboration Techniques, (MIRAGE 2007). LNCS Series, vol. 4418, pp. 283–294, March 2007
9. Delong, A., Osokin, A., Isack, H., Boykov, Y.: Fast approximate energy minimization with label costs. In: IEEE Conference on Computer Vision and Pattern Recognition (CVPR), pp. 2173–2180 (2010)
10. Lellmann, J., Kappes, J., Yuan, J., Becker, F., Schnörr, C.: Convex multi-class image labeling by simplex-constrained total variation. In: Tai, X.-C., Mórken, K., Lysaker, M., Lie, K.-A. (eds.) Scale Space and Variational Methods in Computer Vision (SSVM 2009). Volume 5567 of LNCS, pp. 150–162. Springer, Berlin/Heidelberg (2009)

11. Lellmann, J., Breitenreicher, D., Schnörr, C.: Fast and exact primal-dual iterations for variational problems in computer vision. In: European Conference on Computer Vision (ECCV). LNCS vol. 6312, pp. 494–505 (2010)
12. Lempitsky, V., Blake, A., Rother, C.: Image segmentation by branch-and-mincut. In: Proceedings of the 10th European Conference on Computer Vision: Part IV, pp. 15–29. Springer, Berlin, Heidelberg (2008)
13. Mumford, D., Shah, J.: Optimal approximation by piecewise smooth functions and associated variational problems. Commun. Pure Appl. Math. **42**, 577–685 (1989)
14. Pock, T., Chambolle, A., Bischof, H., Cremers, D.: A convex relaxation approach for computing minimal partitions. In: IEEE Conference on Computer Vision and Pattern Recognition (CVPR), Miami, Florida (2009)
15. Pock, T., Cremers, D., Bischof, H., Chambolle, A.: An algorithm for minimizing the piecewise smooth mumford-shah functional. In: IEEE International Conference on Computer Vision (ICCV), Kyoto, Japan (2009)
16. Potts, R.B.: Some generalized order-disorder transformations. In: Proceedings of the Cambridge Philosophical Society, vol. 48, pp. 106–109 (1952)
17. Strandmark, P., Kahl, F., Overgaard, N.C.: Optimizing parametric total variation models. In: IEEE 12th International Conference on Computer Vision, pp. 2240–2247, pp. 26–33 (2009)
18. Yuan, J., Boykov, Y.: Tv-based image segmentation with label cost prior. In: BMVC, Article no 101, pp. 101:1–101:12. BMVA Press, Sept 2010
19. Yuan, J., Bae, E., Tai, X.-C., Boykov, Y.: A continuous max-flow approach to potts model. In: ECCV. Lecture Notes in Computer Science, vol. 6316, pp. 379–392 (2010)
20. Yuan, J., Shi, J., Tai, X.-C.: A convex and exact approach to discrete constrained tv-l1 image approximation. Technical report CAM-10–51, UCLA, CAM, UCLA (2010)
21. Yuan, J., Bae, E., Boykov, Y., Tai, X.C.: A continuous max-flow approach to minimal partitions with label cost prior. In: Bruckstein, A.M., ter Haar Romeny, B.M., Bronstein, A.M., Bronstein, M.M. (eds.) Scale Space and Variational Methods in Computer Vision. Lecture Notes in Computer Science, vol. 6667, pp. 279–290 (2012)
22. Zach, C., Gallup, D., Frahm, J.-M., Niethammer, M.: Fast global labeling for real-time stereo using multiple plane sweeps. In: Deussen, O., Keim, D.A., Saupe, D. (eds.) Proceedings of the Vision, Modeling and Visualization Conference (VMV), pp. 243–252 (2008)
23. Zhu, S.C., Yuille, A.: Region competition: Unifying snakes, region growing, and bayes/mdl for multi-band image segmentation. IEEE Trans. Pattern Anal. Mach. Intell. **18**, 884–900 (1996)

Chapter 20
Fluctuating Distance Fields, Parts, Three-Partite Skeletons

Sibel Tari

So the nature of a visual experience can not be described in terms of inches of size and distance.
Rudolf Arnheim

Abstract Shapes are continuous objects, even when they are drawn on digital media or composed using finite number of elements. As such, they defy analytic approach; explicitization of their parts, hierarchies, skeletons, or even centroids is ill-posed. I describe a novel approach to perceptually organize shapes and explicate their features without being negligent of their continuous nature. The basic construct is an unusual phase field that can be conceived in a number of varying ways using varying mathematical machinery, so highlighting the field itself rather than how it is being computed. Connections among the field, Mumford-Shah and Tari-Shah-Pien models, and reaction-diffusion equation suggest that the field may bridge low-level and high-level visual processing.

20.1 Shapes Are Continuous

Shapes are continuous objects, even when they are drawn on digital media or composed using finite number of elements. In favor of the argument, Gestalt school produced numerous illustrations, e.g., (1) a finite set of isolated dots placed at a fixed distance from a fixed location creates a circle percept; (2) a peanut shape prevailingly reveals two discrete blobs or approaches to a blob, as its neck thins or thickens; (3) two squares juxtaposed to form an eight-pointed star gives way to new parts, namely eight triangles and an octagon.

S. Tari (✉)
Middle East Technical University, 06531, Ankara, Turkey
e-mail: stari@metu.edu.tr

M. Breuß et al. (eds.), *Innovations for Shape Analysis*, Mathematics and Visualization, DOI 10.1007/978-3-642-34141-0_20,

In a similar vein but surpassing Gestalt, Stiny [50] speaks for ambiguity (i.e. multiple interpretations) in part perception, convincing the reader to place part boundaries wherever and whenever desired. Such a liberal view of shape topology certainly stands in strong contrast to prevalent theories on part perception exemplified by seminal works of Biederman [9] (imposing regularity on parts) and Hoffman and Richards [23] (imposing regularity on break points); nevertheless, contemporary work in neuroscience support the continuity of the process, too [37].

Additionally, shapes live in continua as opposed to standing in isolation. The basis of this argument is two-fold: (1) perception of a shape is influenced by nearby shapes (underlying category structures); indeed, several papers, e.g. [17,34], draw parallels between this contextual effect and experimentally observed non-metric characteristics (asymmetry and violation of triangle inequality) of pairwise dissimilarities; (2) shapes naturally deform into other shapes, say, a peanut changes to a blob-like shape or a pair of disjoint blobs as its neck thickens or thins, respectively.

To a great extent, it is the continuous nature of shapes that hinders their comprehension and description using analytic methods; thus, it is only natural that several chapters from this book present state of the art techniques for situating shapes in continua by constructing shape dynamics, shape spaces etc.

In pursuit of the goal, the presented approach is an attempt to perceptually organize shapes and explicate their features in the form of skeletons without being negligent of their continuous nature. At the core of constructions is an unusual distance-like, scalar field, ω, that implicitly codes interactions among shape elements, both over the region and along the boundary and both local and non-local.

Overview of the New Distance Field The most distinct feature of the new field is that it fluctuates, i.e., unlike previous ones [4, 22, 28, 42, 54], the new one is a non-monotone function of the distance to a set of input points. It is non-monotone because the value of the field at each point mimics the distance (a *diffuse* one [54, 55]) from the point to the union of (1) a set of input points and (2) *a yet to be determined curve.*

The set of input points specifies the shape boundary either incompletely (in cases of open contours, illusory contours or outcomes of edge detectors) or completely as the bounding curve of a connected bounded open set, say $\Omega \subset \mathbf{R}^2$. In either case, the shape boundary is indicated by the set of input points, denoted by $\partial\Omega$ and subjected to homogeneous Dirichlet type boundary condition.

The construction of the ω is valid regardless of the form the boundary is specified. If it is completely specified with a clear distinction between interior and exterior, the field can be computed only on the interior. If, however, the boundary is incompletely specified, the field needs to be computed on a wider scope R which is a simply connected subset of $\mathbf{R}^2$ containing the input set. Of course, a suitable boundary condition is needed on ∂R.

The new field has interesting emergent features; for example, it attains value *zero* not only on the shape boundary (which is due to homogeneous Dirichlet boundary condition on $\partial\Omega$) but also on a curve bounding a central cut. The latter locus is an emergent feature and divides the shape into central and peripheral regions.

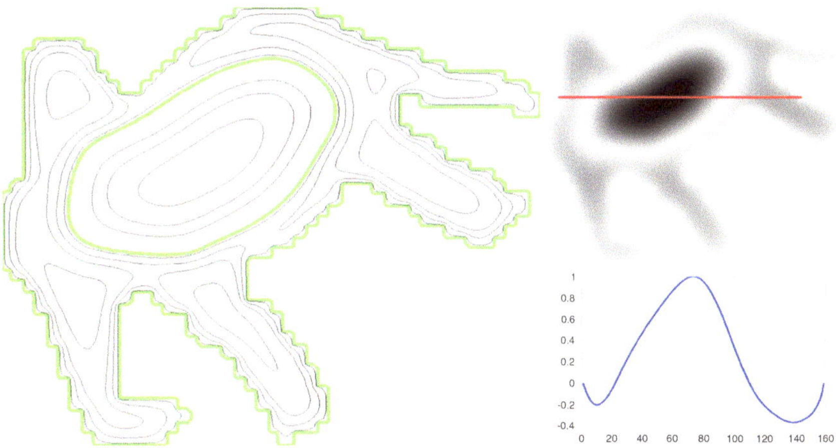

Fig. 20.1 The values of the new field fluctuate between positive and negative. The iso-level *curves* reflect the minimum distance to the union of the shape boundary and an emergent *curve* (shown in *green*)

Roughly speaking, the central region is the least deformable and coarse structure of the shape; moreover, the field inside it is a flexible indicator of the topology and centrality, not necessarily bound to a unique interpretation.

A sample field is depicted in Fig. 20.1. The leftmost figure displays some iso-level curves of the field and the top right one displays its absolute value. Finally, the bottom right figure is the plot of the values at the cross-section marked by the red line in the top right. Observe that the field exhibits interesting features in addition to the emergent zero-level curve: lower level sets in the peripheral region code appendages, the level curves neighboring the boundary resemble erosions of the shape, and the level curves neighboring the zero-level curve resemble erosions and dilations of the central region. All create a dynamic view of the shape.

What's more, the geometry of the level curves is intricately related to that of a well studied diffuse distance field [3, 4, 47, 54, 55] which is the solution of the screened Poisson PDE and has close connection to the Ambrosio-Tortorelli approximation [1] of the Mumford-Shah functional [35] (more on this in Sect. 20.2.3).

The intricate relation between the two fields facilitates an unorthodox skeleton abstraction that I call Three-Partite, due to its being composed of three distinct groups providing a separation into (1) gross structure; (2) limbs and (3) boundary texture or noise. Two examples are shown in Fig. 20.2 where the three skeleton groups are respectively shown in red, blue and green. The first group (red) consists of nested closed curves and abstracts the topology as well as the centrality of the coarse structure. It has two distinct layers: the inner (dark red) and the outer (light red). Both layers are coarser forms of the boundary. The outer layer, which is finer than the inner one, can be considered as the coarsest and least articulable structure still indicating the appendages. The inner layer is the emergent zero-level curve, bounding the central region. The second group (blue) consists of disjoint branches

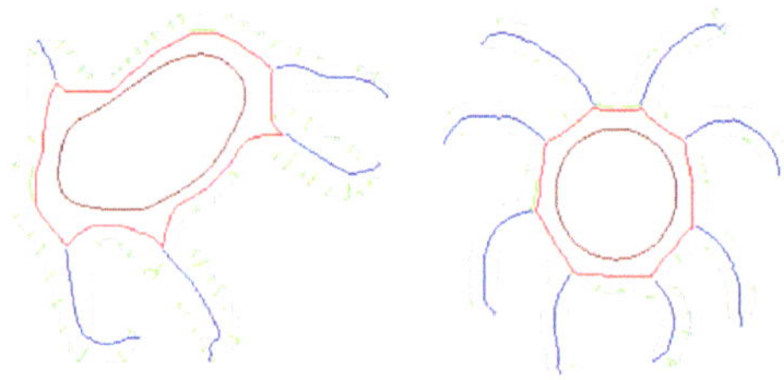

Fig. 20.2 The three skeleton groups are respectively shown in *red*, *blue* and *green* colors

(or loops in case of holes in the periphery) that terminate upon reaching the outer layer of the first group. Each branch of the second group captures a ribbon-like section, a limb or protrusion. The third group (green) consists of disjoint branches, too; but these branches correspond to secondary details.

Incentives for the Three-Partite Skeleton are given in Sect. 20.3 after the new field is explored in Sect. 20.2.

20.2 Fluctuating Distance Field ω

The new field is formulated firstly in a discrete setting, and then re-defined in a continuous setting to better understand the geometry of its level curves by relating it to a previously studied one. In both settings, the field is denoted by ω.

In the discrete setting, ω is the minimizer of an energy which reflects both local and global interactions in the shape interior and along the shape boundary.

On one hand, the boundary and the interior are interchangeable indicators for a shape but each explicates a different set of features. Though it is customary to classify methods as either boundary or region based [21], complementarities between the two, as an experimental guess since Gestalt, are exploited in established mathematical models [33, 35]. Likewise, in the proposed computational model, boundary and interior terms are kept distinct.

As the complementarity between the region and the boundary, the complementarity between the local and the global is an experimental guess: On the computer vision side, several working techniques, e.g. [8,13,32,57,58], utilize local and global features simultaneously; on the human vision side, theories that differ in details agree on that the shape perception is guided by both local and global factors [7, 14–16, 36, 49].

I remark that the choice of terms in the proposed model are driven by experimental guess as well as a bias towards simple linear formulations, since precise physics of the problem is yet unclear.

20.2.1 Formulation

In the discrete setting, the field $\omega : \Omega \to \mathbf{R}$ is a real valued function on a discrete set of sites, $\Omega \subset \mathbf{Z} \times \mathbf{Z}$, equipped with a neighborhood system. The energy $E(\omega)$

is taken as a linear combination of a region based energy $E_{Reg}(\omega)$ defined on Ω and a boundary based energy $E_{Bdy}(\omega)$ defined on $\partial\Omega$, both of which is a sum of pixel-wise terms:

$$E(\omega) = \sum_{(i,j)\in\Omega} E_{Reg}(\omega_{i,j}) + \lambda \sum_{(i,j)\in\partial\Omega} E_{Bdy}(\omega_{i,j}) \tag{20.1}$$

where $\lambda > 0$ is a constant. The additive form is for computational simplicity.

Region Energy The region energy composed of both local and global terms is expressed as a linear combination, too:

$$E_{Reg} = \underbrace{\sum_{(i,j)\in\Omega} E^G_{Reg}(\omega_{i,j})}_{E^G_{Reg}\,:\,\text{global}} + \beta \underbrace{\sum_{(i,j)\in\Omega} E^L_{Reg}(\omega_{i,j})}_{E^L_{Reg}\,:\,\text{local}} \tag{20.2}$$

where $\beta > 0$ is a constant. In the absence of a specific bias towards local or global, a natural choice may be $\beta = 1$.

Global Component Consider the squared average, the simplest quadratic expression linking all the nodes, as the global energy:

$$E^G_{Reg}(\omega_{i,j}) := \left(\frac{1}{|\Omega|} \sum_{(k,l)\in\Omega} \omega_{k,l} \right)^2$$

Differentiating its sum over Ω w.r.t. $\omega_{i,j}$:

$$\begin{aligned} \frac{\partial}{\partial\,\omega_{i,j}} E^G_{Reg} &= \frac{\partial}{\partial\,\omega_{i,j}} \sum_{(i,j)\in\Omega} \left(\frac{1}{|\Omega|} \sum_{(k,l)\in\Omega} \omega_{k,l} \right)^2 \\ &= \frac{2}{|\Omega|} \sum_{(k,l)\in\Omega} \omega_{k,l} \end{aligned} \tag{20.3}$$

Notice that the considered energy is minimized by either the zero function or a fluctuating function of which positive and negative values cancel each other.

Local Component The simplest local interaction model appears to be the sum of pairwise products of ω value at a site with ω values at the site's nearest neighbors. Such a component in the energy imposes regularity by favoring configurations in which each node agrees with its immediate neighbors. Assuming the usual four-connectivity, such that $\{(i+1,j),(i-1,j),(i,j+1),(i,j-1)\}$ defines the nearest neighbors of site (i,j), local interaction energy takes the form:

$$E^L_{Reg}(\omega_{i,j}) := -\left(\omega_{i,j}\cdot\omega_{i+1,j} + \omega_{i,j}\cdot\omega_{i-1,j} + \omega_{i,j}\cdot\omega_{i,j+1} + \omega_{i,j}\cdot\omega_{i,j-1}\right)$$

Differentiating w.r.t. $\omega_{i,j}$:

$$\frac{\partial E^L_{Reg}}{\partial \omega_{i,j}} = \frac{\partial}{\partial \omega_{i,j}} \sum_{(i,j)\in\Omega} -\left(\omega_{i,j}\cdot\omega_{i+1,j} + \omega_{i,j}\cdot\omega_{i-1,j} + \omega_{i,j}\cdot\omega_{i,j+1} + \omega_{i,j}\cdot\omega_{i,j-1}\right)$$

$$= -2\left(\omega_{i+1,j} + \omega_{i-1,j} + \omega_{i,j+1} + \omega_{i,j-1}\right) \tag{20.4a}$$

$$= -2\left[\mathbb{L}(\omega_{i,j}) + 4\omega_{i,j}\right] \tag{20.4b}$$

where $\mathbb{L}$ denotes the usual five-point discretization of the Laplace operator $\triangle := \frac{\partial^2}{\partial x^2} + \frac{\partial^2}{\partial y^2}$, using step size *one*. Notice that (20.4a) is nothing but the sum of the four neighbors, and (20.4b) is obtained by re-arranging the terms after adding and subtracting $4\,\omega_{i,j}$.

The scope of the regularity induced by the local energy can be extended by considering pairwise interactions in a larger neighborhood $\mathcal{N}_{ij}$:

$$E^L_{Reg}(\omega_{i,j}) := -\sum_{(k,l)\in\mathcal{N}_{ij}} a_{i,j;k,l}\cdot\omega_{i,j}\cdot\omega_{k,l} \tag{20.5}$$

where $a_{i,j;k,l} = a_{k,l;i,j}$ are suitably chosen positive weights, say, inversely proportional to the squared distances between the pairs as defined by the neighborhood system. Then the derivative of E^L_{Reg} w.r.t. $\omega_{i,j}$ is nothing but a weighted sum of the neighbors:

$$\frac{\partial E^L_{Reg}}{\partial \omega_{i,j}} = \frac{\partial}{\partial \omega_{i,j}} \sum_{(i,j)\in\Omega} -\left(\sum_{(k,l)\in\mathcal{N}_{ij}} a_{i,j;k,l}\cdot\omega_{i,j}\cdot\omega_{k,l}\right)$$

$$= -2\sum_{(k,l)\in\mathcal{N}_{ij}} a_{i,j;k,l}\cdot\omega_{k,l} \tag{20.6}$$

As a result, similar to (20.4b), it can be be re-written as

$$-2\cdot\left[\mathbb{L}_{\mathcal{G}}(\omega_{i,j}) + \mu_{\mathcal{G}}\,\omega_{i,j}\right] \tag{20.7}$$

where $\mathbb{L}_{\mathcal{G}}(.)$ denotes the graph Laplacian and $\mu_{\mathcal{G}} > 0$ a constant that depends on the neighborhood system.

Of course, the scope of the local regularity may be further extended by including all of the pairwise interactions among the elements in $\mathcal{N}_{ij}$:

$$\sum_{\substack{(k,l)\in\mathcal{N}_{ij}\\(m,n)\in\mathcal{N}_{ij}\\(k,l)\neq(m,n)}} a_{k,l;m,n}\cdot\omega_{k,l}\cdot\omega_{m,n} \tag{20.8}$$

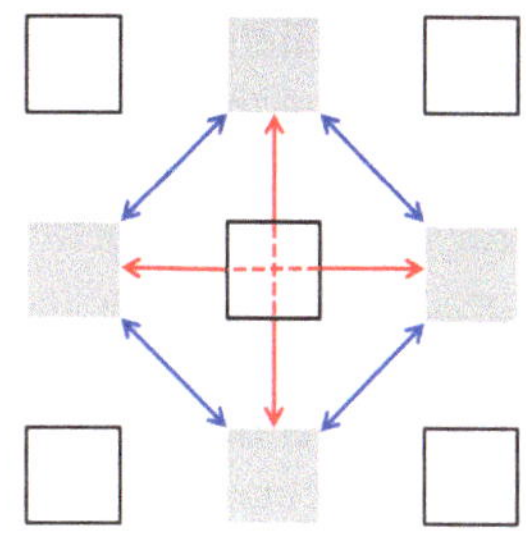

Fig. 20.3 The distance between a pair of nodes joined by a *blue arrow* is either 2 (the Manhattan) or $\sqrt{2}$ (the Euclidean)

with the weights reflecting the distances (Fig. 20.3). Then the derivative w.r.t. $\omega_{i,j}$ is either of the following nine-point weighted averages in a 5×5 window centered at (i, j):

$$\frac{1}{4}\begin{bmatrix} \cdot & \cdot & 1 & \cdot & \cdot \\ \cdot & 2 & \cdot & 2 & \cdot \\ 1 & \cdot & \cdot & \cdot & 1 \\ \cdot & 2 & \cdot & 2 & \cdot \\ \cdot & \cdot & 1 & \cdot & \cdot \end{bmatrix} \qquad \frac{1}{4}\begin{bmatrix} \cdot & \cdot & 1 & \cdot & \cdot \\ \cdot & 4 & \cdot & 4 & \cdot \\ 1 & \cdot & \cdot & \cdot & 1 \\ \cdot & 4 & \cdot & 4 & \cdot \\ \cdot & \cdot & 1 & \cdot & \cdot \end{bmatrix}$$

The one the left is obtained if the weights are calculated according to the Manhattan, the natural metric for the 4-connected grid, and the one on the left according to the Euclidean. Upon subtracting a suitable multiple of $\omega_{i,j}$, both stencils become a realization of the nine-point discrete Laplacian on a redefined grid. That is, as long as local interactions are restricted to pairwise products (linear combinations of (20.5) and (20.8)), the derivative of the local energy w.r.t. $\omega_{i,j}$ is of the form:

$$\frac{\partial E^L_{Reg}}{\partial \omega_{i,j}} = -2\mu_1 \Big[\mathbb{L}_*\big(\omega_{i,j}\big) + \mu_2 \omega_{i,j} \Big] \tag{20.9}$$

where $\mathbb{L}_*$ is some approximation to the discrete Laplacian.

Adding Local and Global Components Combining (20.9) and (20.3) according to the additive model (20.2) yields

$$\frac{\partial E_{Reg}}{\partial \omega_{i,j}} = -2 \Big[\mathbb{L}_*\big(\omega_{i,j}\big) + \bar{\mu}\, \omega_{i,j} - \frac{1}{O\big(|\Omega|\big)} \sum_{(k,l)\in\Omega} \omega_{k,l} \Big] \tag{20.10}$$

where $O\big(|\Omega|\big)$ denotes a positive constant on the order of the domain size and $\bar{\mu} > 0$ depends on assumed local connectivities as well as the parameter β in (20.2). Later on, we will see that the choice of $\bar{\mu}$ has no bearing on the result.

Boundary Energy Formulating interactions along a contour is not as simple as formulating them over a region. The task gets further complicated in the case of long-range interactions because critical ones occur among specific pairs of boundary points, e.g., two opposing indentation points on a neck region.

Fortunately, the usual distance transform provides a way out. At the outset, it is a transform that assigns to each point on a domain the minimum distance from the point to the domain boundary, yet it is at the same time a compact representation for boundary interactions. Indeed, its construction can be imagined as a dynamic process, a time-dependent evolution of the shape boundary such that each point on it moves with a unit speed in the direction of the unit normal; so the distance field t as a function over the digital plane is defined by setting $t_{i,j}$ = the time when the evolving curve passes through the point (i,j). Through the course of the evolution, singularities form as opposing pairs of boundary points meet and characterize the shape as its parts and symmetries are revealed [10].

The key point is that once a bridge between boundary interactions and the distance transform is established, E_{Bdy} is easily defined as a quadratic expression over the region Ω:

$$\sum_{(i,j)\in\partial\Omega} E_{Bdy}(\omega_{i,j}) := \sum_{(i,j)\in\Omega} \left(\omega_{i,j} - t_{i,j}\right)^2 \tag{20.11}$$

As a result,

$$\begin{aligned} \frac{\partial E_{Bdy}}{\partial \omega_{i,j}} &= \frac{\partial}{\partial \omega_{i,j}} \sum_{(i,j)\in\partial\Omega} E_{Bdy}(\omega_{i,j}) \\ &= 2\left(\omega_{i,j} - t_{i,j}\right) \end{aligned} \tag{20.12}$$

That is, $\omega = t$.

Complete Energy Combining (20.12) and (20.10) according to the additive model (20.1):

$$\frac{\partial E}{\partial \omega_{i,j}} = -\mathbb{L}_*\left(\omega_{i,j}\right) + \frac{1}{O\left(|\Omega|\right)} \sum_{(k,l)\in\Omega} \omega_{k,l} + (\lambda - \bar{\mu})\,\omega_{i,j} - \lambda t_{i,j} \tag{20.13}$$

Then the condition satisfied by the minimizer is obtained by setting the expression in (20.13) to *zero*, which is in matrix notation given by

$$\left(-\mathbf{L} + \frac{1}{O\left(|\Omega|\right)}\mathbf{J} + (\lambda - \bar{\mu})\,\mathbf{I}\right)\mathbf{w} = \lambda\,\mathbf{t} \tag{20.14}$$

where

- **w** and **t** are vector representations for the field ω and the distance transform t;
- **L** is the $|\Omega| \times |\Omega|$ matrix representation for a discrete Dirichlet Laplacian;
- **J** is the $|\Omega| \times |\Omega|$ matrix of ones;
- **I** is the $|\Omega| \times |\Omega|$ identity matrix.

Firstly, notice that the uniform scaling of the right hand side term with λ has no effect on the geometry of ω; it only scales the values, leaving the shape of the level curves intact. As a result, the geometric features of ω depends only on the choice of $(\lambda - \bar{\mu})$. Secondly, notice that $\left(-\mathbf{L} + \frac{1}{O(|\Omega|)} \cdot \mathbf{J} + (\lambda - \bar{\mu}) \cdot \mathbf{I}\right)$ is a positive definite matrix if $(\lambda - \bar{\mu}) \geq 0$. The difference between the strict positivity and zero, from the linear algebraic point of view, is that the former one has a better condition number so strict positivity of $(\lambda - \bar{\mu})$ may be preferable. But if $(\lambda - \bar{\mu})$ exceeds $\frac{1}{O(|\Omega|)}$, the non-spatial term dominates over the spatial ones. Such intuition suggest that $(\lambda - \bar{\mu})$, the only parameter, should be approximately $\frac{1}{|\Omega|}$. Later on this choice is supported in the continuous domain model as well.

Implementation Based on the above discussion the parameters $\bar{\mu} - \lambda$ and λ are eliminated: Firstly, $\bar{\mu} - \lambda$ is selected as inversely proportional to the domain size. Secondly, t is replaced by $\bar{t}$ denoting an arbitrary scaling of the distance transform because the right hand side scaling by λ does not affect the geometry of the solution.

Using the method of gradient descent, computing ω is cast as computing the steady state solution of the following differential-difference equation:

$$\begin{aligned} \frac{\mathrm{d}\omega_{i,j}(\tau)}{\mathrm{d}\tau} &= -\frac{\partial E}{\partial \omega_{i,j}} \qquad (20.15) \\ &= \mathbb{L}_*(\omega_{i,j}) - \frac{1}{O(|\Omega|)} \sum_{(k,l)\in\Omega} \omega_{k,l} - \frac{1}{O(|\Omega|)} \omega_{i,j} + \bar{t}_{i,j} \end{aligned}$$

Discretizing τ as

$$\begin{aligned} \frac{\omega_{i,j}^{n+1} - \omega_{i,j}^{n}}{\Delta\tau} &= \left(\omega_{i,j+1}^{n} + \omega_{i,j-1}^{n} + \omega_{i+1,j}^{n} + \omega_{i-1j,}^{n} - 4\omega_{i,j}^{n}\right) \qquad (20.16) \\ &\quad - \frac{1}{|\Omega|} \sum_{(k,l)\in\Omega} \omega_{k,l}^{n} - \frac{1}{|\Omega|} \omega_{k,l}^{n+1} + \bar{t}_{i,j} \end{aligned}$$

where $n = 0, 1, 2, \cdots$ is the iteration variable yields an iterative scheme once the terms in (20.16) are re-arranged. Note that $\mathbb{L}_*$ is simply replaced with the usual five point discrete Laplacian, and $O(|\Omega|)$ with $|\Omega|$.

Behavior of ω The solution satisfies three competing criteria:

– having low expected value (20.17a)

– being smooth (20.17b)

– staying close to the usual distance transform (20.17c)

At a first glance, the first criterion seems to favor flatness by forcing the minimizer to attain values close to zero; yet this criterion is also satisfied when the values of

ω fluctuate between positive and negative. The other two criteria impose additional regularities so the fluctuations form a certain spatial pattern as opposed to being random localization of positive and negative values. Firstly, the locations of identical sign tend to form spatial proximity groups as defined by the graph neighborhood system. Secondly, the positive clusters tend to occur in the central region. This is due to the third criterion: Whenever ω is negative at a point (i, j), a penalty larger than $t_{i,j}^2$ is incurred; therefore, ω has a stronger tendency to be positive at locations where the value of distance transform is higher in order to minimize the total cost of deviating from the distance transform. Consequently, ω attains positive values in central locations where the value of the distance transform is higher and negative values at peripheral locations where the value of the distance transform is lower.

20.2.2 Illustrative Results

Few illustrative results are depicted in Figs. 20.4–20.6. In color illustrations, ω values are normalized to [0, 1] within regions of identical sign for convenience of visualization and the zero-level is depicted in dark blue.

Central and Peripheral Separation In each case, the zero-level set partitions the shape domain into upper and lower zero level sets, Ω^+ and Ω^-, denoting respectively central and peripheral regions.

Ω^- contains all the detail: limbs, protrusions, and boundary texture or noise. In contrast, Ω^+ is coarse blob-like form, which can even be considered as an interval estimate of the center; most commonly, it is a simply connected set. Of course, it may also be either disconnected or multiply connected. For instance, it is disconnected for a dumbbell-like shape (two blobs of comparable radii combined through a thin neck) whereas it is multiply connected for an annulus formed by two concentric circles. Indeed, the annulus gets split into three concentric rings. Whereas the middle ring is the Ω^+, the union of the outermost and the innermost rings form a multiply-connected set corresponding to the Ω^-, separated by the Ω^+.

For quite a many shapes, however, Ω^+ is a simply connected set. Firstly, shapes obtained by protruding a blob as well as shapes whose peripheral parts are smaller or thinner than their main parts always have a simply connected Ω^+. This is expected: When the width of a part is small, the highest value of the distance transform t inside the part is small. That is, the local contribution to $(\omega - t)^2$ incurring due to negative values is less significant for such a part as compared to locations with higher positive values of t. Consequently, ω tends to attain negative values on narrow or small parts as well as on protrusions. Shapes with holes also have a simply connected Ω^+ as long as the holes are far from the center. Secondly, even a dumbbell-like shape may have a simply connected Ω^+. This happens if the join area, namely the neck, is wide enough, e.g., the top right in Fig. 20.4. Nevertheless, this does not cause any representational instability: Whereas the Ω^+ for a blob-like shape has a unique maximum located roughly at its centroid, the Ω^+ for a

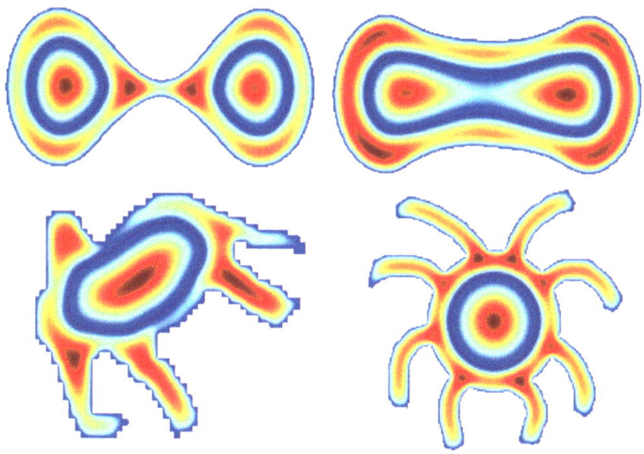

Fig. 20.4 Absolute value of ω. For ease of color visualization, values of ω are normalized within regions of identical sign. *Dark blue* color indicates zero

dumbbell-like shape has two local maxima indicating two bodies. Each body is captured by a connected component of an upper level set whose bounding curve passes through a saddle point. At a saddle point $\mathbf{p}$, such that $\omega(\mathbf{p}) = s$, the $s - level$ curve has a double point singularity, *i.e.* it forms a cross. As a result, the upper level set $\{\Omega^s = (x, y) \in \Omega^+ : \omega(x, y) > s\}$ yields two disjoint connected components capturing the two parts of the central region.

In contrast to Ω^+, the peripheral region, Ω^-, is multiply connected, even if the shape is simply connected. Most typically it is a ring-like domain whose holes are carved by Ω^+. It is also possible that Ω^- is disconnected. For instance, it is two concentric rings for an annulus. Additionally, Ω^- may be disconnected when there are several elongated limbs organized around a rather small central body, e.g., a palm tree. Ω^+, being small, is tolerated to grow and reach to the most concave parts of the shape boundary creating a split of Ω^- by the zero-level curve. The emergent zero-level curve bounds both Ω^+ and Ω^-.

Gap Closure An application in the case of incompletely specified boundaries is depicted in Fig. 20.5. The small figure on the left is the input, which consists of twelve isolated dots organized around a center point. The field ω computed in a rectangular region is depicted in the right. The gray is the central region; it is a circle. Notice that the level curves in the vicinity of the input points enclose firstly the individual dots and secondly the pairs of dots. Later the outer level curves enclose the entire set and the inner level curves become circles.

Restrictions of ω Each restriction of ω to either Ω^+ or Ω^- emulates a diffuse distance transform applied to the respective domain. The analogy is surely over the geometry of the level curves rather than actual ω values. More precisely, the level curves of ω in either domain resemble curvature dependent evolution of the boundary of the respective domain.

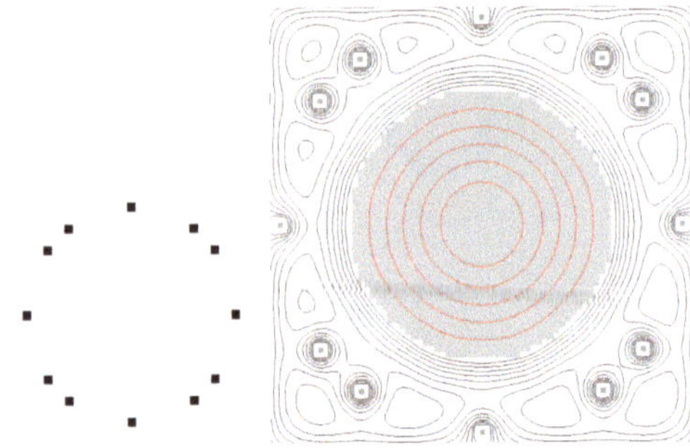

Fig. 20.5 Gap closure. (*left*) The input. (*right*) The field computed over a *rectangular* region; the central region is shown in *gray*

The qualitative resemblance of the level curves of ω to fronts evolving with curvature dependent speed triggers the observation that the condition satisfied by the minimizer of the discrete energy coding local/non-local and region/boundary interactions is indeed a central difference approximation for an integro-differential equation:

$$\triangle\omega(x,y) - \frac{1}{|\Omega|}\iint \omega(\alpha,\beta)\,\mathrm{d}\alpha\,\mathrm{d}\beta - \frac{1}{|\Omega|}\omega(x,y) \;=\; -\bar{t}(x,y) \qquad (20.18)$$

where $\triangle$ stands for the Laplace operator. Recall that $\bar{t}$ denotes any uniform scaling of t. Interestingly, when the global component of the energy which is responsible for Ω^+/Ω^- separation is omitted, (20.18) reduces to the screened Poisson PDE:

$$\left(\triangle - \frac{1}{|\Omega|}\right)\omega(x,y) \;=\; -\bar{t}(x,y) \qquad (20.19)$$

The screened Poisson PDE with a constant right hand side has been previously proposed by Tari et al. [54, 55] as the governing equation for a diffuse distance field, with a motivation to bridge the higher level process of shape abstraction and the lower level processes of image smoothing and boundary extraction. Tari, Shah and Pien (TSP) model is closely related to a famous by-product of variational segmentation models: the Ambrosio-Tortorelli (AT) phase field.

In Sect. 20.2.3, I relate (20.18) to a modification of the previous field which I call as the AT/TSP field. The connection between the two fields (AT/TSP and ω) improves our understanding of the new one. For example, the restriction of ω to either the central or peripheral region is geometrically equivalent to the AT/TSP field constructed for that region. Hence, the visual observation that the level curves in the vicinities of $\partial\Omega$ and the emergent zero-level curve are mimicking a curvature dependent motion is substantiated.

The curvature dependent behavior of the level curves, however, deteriorates away from the boundary; eventually, even initially parallel boundaries form a cross. This formation of saddle points is a consequence of increased interaction and it has been discussed previously by Tari et al. [55] (see Fig. 1 in [55]).

Fig. 20.6 Level *curves* of ω. For visualization purposes, the values are normalized within central and peripheral regions; the zero level is depicted in *dark blue*

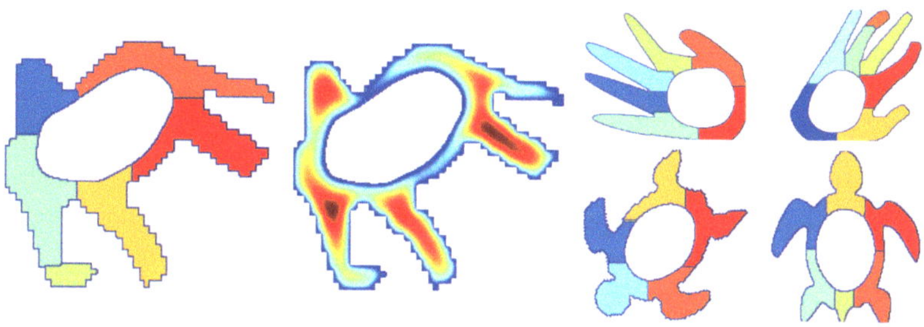

Fig. 20.7 Formation of watershed zones in the peripheral region is a consequence of the relation of the new field to the AT/TSP field

Fortunately, in the new framework, this behavioral deviation provides considerable convenience by assuring that the peripheral region is always partitioned, unless it is a perfect annulus. Figure 20.6 depicts level curves of ω for the octopus. Notice the formation of eight saddle points separating the arms of the octopus. Indeed the zones separated by the saddle points are easily extracted as watershed zones [31] of the peripheral region (Fig. 20.7). Of course, raw watershed partitions are subject to topologic instabilities. Obtaining topologically stable partitions is a separate issue which is tackled in a recent work by Tari and Genctav [51] by constructing a probabilistic part tree; nevertheless, watershed zones surrounding the central region are perfectly sufficient as rough part estimates.

20.2.3 ω and the AT/TSP Field

In vision applications, the AT (Ambrosio-Tortorelli) phase field appeared first as a technical device to apply gradient descent to the Mumford-Shah functional [35], where it served as a continuous indicator for boundary/not-boundary phases at each domain point [44]. The field is the minimizer of an energy composed of two competing terms: One term favors configurations that take values close to either 0 or 1 (separation into boundary/not-boundary phases) and the other term encourages local interaction in the domain by penalizing spatial inhomogeneity. A parameter controls the relative influence of these two terms hence the interaction. As this "interaction" parameter tends to 0, the separation term is strongly emphasized; consequently, the field tends to the characteristic function $1-\chi_S$ of the boundary set S and the respective energy tends (following the Γ convergence framework [11]) to the boundary length.

Over the years, the AT field has been used in numerous applications addressing a rich variety of visual problems, cf. Shah and colleagues [39, 44, 45, 48], March and Dozio [29], Proesman et al. [41], Teboul et al. [56], Bar et al. [6], Droske and Rumpf [40], Erdem and Tari [18], Patz and Preusser [38], and Jung and Vese [24]. In all these applications, because the field serves as an auxiliary variable to facilitate discontinuity-preserving smoothing and boundary detection, the interaction parameter ρ is chosen sufficiently small, i.e., $\rho \to 0$, to localize boundaries accurately.

Increased Interaction: the AT/TSP Field In the late 1990s, Tari, Shah and Pien (TSP) re-casted the AT field as a compact representation of the morphology of regions implied by putative edges [54, 55]. To such end, they weakened boundary/not-boundary separation by letting the interaction parameter be sufficiently large, in contrast to the prevalent applications of the AT field exemplified by the above referenced works. Furthermore, they examined the geometry of the level curves and then devised robust schemes that explicate curvature related criticalities such as skeletons.

On the technical side, they used standard coupled equations obtained by applying gradient descent to the AT approximations of available segmentation energies [44, 45]. For example, given an image $g(x, y)$ defined on a region $R \subset \mathbf{R}^2$, they computed a de-noised image $u(x, y)$ and the AT field $v(x, y)$ by solving

$$\frac{\partial u}{\partial \tau} = \nabla \cdot (v^2 \nabla u) - \tfrac{\beta}{\alpha}(u - g) \tag{20.20}$$

$$\frac{\partial v}{\partial \tau} = \Delta v - \tfrac{2\alpha|\nabla u|^2 v}{\rho} - \tfrac{(v-1)}{\rho^2} \tag{20.21}$$

subject to homogeneous Neumann conditions on the image boundary:

$$\left.\frac{\partial u}{\partial n}\right|_{\partial R} = 0, \quad \left.\frac{\partial v}{\partial n}\right|_{\partial R} = 0 \tag{20.22}$$

where ∂R is the image boundary and n is the outer unit normal vector to ∂R, while α and β are Mumford-Shah related parameters determining contrast and scale [45, 55], ρ is the interaction parameter. The key trick is to choose ρ sufficiently large, typically at the order of the radius of the maximal circle that can be fit inside putative shapes – regions implied by putative boundaries.

That is, in the TSP approach, boundary localization is traded with shape abstraction, thus, the role of AT field changed from being a fixed property detector (boundary/no-boundary) to being an implicit shape coder. From now on, I refer to the v function as AT/TSP field if local interactions are emphasized during its computation.

When the computation of the AT/TSP field is restricted to shape interior (indicated by $\Omega \subset \mathbf{R}^2$, a connected bounded open set with a boundary $\partial\Omega$), the field satisfies on Ω a screened Poisson PDE with a constant source term:

$$\left(\triangle - \frac{1}{\rho^2}\right) v(x,y) = -\frac{1}{\rho^2} \tag{20.23}$$

subject to homogenous Dirichlet condition. The solution to (20.23) is the unique minimizer of the AT energy:

$$\iint_{\Omega} \frac{1}{\rho} \underbrace{\Big[v(x,y) - \chi_{\Omega}(x,y)\Big]^2}_{\substack{\text{boundary/interior}\\ \text{separation}}} + \rho \underbrace{\big|\nabla v(x,y)\big|^2}_{\text{local interaction}} \, \mathrm{d}x\, \mathrm{d}y$$

$$\text{with } v(x,y) = 0 \text{ for } (x,y) \in \partial\Omega \tag{20.24}$$

where $\chi_{\Omega}(x,y)$ is the shape indicator that attains 1 in Ω and 0 on $\partial\Omega$.

There are two important practical implications of the AT/TSP model:

1. v is a smooth distance transform of which gradient at a point exponentially depends on the medialness of the point; this leads to a robust procedure for extracting curvature based criticalities [53, 55];
2. The level curves of v approximate evolution of the shape boundary such that each point on it moves with a curvature-dependent speed in the direction of the unit normal; that is, a nonlinear wave-like phenomena can be modeled using a linear diffusion operator; as a result, the reaction-diffusion scale space of Kimia et al. [25] can be easily constructed [54].

The right hand side of (20.23) may be scaled if the interest is purely on the geometry of the level curves. Then *infinite* interaction limit $\rho \to \infty$ can be considered, giving the Poisson PDE [22] or approximating the Aslan-Tari model [3, 4]; but then the exponential dependence on medialness [55] is lost. In the original paper by Tari, Shah and Pien, there are several experiments illustrating the role of ρ. In what follows, ρ is assumed to be at least on the order of the radius of the maximal circle that can be fit inside the shape for the diffusive effect of $|\nabla.|^2$ to influence the entire shape.

Now I modify (20.24) in a way that its minimizer is geometrically equivalent to the ω function defined earlier in Sect. 20.2.1 as a balance among different types of interactions. Take a note that the AT/TSP field which is the solution of the screened Poisson equation with a constant source term (20.23) is geometrically equivalent to the solution of a screened Poisson equation for which the source term is a monotonic function of the distance (20.19); the change of the source term affects the actual values while leaving the geometry of the level curves almost intact.

Fluctuating AT/TSP The key idea is to break down the monotone behavior of the AT/TSP to partition the shape into central and peripheral regions within which the field remains qualitatively similar to the AT/TSP. This can be achieved by additively augmenting the energy in (20.24) with a term minimized by a fluctuating function, e.g. a multiple of $\left(\iint \omega(x,y)\,\mathrm{d}x\,\mathrm{d}y\right)^2$, but one has to make sure that locations of identical sign form meaningful spatial proximity groups. The term $\left|\nabla v(x,y)\right|^2$ prevents wild fluctuations and ensures that nearby locations have similar values, but it can not influence where these groups form. An obvious way of obtaining central versus peripheral separation is to increase the residual $\left[v(x,y) - \chi_\Omega(x,y)\right]^2$ at central regions to penalize discrepancies more, hence, forcing the function to be positive in a central area while attaining the opposite sign in remaining locations to minimize penalty incurring due to the new term $\left[\iint \omega(x,y)\,\mathrm{d}x\,\mathrm{d}y\right]^2$.

There are two possible strategies: Either making the phase separation term space variant by multiplying it with a medialness dependent coefficient or using a medialness dependent indicator function. In both cases, medialness dependence can be modeled via any monotonic function of the distance. Using the second strategy and letting the distance transform $t(x,y)$ serve as the weighted indicator yields

$$\iint\limits_\Omega \frac{1}{\rho}\left[\omega(x,y) - t(x,y)\right]^2 + \rho\left[\left|\nabla\omega(x,y)\right|^2 + \left(\frac{1}{|\Omega|}\iint\limits_\Omega \omega(\alpha,\beta)\,\mathrm{d}\alpha\,\mathrm{d}\beta\right)^2\right]\mathrm{d}x\,\mathrm{d}y,$$

$$\text{with } \omega(x,y) = 0 \text{ for } x,y = (x,y) \in \partial\Omega. \tag{20.25}$$

Similar to the AT/TSP field, the minimizer of (20.25) is a compromise between inhomogeneity and homogeneity though the inhomogeneity is forced both externally (by t) and internally (by the third term). Unlike the phases defined by the AT/TSP field (its level curves), those defined by ω do not necessarily mimic erosions of the shape boundary but they have a richer meaning that facilitates extraction of a novel skeleton as detailed in the next section.

To find the minimizer of (20.25), let us consider the Gâteaux variation

$$\delta E(\omega,\nu) := \lim_{\epsilon\to 0}\frac{E(\omega+\epsilon\nu) - E(\omega)}{\epsilon} = \frac{\partial}{\partial\epsilon}E(\omega+\epsilon\nu)\Big|_{\epsilon=0} \tag{20.26}$$

for all test functions ν properly prescribed on the boundary. Setting $\delta E(\omega,\nu) = 0$ yields

$$\iint\limits_\Omega \left\{\frac{1}{\rho^2}\Big(\omega - t\Big) - \Delta\omega\right\}\nu + \mathbb{E}[\omega]\mathbb{E}[\nu] = 0 \tag{20.27}$$

where

$$\mathbb{E}[f] = \frac{1}{|\Omega|} \iint\limits_{\Omega} f(x, y) \mathrm{d}x\, \mathrm{d}y.$$

The condition is not intrinsic, but it is is satisfied for all test functions if both of the following holds:

$$\frac{1}{|\Omega|} \iint \omega(\alpha, \beta)\, \mathrm{d}\alpha\, \mathrm{d}\beta = 0 \qquad (20.28a)$$

$$\triangle\, \omega(x, y) - \frac{1}{\rho^2} \omega(x, y) = -\frac{1}{\rho^2} t(x, y) \qquad (20.28b)$$

Even though the above pair of equations and (20.18) are not identical mathematically, they are equivalent conceptually, i.e., constraints induced by the pair (20.28a) and (20.28b) are equivalent to those in the discrete model:

Firstly, (20.28a) implies that the solution integrates to *zero* (compare to (20.17a) in the discrete case). Secondly, (20.28b) implies that the desired solution is a smooth approximation of the distance transform (compare to the pair (20.17b) and (20.17c) in the discrete case). Selecting the smoothing radius as consistent with the suggested practice in the AT/TSP model, i.e., $\rho = O\left(\sqrt{|\Omega|}\right)$, is in agreement with the choice of $\lambda - \bar{\mu}$ in the discrete model. (Recall that ρ needs to be approximately equal to the radius of the maximal circle that can be fit inside the shape for the diffusive effect of $|\nabla .|^2$ to influence the entire shape; see also [3]).

20.3 Three Partite Skeletons

The invention of skeletons is one of the most influential outcomes of the search for a mathematics of shape:

In the mid 1960s the school of mathematical morphology [43] brilliantly opted to understand a shape merely by observing how it interacts with template shapes called structuring elements using two basic processes called erosion and dilation. Eroded and dilated shapes define a *time* evolution in which the time is the size of the structuring element, hence, shapes are equipped with dynamics.

Independent from the school of mathematical morphology, Blum [10] proposed to code topologic and geometric properties implicitly using the Euclidean distance field $t(x, y)$. The relevant link is that successive level sets of the distance transform correspond to erosions and dilations of the shape with discs of increasing radius making the Euclidean distance field a compact representation of morphology. The level curves of the Euclidean distance transform are analogous to a time evolution of the shape boundary such that each point on it moves with a unit speed in the direction of the unit normal, so the distance field t as a function over the plane is defined by setting $t(x, y)$ = the time when the evolving curve passes through the point (x, y). During the evolution opposing pairs of boundary points meet, each collapsing to a single point inside the shape. The locus of meeting points is called the skeleton. Each point on the skeleton is equidistant from at least two boundary points.

In the following years, several practitioners conceptualized skeletons strictly as thin versions of shapes with equidistance property; vast literature is dedicated to thinning and solving consequently arising problems; axioms that are necessitated by chosen algorithms rather than being induced by theory (e.g. the connectivity requirement) are introduced; holding back nonconforming variants.

Nevertheless, significant liberation is gained in the 1990s when vision researchers started to reveal connections among different computational models and algorithms to visual processing [12, 33, 45, 46]. These works contributed towards a better understanding of the theory level of visual processing emphasizing *what the computational problem is rather than how it is being solved* [30].

In particular, mathematical morphology, skeletons, curve evolution approaches, diffusion PDEs and variational methods are naturally tied by distance-like fields [28], reviving the Gestalt school's potential field theory. The Euclidean Distance Transform is given by the solution to the Eikonal equation $|\nabla t(x, y)| = c(x, y)$ with the shape boundary as the initial front and $c(x, y) = 1$ and new distance fields can be constructed by varying $c(x, y)$ resulting in weighted distance transforms or augmenting the initial front with seed points resulting in seeded distance transforms [28]. Of course a weighted distance transform of which level sets emulating curvature dependent motion is readily constructed via an elliptic PDE, namely the Poisson equation, be it screened [55] or not [22].

Indeed the ω function is equivalent to a seeded weighted distance transform: Consider the shape boundary depicted by the red curve in Fig. 20.8a, and imagine that the initial front is its augmentation with two seed points. During the course of the evolution, fronts growing out of each of the two seeds merge forming the blue curve in (a); then the blue curve propagating outward and the shape boundary (the red) propagating inward meet at the zero level curve (the pink). In Fig. 20.8b more level curves in the region where ω is negative, namely Ω^-, are depicted.

New distance fields give way to new skeleton models, one of which is the Three-Partite Skeleton.

The Three-Partite Skeleton consists of three groups. Recall Fig. 20.2. The first group S_G (shown in red) is an abstraction of the central or gross structure. It consists of two components: the inner component S_G^{in} (the dark red) coinciding with the zero-level curve and the outer component S_G^{out} (the light red). The second group S_R (shown in blue) abstracts ribbon-like sections corresponding to limbs and significant protrusions. Finally the third group S_T (shown in green) captures secondary details typically arising due to boundary texture or noise.

Conceiving the Three Partite Skeleton The relation between the level curves of ω and those of the AT/TSP field makes conceiving the Three-Partite Skeleton quite natural: With the exception of S_G^{in}, which is readily provided by the zero level curve of ω, all of the points of the Three-Partite Skeleton are in a generalized ridge set

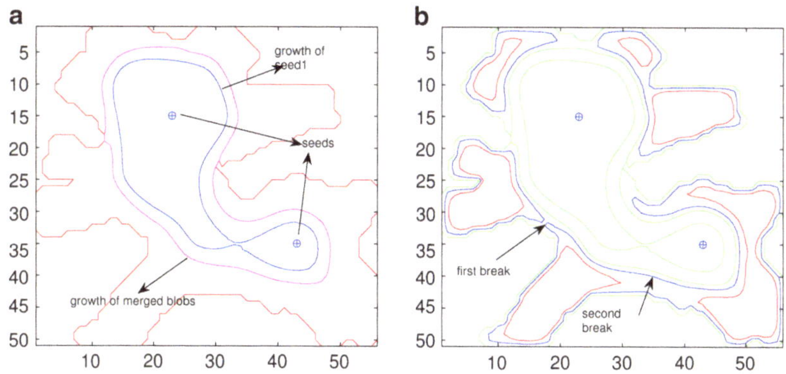

Fig. 20.8 Seeded distance transform analogy for ω

defined by the zero-crossings of the flowline curvature, i.e., $\frac{d|\nabla\omega|}{ds} = 0$ where s is the arc-length along level curves. Moreover, the ridge points can be split into maximum and minimum convexity groups depending on the sign of the second derivative $\frac{d^2|\nabla\omega|}{ds^2}$ [55]. In Fig. 20.9, the points are depicted in black if $\frac{d^2|\nabla\omega|}{ds^2} < 0$ and in light gray if $\frac{d^2|\nabla\omega|}{ds^2} > 0$; skeleton points tend to be on the first group.

Branches tracking discretization artifacts are clearly visible in the form of disconnected short segments; they belong to S_T. Branches tracking main limbs are clearly visible too; they belong to S_R. Each S_R branch starts from a curvature maximum of the shape boundary $\partial\Omega$ and comes to an end at a point where the front initiated at $\partial\Omega$ meets the front initiated at the emergent zero level curve $\partial\Omega^+$, during the course of propagation. The termination points are in a sub-locus composed of a circular sequence of saddles and minima so defining a numerical realization of a parabolic locus [52]. This locus is S_G^{out}. All points in S_G^{out} are simultaneously reachable from $\partial\Omega$ and $\partial\Omega^+$; as such, S_G^{out} defines a *minimal* skeleton for Ω^-, the ring-like region bounded by the shape boundary $\partial\Omega$ and the emergent zero level curve $\partial\Omega^+$.

It is important that the level curves of the restriction of ω to Ω^- behave similar to the level curves of a highly diffused distance transform: Only if the tip of a protrusion moves sufficiently faster than the opposing point on $\partial\Omega^+$, the respective fronts meet at a point that is sufficiently away from the tip, leaving enough room for the respective branch to grow. As a result, a branch tracking a limb terminates when the limb connects to a body, making the branch length a good correspondent of the limb length.

The ridge locus takes an irregular appearance so becomes difficult to interpret as the level curves get rounder. Fortunately, the irregularities are to a large extent within the confines of the central region Ω^+, thus, can be cut out (Fig. 20.10).

Grouping the set of ridge points in the form of segments (branches) is an implementation issue, typically subject to several ad-hoc decisions. In the case of skeletons from ω, the implementation can be considerably simplified by confining

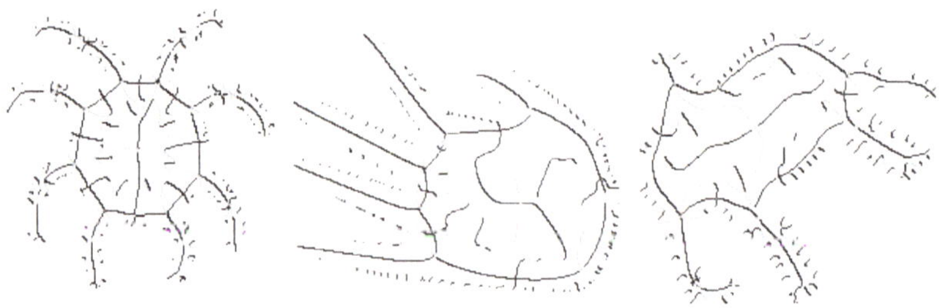

Fig. 20.9 Ridge points extracted by applying Tari et al. method [55] on ω

Fig. 20.10 Removing the ridge points that are inside Ω^+ (the *light gray* region) significantly reduces the set of candidate points. The *thick red curve* is S_G^{out}

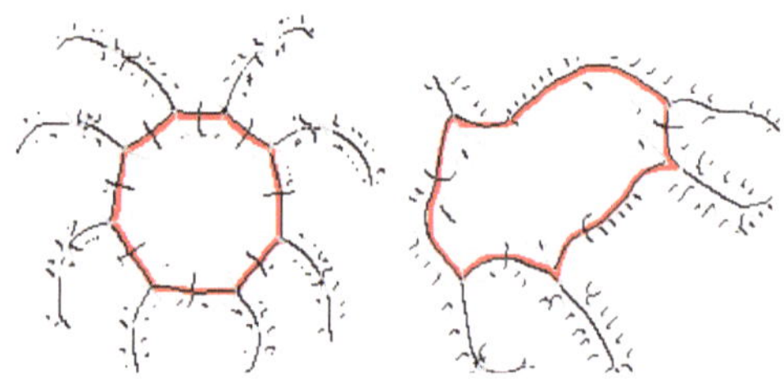

the grouping process to individual zones (recall that the peripheral region, unless it is a perfect annulus, always breaks into parts). Then grouping within each zone may be cast as a curve tracing task, performed by starting either from the boundary or the interior [2,20]. Unfortunately both cases necessitates too many ad-hoc decisions. A robust formulation via energy minimization is under development. In the mean time, leaving the grouping task as a technical issue that requires further attention, a sample set of branches is depicted in Fig. 20.11 in order to convey the concept. Typically the branches in S_R and S_T are linear disconnected elements, but there may be loop-like branches too, which occur if the shape has a hole in its peripheral region (Fig. 20.12).

Whether a branch belongs to S_R or S_T becomes unclear if limbs have limbs too (e.g. fingers of an arm). On one hand, it may be natural to put such branches in S_T since they track secondary details. On the other hand, it may be critical to distinguish them from noise, discretization artifacts and boundary texture. A generic strategy could be to assign fuzzy membership to such branches. This all depends on the task at hand. (The more important point is not to prune any branch.)

If the task at hand necessitates distinguishing the limbs of limbs from other secondary details, proper criteria need to be defined. Typically, within each zone, S_R branches are distinguished by the proximity of their termination point to a fixed locus, i.e. S_G^{out}. Likewise, a branch denoting a limb of a limb may be distinguished by the proximity of its termination point to a branch in S_R. A better strategy, however, is to consider the contribution of each branch to the reconstruction of the shape, i.e. the effective coverage of a branch, defined as the ratio of the area exclusively covered by the branch to its total coverage. Note that the region covered by a skeleton branch is the envelope of all circles such that each circle is centered at a skeleton point on

Fig. 20.11 Sample branches

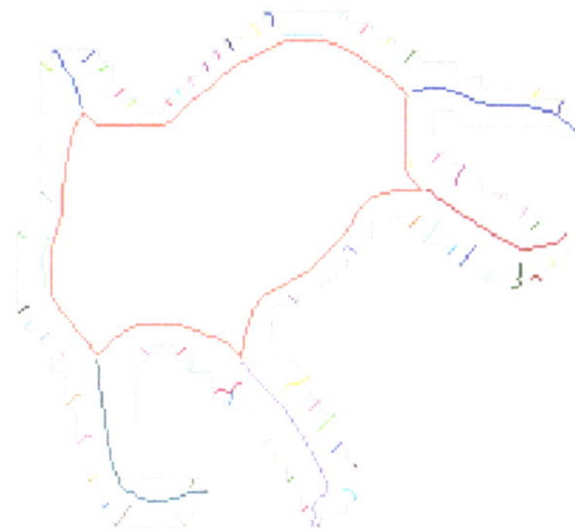

Fig. 20.12 Shape with a hole

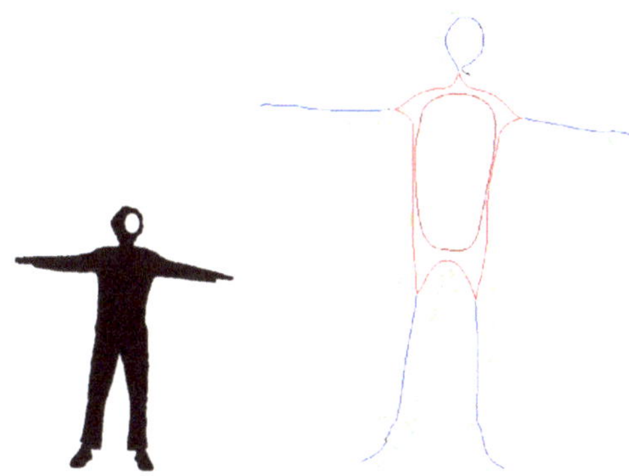

the branch, and has a radius given by the value of the distance transform at that point [10].

In general, computing the effective coverage is difficult [19] but becomes easy in our case because the branches can be put into a priority order, at least within a zone. This order is based on the branch length and proximity to a fixed locus, be it S_G^{out} or members of S_R.

Effective coverages of the branches for the elephant shape are depicted in Fig. 20.13. The gray toning of the covered regions (left) reflects the magnitudes of the effective coverage ratios of the corresponding branches. On the right, the five branches forming S_R covers almost the entire peripheral region.

20.3.1 Why Three Partite Skeletons?

I make four arguments in favor of the new skeleton model:

Boundary Details The most frequently reported problem of the classical skeleton is its sensitivity to boundary perturbations: small indentations or protrusions introduce spurious branches. A classical connectivity-preserving skeleton for the

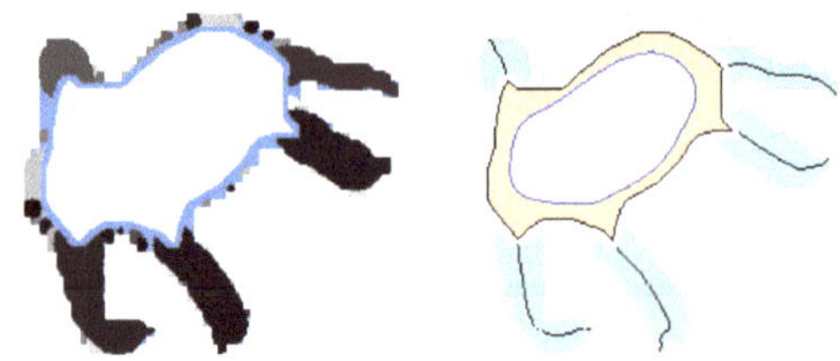

Fig. 20.13 Coverage of branches in S_R and S_T. The *gray* toning of the covered regions (*left*) reflects the magnitudes of the effective coverage ratios of the corresponding branches

elephant shape is shown in Fig. 20.14; compare it to the Three-Partite one of the same shape (Fig. 20.2).

Both skeletons contain spurious branches due to the wiggly boundary. Nevertheless, in the new model, they remain short and isolated, hence easily distinguishable, from the main branches. In contrast, in the classical model, it is difficult to identify spurious branches; consequently, there has been an extensive research on defining criteria for branch pruning [5, 19, 27]. Pruning criteria and methods differ, but almost all of the working skeleton methods reduce skeletons by eliminating those branches that contribute little to the reconstruction of the shape.

On one hand, reduced skeletons decrease classification errors when experimented using typical shape databases because secondary details most of the time act like noise [4]. On the other hand, in a different context, secondary details may be characterizing features: Observe that separation of the shapes shown in Fig. 20.15 into two perceptual proximity groups, as depicted in Fig. 20.16, is based on the boundary texture, a secondary detail subject to pruning.

Thus, it is essential not to throw away spurious branches but to separate them from primary ones, keeping in mind that the purpose of perceptual organization is to organize and gradually explicate relations. In the Three-Partite Skeleton the desired separation is achieved by letting each branch to be disconnected as a result of extensive smoothing. Disconnected skeletons may be uncommon but have always been around: Earlier examples are Leyton [26], Burbeck and Pizer [14], and Tari et al. [55], and more recent examples are Aslan and Tari [3,4] and Macrini et al. [27].

Perceived Topology Preserving topology is the common promise of skeleton models, but it may cause instability because topologic equivalence is not perceptual equivalence. Let us consider a set of peanuts and blobs (Fig. 20.17a). Where should we place category boundaries? Is the fifth shape more similar to the second or to the last one? How different are the first two shapes?

It seems natural to consider multiple abstractions of the topology. In the case of Three-Partite skeletons, this is achieved by the layered structure of S_G considered together with the critical points of ω inside the region enclosed by S_G^{in}. The following is important: If a simple closed curve in S_G^{in} is blob-like, ω has a unique maximum located roughly at the centroid of the region it encloses; but if it has indentations, ω has at least two local maxima roughly located at the centroids of implied blobs.

Let us consider the four Three-Partite Skeletons depicted in Fig. 20.17b, corresponding to the four peanuts shown in the middle group in (a). As before, S_G^{out} and

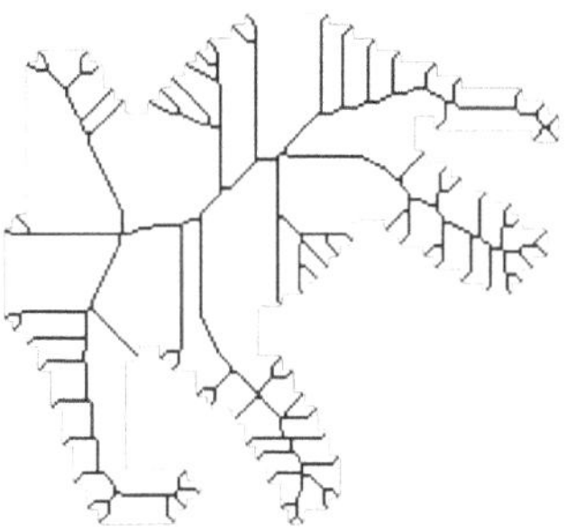

Fig. 20.14 A connectivity preserving skeleton. The branch length is not a good indicator of significance

Fig. 20.15 A collection of shapes

Fig. 20.16 Depending on the context, a secondary detail (e.g. boundary texture) may be the key feature

S_G^{in} are respectively shown as light and dark red; local maxima pairs are combined via a magenta line whenever applicable. The S_G of the leftmost shape consists of a pair of nested closed curves; the inner component S_G^{in} is blob-like so encloses the unique maximum of ω; signaling that the shape is a single blob. As the neck of the shape thins giving the second shape, S_G^{in} develops indentations, so the maximum becomes the saddle separating two local maxima, indicating that the shape is a composition of two blobs. As the neck thins further giving the third shape, S_G^{in} breaks into two curves, each of which is blob-like; the outer layer S_G^{out} still remains connected. On one hand, the second shape is similar to the first shape because its skeleton is a single pair of nested closed curves; on the other hand, S_G^{in} encloses two local maxima signaling that the shape is a composition of two blobs. That is, the second shape is a borderline shape for which two abstractions of the topology should be retained. Finally, as the neck sufficiently thins to yield the rightmost shape, the outer loop also splits; signaling even more strongly that the shape is a composition of two blobs. The fourth peanut (the fifth in Fig. 20.17a) is another borderline shape. The magenta line connecting the pair of local maxima passes though a saddle point region (light blue) which falls outside S_G. Hence the alternative abstraction connects the last shape to two discrete blobs.

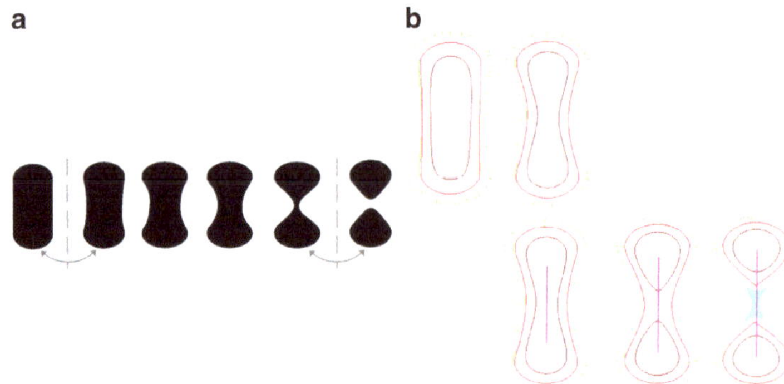

Fig. 20.17 (**a**): Peanuts of varying neck thickness and blobs. (**b**): The Three-Partite Skeletons of the four shapes depicted in the middle group in (**a**). The new model enables multiple interpretations of the shape topology. Transition shapes provide paths between blobs and peanuts

Fig. 20.18 Centrality

Centrality for Nearly Symmetric Shapes The classical skeleton of a circle is a single point located at its center reflecting the shape's symmetry. Likewise the branches of the skeleton of n-fold symmetric stars (or n-fold symmetric polygons) meet at the respective shape centroid; unfortunately, a slight deviation from the symmetry changes the connectivity pattern, e.g., the branches tracing the eight arms of the octopus in Fig. 20.18 do not all meet at a single point. This means that skeleton based abstractions fail to capture *centrality*, a natural consequence of fold symmetry (e.g., Figs. 20.18 and 20.5). This is rather ironic because the points of the classical skeleton are the medial ones.

In the new model, the notion of medialness is revised. S_G is a robust explication of the centrality. Indeed, the region bounded by S_G^{in}, namely Ω^+, can be considered as an interval estimate of the center, a robust alternative to a point estimate. The problem associated to connectivity pattern is solved by considering the branches outside S_G as they remain disjoint.

Notice that in Fig. 20.2, the eight disjoint branches in S_R (blue) tracing the eight arms of the octopus terminate without distorting the central part, which is a *perfect* circle. The S_G^{in} of the prickly ball is also a circle whereas the S_G^{in} of the triskelion is a triangle. For both shapes, S_R is empty; the centrality of the triskelion is emphasized (Fig. 20.19) consistent with the historic meaning of this shape.

Ligature Problem The fragility of the branch connectivity pattern is not specific to nearly symmetric shapes. Instabilities near junctions almost always occur; this is commonly addressed as the *ligature* problem [10]. One way to deal with the ligature problem is to restrict skeleton extraction to ribbon-like sections where the

Fig. 20.19 Consistent with the historic meaning of the triskelion, centrality is emphasized

procedure is stable: While Shah [47] completed missing parts of the skeleton via interpolation, Aslan and Tari [3] kept the skeleton disconnected. Recently Macrini et al. [27] also proposed a disconnected skeleton by removing the ligature part of the classical skeleton.

Ligature problem is all the more reason to cut out a central part.

20.4 Summary and Concluding Remarks

A distance-like field that attains both positive and negative values is described. It emerges out of simple spatial interactions, both local and global. It is a feature conscience field: A hierarchy of parts starting with peripheral versus gross structure separation is readily provided by its level curves, which mimic curvature dependent motion (so code curvature), and its generalized ridge points give rise to a generalized medial locus that is organized in the form of an unconventional skeleton. The new skeleton model suggests, among other things, (1) multiple interpretation of the shape topology, and (2) *gross structure* + *texture* + *part* separation. Incentives for the new skeleton model are discussed.

If the global interaction is ignored, the set of linear equations defining the field reduces to a discrete implementation of a screened Poisson PDE, a computationally advantageous alternative model for constructing reaction-diffusion scale space for shapes. Exploiting a connection between the screened Poisson PDE and the famous phase field of Ambrosio and Tortorelli, a connection between the new field and a modification to the Ambrosio-Tortorelli model is established. Implication of this connection is two fold: (1) a novel skeleton model, for which strong perceptual incentives exist, is conceived, (2) the issue of bridging image segmentation and shape analysis is raised, following former work by Tari, Shah and Pien.

Acknowledgements The author thanks to the Alexander von Humboldt Foundation for a generous financial support and extends her gratitude to Folkmar Bornemann, Sci. Comp. Dept. of Tech. Universität München for providing a wonderful sabbatical stay during which this work has been completed. She also thanks to anonymous reviewers and the editors, A. Bruckstein, M. Breuß and P. Maragos, for their meticulous editing.

References

1. Ambrosio, L., Tortorelli, V.: On the approximation of functionals depending on jumps by elliptic functionals via Γ-convergence. Commun. Pure Appl. Math. **43**(8), 999–1036 (1990)
2. Aslan, C.: Disconnected skeletons for shape recognition. Master's thesis, Department of Computer Engineering, Middle East Technical University (2005)
3. Aslan, C., Tari, S.: An axis-based representation for recognition. In: Proceedings of the ICCV, pp. 1339–1346. Springer, Berlin/New York (2005)
4. Aslan, C., Erdem, A., Erdem, E., Tari, S.: Disconnected skeleton: shape at its absolute scale. IEEE Trans. Pattern Anal. Mach. Intell. **30**(12), 2188–2203 (2008)
5. Bai, X., Latecki, L., Liu, W.Y.: Skeleton pruning by contour partitioning with discrete curve evolution. IEEE Trans. Pattern Anal. Mach. Intell. **29**, 449–462 (2007)
6. Bar, L., Sochen, N., Kiryati, N.: Image deblurring in the presence of impulsive noise. Int. J. Comput. Vis. **70**(3), 279–298 (2006)
7. Barenholtz, E., Feldman, J.: Visual comparisons within and between object-parts: evidence for a single- part superiority effect. Vis. Res. **43**, 1655–1666 (2003)
8. Belongie, S., Malik, J., Puzicha, J.: Shape matching and object recognition using shape contexts. IEEE Trans. Pattern Anal. Mach. Intell. **24**, 509–522 (2002)
9. Biederman, I.: Recognition-by-components: a theory of human image understanding. Psychol. Rev. **94**(2), 115–117 (1987)
10. Blum, H.: Biological shape and visual science. J. Theor. Biol. **38**, 205–287 (1973)
11. Braides, A.: Approximation of Free-Discontinuity Problems. Lecture Notes in Mathematics, vol. 1694. Springer, Berlin/New York (1998)
12. Brockett, R., Maragos, P.: Evolution equations for continuous–scale morphology. In: Proceedings of the ICASSP, vol. 3, pp. 125–128. IEEE, Piscataway (1992)
13. Buades, A., Coll, B., Morel, J.M.: A non-local algorithm for image denoising. In: Proceedings of the CVPR, pp. 60–65. IEEE Computer Society, Los Alamitos (2005)
14. Burbeck, C.A., Pizer, S.M.: Object representation by cores: identifying and representing primitive spatial regions. Vis. Res. **35**, 1917–1930 (1995)
15. Cohen, E.H., Singh, M.: Geometric determinants of shape segmentation: tests using segment identification. Vis. Res. **47**, 2825–2840 (2007)
16. de Winter, J., Wagemans, J.: Segmentation of object outlines into parts: a large-scale integrative study. Cognition **99**, 275–325 (1999)
17. Edelman, S., Cutzu, F., Duvdevani-Bar, S.: Similarity to reference shapes as a basis for shape representation. In: Proceedings of the COGSCI, San Diego (1996)
18. Erdem, E., Tari, S.: Mumford-Shah regularizer with contextual feedback. JMIV **33**, 67–84 (2009)
19. Feldman, J., Singh, M.: Bayesian estimation of the shape skeleton. PNAS **103**(47), 18014–18019 (2006)
20. Genctav, M.: Matching global skeleton. Master's thesis, Department of Computer Engineering, Middle East Technical University (2010)
21. Gonzalez, R., Woods, R.: Digital Image Processing, 2nd edn. Addison-Wesley Longman Publishing Co., Inc., Boston (2001)
22. Gorelick, L., Galun, M., Sharon, E., Basri, R., Brandt, A.: Shape representation and classification using the poisson equation. IEEE Trans. Pattern Anal. Mach. Intell. **28**(12), 1991–2005 (2006)
23. Hofmann, D., Richards, W.: Parts of recognition. Cognition **18**, 65–96 (1984)
24. Jung, M., Vese, L.: Nonlocal variational image deblurring models in the presence of gaussian or impulse noise. In: Proceedings of the SSVM, pp. 401–412. Springer, Berlin/New York (2009)
25. Kimia, B., Tannenbaum, A., Zucker, S.: Shapes, shocks, and deformations I: the components of two-dimensional shape and the reaction-diffusion space. Int. J. Comput. Vis. **15**(3), 189–224 (1995)

26. Leyton, M.: A process-grammar for shape. Art. Intell. **34**(2), 213–247 (1988)
27. Macrini, D., Dickinson, S., Fleet, D., Siddiqi, K.: Bone graphs: medial shape parsing and abstraction. Comput. Vis. Image Underst. **115**, 1044–1061 (2011)
28. Maragos, P., Butt, M.A.: Curve evolution, differential morphology and distance transforms as applied to multiscale and eikonal problems. Fundam. Inf. **41**, 91–129 (2000)
29. March, R., Dozio, M.: A variational method for the recovery of smooth boundaries. Image Vis. Comput. **15**(9), 705–712 (1997)
30. Marr, D.: Vision: A Computational Investigation into the Human Representation and Processing of Visual Information. W.H. Freeman, San Francisco (1982)
31. Meyer, F.: Topographic distance and watershed lines. Signal Process. **38**, 113–125 (1994)
32. Mi, X., DeCarlo, D.: Separating parts from 2d shapes using relatability. In: Proceedings of the ICCV, pp. 1–8. IEEE Computer Society, Los Alamitos (2007)
33. Morel, J.-M., Solimini, S.: Variational Methods in Image Segmentation. Birkhäuser, Boston (1995)
34. Mumford, D.: Mathematical theories of shape: do they model perception? In: Proceedings of the SPIE, vol. 1570. SPIE, Bellingham (1991)
35. Mumford, D., Shah, J.: Optimal approximations by piecewise smooth functions and associated variational problems. Commun. Pure Appl. Math. **42**, 577–685 (1989)
36. Navon, D.: Forest before trees: the precedence of global features in visual perception. Cogn. Psychol. **9**, 355–383 (1977)
37. Pasupathy, A., Connor, C.: Population coding of shape in area V4. Nat. Neurosci. **5**(2), 1332–1338 (2002)
38. Patz, T., Preusser, T.: Ambrosio-Tortorelli segmentation of stochastic images. In: Proceedings of the ECCV, pp. 254–267. Springer, Berlin (2010)
39. Pien, H.H., Desai, M., Shah, J.: Segmentation of mr images using curve evolution and prior information. IJPRAI **11**(8), 1233–1245 (1997)
40. Preußer, T., Droske, M., Garbe, C., Rumpf, M., Telea, A.: A phase field method for joint denoising, edge detection and motion estimation. SIAM J. Appl. Math. **68**(3), 599–618 (2007)
41. Proesman, M., Pauwels, E., van Gool, L.: Coupled geometry-driven diffusion equations for low-level vision. In: Romeny, B. (ed.) Geometry Driven Diffusion in Computer Vision. Lecture Notes in Computer Science. Kluwer, Dordrecht/Boston (1994)
42. Rosenfeld, A., Pfaltz, J.L.: Distance functions on digital pictures. Pattern Recognit. **1**, 33–61 (1968)
43. Serra, J.: Image Analysis and Mathematical Morphology. London Academic, Orlando (1982)
44. Shah, J.: Segmentation by nonlinear diffusion. In: Proceedings of the CVPR, pp. 202–207. IEEE Computer Society, Los Alamitos (1991)
45. Shah, J.: A common framework for curve evolution, segmentation and anisotropic diffusion. In: Proceedings of the CVPR, pp. 136–142 (1996)
46. Shah, J.: Riemannian drums, anisotropic curve evolution and segmentation. In: Proceedings of the Scale-Space, pp. 129–140. Springer, Berlin/New York (1999)
47. Shah, J.: Skeletons and segmentation of shapes. Technical report, Northeastern University. http://www.math.neu.edu/~shah/publications.html (2005)
48. Shah, J., Pien, H.H., Gauch, J.: Recovery of shapes of surfaces with discontinuities by fusion of shading and range data within a variational framework. IEEE Trans. Image Process. **5**(8), 1243–1251 (1996)
49. Siddiqi, K., Tresness, K.J., Kimia, B.: Parts of visual form: ecological and psychophysical aspects. Perception **25**, 399–424 (1996)
50. Stiny, G.: Shape: Talking about Seeing and Doing. MIT, Cambridge (2006)
51. Tari, S., Genctav, M.: From a modified Ambrosio-Tortorelli to a randomized part hierarchy tree. In: Bruckstein, A.M., ter Haar Romeny, B.M., Bronstein, A.M., Bronstein, M.M. (eds.) Scale Space and Variational Methods. Lecture Notes in Computer Science, vol. 6667, pp. 267–278. Springer, Berlin/Heidelberg (2011)
52. Tari, S., Shah, J.: Simultaneous segmentation of images and shapes. In: Proceedings of the SPIE, San Diego, vol. 3168, San Diego pp. 88–94 (1997)

53. Tari, S., Shah, J.: Local symmetries of shapes in arbitrary dimension. In: Proceedings of the ICCV, pp. 1123–1128. Narosa, New Delhi (1998)
54. Tari, S., Shah, J., Pien, H.: A computationally efficient shape analysis via level sets. In: Proceedings of the MMBIA, pp. 234–243. IEEE Computer Society, Los Alamitos (1996)
55. Tari, S., Shah, J., Pien, H.: Extraction of shape skeletons from grayscale images. CVIU **66**(2), 133–146 (1997)
56. Teboul, S., Blanc-Féraud, L., Aubert, G., Barlaud, M.: Variational approach for edge preserving regularization using coupled PDE's. IEEE Trans. Image. Process. **7**, 387–397 (1998)
57. Xu, C., Liu, J., Tang, X.: 2d shape matching by contour flexibility. IEEE Trans. Pattern Anal. Mach. Intell. **31**(1), 180–186 (2009)
58. Zeng, J.T., Lakaemper, R., Wei, X., Li, X.: 2d shape decomposition based on combined skeleton-boundary features. In: Proceedings of the Advances in Visual Computing, 4th International Symposium, ISVC, pp. 682–691. Springer, Berlin/New York (2008)

Chapter 21
Integrated DEM Construction and Calibration of Hyperspectral Imagery: A Remote Sensing Perspective

Christian Wöhler and Arne Grumpe

Abstract In this study, we present from a remote sensing perspective a method for the combination of surface gradient information obtained by photoclinometry and shape from shading with absolute depth data (here: light detection and ranging (LIDAR) data) by exploiting their respective advantages, regarding distinctly non-Lambertian surfaces with non-uniform albedos. While photometry-based 3D reconstruction methods yield reliable small-scale surface gradient information for each image pixel, absolute depth data which are typically noisy on small scales but reliable on large scales are provided by LIDAR techniques. The first step of the proposed algorithm consists of an extended photoclinometry approach which takes into account both image and LIDAR data. In a second step the reconstructed surface is refined based on an iterative scheme relying on the minimisation of a global error functional, thus compensating the inaccuracies of the measured surface gradients and the LIDAR data on the respective scales. The surface shape and non-uniform albedo map represent the best fit to the observed image radiances and LIDAR data. We apply our framework to the construction of digital elevation models (DEM) of lunar surface regions. We use hyperspectral imagery in order to employ the DEM to normalise the wavelength-dependent surface reflectance to a standard illumination and viewing geometry. Although we employ a highly realistic, physically motivated reflectance model (the Hapke model), systematic topography-dependent distortions of the pixel spectra occur, which lead to errors in the extracted spectral parameters (e.g. the absorption wavelength, depth, and width of prominent absorption troughs) and for which we propose an empirical, PCA-based correction approach. Based on the correspondingly corrected surface reflectances we obtain a refined DEM along with spectral parameter maps in which (except for the hydroxyl absorption) topographic effects are nearly completely removed.

C. Wöhler (✉) · A. Grumpe
Image Analysis Group, Dortmund University of Technology, Otto-Hahn-Str. 4, 44227 Dortmund, Germany
e-mail: christian.woehler@tu-dortmund.de; arne.grumpe@tu-dortmund.de

M. Breuß et al. (eds.), *Innovations for Shape Analysis*, Mathematics and Visualization, DOI 10.1007/978-3-642-34141-0_21,

21.1 Introduction

The concept of photoclinometry is introduced by Wilhelms [52] for the image-based measurement of lunar surface slopes. The concepts of shape from shading and photometric stereo are introduced by Horn [20] and Woodham [54], respectively. The method of characteristic strip expansion by Horn [21] and also the method by Kozera [29] directly yield a depth map of the surface. In contrast, other methods by Horn [20, 22] estimate the gradient field of the surface and infer the depth map by integration. The computational efficiency of such gradient-based approaches is increased e.g. by Simchony et al. [47] by an analytical approach. An extensive survey is provided in [1].

Improved results can be achieved by combining photometrically determined surface gradient data with independently measured absolute depth data. General solutions for this problem are proposed e.g. in [7] and [40], where a highly detailed 3D surface reconstruction is obtained by a combination of the low-frequency part of the absolute depth data with the high-frequency component of the photometrically determined surface gradients. Non-Lambertian metallic surfaces are reconstructed e.g. in [53], multiple views are taken into account in [25,32,55], and moving objects are analysed in [46].

In the context of lunar remote sensing, a global digital elevation model (DEM) of the Moon of 100 m nominal lateral resolution with a typical vertical accuracy of 23 m is described in [44], which is termed "GLD100" and has been constructed based on the stereoscopic evaluation of images acquired by the LROC WAC (Lunar Reconnaissance Orbiter Wide Angle Camera). LROC NAC (Narrow Angle Camera) based stereo images analysis has led to the construction of local lunar DEMs of very high lateral resolution [31, 36]. Detailed information about the LROC WAC and NAC instruments can be found in [43]. A nearly global lunar hyperspectral image mosaic of typically 140 m resolution has been acquired by the Moon Mineralogy Mapper (M^3) instrument on board the Indian spacecraft Chandrayaan-1 [42] (http://m3.jpl.nasa.gov/m3data.html).

A combination of stereo image analysis and shape from shading is used in [28] to construct DEMs of high lateral resolution of the surface of Mars. Lunar DEMs are generated in [33] based on images acquired by the Clementine spacecraft using a technique termed "multi-image shape from shading", where the elevation values and the parameters of the surface reflectance function are adapted to the image radiance information, assuming a uniform surface albedo. A photometric stereo method is used for DEM construction in [10] based on overlapping images of the lunar southern polar region acquired by the SMART-1 Advanced Moon micro-Imager Experiment (AMIE) camera. Detailed information about the AMIE instrument can be found in [26]. The properties of the recently acquired lunar orbital image data sets are summarised in Table 21.1.

The global DEMs obtained by the LIDAR instrument LALT on board the Japanese Kaguya spacecraft and the Lunar Orbiter Laser Altimeter (LOLA) instrument on board the LRO spacecraft have vertical accuracies of a few metres (cf. [4] and [31] for a short description of LALT and LOLA data, respectively). The nominal

Table 21.1 Properties of recently acquired lunar orbital imagery. The numbers in brackets denote the M^3 target mode applied to selected regions only

Instrument	Bands	Wavelength (nm)	Resolution (m/pixel)	Reference
SMART-1 AMIE	No filter + 3	Visible, 750–960	27–270	[26]
Chandrayaan-1 M^3	85 (260)	461–2,976	140 (70)	[42]
LROC WAC	7	320–690	100	[43]
LROC NAC	1	Visible	0.5	[43]

lateral resolution of the gridded LOLA DEM (available at http://pds-geosciences.wustl.edu/missions/lro/lola.htm) corresponds to 30 m at the lunar equator, but its effective lateral resolution is several times lower in most regions due to missing data and the resulting interpolation artifacts.

This study presents a method for the integration of surface gradient data obtained by shape from shading with absolute depth data inferred from laser altimetry measurements, relying on M^3 orbital images and the global LOLA DEM. This framework is applied to the construction of DEMs of the lunar surface, which displays a non-Lambertian reflectance behaviour and a non-uniform albedo. The resulting DEMs have a high effective lateral resolution that comes close to the pixel resolution of the utilised images. Furthermore, based on the resulting DEM data we introduce the first procedure (to our knowledge) that normalises wavelength-dependent lunar surface reflectance values to a standard illumination and viewing geometry based on the topography on large and small spatial scales (given the corresponding values of the pixel resolution, large spatial scales will in the following refer to kilometres and small scales to pixels).

21.2 Reflectance Modelling

The light power emitted by a unit surface into a unit solid angle is termed "radiance" and is measured in $\mathrm{W\,m^{-2}\,sr^{-1}}$. The incoming light power per apparent unit surface is termed "irradiance" and has the physical unit $\mathrm{W\,m^{-2}}$. The so-called "reflectance" is defined as the ratio between radiance and irradiance [21]. It usually depends on the incidence angle θ_i between the surface normal $\mathbf{n}$ and the direction $\mathbf{l}$ to the light source, the emission angle θ_e between $\mathbf{n}$ and the direction $\mathbf{v}$ to the camera, and on the phase angle α between the vectors $\mathbf{l}$ and $\mathbf{v}$ [14] (Fig. 21.1). The radiances and irradiances regarded in the experiments of this study are additionally normalised to wavelength ("spectral radiance" and "spectral irradiance" [14, 27]).

A reflectance model based on the theory of radiative transfer [5] has been proposed by Hapke [14–17], which in its general form according to [17] can be written as

$$R_{\mathrm{AMSA}}(\mu_0, \mu, \alpha) = \frac{w}{4\pi}\frac{\mu_0}{\mu_0 + \mu}\left[p(\alpha)B_{\mathrm{SH}}(\alpha) + M(\mu_0, \mu)\right]B_{\mathrm{CB}}(\alpha) \qquad (21.1)$$

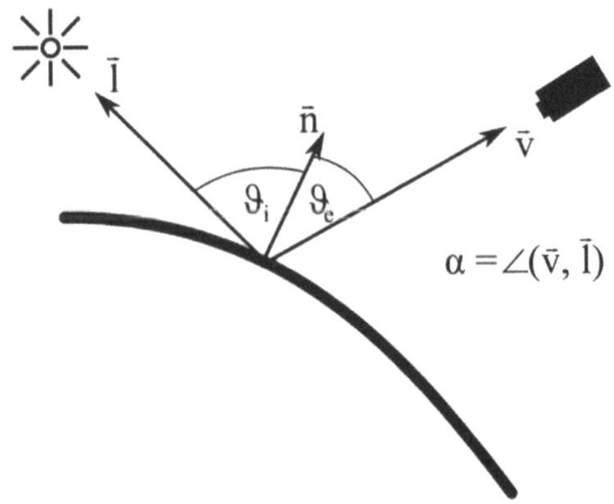

Fig. 21.1 Illustration of the illumination and observation geometry

with $\mu_0 = \cos\theta_i$ and $\mu = \cos\theta_e$. According to [14–17], the intrinsic reflectivity of a surface particle is denoted by w and is termed "single-scattering albedo". Typical planetary surfaces display a peak of the reflectance function around $\alpha = 0°$. This so-called "opposition effect" is composed of a shadow hiding [16, 17] and a coherent backscatter [17] component, described by the functions $B_{\mathrm{SH}}(\alpha)$ and $B_{\mathrm{CB}}(\alpha)$ in Eq. (21.1), respectively. Each of these functions is governed by two parameters describing the strength and width of the respective component [17]. The function $p(\alpha)$ is the so-called "single-particle scattering function" describing first-order scattering processes at the particles [14]. The term $M(\mu_0, \mu)$ according to [17] describes multiple scattering processes, which are assumed to be isotropic in [14–16], leading to the "isotropic multiple scattering approximation" (IMSA), while $p(\alpha)$ is expanded as a series of Legendre polynomials in [17], resulting in the "anisotropic multiple scattering approximation" (AMSA). An extension that takes into account the roughness of the surface is introduced in [15].

The so-called "lunar-Lambert" function, a simplified, empirical reflectance function for planetary surfaces, is proposed in [38] according to

$$R_{\mathrm{LL}}(\mu_0, \mu, \alpha) = \rho \left[2L(\alpha) \frac{\mu_0}{\mu_0 + \mu} + (1 - L(\alpha))\, \mu_0 \right]. \tag{21.2}$$

The factor ρ depends on the reflectivity of the surface but also on camera-specific factors such as the sensor sensitivity. The empirical function $L(\alpha)$ is determined in [38] based on a phase angle specific maximisation of the similarity between the lunar-Lambert reflectance $R_{\mathrm{LL}}(\mu_0, \mu, \alpha)$ and the Hapke IMSA reflectance [14–16] across a broad interval of incidence and emission angles for a variety of parameter configurations of the Hapke IMSA model.

The lunar-Lambert model is commonly used as the basis for the normalisation of reflectance data to a standard illumination and viewing geometry, as outlined e.g. in [39]. It is favourably used for DEM construction if relative pixel intensities rather than absolute radiance data are available.

In the DEM construction framework proposed in Sect. 21.3, radiance images are evaluated, where the generally non-uniform value of the single-scattering albedo w is recovered for each surface location simultaneously with the DEM, while the other parameters of the Hapke model [14–17] are adopted from [51] (Moon solution 1 therein) and are assumed to be constant.

21.3 DEM Construction

In the following, the surface gradients are denoted by p and q, implying a local surface normal vector $\mathbf{n} = (-p, -q, 1)^T$, the surface by z, the observed radiance by L^{obs}, the solar irradiance by E^{sun}, the modelled reflectance by R, and the lateral coordinates by x and y. The value of E^{sun} depends on the wavelength range in which the image has been acquired and is assumed to be known. In our remote sensing scenario, the observed reflectance is computed by a pixel-wise division of L^{obs} by E^{sun}. For brevity, the dependence on x and y is omitted where possible.

21.3.1 The Error Functional

The description in this section and in Sect. 21.3.2 has been adapted from [18]. Horn [22] proposes a minimisation of a global error functional composed of the reflectance error

$$E_R = \iint \left(\frac{L^{\text{obs}}(x,y)}{E^{\text{sun}}} - R\left(p(x,y), q(x,y)\right) \right)^2 dx\, dy \tag{21.3}$$

and the integrability error

$$E_{\text{int}} = \iint \left[(z_x - p)^2 + (z_y - q)^2 \right] dx\, dy. \tag{21.4}$$

We extend this framework by adding the error term

$$E_{\text{DEM}} = \iint \left[f_{\sigma_{\text{DEM}}} \left(\frac{\partial z_{\text{DEM}}}{\partial x} \right) - f_{\sigma_{\text{DEM}}}(p) \right]^2 + \left[f_{\sigma_{\text{DEM}}} \left(\frac{\partial z_{\text{DEM}}}{\partial y} \right) - f_{\sigma_{\text{DEM}}}(q) \right]^2 dx\, dy \tag{21.5}$$

to the error functional $E = E_R + \gamma E_{\text{int}}$ proposed in [22], such that absolute depth data z_{DEM} like the LOLA DEM can be taken into account. The term E_{DEM} takes into account the deviation between the large-scale gradients of the depth data and the surface z. The noisy high-frequency component of the absolute depth data is suppressed by the lowpass filter function $f_{\sigma_{\mathrm{DEM}}}$, which is implemented as a Gaussian filter of width σ_{DEM}. The low-frequency component of the absolute depth data can be expected to be accurate on large scales. The reconstruction error E_{Σ} to be minimised thus corresponds to

$$E_{\Sigma} = E_R + \gamma E_{\mathrm{int}} + \delta E_{\mathrm{DEM}} \tag{21.6}$$

with γ and δ as weight parameters.

At this point one might argue that it would be more straightforward to take into account the low-frequency component of z_{DEM} directly instead of the corresponding surface gradients. However, a computationally efficient formulation comparable to the gradient-based numerical scheme presented in Sect. 21.3.2 is not available, such that an error term based on z_{DEM} would be much more intricate to minimise. Furthermore, the normalisation of the hyperspectral data described in Sect. 21.5 requires accurate values of the surface gradients rather than the surface itself, such that we expect an explicit determination of the surface gradients to be beneficial in that context.

21.3.2 Variational Optimisation Scheme

The minimisation of E_{Σ} with a nonzero value of γ aims for generating an integrable surface gradient field. For this purpose, we use a variational scheme according to [22], leading to

$$\begin{aligned} p^{(n+1)} &= z_x^{(n)} + \frac{1}{\gamma}\left[\frac{L^{\mathrm{obs}}}{E^{\mathrm{sun}}} - R\left(z_x^{(n)}, z_y^{(n)}\right)\right] \left.\frac{\partial R}{\partial p}\right|_{z_x^{(n)}} \\ &\quad + \frac{\delta}{\gamma} \iint \left[f_{\sigma_{\mathrm{DEM}}}\left(\frac{\partial z_{\mathrm{DEM}}}{\partial x}\right) - f_{\sigma_{\mathrm{DEM}}}(z_x^{(n)}) \right] \left.\frac{\partial f_{\sigma_{\mathrm{DEM}}}(p)}{\partial p}\right|_{z_x^{(n)}} dx\,dy \\ z^{(n+1)} &= \bar{z}^{(n)} - \frac{\varepsilon^2}{\kappa}\left[p_x^{(n+1)} + q_y^{(n+1)}\right]. \end{aligned} \tag{21.7}$$

The corresponding equation for q is obtained analogously. In Eq. (21.7), the average over the κ nearest neighbouring pixels is denoted by $\bar{z}$, and the lateral offset between neighbouring pixels is given by ϵ. The surface may be initialised e.g. by the LOLA DEM.

The double integral in Eq. (21.7) corresponds to $\frac{\partial E_{\mathrm{DEM}}(p,q)}{\partial p}$. A computationally efficient implementation is obtained for the complete image by two consecutive linear filtering operations if the filtering function $f_{\sigma_{\mathrm{DEM}}}$ is implemented as a linear filter. According to [13], the double integral is written as a sum over $U \times V$ discrete cells, where the filter function $f_{\sigma_{\mathrm{DEM}}}$ is defined as a $K \times L$ filter matrix $F_{\sigma_{\mathrm{DEM}}}$, leading to

$$\frac{\partial E_{\mathrm{DEM}}(p,q)}{\partial p} = -\sum_{i=1}^{U}\sum_{j=1}^{V} A(i,j)\left[\sum_{k=-K/2}^{K/2}\sum_{l=-L/2}^{L/2} F_{\sigma_{\mathrm{DEM}}}(k,l)\left(\left.\frac{\partial z_{\mathrm{DEM}}}{\partial x}\right|_{i+k,j+l} - p(i+k,j+l)\right)\right]\left[\sum_{k=-K/2}^{K/2}\sum_{l=-L/2}^{L/2} F_{\sigma_{\mathrm{DEM}}}(k,l)\left.\frac{\partial p(i+k,j+l)}{\partial p}\right|_{u,v}\right]. \quad (21.8)$$

In Eq. (21.8), the filter $f_{\sigma_{\mathrm{DEM}}}$ is correlated with the discrete data. The areal extent $A(i,j)$ of a discrete cell is defined such that it denotes the surface area covered by the respective pixel. This quantity is not necessarily identical for all pixels. The summands in Eq. (21.8) are only non-zero for $u = l+k$ and $v = j+l$. Omitting the zero-valued summands and shifting the indices i and j such that they are centred at u and v yields

$$\frac{\partial E_{\mathrm{DEM}}(p,q)}{\partial p} = -\sum_{i=-K/2}^{K/2}\sum_{j=-L/2}^{L/2} F_{\sigma_{\mathrm{DEM}}}(-i,-j)\left[\sum_{k=-K/2}^{K/2}\sum_{l=-L/2}^{L/2} F_{\sigma_{\mathrm{DEM}}}(k,l) \times \left(\left.\frac{\partial z_{\mathrm{DEM}}}{\partial x}\right|_{u+k+i,v+l+j} - p(u+k+i,v+l+j)\right)\right] A(u+i,v+j) \quad (21.9)$$

$$\frac{\partial E_{\mathrm{DEM}}(p,q)}{\partial q} = -\sum_{i=-K/2}^{K/2}\sum_{j=-L/2}^{L/2} F_{\sigma_{\mathrm{DEM}}}(-i,-j)\left[\sum_{k=-K/2}^{K/2}\sum_{l=-L/2}^{L/2} F_{\sigma_{\mathrm{DEM}}}(k,l) \times \left(\left.\frac{\partial z_{\mathrm{DEM}}}{\partial y}\right|_{u+k+i,v+l+j} - q(u+k+i,v+l+j)\right)\right] A(u+i,v+j). \quad (21.10)$$

According to Eq. (21.9), $\frac{\partial E_{\mathrm{DEM}}(p,q)}{\partial p}$ is obtained by a correlation between the filter matrix $F_{\sigma_{\mathrm{DEM}}}$ and the difference between the surface gradient $\frac{\partial z_{\mathrm{DEM}}}{\partial x}$ of the absolute depth map and the surface gradient p to be optimised, and in a second step a convolution of the correlation result with the filter matrix $F_{\sigma_{\mathrm{DEM}}}$ weighted by the areal extent $A(i,j)$. The value of $\frac{\partial E_{\mathrm{DEM}}(p,q)}{\partial q}$ is obtained in an analogous manner (cf. Eq. (21.10)). The order of correlation and convolution in Eqs. (21.9) and (21.10) is exchanged if the filter function $f_{\sigma_{\mathrm{DEM}}}$ is defined by a convolution rather than a correlation.

For a simultaneous reconstruction of the surface and the spatially non-uniform albedo, the optimisation according to Eq. (21.7) is incorporated into a further iterative scheme. The angles $\theta_i^{(m)}$ and $\theta_e^{(m)}$ and their cosines $\mu_0^{(m)}$ and $\mu^{(m)}$are determined based on the surface $z(m)$ and the corresponding gradients $p(m)$ and $q(m)$ for each iteration m. The non-uniform single-scattering albedo $w^{(m)}(x,y)$ is obtained by solving

$$R\left(\mu_0^{(m)}, \mu^{(m)}, \alpha, w^{(m)}(x,y)\right) = \frac{L^{\mathrm{obs}}(x,y)}{E^{\mathrm{sun}}} \quad (21.11)$$

with respect to $w^{(m)}(x, y)$ for each pixel (x, y). A Gaussian low-pass filter $G_{\sigma_w^{(m)}}$ of width $\sigma_w^{(m)}$ is then convolved with $w^{(m)}(x, y)$. The surface gradients $p(m+1)$, $q(m+1)$, and the surface $z(m+1)$ of the subsequent step $(m+1)$ are computed based on the filtered albedo map $\left[G_{\sigma_w^{(m)}} * w^{(m)}(x, y)\right]$. The width parameter $\sigma_w^{(m)}$ is decreased in the course of the iteration process such that an albedo map of increasing spatial resolution is obtained.

21.3.3 Initialisation by an Extended Photoclinometry Scheme

Photoclinometry denotes the pixel-wise determination of the surface gradients according to the observed image reflectances by setting the reflectance error E_R (cf. Eq. (21.3)) to zero for each pixel [22]. However, if only a single image is available but two surface gradients and (in the case of a non-uniform albedo map) the surface albedo need to be estimated, this ill-posed problem has an infinite number of solutions. Commonly, assumptions are made that regularise the problem and reduce the solution space to a single solution. For example, the albedo might be assumed to be constant over the surface or line-wise constant in the azimuthal direction of incident light. Additionally, the gradient perpendicular to the azimuthal direction of the incident light is difficult to estimate accurately; if it is small, it may be set to zero [22]. A constant or line-wise constant albedo can be estimated when the mean surface slope on large spatial scales is known. A favourable property of the surface gradients estimated by photoclinometry is that they are not affected by regularisation conditions such as smoothness of the surface. On the other hand, spurious artifacts tend to occur perpendicular to the direction of incident light.

However, such assumptions are obsolete in the case of independent depth information available on large spatial scales. Hence, instead of setting the reflectance error E_R to zero independently for each pixel, we minimise a weighted sum of the reflectance error E_R according to Eq. (21.3) and the DEM error E_{DEM} according to Eq. (21.5). Therefore, in analogy to classical photoclinometry, our extended photoclinometry scheme does not take into account a regularisation term such as smoothness or integrability of the surface gradients. Rather, we perform an independent pixel-wise minimisation of E_Σ according to Eq. (21.6) with $\gamma = 0$ [11]. In order to examine the behaviour of this minimisation, the Hessian matrix of E_{DEM} is required. For its first diagonal element we obtain the expression

$$\frac{\partial^2 E_{\text{DEM}}}{\partial p^2} = \sum_{i=-K/2}^{K/2} \sum_{j=-L/2}^{L/2} F_{\sigma_{\text{DEM}}}(-i,-j)\, A(u+i, v+j) \left[\sum_{k=-K/2}^{K/2} \sum_{l=-L/2}^{L/2} F_{\sigma_{\text{DEM}}}(k,l) \times \frac{\partial}{\partial p} \left(- \left.\frac{\partial z_{\text{DEM}}}{\partial x}\right|_{u+k+i,v+l+j} + p(u+k+i, v+l+j) \right) \right]$$

$$= \sum_{i=-K/2}^{K/2} \sum_{j=-L/2}^{L/2} F_{\sigma\mathrm{DEM}}(-i,-j)\, A(u+i, v+j) \left[\sum_{k=-K/2}^{K/2} \sum_{l=-L/2}^{L/2} F_{\sigma\mathrm{DEM}}(k,l) \times \frac{\partial}{\partial p}(p(u+k+i, v+l+j)) \right]. \quad (21.12)$$

In Eq. (21.12), $k + i = 0$ and $l + j = 0$ must hold for all non-zero summands, and the equation reduces to

$$\frac{\partial^2 E_{\mathrm{DEM}}}{\partial p^2} = \sum_{i=-K/2}^{K/2} \sum_{j=-L/2}^{L/2} F^2_{\sigma\mathrm{DEM}}(-i,-j) \cdot A(u+i, v+j). \quad (21.13)$$

The second derivative with respect to q has the same value. Since neither Eq. (21.9) depends on q nor Eq. (21.10) depends on p, the mixed derivatives are given by

$$\frac{\partial^2 E_{\mathrm{DEM}}}{\partial p \partial q} = \frac{\partial^2 E_{\mathrm{DEM}}}{\partial q \partial p} = 0. \quad (21.14)$$

Therefore, the Hessian matrix of E_{DEM} reduces to a diagonal matrix. Due to the fact that not all filter coefficients of $F_{\sigma\mathrm{DEM}}$ are zero and the pixel area A is positive by definition, all eigenvalues of the Hessian matrix are positive for all pixels. This implies that E_{DEM} possesses only one global minimum. Assuming a minimisation of E_{DEM} with the soft constraint E_R weighted by $1/\delta$, convergence is thus guaranteed for sufficiently large δ.

The resulting surface gradients p and q may be quite far from being integrable, such that we determine the "nearest" surface z based on the approach introduced in [21]. As an alternative to the LOLA DEM, the resulting surface can be used as an initialisation of the variational optimisation scheme described in Sect. 21.3.2. It will be shown in Sect. 21.4 that this initialisation results in a significantly higher absolute reconstruction accuracy.

The albedo map is computed pixel-wise using the initial surface and is then low-pass filtered as explained in Sect. 21.3.2. Since the method by Horn [21] of estimating the surface z from the gradients p and q does not allow large variations of z on large spatial scales, we apply a pyramidal approach, where both the image and the LOLA DEM are downscaled n_{pyr} times by a factor of 2. Starting from the lowest resolution level, the surface and the albedo map are computed and passed on to the next pyramid level until the full image resolution is reached.

21.4 Results of DEM Construction

For DEM construction, we utilise imagery acquired with the Moon Mineralogy Mapper (M^3) on board the Indian Chandrayaan-1 spacecraft. The M^3 instrument is a hyperspectral pushbroom sensor with (in global mode) 85 pixel-synchronous spectral channels with 20–40 nm per channel with centre wavelengths between 461 and 2,976 nm, where the image resolution typically corresponds to 140 m per pixel [42] (cf. also Table 21.1). The M^3 data archive accessible at http://m3.jpl.nasa.gov/m3data.html comprises radiance images for each wavelength channel and provides for each image pixel the selenographic longitude and latitude as well as the incidence angle $\theta_i^{(0)}$, the emission angle $\theta_e^{(0)}$, and the phase angle α, where topography is neglected and a smooth spherical surface is assumed for the given values of $\theta_i^{(0)}$ and $\theta_e^{(0)}$. We compute the reflectance error E_R as the sum of squared channel-wise errors according to Eq. (21.3) for channels 50–60, covering the wavelength range from $1,579$ to $1,978$ nm. As the lunar surface temperature may reach up to 396 K [50], we do not use channels with centre wavelengths beyond 2,000 nm in order to avoid the resulting thermal radiance component. The DEM error according to Eq. (21.5) is computed based on the gridded DEM derived from LOLA data (http://pds-geosciences.wustl.edu/missions/lro/lola.htm), which provides absolute elevation values relative to the mean lunar radius of 1,738 km and is georeferenced to the same coordinate system as the M^3 images.

Our first example DEM covers the eastern half of the lunar crater Alphonsus, which shows a distinctly non-uniform surface albedo due to the presence of pyroclastic deposits on the crater floor which consist of dark volcanic ash [9]. Bright highland material appears as high-albedo regions. This crater is well suited for demonstrating the capability of the DEM construction algorithm proposed in Sect. 21.3 to separate radiance variations due to topography from those due to non-uniform albedo. The M^3 image of Alphonsus, the corresponding LOLA DEM, the refined DEM and the albedo map are shown in Fig. 21.2. Another lunar region displaying distinct albedo variations is the crater Bullialdus (cf. Fig. 21.3). Our refined DEM reveals the terraced structure of the inner crater walls and a multitude of small craters which are invisible in the LOLA DEM. Similarly, we found that despite the high nominal lateral resolution of the GLD100 according to [44], our constructed DEMs reveal much finer details than the GLD100 (cf. also [12]).

No topographic map of sufficiently high lateral resolution is available for the examined regions as ground truth. In order to nevertheless obtain an estimate of the absolute accuracy, a synthetic image of Bullialdus crater was created by illuminating the constructed DEM using the inferred albedo map. The illumination and viewing direction were adopted from the original M^3 image. This synthetic image was used as an input image for our DEM construction algorithm. When the variational optimisation scheme is initialised with the LOLA DEM, the reconstruction error (RMSE) corresponds to 54 m, while it becomes as low as 9 m when the result of the extended photoclinometry scheme is used for initialisation. The RMSE of the inferred albedo map at 1,978 nm wavelength amounts to 0.007 and 0.005, corresponding to 1.8 and 1.3 % of the mean albedo of 0.391, respectively.

These results were obtained by performing 8 iterations of the variational approach (cf. Sect. 21.3.2) with σ_w decreasing from 11 to 7 pixels, $\sigma_{\mathrm{DEM}} = 7$ pixels, $\gamma = 9 \times 10^{-4}$, and $\delta = 2.5 \times 10^{-5}$. The values of γ and δ have been chosen as small as possible for the optimisation scheme to converge in order to maximise

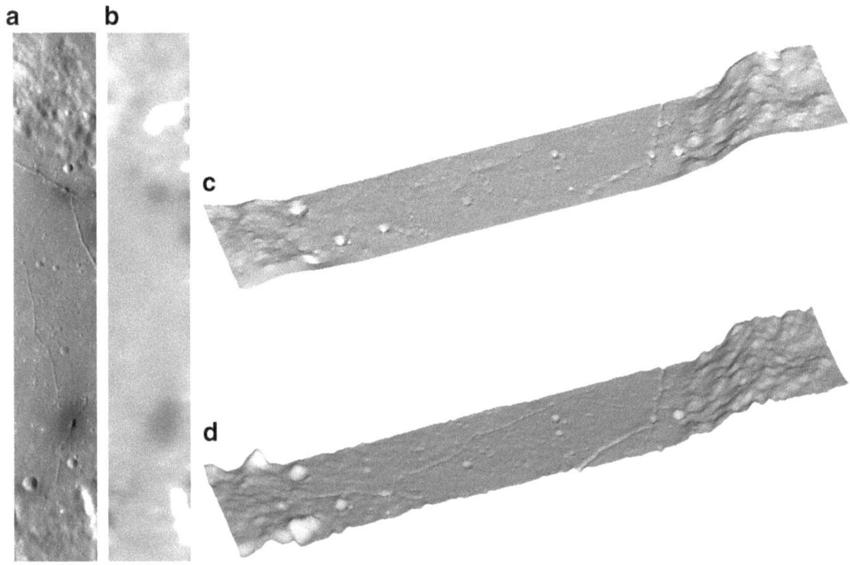

Fig. 21.2 Part of the floor of the lunar crater Alphonsus. (**a**) M^3 radiance image at 1,978 nm (Courtesy NASA/JPL-Caltech). (**b**) Map of the single-scattering albedo w at 1,978 nm, obtained with $\sigma_w = 5$ pixels. The grey value range is 0–0.6. (**c**) LOLA DEM. (**d**) Refined DEM. In the shaded perspectivical views of the DEMs, the vertical axis is three times exaggerated and the albedo is assumed to be constant

the influence of the reflectance error E_R. The variational optimisation scheme was initialised by the result of the extended photoclinometry method outlined in Sect. 21.3.3, where we have set $\sigma_w = 16$ pixels due to the low effective lateral resolution of the LOLA DEM, $\delta = 0.05$, and $n_{\text{pyr}} = 4$. Furthermore, we achieved a favourable convergence behaviour when applying the pyramidal approach also in the variational optimisation. As the photoclinometry result used for initialisation is already close to the final result, we set $n_{\text{pyr}} = 1$ in the variational scheme.

21.5 Calibration of Hyperspectral Imagery

This section describes the preprocessing and normalisation of the hyperspectral image data used for DEM construction. Furthermore, the topography dependence of the spectral parameters extracted from the pixel spectra is regarded, and an empirical topography correction scheme based on a principal component analysis (PCA) is proposed. The topography dependence of spectral ratios extracted from Clementine multispectral imagery and the correspondingly estimated abundances of chemical compounds in the lunar soil is shown in [24]. As our method is based on the ratios of spectra acquired at almost identical phase angles α (the variation of α within an

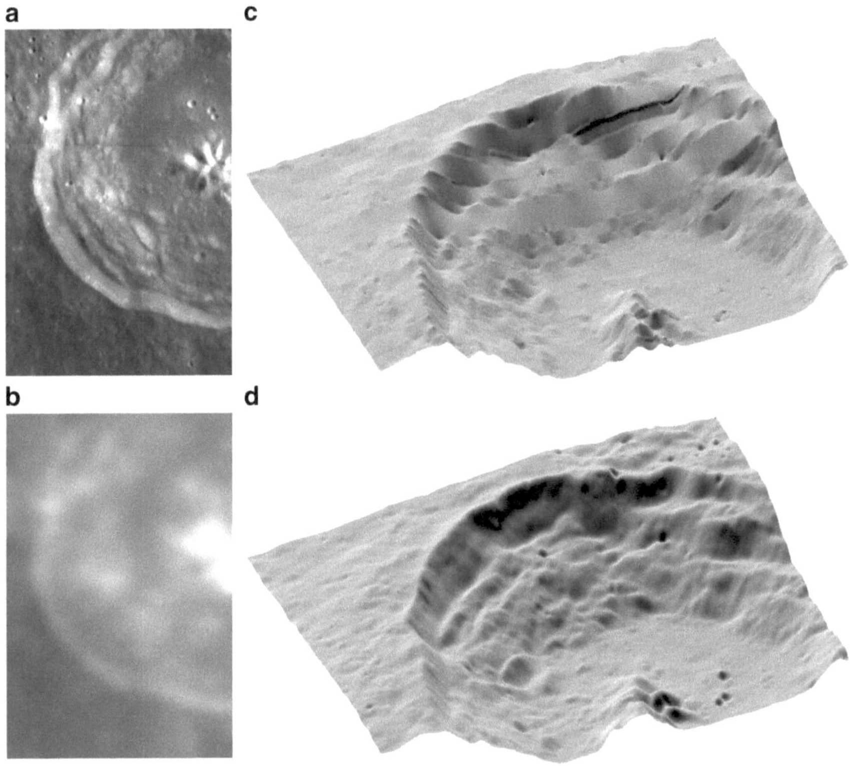

Fig. 21.3 Lunar crater Bullialdus. (**a**) M^3 radiance image at 750 nm (Courtesy NASA/JPL-Caltech). (**b**) Map of the single-scattering albedo w at 750 nm, obtained with $\sigma_w = 9$ pixels. The grey value range is 0–0.35. (**c**) LOLA DEM. (**d**) Refined DEM. In the shaded perspectivical views of the DEMs, the vertical axis is three times exaggerated and the albedo is assumed to be constant

image is always smaller than a few degrees), the approach based on an empirical phase angle dependent photometric function suggested in [19] for M^3 data is not able to perform such a correction of the spectra with respect to local topography.

21.5.1 Preprocessing and Normalisation of Reflectance Spectra

In our spectral analyses we omit M^3 channels 1–5 (which are known to be poorly calibrated [23]) and 85 (this channel is also omitted in [19]) and only use channels 6–84, covering the wavelength range from 661 to 2,936 nm (cf. Fig. 21.4 for a typical observed radiance spectrum). In remote sensing, it is a standard procedure to take into account the thermal emission component of the observed spectra [45].

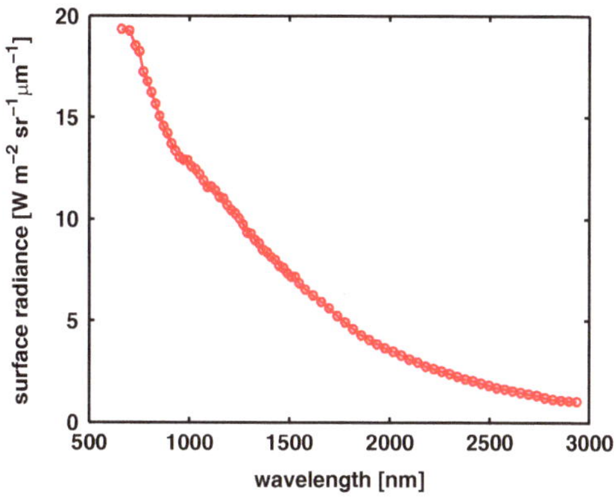

Fig. 21.4 Typical M^3 radiance spectrum (channels 6–84 covering the wavelength range 661–2,936 nm)

At this point, we make the assumption that the observed radiance spectrum L_λ^{obs} between about 2,400 and 3,000 nm used for temperature estimation can be modelled as the sum of a reference spectrum and the emission spectrum $\beta B_\lambda(T)$ of a black body of temperature T and emissivity β [45], which we assume to be wavelength-independent. Our reference spectrum R_λ^{62231} is the extrapolation of a linear fit to the laboratory reflectance spectrum of returned lunar sample no. 62231 (http://www.planetary.brown.edu/pds/AP62231.html). The linear fit is performed in the wavelength range 2,400–2,600 nm of the laboratory spectrum and yields the linear reference spectrum

$$R_\lambda^{62231} = 0.081151\,\lambda[\mu m] + 0.1423. \qquad (21.15)$$

The corresponding radiance spectrum is readily obtained by multiplying R_λ^{62231} with the solar irradiance spectrum E_λ^{sun}. Hence, the model equation

$$L_\lambda^{model} = aR_\lambda^{62231}E_\lambda^{sun} + \beta B_\lambda(T) \qquad (21.16)$$

is fitted to the observed radiance spectrum L_λ^{obs} of each M^3 pixel in the wavelength range 2,377–2,936 nm, where a is a mean surface reflectivity parameter. This procedure yields maps of the parameters a, β, and T. We always obtained values of the emissivity β close to 1. Notably, due to the limited wavelength coverage of the M^3 spectra, it is not possible to reliably estimate temperature values below about 300 K. Finally, the thermally corrected radiance spectrum L_λ^{corr} is obtained according to

$$L_\lambda^{corr} = L_\lambda^{obs} - \beta B_\lambda(T). \qquad (21.17)$$

Figure 21.5 shows a comparison between the apparent reflectance $L_\lambda^{obs}/E_\lambda^{sun}$ obtained without thermal correction and the thermally corrected reflectance $L_\lambda^{corr}/E_\lambda^{sun}$ for an inferred surface temperature of $T = 368$ K. The deviation rapidly increases for wavelengths beyond 2,200 nm.

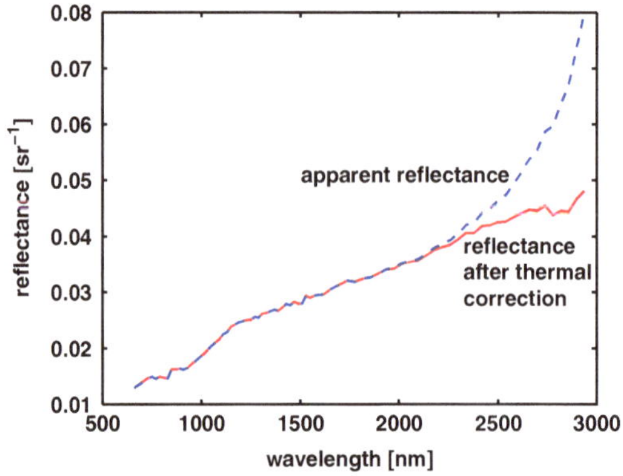

Fig. 21.5 Comparison between apparent and thermally corrected reflectance for an inferred surface temperature of $T = 368$ K

At this point, we compute for each pixel the true incidence angle θ_i and emission angle θ_e based on the constructed DEM and the angles $\theta_i^{(0)}$ and $\theta_e^{(0)}$ provided with the M^3 image data under the assumption of a smooth spherical surface. The single-scattering albedo $w(x, y)$ obtained based on Eq. (21.11) during the DEM construction process is then inserted into the utilised reflectance model (Hapke IMSA [14–16] or AMSA [17]), using the standard illumination and viewing configuration with $\theta_i^{\mathrm{ref}} = 30°$, $\theta_e^{\mathrm{ref}} = 0°$, and $\alpha^{\mathrm{ref}} = 30°$, which yields the normalised reflectance $R_\lambda^{\mathrm{norm}}$ according to

$$R_\lambda^{\mathrm{norm}}(x, y) = R_{\mathrm{IMSA/AMSA}}\left(\mu_0^{\mathrm{ref}}, \mu^{\mathrm{ref}}, \alpha^{\mathrm{ref}}, w(x, y)\right) \qquad (21.18)$$

with $\mu_0^{\mathrm{ref}} = \cos(\theta_i^{\mathrm{ref}})$ and $\mu^{\mathrm{ref}} = \cos(\theta_e^{\mathrm{ref}})$. The other parameters of the Hapke model are chosen according to [51] (Moon solution 1 therein).

In this study we use the Hapke IMSA [14–16] and AMSA [17] models for reflectance normalisation. However, the lunar-Lambert model according to Eq. (21.2) is commonly used instead of the Hapke model as the basis for the normalisation of reflectance data [39]. Since this results in a multiplication of the original reflectances by a wavelength-independent factor, wavelength-dependent effects then have to be taken into account by additional correction functions with empirically determined parameters (cf. e.g. [19, 39]).

In the normalised reflectance spectra, one would expect topography effects to be absent. We will see in Sect. 21.5.3, however, that even after normalisation to standard illumination and viewing geometry subtle topography effects can still be observed which may have a significant influence on the extracted spectral parameters.

21.5.2 Extraction of Spectral Parameters

A typical thermally corrected lunar reflectance spectrum is shown in Fig. 21.6. The reflectance strongly increases from visible towards near-infrared wavelengths. The absorption trough around 1,000 nm is due to minerals containing iron in the form of

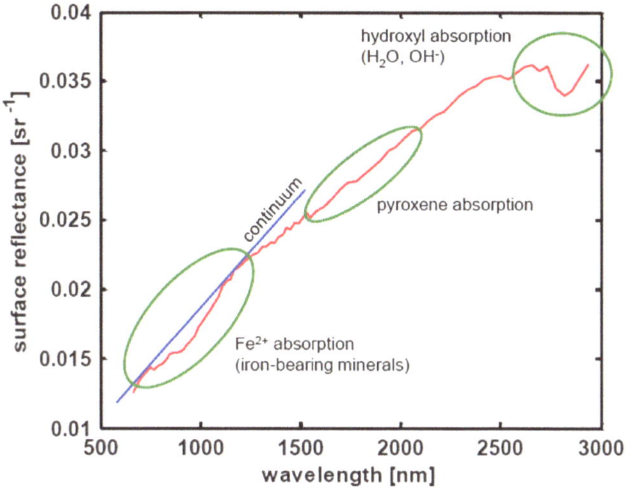

Fig. 21.6 Typical thermally corrected lunar reflectance spectrum, illustrating the primary absorption troughs

Fe^{2+} ions [3,35,48]. A second absorption trough around 2,000 nm can be attributed to the mineral pyroxene [3]. The absorption trough around 2,800 nm is described e.g. in [6], where it is attributed to the presence of water molecules and OH^- ions in the lunar soil. According to [37], they originate from chemical reactions of oxygen contained in the minerals of the lunar surface with protons of the solar wind.

Easily obtainable quantities to characterise the depths of the ferrous absorption trough around 1,000 nm and the hydrous absorption trough around 2,800 nm are the diagnostic spectral ratios $R_{950}^{\mathrm{norm}}/R_{750}^{\mathrm{norm}}$ (cf. e.g. [9]) and $R_{2817}^{\mathrm{norm}}/R_{2657}^{\mathrm{norm}}$, respectively. Additionally, the detailed composition of lunar iron-bearing minerals causes subtle variations of the shape of the ferrous absorption trough, which can be examined based on the continuum-removed spectrum [35,48].

To compute the continuum-removed reflectance spectrum, the normalised reflectance spectrum is smoothed by adapting a curve for which the mean squared deviation from the measured reflectance values and the mean squared second derivative are minimised simultaneously [34]. The spectral continuum is then approximated by a straight line between the smoothed reflectances at 701 and 1,249 nm (cf. [30]). Division of the smoothed reflectance spectrum by the linear continuum then yields the continuum-removed reflectance spectrum [30], to which an interpolation based on the Akima method [2] is applied (cf. Fig. 21.7).

The extracted spectral parameters of the absorption trough around 1,000 nm are (as illustrated in Fig. 21.7) the absorption wavelength λ_{abs}, the full width at half maximum (FWHM), the absorption depth δ (cf. [30]), and the integrated band depth (IBD) [8], corresponding to the area under the trough in Fig. 21.7. The absorption wavelength e.g. allows to differentiate between pyroxenes containing high and low abundances of calcium [35, 48]. The absorption depth depends on the amount of

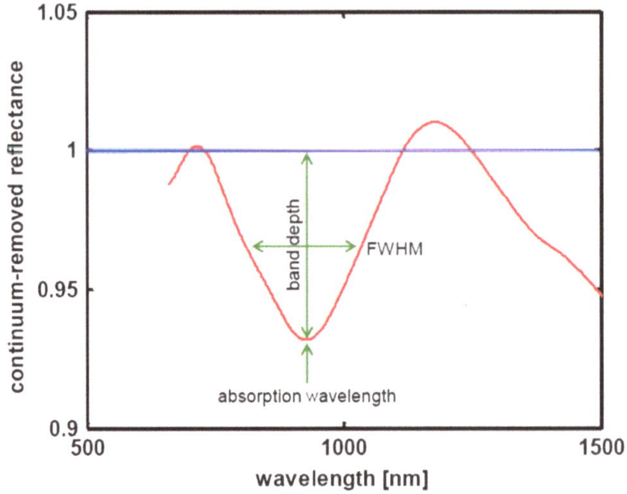

Fig. 21.7 Typical smoothed and continuum-removed reflectance spectrum showing the ferrous absorption trough around 1,000 nm

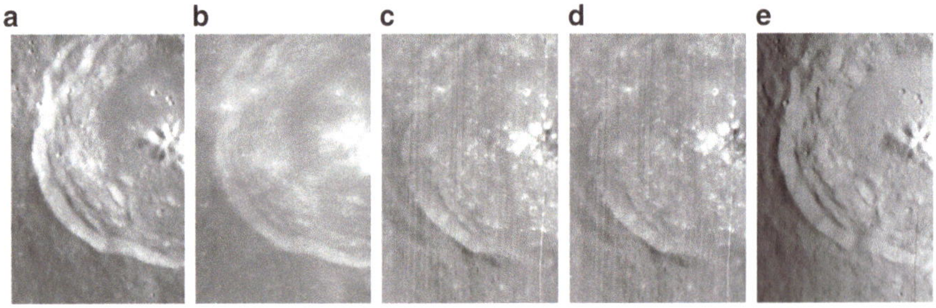

Fig. 21.8 (**a**) M^3 750 nm radiance image of the crater Bullialdus (Courtesy NASA/JPL-Caltech). (**b**) Single-scattering albedo w at 750 nm. (**c**) and (**d**) Integrated band depth (IBD) of the ferrous absorption trough around 1,000 nm inferred based on the lunar-Lambert and on the Hapke AMSA reflectance function, respectively (grey value range 0–50). (**e**) Relative difference of the IBD values (grey value range 0–0.15)

mafic minerals [48] and also on the so-called "optical maturity" of the soil [41], i.e. the period of time the surface has been exposed to high-energy solar radiation. A broad absorption, i.e. a large FWHM value, indicates that olivine occurs in the soil [35,48]. The IBD of the ferrous absorption trough is correlated with the abundance of pyroxene [8].

Figure 21.8 illustrates the difference between the values of the IBD of the ferrous absorption trough inferred based on the wavelength-independent lunar-Lambert reflectance function [38] according to Eq. (21.2) and on the Hapke AMSA reflectance function [17] according to Eq. (21.1), respectively.

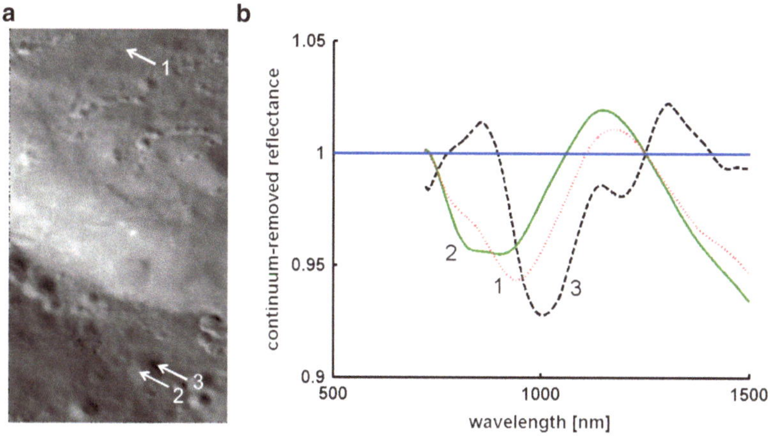

Fig. 21.9 (**a**) M³ 750 nm radiance image of the Huggins test region (Courtesy NASA/JPL-Caltech). A location on the flat floor of the crater Huggins and two locations inside a small crater just south of Huggins are indicated by *arrows*. (**b**) Continuum-removed spectra of the indicated locations. Although from a geological point of view, spectra 2 and 3 from the inner wall of the small crater should be identical, a strong dependence on surface orientation is apparent

21.5.3 Topography Dependence of Spectral Parameters

Figure 21.9 shows continuum-removed spectra of a location on the flat floor of the crater Huggins and two locations inside a small crater just south of Huggins. Although from a geological point of view, spectra 2 and 3 from the inner wall of the small crater should be identical, the spectra appear to be distorted and a strong dependence on surface orientation is visible. In contrast, the spectrum 1 of the flat crater floor displays a regular, symmetric absorption trough. Similar effects are observed for all regions examined in this study. They may in part result from radiance calibration errors e.g. due to a nonlinear behaviour of some M³ channels. However, one would expect such nonlinearities to result in a radiance dependence of the distortions, whereas in Fig. 21.9 the average radiances of locations 1 and 2 are nearly identical while spectrum 2 displays a distortion but spectrum 1 does not. Hence, inaccuracies of the reflectance model probably also contribute to the topographic effects.

Regarding the spectral parameters, the absorption wavelength λ_{abs} shows the strongest dependence on topography. For a part of the floor of the crater Purbach, Fig. 21.10a, b shows that the slopes inclined towards the sun have much lower absorption wavelengths λ_{abs} than the slopes inclined away from the sun. This systematic effect may lead to geologic misinterpretations as it is very unlikely that e.g. the western and the eastern flank of the mountain range in the image centre consist of strongly different materials, as it would appear from a naive interpretation of the λ_{abs} values.

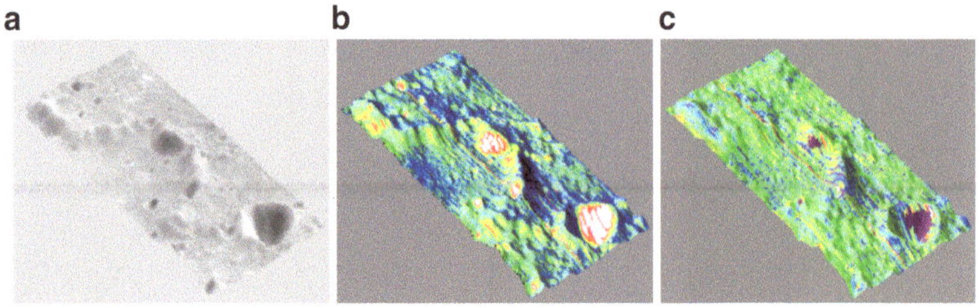

Fig. 21.10 (**a**) DEM of the floor of the crater Purbach with M^3 750 nm radiance image as overlay (Radiance image: courtesy NASA/JPL-Caltech). (**b**) Topography dependence of the absorption wavelength λ_{abs} of the ferrous absorption trough. (**c**) Result of the correction of the topography dependence of λ_{abs} according to Sect. 21.5.4. Low-radiance pixels are masked out in *dark purple* colour. The range of the colour-coded spectral parameter maps is 880–1,000 nm

21.5.4 Empirical Topography Correction

This section describes an empirical approach to the correction of the effect of topography on the M^3 spectra. The basic assumption, which is geologically reasonable, is that all pixels on the inclined inner wall of a small crater should have identical reflectance spectra, while inclined and even surface parts may possibly display systematically different spectra.

Hence, the inclined inner wall of a small crater in the region under study is used as a reference region, where the reference spectrum $S_\lambda = \langle R_\lambda \rangle_{\text{ref}}$ is taken to be the average of all spectra of the reference region. For each pixel position (x, y), the normalised ratio spectrum is then defined as

$$Q_\lambda(x, y) = \frac{R_\lambda(x, y)/S_\lambda}{\langle R_\lambda(x, y)/S_\lambda \rangle_\lambda}. \tag{21.19}$$

A principal component analysis (PCA) of all normalised ratio spectra of the reference region yields a set of PCA components $P_\lambda^{(i)}$ and, for each pixel, a set of PCA coefficients a_i. The DEM constructed according to Sect. 21.3 allows to compute a pixel-wise unit normal vector **n**, which in turn yields the surface inclination angle $\sigma = \arccos(n_z)$ and the azimuth angle $\psi = \text{atan2}\left(n_y, n_x\right) =: \arctan\left(n_y/n_x\right)$, where $\sigma \in [0°, \ldots, 90°]$ and $\psi \in [0°, \ldots, 360°]$. A polynomial function of second order in σ and eighth order in ψ is fitted to the PCA coefficients extracted from the reference region. For a pixel located at (x, y) outside the reference region, the PCA coefficients $a_i\,(\sigma(x, y), \psi(x, y))$ are then computed according to the DEM, which yields a corrected normalised ratio spectrum

$$Q_\lambda^{\text{corr}}(x, y) = \frac{R_\lambda}{S_\lambda} - \sum_{i=1}^{K} a_i\,(\sigma(x, y), \psi(x, y))\; P_\lambda^{(i)} \tag{21.20}$$

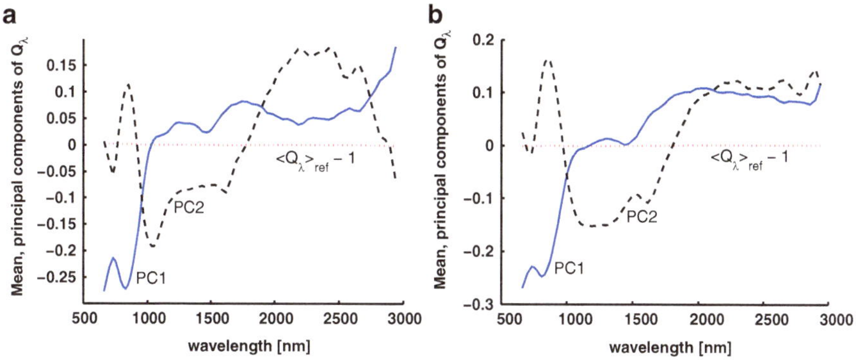

Fig. 21.11 Average value of $Q_\lambda(x, y)$ according to Eq. (21.19) and the first two principal components (denoted by PC1 and PC2), extracted from the pixels belonging (**a**) to the inner walls of two small craters just south of Huggins and (**b**) to those of the small crater Purbach A, respectively

and a corrected reflectance spectrum

$$R_\lambda^{\mathrm{corr}}(x, y) = Q_\lambda^{\mathrm{corr}}(x, y) S_\lambda. \tag{21.21}$$

For our two test regions, the first four principal components comprise 98 % of the information, such that we always set $K = 4$ in Eq. (21.20). The average value of $Q_\lambda(x, y)$ is shown along with the first two PCA components for the Huggins and Purbach test regions in Fig. 21.11. The differences between the test regions are due to the higher southern selenographic latitude of the Huggins region, resulting in a more oblique solar illumination angle.

Based on the corrected reflectances, a refined DEM is constructed according to Sect. 21.3.2, where the Hapke IMSA reflectance model is used due to convergence problems with the AMSA model. Finally, the reflectance normalisation and PCA-based topography correction is repeated using the Hapke AMSA model, where it is favourable to neglect low-radiance pixels (e.g. shaded crater walls or mountain flanks) in the analysis due to their very low signal-to-noise ratio.

21.6 Results of Topography Correction and Final DEM Construction

The final DEMs of the Huggins and Purbach test regions are shown in Fig. 21.12, showing a high amount of small-scale surface detail. The spectral parameters extracted from the uncorrected and the final corrected reflectance spectra of these regions are shown in Figs. 21.13 and 21.14, respectively, where dark pixels with 750 nm radiances below $7\,\mathrm{W\,m^{-2}\,\mu m^{-1}\,sr^{-1}}$ are masked out. Figure 21.10c shows

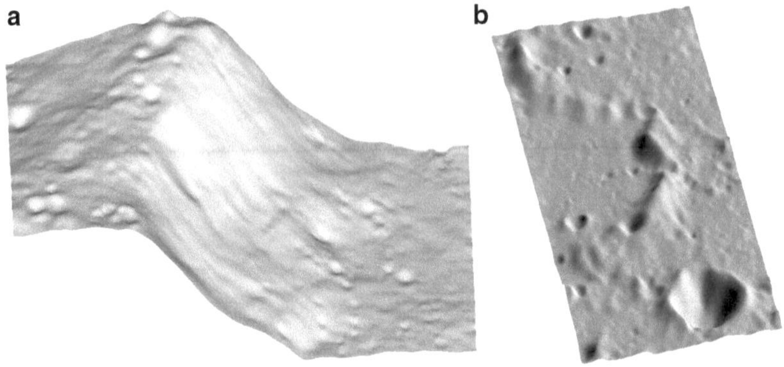

Fig. 21.12 Final DEMs of the test regions (**a**) Huggins and (**b**) Purbach. The vertical axis is three times exaggerated. The shaded DEM rather than the original image is used as an overlay, such that all visible small-scale detail is actually present in the DEM and not only in the original image

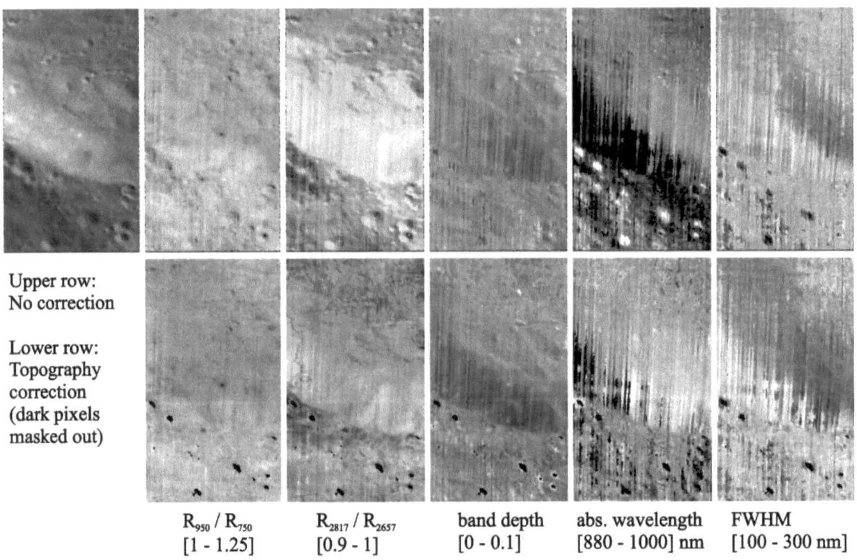

Fig. 21.13 M^3 750 nm radiance image of the Huggins test region (Courtesy NASA/JPL-Caltech) (*upper left*) and spectral parameters extracted from the uncorrected and from the final corrected reflectance spectra (low-radiance pixels are masked out in *black*)

the final DEM of the Purbach test region with false-colour overlay of the absorption wavelength λ_{abs} extracted from the uncorrected and final corrected reflectance spectra, respectively. For all examined spectral parameters, the PCA-based correction according to Sect. 21.5.4 is able to eliminate most topographic effects.

As an independent test example, we examine the crater Bullialdus. This crater is located at approximately the same selenographic latitude as the Purbach test region,

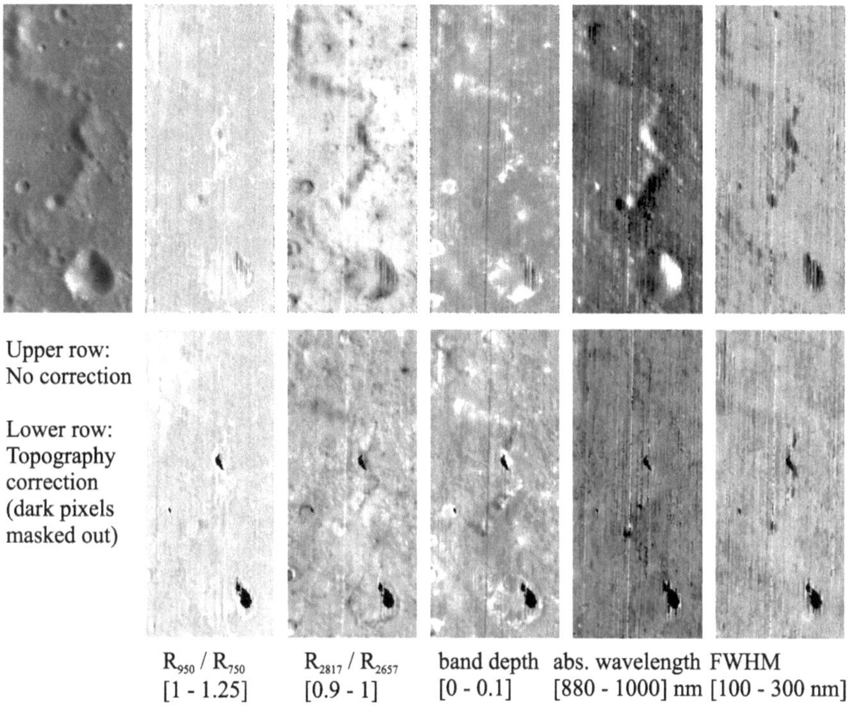

Fig. 21.14 M^3 750 nm radiance image of the Purbach test region (Courtesy NASA/JPL-Caltech) (*upper left*) and spectral parameters extracted from the uncorrected and from the final corrected reflectance spectra (low-radiance pixels are masked out in *black*)

resulting in similar illumination conditions in the corresponding M^3 data. Hence, we use the PCA components $P_\lambda^{(i)}$ and the coefficient functions $a_i\,(\sigma, \psi)$ inferred from the Purbach region in order to compute corrected reflectances according to Eq. (21.21) and extract the corresponding corrected spectral parameters. The spectral parameters extracted from the uncorrected and corrected reflectances are shown in Fig. 21.15 (cf. Fig. 21.3 for the final DEM). Again, the topography correction is most obvious in the map of the absorption wavelength λ_{abs}, which is used as an overlay of the final DEM of Bullialdus in Fig. 21.16. After correction, the crater walls, most of the crater floor, and the even surface outside the crater display similar values of λ_{abs}, only the central peaks still display excessively low absorption wavelengths. These low λ_{abs} values are much more clearly confined to the central peaks than without correction. In [49], the spectral characteristics of the central peaks of Bullialdus are attributed to the occurrence of the mineral norite.

Figure 21.15 shows that the cold, shaded flanks of the central peak of Bullialdus display a stronger hydroxyl absorption (i.e. lower R_{2817}/R_{2657} ratio) than the warmer crater floor. This effect is not eliminated by the empirical correction approach. Temperatures below 250–300 K cannot be estimated reliably due to the

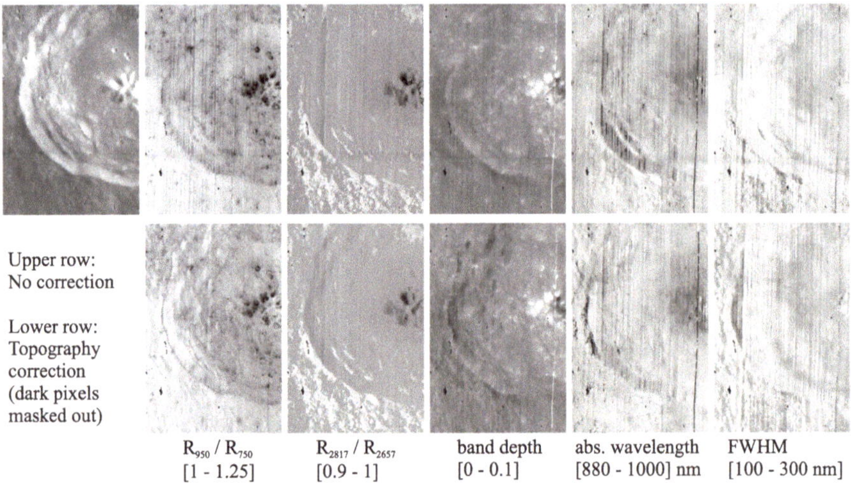

Fig. 21.15 M[3] 750 nm radiance image of Bullialdus crater (Courtesy NASA/JPL-Caltech) (*upper left*) and spectral parameters extracted from the uncorrected and from the final corrected reflectance spectra

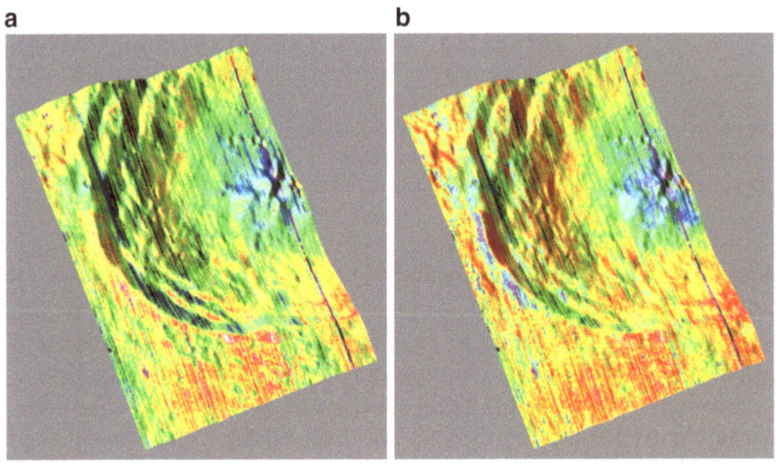

Fig. 21.16 Final DEM of Bullialdus crater with false-colour overlay of the absorption wavelength λ_{abs} (range: 880–1,000 nm) extracted (**a**) from the uncorrected and (**b**) from the final corrected reflectance spectra, respectively

limited spectral range, such that for these pixels no thermal correction can be performed. The high R_{2817}/R_{2657} values southwest of the crater appear to be due to cool surface parts with temperatures just below the detection threshold, for which the thermal correction is inaccurate. Generally spoken, it is not possible based on the available data to distinguish topography effects on the R_{2817}/R_{2657} ratio (cf. also Figs. 21.13 and 21.14) from thermal effects.

21.7 Summary and Conclusion

A method for the construction of lunar DEMs which combines surface gradients obtained by photoclinometry and shape from shading with absolute depth data (here: lunar orbital LIDAR data) by exploiting their respective advantages has been presented. The lunar surface has a non-Lambertian reflectance behaviour and a non-uniform albedo. In a first step, the surface gradients are estimated based on an extended photoclinometry approach which takes into account both image and LIDAR data, and the surface is reconstructed by integration. The second step consists of the minimisation of a global error functional based on a variational approach in order to combine the surface gradient information on small spatial scales and the LIDAR data on large spatial scales, where the result of the first step is used for initialisation. This framework has been applied to the construction of DEMs of lunar surface regions. Based on the available hyperspectral imagery, the obtained DEMs have been used to normalise the wavelength-dependent surface reflectance to a standard illumination and viewing geometry. In this context, an empirical, PCA-based correction approach has been proposed to compensate the detected systematic topography-dependent distortions of the pixel spectra which affect the extracted spectral parameters. Relying on the corrected surface reflectance data, refined DEMs have been constructed and spectral parameter maps have been obtained in which (except for the hydroxyl absorption) topographic effects are nearly completely removed.

References

1. Agrawal, A., Raskar, R., Chellappa, R.: What is the range of surface reconstructions from a gradient field? In: Leonardis, A., Bischof, H., Pinz, A. (eds.) Proceedings of European Conference on Computer Vision. Lecture Notes in Computer Science, vol. 3951, pp. 578–591. Springer (2006)
2. Akima, H.: A new method of interpolation and smooth curve fitting based on local procedures. J. Assoc. Comput. Mach. **17**(4), 589–602 (1970)
3. Burns, R.G., Abu-Eid, R.M., Huggins, F.E.: Crystal field spectra of lunar pyroxenes. In: Proceedings of Lunar Science Conference, vol. 2, pp. 533–543. Lunar and Planetary Institute, Houston (1972)
4. Bussey, D.B.J., McGovern, J.A., Spudis, P.D., Neish, C.D., Sörensen, S.A.: Lunar polar illumination conditions derived using Kaguya laser data. In: Proceedings of Annual Meeting of LEAG. Lunar and Planetary Institute, Houston (2009)
5. Chandrasekhar, S.: Radiative Transfer. Dover, New York (1960)
6. Clark R., Pieters, C.M., Green, R.O., Boardman, J., Buratti, B.J., Head, J.W., Isaacson, P.J., Livo, K.E., McCord, T.B., Nettles, J.W., Petro, N.E., Sunshine, J.M., Taylor, L.A.: Water and Hydroxyl on the Moon as seen by the Moon Mineralogy Mapper (M^3). In: Proceedings of Lunar Planetary Science XXXXI, abstract #2302. Lunar and Planetary Institute, Houston (2010)
7. Cryer, J.E., Tsai, P.-S., Shah, M.: Integration of shape from shading and stereo. Pattern Recognit. **28**(7), 1033–1043 (1995)

8. Dhingra, D., Pieters, C.M., Isaacson, P., Staid, M., Mustard, J., Klima, R., Taylor, L.A., Kramer, G., Nettles, J., M^3 team: Spectroscopic signature of the high titanium basalts at mare tranquillitatis from Moon Mineralogy Mapper (M^3). In: Proceedings of Lunar Planetary Science XXXXI, abstract #2494. Lunar and Planetary Institute, Houston (2010)
9. Gaddis, L.R., Staid, M.I., Tyburczy, J.A., Hawke, B.R., Petro, N.E.: Compositional analyses of lunar pyroclastic deposits. Icarus **161**, 262–280 (2003)
10. Grieger, B., Beauvivre, S., Despan, D., Erard, S., Josset, J.-L., Koschny, D.: Investigating a peak of (almost) eternal light close to the lunar south pole with SMART-1/AMIE images. In: Proceedings of European Planetary Science Congress, EPSC2008-A-00205, Münster, Germany (2008)
11. Grumpe, A., Wöhler, C.: DEM construction and calibration of hyperspectral image data using pairs of radiance images. In: Proceedings of IEEE International Symposium on Image and Signal Processing and Analysis, Dubrovnik, Croatia, pp. 609–614 (2011)
12. Grumpe, A., Wöhler, C.: Image-based construction of lunar digital elevation models of very high lateral resolution. In: Proceedings of Lunar Planetary Science XXXXIII, abstract #2597. Lunar and Planetary Institute, Houston (2012)
13. Grumpe, A., Herbort, S., Wöhler, C.: 3D reconstruction of non-Lambertian surfaces with non-uniform reflectance parameters by fusion of photometrically estimated surface normal data with active range scanner data. In: Proceedings of Oldenburger 3D-Tage, Oldenburg, Germany, pp. 54–61 (2011)
14. Hapke, B.W.: Bidirectional reflectance spectroscopy 1: theory. J. Geophys. Res. **86**, 3039–3054 (1981)
15. Hapke, B.W.: Bidirectional reflectance spectroscopy 3: correction for macroscopic roughness. Icarus **59**, 41–59 (1984)
16. Hapke, B.W.: Bidirectional reflectance spectroscopy 4: the extinction coefficient and the opposition effect. Icarus **67**, 264–280 (1986)
17. Hapke, B.W.: Bidirectional reflectance spectroscopy 5: the coherent backscatter opposition effect and anisotropic scattering. Icarus **157**, 523–534 (2002)
18. Herbort, S., Grumpe, A., Wöhler, C.: Reconstruction of non-Lambertian surfaces by fusion of shape from shading and active range scanning. In: Proceedings of IEEE International Conferences on Image Processing, Brussels, Belgium, pp. 17–20 (2011)
19. Hicks, M.D., Buratti, B.J., Nettles, J., Staid, M., Sunshine, J., Pieters, C.M., Besse, S., Boardman, J.: A photometric function for analysis of lunar images in the visual and infrared based on Moon Mineralogy Mapper observations. J. Geophys. Res. **116**, E00G15 (2011). doi:10.1029/2010JE003733
20. Horn, B.K.P.: Shape from shading: a method for obtaining the shape of a smooth opaque object from one view. MIT Technical Report 232, Massachusetts Institute of Technology (1970)
21. Horn, B.K.P.: Robot Vision. MIT Press, Cambridge, MA (1986)
22. Horn, B.K.P.: Height and gradient from shading. AI Memo 1105A, MIT AI Lab (1989)
23. Isaacson, P.: M^3 overview and working with M^3 data. M^3 Data Tutorial at Lunar Planetary Science XXXXII. http://m3.jpl.nasa.gov/pubs/Isaacson_M3DataWorkshop_LPSC2011.pdf (2011)
24. Jolliff, B.L.: Clementine UVVIS multispectral data and the Apollo 17 landing site: what can we tell and how well? J. Geophys. Res. **104**(E6), 14123–14148 (1999)
25. Joshi, N., Kriegman, D.J.: Shape from varying illumination and viewpoint. In: Proceedings of the International Conference on Computer Vision. IEEE, New York (2007)
26. Josset, J.-L., et al.: Science objectives and first results from the SMART-1/AMIE multicolour micro-camera. Adv. Space Res. **37**, 14–20 (2006)
27. Kieffer, H.H., Stone, T.C.: The spectral irradiance of the Moon. Astron. J. **129**, 2887–2901 (2005)
28. Kirk, R.L., Soderblom, L.A., Howington-Kraus, E., Archinal, B.: USGS high-resolution topomapping of Mars with Mars orbiter camera narrow-angle images. In: Proceedings of ISPRS Symposium on Geospatial Theory, Processing and Applications. International Society for Photogrammetry and Remote Sensing (2002)

29. Kozera, R.:. Existence and uniqueness in photometric stereo. Appl. Math. Comput. **44**(1), 1–103 (1991)
30. Le Mouélic, S., Lucey, P.G., Langevin, Y., Hawke, B.R.: Calculating iron contents of lunar highland materials surrounding Tycho crater from integrated Clementine UV-visible and near-infrared data. J. Geophys. Res. **107**(E10), 5074 (2002). doi:10.1029/2000JE001484
31. Li, R., Wang, W., He, S., Yan, L., Meng, X., Crawford, J., Robinson, M.S., Tran, T., Archinal, B.A., the LROC Team: Latest results of 3D topographic mapping using lunar reconnaissance orbiter narrow-angle camera data. In: Proceedings of Lunar Planetary Science XXXXII, abstract #2010. Lunar and Planetary Institute, Houston (2011)
32. Lim, J., Ho, J., Yang, M.-H., Kriegman, D.: Passive photometric stereo from motion. In: Proceedings of International Conference on Computer Vision, vol. 2, pp. 1635–1642. IEEE, New York (2005)
33. Lohse, V., Heipke, C., Kirk, R.L.: Derivation of planetary topography using multi-image shape-from-shading. Planet. Space Sci. **54**, 661–674 (2006)
34. Marsland, S.: Machine Learning: An Algorithmic Perspective. Chapman & Hall/CRC Machine Learning and Pattern Recognition Series. CRC, Boca Raton (2009)
35. Matsunaga, T., et al.: Discoveries on the lithology of lunar crater central peaks by SELENE Spectral Profiler. Geophys. Res. Lett. **35**, L23201 (2008). doi:10.1029/2008GL035868
36. Mattson, S., Ojha, A., Ortiz, A., McEwen, A.S., Burns, K.: Regional digital terrain model production with LROC-NAC. In: Proceedings Lunar Planetary Science XXXXIII, abstract #2630. Lunar and Planetary Institute, Houston (2012)
37. McCord, T.B., Taylor, L.A., Orlando, T.M., Pieters, C.M., Combe, J.-Ph., Kramer, G., Sunshine, J.M., Head, J.W., Mustard, J.F.: Origin of OH/Water on the lunar surface detected by the Moon Mineralogy Mapper. In: Proceedings of Lunar Planetary Science XXXXI, abstract #1860. Lunar and Planetary Institute, Houston (2010)
38. McEwen, A.S.: Photometric functions for photoclinometry and other applications. Icarus **92**, 298–311 (1991)
39. McEwen, A.S., Eliason, E., Lucey, P., Malaret, E., Pieters, C., Robinson, M., Sucharski, T.: Summary of radiometric calibration and photometric normalization steps for the Clementine UVVIS images. In: Proceedings Lunar Planetary Science XXIX, abstract #1466. Lunar and Planetary Institute, Houston (1998)
40. Nehab, D., Rusinkiewicz, S., Davis, J., Ramamoorthi, R.: Efficiently combining positions and normals for precise 3D geometry. ACM Trans. Graph. (Proc. SIGGRAPH) **24**(3), 536–543 (2005)
41. Nettles, J.W., Besse, S., Boardman, J., Combe, J.-P., Clark, R., Dhingra, D., Isaacson, P., Klima, R., Kramer, G., Petro, N.E., Pieters, C.M., Staid, M., Taylor, L.A.: Progress toward a new lunar optical maturity measure based on Moon Mineralogy Mapper (M^3) data. In: Proceedings of Lunar Planetary Science XXXXI, abstract #2217. Lunar and Planetary Institute, Houston (2010)
42. Pieters, C.M., et al.: The Moon Mineralogy Mapper (M^3) on Chandrayaan-1. Curr. Sci. **96**(4), 500–505 (2009)
43. Robinson, M.S., et al.: Lunar reconnaissance orbiter camera (LROC) instrument overview. Space Sci. Rev. **150**, 81–124 (2010)
44. Scholten, F., Oberst, J., Matz, K.-D., Roatsch, T., Wählisch, M., Robinson, M.S., the LROC Team: GLD100 – the global lunar 100 meter raster DTM from LROC WAC Stereo models. In: Proceedings of Lunar Planetary Science XXXXII, abstract #2046. Lunar and Planetary Institute, Houston (2011)
45. Schowengerdt, R.A.: Remote Sensing: Models and Methods for Image Processing. Academic Press, Burlington (2006)
46. Simakov, D., Frolova, D., Basri, R.: Dense shape reconstruction of a moving object under arbitrary, unknown lighting. In: Proceedings of International Conferences on Computer Vision, vol. 2, pp. 1202–1209. IEEE, New York (2003)
47. Simchony, T., Chellappa, R., Shao, M.: Direct analytical methods for solving poisson equations in computer vision problems. IEEE Trans. Pattern Anal. Mach. Intell. **12**(5), 435–446 (1990)

48. Smrekar, S., Pieters, C.M.: Near-infrared spectroscopy of probable impact melt from three large lunar highland craters. Icarus **63**, 442–452 (1985)
49. Tompkins, S., Pieters, C.M., Mustard, J.F., Pinet, P., Chevrel, S.D.: Distribution of materials excavated by the lunar crater Bullialdus and implications for the geologic history of the Nubium region. Icarus **110**(2), 261–274 (1994)
50. Vaniman, D., Reedy, R., Heiken, G., Olhoeft, G., Mendell, W.: The lunar environment. In: Heiken, G., Vaniman, D., French, B.M. (eds.) Lunar Sourcebook. Cambridge University Press, Cambridge, UK (1991)
51. Warell, J.: Properties of the Hermean regolith: IV. Photometric parameters of Mercury and the Moon contrasted with Hapke modelling. Icarus **167**(2), 271–286 (2004)
52. Wilhelms, D.E.: A photometric technique for measurement of lunar slopes. In: Astrogeologic Studies, Annual Progress Report, Part D: Studies for Space Flight Program, USGS preliminary report, United States Geological Survey, pp. 1–12 (1964)
53. Wöhler, C., d'Angelo, P.: Stereo image analysis of non-Lambertian surfaces. Int. J. Comput. Vis. **81**(2), 172–190 (2009)
54. Woodham, R.J.: Photometric method for determining surface orientation from multiple images. Opt. Eng. **19**(1), 139–144 (1980)
55. Zhang, L., Curless, B., Hertzmann, A., Seitz, S.M.: Shape and motion under varying illumination: unifying structure from motion, photometric stereo, and multi-view stereo. In: Proceedings of International Conference on Computer Vision, vol. 1, pp. 618–626. IEEE, New York (2003)

Index

M. Breuß et al. (eds.), *Innovations for Shape Analysis*, Mathematics and Visualization,
DOI 10.1007/978-3-642-34141-0, © Springer-Verlag Berlin Heidelberg 2013

MIX
Papier aus verantwortungsvollen Quellen
Paper from responsible sources
FSC® C105338

If you have any concerns about our products,
you can contact us on
ProductSafety@springernature.com

In case Publisher is established outside the EU,
the EU authorized representative is:
Springer Nature Customer Service Center GmbH
Europaplatz 3, 69115 Heidelberg, Germany

Printed by Libri Plureos GmbH
in Hamburg, Germany